Dave Cisco
FRZ-8593

DM Bailey
Prof. Emp Rep
2000N

Methods for Solving Engineering Problems

Using Analog Computers

McGraw-Hill Series in Information Processing and Computers

J. P. Nash, Consulting Editor
Richard W. Hamming, Associate Consulting Editor

DAVIS: *Computability and Unsolvability*

EVANS AND PERRY: *Programming and Coding for Automatic Digital Computers*

GOTLIEB AND HUME: *High-speed Data Processing*

KARPLUS: *Analog Simulation*

LEDLEY: *Programming and Utilizing Digital Computers*

LEVINE: *Methods for Solving Engineering Problems Using Analog Computers*

MCCORMICK: *Digital Computer Primer*

OAKFORD: *Introduction to Electronic Data Processing Equipment*

ROGERS AND CONNOLLY: *Analog Computation in Engineering Design*

SCOTT: *Analog and Digital Computer Technology*

WILLIAMS: *Digital Computing Systems*

WRUBEL: *A Primer of Programming for Digital Computers*

Methods for Solving Engineering Problems

Using Analog Computers

LEON LEVINE

SENIOR STAFF ENGINEER
HUGHES AIRCRAFT COMPANY
CULVER CITY, CALIFORNIA

CHAPTERS 9 AND 10 CONTRIBUTED BY

Arnold Levine

STAFF ENGINEER
HUGHES AIRCRAFT COMPANY
CULVER CITY, CALIFORNIA

McGraw-Hill Book Company

NEW YORK SAN FRANCISCO TORONTO LONDON

Methods for Solving Engineering Problems Using Analog Computers

Library of Congress Catalog Card Number 63-17954

37373

To my wife, Joan

To my children, Beth Anne, Eve Linda

To my parents, Jacob, Mary

all of whom in their different ways gave me so much to help make this book possible

Preface

The very success of the analog computer as a calculating device has tended to obscure two of its most important characteristics:

1. *The computer is a tool which can be used to provide numerical solutions which show functional dependence on problem parameters.* The computer provides information blindly at the behest of the programmer. It is the duty of the user to schedule runs which give him the data he requires with a minimum of computer effort. Then the user must interpret the data. A relatively small amount of computer results wisely applied is far more useful than a greater amount of data, inadequately understood.

2. *Computer operations are inexactly performed.* Hence, there is always an error inherent in the solution. The more skillful the operator, the better he can program the problem and reduce the effect of these errors on the solution.

The degree to which the computer user understands these characteristics will largely determine how effectively the computer assists him in solving his problem. My aim in this book has been to discuss these characteristics thoroughly, so that the reader may comprehend how the analog computer can serve as an important means for aiding the problem analyst. To this end I devote most of the book to a discussion of methods for utilization of the computer for solving problems. Thus, the detailed circuitry of the particular components is omitted except where such knowledge is essential for shedding light on performance, e.g., the nature and magnitude of component errors.

The book should be useful either for industrial computer users or as a graduate text, alone or in conjunction with another book.

Chapter 2 covers the subject of ordinary differential equations, treating mainly their properties and methods of solution. Although this material may be found in good classical texts like References 24 and 27, its value to the reader is that, unlike the mathematical texts, the chapter emphasizes those properties which are important for computer users. A capable user of the computer must understand thoroughly the content of this chapter.

Chapter 3 treats the various methods available for programming the computer from differential equations. I place particular emphasis on initial conditions, normalized equations, and adjoint equations. The latter two subjects deal with methods of minimizing the number of computer runs for recording all the essential information about linear systems.

Chapter 4 is devoted to techniques for manipulating block diagrams and for performing time and amplitude scaling. These methods allow the operator to modify an existing program easily and with little likelihood of error. In particular, I describe an original approach to time and amplitude scaling which, experience has shown, makes elementary an otherwise obscure subject.

Chapters 5 and 6 discuss the generations of explicit and implicit functions, respectively. Except for the discussion of curve fitting and scaling, the material of Chapter 5 can be found in other good books on analog computing. Chapter 6, however, covers many more of the ramifications of programming implicit-function equations than the reader is apt to find elsewhere.

Chapter 7 emphasizes various methods at the programmer's disposal to reduce the effects of component errors. An application of the information (much of which is nowhere else available) in this chapter will allow the user to solve problems whose accuracy requirements are otherwise considered beyond the capability of the computer.

Chapter 8 describes various ways in which the computer may be programmed to optimize solutions. I have provided a detailed mathematical and heuristic explanation of gradient optimization procedures, as well as several pertinent examples. The present surge of interest by engineers in optimization should see further developments in this field during the next few years. The chapter should provide a suitable foundation for people interested in the subject of optimization.

Chapters 9, 10, and 11 are interrelated. *Together they constitute by far the most extensive discussion of statistics and the computer implementation of statistical problems yet found in any book on analog computers.* The subject matter is geared to the engineer.

Chapter 9 discusses some important aspects of statistics and shows how these properties may be used to solve statistical problems. Chapter 10 considers how much data is necessary to meet the accuracy requirements of a statistical problem, and also discusses techniques which can be used to detect computer malfunctions. Chapter 11 covers numerous methods for simulating statistical problems on the computer.

This book has evolved from a set of notes which I wrote for a one-year course on analog computers. This course has been taught in the fifth-year curriculum of the Electrical Engineering Department at the University of Southern California. The first six chapters comprise the first term and Chapters 7 to 11 the second term. The material through Chapter 8 can be understood by anyone with a good background in differential equations. Chapters 9 to 11 require a basic course in probability and statistics.

During my several years in the field I have met several people who have influenced my thinking on computers. Above all others, I would single out Hans Meissinger, who for more than nine years has been a friend and colleague. Several ideas in this book were suggested by him in the course of our discussions.

I would also like to acknowledge the help provided by Hughes Aircraft

Company and, particularly, Dr. Fred P. Adler, Manager of the Space Systems Division, in making available to me so many company facilities. For special attention I would single out Iris Fullerton, head of the Publications Department, and some of her workers: Sheila Kirkland, Joy Garcia, Carol Grossman, and Claire Tebbetts.

My secretary, Barbara Woodland, worked so hard for me that I simply cannot indicate in words my thanks for all she has done. To my wife I say, "I now know, with you as an example, why so many authors dedicate their books to their wives." Without her ebullient spirit to buoy me up and her copyreading efforts to polish my writing, I cannot conceive how I would ever have finished this book.

Leon Levine

Contents

1 *Perspective*

1.1 Scope of the Chapter

Some general concepts of analog computing will be discussed in this first chapter to provide a basic preparation for the following chapters, each of which details a particular aspect of the subject. We begin by defining the analog computer and contrast its behavior with the digital computer. Next we discuss the present impact of the computer on the scientific community and follow this by a history of analog computing from antiquity to the most recent past.

We then consider the electronic differential analyzer in some detail. This is the most important indirect analog computer, and it solves problems by combining elements to satisfy the same mathematical equations. It is the principal computer considered in this book, and we spend the rest of the chapter discussing its properties.

These subjects include (1) the mathematical operations it performs, (2) its speed and accuracy, (3) the types of problems it solves (which we compare with those solved by the digital computer). We discuss the extent to which the electronic differential analyzer simulates human thinking, and the manner in which the computer should be used to extract information about a problem. We then proceed to describe how the computer can be used to find a set of equations which correctly represent a physical system, a process we call model building.

We close the chapter with a brief description of physical simulation, which is a combination of some actual system components with a computer that represents the mathematical characteristics of the rest of the system. Sometimes, as we show, this is the only way a problem can be solved.

1.2 Definition

The ancient Greeks, who had a word for almost everything, used the word *analogia* to mean proportion. Derived from this source, *analogy* is defined as a relation of likeness between two things consisting of a resemblance not of the things themselves but of two or more circumstances, or effects. Hence, if a device solves a problem by resembling the attributes of the problem, it may be said to be computing by analogy.[1] In its present usage the term *analog computer* applies to a device in which quantities usually vary continuously. However, problems involving finite discontinuities, like those encountered in sampled-data systems, can also be solved by the analog computer.

If we consider them from the standpoint of how they solve problems, analog computers divide into two categories—direct and indirect. The *direct analog computer* is a device which in its total structure resembles some physical system. One example is a resistive paper sheet whose electrical current behaves like a laplacian field. In another example a network analyzer resembles a mechanical system in detail where in one representation the following electrical and mechanical quantities are analogous: voltage and force, inductance and mass, resistance and damping, current and velocity, capacitance and compliance, charge and displacement.[2]

The *indirect analog computer* consists of a set of separate elements, each of which modifies some measurable quantity like voltage or length in accordance with certain mathematical rules, e.g., addition and integration. The elements are combined so that their composite behavior is represented by the same set of mathematical equations as the problem being solved. The analogy between the indirect analog computer and physical system is in the equation being solved. In the direct analog there is a direct physical analogy between the direct analog computer and the physical system.

In contrast to the analog computer the digital computer has information presented in terms of discrete numbers. A logical sequence of operations is performed whose end result is the solution of the problem. However, in general, this sequence bears no direct relation to the system under study.

1.3 Significance of the Computer

The analog computer became an important instrument for solving mathematical problems when the mechanical differential analyzer was developed in the 1930s. During the 1950s its successor, the electronic differential analyzer, came into use by engineers, scientists, and mathematicians to solve a wide variety of problems. Judiciously used, these machines give accurate, rapid solutions, which are presented in understandable form.

Although these computers have been in use a relatively short time, they have already made a significant imprint on our society. Ehricke [4] believes that the accelerated development of American missiles would not have been

[1] The use and importance of analogy in mathematical reasoning is treated in an interesting manner by Polya [1]. (Bracketed numeral is keyed to References at the end of the book.)

[2] Since the direct analog computer will not be considered in this book, the reader should consult Refs. 2 and 3.

possible without computers.[1] This statement is not difficult to understand. The equations describing missile behavior cannot be solved analytically—especially when the noise effects are included. Lacking a computer, the designer would be forced to go through an elaborate series of ground and flight tests to observe the shortcomings which need to be corrected.[2] Such a process is costly and time-consuming. Furthermore, there is no guarantee that a given flight might produce any useful data at all (the missile could blow up at launch). By contrast, the computer determines the missile behavior under controlled laboratory conditions where every variable is easily measured. Whereas a flight test may take months of preparation and cost more than $1,000,000, the computer can get the same data in a few seconds at minuscule cost (about $100).

Flight tests cannot be entirely dispensed with. The physical characteristics of the missile in flight may not be known exactly; hence the computer obtains only an approximate solution. However, the designer can usually specify his missile accurately enough so that the major design is done on the computer and relatively minor (but important) modifications are made as a result of information gained during flight tests.

The solution of physical problems on the analog computer is called *simulation*. The representation of the equations of the physical system is mathematical model building, an important area of computer analysis of problems. We treat this subject in Sec. 1.8.

The computer, in solving certain problems, can open new areas for study. Computer problems are becoming increasingly complex even as the pressure always exists to obtain solutions in shorter time intervals. There will be a continuing need to develop new techniques and equipment if the computer is to grapple successfully with the new problems of the coming decade.

In performing its assigned role the machine relieves competent scientists from the tedious task of solving routine[3] problems. As a result, these talented people are given added leisure to utilize to the fullest their capability to think. In retrospect we can sympathize with Johannes Kepler (1571–1630), who spent years monotonously performing calculations without a computer or even logarithms to ease the burden [6]. Sustained by the faith that order exists in the universe, he discovered the first basic laws of celestial motion. How much more might this genius have found if he had had the time?

By increasing intellectual productivity, computers will have a great

[1] Digital computers have also played an important role in this development. Except where otherwise qualified, the term computer will refer to an analog computer. However, in certain instances like this one, it may apply to a digital computer as well.

[2] Scott [5, p. 2] describes how the aircraft industry used such a design procedure in the 1930s. The results were obtained in a laborious, inefficient manner which was often dangerous to the personnel involved.

[3] Routine in the sense that the manner of solving the problem is known—although the solution may be obtained only by a complicated operation involving much time.

impact first on the scientific community and ultimately on the entire society. We may expect such areas as the national economy, higher education, scientific research, and international politics to be materially affected by these computers. These subjects, important as they admittedly may be, are beyond the scope of this book [7].

1.4 History

In focusing our attention on the explosive increase in the use of computers since the end of World War II, we are apt to conclude that analog computing devices are of recent origin. In truth several computers have been invented, with the earliest going back to the dawn of history.[1]

As early as 3800 B.C., the Babylonians were using analog computation in surveying and map making. About 80 B.C. the Greeks constructed a planetarium by means of which they computed the position of the sun and planets using a Ptolemaic model of the solar system [10]. In the early fifteenth century al-Kushi, who worked in Samarkand, built a "plate of conjunctions"—a device for calculating the time of day when two planets were at the same earth longitude [11]. Later he built another computer which determined the longitudes at a given time of the sun, moon, and the known planets (the five nearest to earth) [12, 13]. This latter device was based on a Ptolemaic model, like its Greek forebear.

Edmund Gunter, an Englishman, developed the first slide rule in 1620 using the logarithm concept invented by Napier in 1614. About 1814 a Bavarian engineer, J. H. Hermann, designed the first planimeter, which measured area on a plane surface bounded by an irregular curve.

The direct ancestor of the modern mathematical computers stemmed from work reported in England in 1876 by James Thomson and his brother, Lord Kelvin (William Thomson). James Thomson invented a ball-and-disk integrator, and Lord Kelvin conceived the idea of connecting these integrators to solve differential equations [14–17]. The principle of analog computing suggested by Lord Kelvin is the one employed today.[2] Apparently the ingenious ideas of Lord Kelvin were exploited briefly in Europe and then were forgotten.[3] Vannevar Bush built his mechanical differential analyzer in 1931, unaware that Lord Kelvin had done the same work 55 years before [19].

Digital computing also has a record of basic ideas lost and later redis-

[1] Some of the material in this section, as well as a history of digital computers, may be found in Ref. 22, pp. 27–32. References 8 and 9 discuss several devices not covered in this section.

[2] A further discussion of Lord Kelvin's work is found in Sec. 3.1.2.

[3] Apparently without knowing of Lord Kelvin's work, A. M. Krylov designed and built a four-integrator device in 1904–1911 in Russia. The Russians did not build their first differential analyzer until 1938. For additional details see Ref. 18, p. 39.

covered. During the period 1812–1833, Charles Babbage under the sponsorship of the British government devised, but did not construct, an "analytic engine which might solve mathematic equations." When in 1939 Aiken built the Mark I digital computer at Harvard University, he did not know that Babbage had almost the same concept nearly a century before [20]. These examples illustrate that good, even great, scientific ideas which do not move quickly into the mainstream of scientific activity are often forgotten and must be rediscovered in a later generation.

The early twentieth century witnessed extensive effort on polynomial solvers and harmonic analyzers (devices to calculate Fourier coefficients). In the late 1920s General Electric built the first network analyzer, which measured phase and amplitude. As we have already mentioned, Dr. Vannevar Bush of MIT built in 1931 the first mechanical differential analyzer, a highly accurate and reliable device capable of solving problems dealing with functions of several variables. This machine, however, had relatively little impact on scientific research. It was large, slow, and expensive. As a result only a few mechanical differential analyzers were built.

The operational amplifier, the key element of the present indirect electronic analog computer (usually called an electronic differential analyzer), was developed by Philbrick in 1938 and independently by Lovell of Bell Telephone Laboratories during World War II. In 1950 Goldberg of RCA designed the stabilized d-c amplifier, which greatly reduced amplifier drift. This was the all-important step needed to give the electronic differential analyzer (EDA) sufficient accuracy to meet the needs of the scientist and engineer. The rest of the 1950s saw the advent of high-precision multipliers, recorders, and other devices which give the computer the versatility and accuracy needed to solve complex system problems. In this decade new packaging techniques have reduced the size of computer elements while increasing their accuracy. Solid-state components should cause this trend to continue in the 1960s. Also during this period equipment was designed to automatically set potentiometers and functions and automatically read out from any computer element. The control information is stored on punched tape.

There have also been attempts to use digital and analog computers in combined simulation, notably by Convair and Space Technology Laboratories. The purpose was to combine the high accuracy of the digital computer with the speed of the analog computer to solve long-range missile-trajectory problems. This effort has thus far had only limited success. Recently a new device has appeared which is termed a hybrid computer and combines analog and digital operations.[1] The digital operations are used principally to control the analog computer rather than to perform the solution of systems equations as before. It is too early for us to know what

[1] At the present time the two commercial hybrid computers are the HYDAC, built by Electronic Associates, Inc., and the SPAC, built by Beckman Instruments Co.

effect the hybrid device will have on computing. However, it promises to extend greatly the capability of the analog computer.

1.5 Electronic Differential Analyzer (EDA)

The greater speed of the EDA and the greater accuracy and versatility of the digital computer have combined to relegate the mechanical differential analyzer to the museum. Today the most important indirect analog computer is the EDA. The major portion of this book is concerned with methods for solving problems using this machine—and to a lesser degree its faster counterpart, the high-speed repetitive computer.

The primary application of the EDA is in the solution of ordinary differential equations of the first degree. The term *electronic* is used in the name of the EDA because, with the exception of the independent variable which is analogous to the solution time of the computer, all dependent variables of the problem are represented by voltages in the computer. Some operations like switching, multiplication, and function generation may be electromechanical. However, the mechanical aspect of the operation, like the turning of a shaft, is always converted to an equivalent voltage before being used by the computer for other mathematical operations.

Computer users sometimes have access to digital computers, like the IBM 7090, as well as EDAs. Under these conditions they must make a choice between computers on the basis of which most closely meets their problem requirements. If the forte of the EDA is the solution of ordinary differential equations, that of the digital computer is algebraic equations. These include, naturally, solution of simultaneous algebraic equations, the determination of roots of polynomials, inversion of matrices, establishment of least squares, and other related criteria. The digital computer is also used to solve partial differential equations.

These problems are not the exclusive domain of either the analog or the digital computer. Both can solve all the problems mentioned here. Nor can either computer solve all these problems better than the other. The digital computer is used to solve differential equations, which require a solution accuracy beyond the capacity of the analog computer. Otherwise, the EDA has all the advantages. It can solve the problem faster and at less cost. Also modifications in the equations (system structure) are considerably easier to make. Conversely, the analog computer should not be used to solve algebraic problems unless there is a compelling nontechnical reason for avoiding the digital computer.

In the EDA each computer element performs a single mathematical operation and is a separate entity. The problem is programmed by an interconnection of the required elements in proper sequence. The connections are made via wires on a removable patch board, which when inserted into a patch bay completes the connections. With the patch board in place and the parameters set, the computer then solves automatically the programmed problem. A diagram, in which the elements are represented symbolically, is used by the programmer to indicate the sequence of interconnections. A standardized set of these symbols, with a brief description of their function, is shown in Fig. 1.1. The symbols shown in Fig. 1.1 are representations of the most common mathematical operations performed by the computer. There is, however, some redundancy since the italicized elements alone may in proper combination perform all the operations shown. These elements do the essential operations of the computer since with them one can program any type of ordinary differential equation. However, the other elements make programming a much easier task. An important characteristic of the EDA is that the basic mathematical operations are simple and few, yet when connected correctly they are sufficient to solve complex differential equations.

The EDA is a parallel computer because all the mathematical operations are carried out simultaneously. Consequently, the speed of solution and number of elements used bear no direct relationship to one another. Rather, the rapidity with which the problem is solved accurately is limited by the bandwidth of the elements, which is considerably less for electromechanical elements like servomultipliers than for electrical elements like amplifiers.

Two terms, *precision* and *accuracy*, specify the performance of the computer. *Precision* is the degree of fineness or resolution with which a measurable quantity may be set or recorded. A potentiometer is set by a four-phase digital voltmeter to one place in ten thousand. A graphical plot of the solution is read to no better than one part in a thousand. *Accuracy* is the closeness with which the computer solution compares with the correct solution. The computer elements only *approximately* perform the mathematical operations for which they have been designed. In Chap. 7 we show methods which will reduce the effect of these individual errors on the overall solution. Nonetheless, for all our ingenuity some error will exist. The amount of the error is strongly affected by the nature of the problem. In general we expect the solution error to lie between 0.1 and 1.0 per cent of maximum value.

Precision and *accuracy* are separate computer properties which are not directly related. However, the precision should be sufficiently fine so that it does not introduce any errors in the solution. For example, suppose the computer solves the problem with a 0.1 per cent error but the precision is one part in a hundred. Here the accuracy is determined by the computer precision instead of the performance of the elements.

1.6 Do Computers Think?

We define thinking as processes which are performed by the human mind. The provocative question as to the extent to which computers can be

Computing element	Symbol	Explanation
1. *High-gain amplifier*	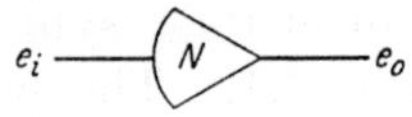	K = open-loop gain
2. Summing amplifier	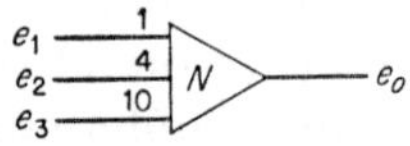	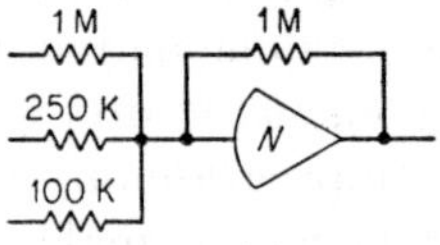
3. Inverting amplifier	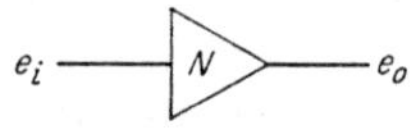	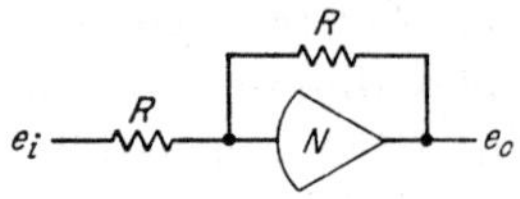
4. *Integrating amplifier*	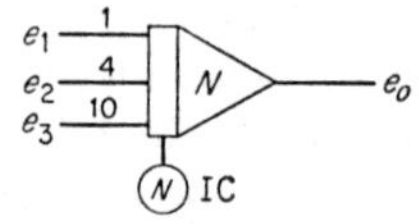	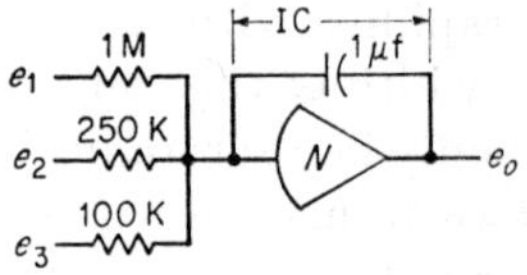
5. Potentiometer (single-sided)	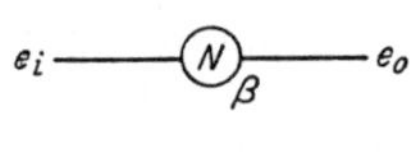	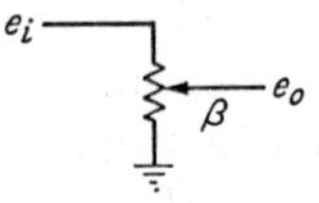
6. *Potentiometer (two-sided)*	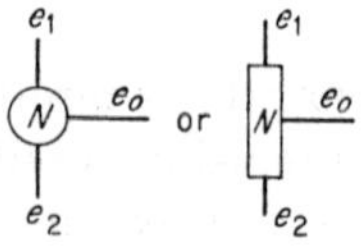	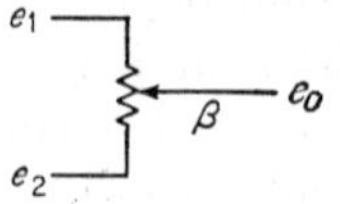
7. *Electronic multiplier*	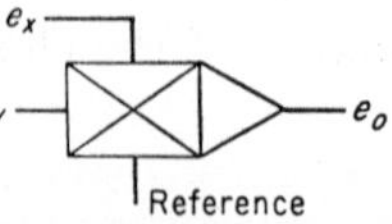	Reference is fixed voltage

FIG. 1.1. Standardized computer-element symbols, based on suggestions of IRE Com

Equation	Comment
$e_o = -Ke_i$	In better computers $K > 10^6$
$e_o = -(e_1 + 4e_2 + 10e_3)$	Consists of high-gain amplifier and passive elements; this also applies to inverting and integrating elements
$e_o = -e_i$	Same as above comment
$e_o = -\left[\int_0^t (e_1 + 4e_2 + 10e_3)\,dt + \mathrm{IC}\right]$	IC = initial condition at time equal to zero; appears as charge across capacitor
$e_o = \beta e_i$	$0 \leq \beta \leq 1$ If total resistance of potentiometer is R and resistance from arm to ground (including any loading by the output) is R_B, then $\beta = R_B/R$; in most computers the ground terminal is permanently connected and inaccessible to operator
$e_o = \beta e_1 + (1 - \beta)e_2$	See above comment for explanation of β
$e_o = \dfrac{e_x e_y}{\text{reference}}$	This text will omit reference voltage and assume it to be unity This symbol will be used to indicate the general operation of multiplication

mittee on Analog Computer Standards. N = number identifying computer element.

Computing element	Symbol	Explanation
8. *Electronic function generator*		. . . .
9. Electrical connection	———————	
10. Mechanical connection	– – – – – – – – – – –	
11. Servomultiplier	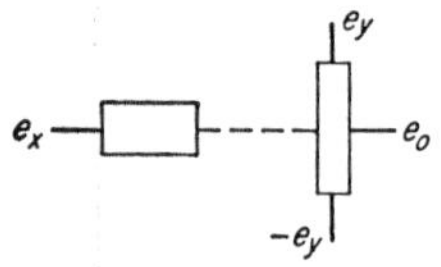	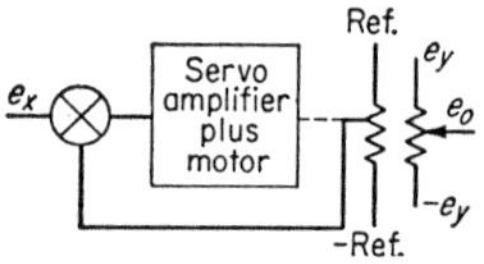
12. *Resolver*	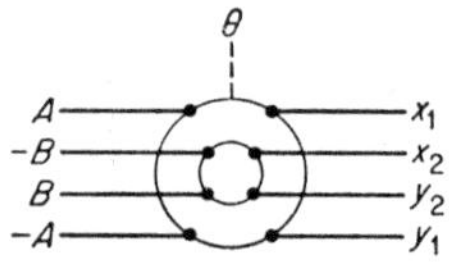	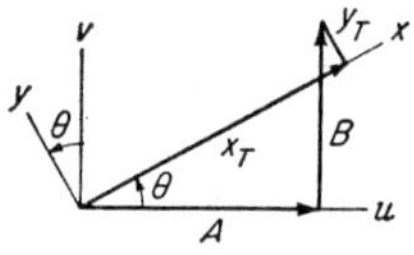
13. Polar resolver	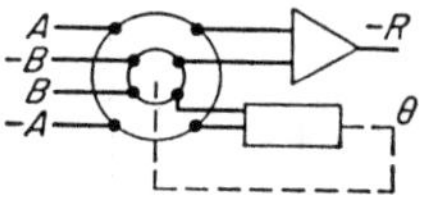	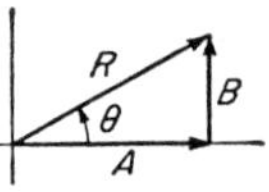
14. *Relay*	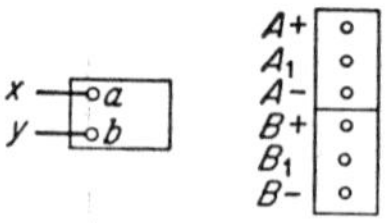	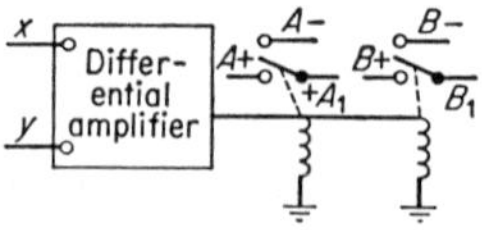

FIG. 1.1 (*Continued*)

designed to think has been explored by several contemporary philosophers and scientists, among them Norbert Wiener and John von Neumann [21]. They have speculated on the ultimate computer, whereas our question is addressed to the EDA. We are interested in making this inquiry because it is a suitable means for discussing the capacity and limitations of the computer. A comparison of the EDA and the human mind reveals that the

Equation	Comment
$e_o = f(e_x)$	$f(e_x)$ is an arbitrary function which can be changed from one problem to another
$e_o = \dfrac{e_x e_y}{\text{reference}}$	1. Unless otherwise specified, the reference will be unity 2. The servopotentiometer has same symbol as two-sided potentiometer 3. Condition on input is $\lvert e_x \rvert \leq$ reference
$x_T = x_1 + x_2 \qquad x_1 = A\cos\theta \qquad x_2 = B\sin\theta$ $y_T = -y_1 + y_2 \qquad y_1 = A\sin\theta \qquad y_2 = B\cos\theta$ or, in matrix notation, $\begin{bmatrix} x_T \\ y_T \end{bmatrix} = \begin{bmatrix} \cos\theta & \sin\theta \\ -\sin\theta & \cos\theta \end{bmatrix} \begin{bmatrix} A \\ B \end{bmatrix}$	This is a transformation of the components of a vector (A, B) in one coordinate system (u, v) to the components (x_T, y_T) of a second coordinate system (x, y) rotated an angle θ relative to the first; when A or B and θ are the inputs, the resolver transforms from polar to rectangular coordinates
$\theta = \tan^{-1}\dfrac{B}{A}$ $R = \sqrt{A^2 + B^2}$	This is a transformation from rectangular coordinates (A, B) to polar coordinates (R, θ) See Chap. 6 for a discussion of problems of polar resolution
When $x > y$, A_1 connects with A^+ and B_1 connects with B^+ When $x < y$, A_1 connects with A^- and B_1 connects with B^-	

two have many attributes in common. Some of the anthropomorphic properties of the EDA are:

1. *Learning,* a subject to which Chap. 8 is devoted
2. *Memory,* accomplished by integrators and function generators
3. *Decision making,* or choosing between alternatives, which is done by a comparator
4. *Solution of mathematical problems,* which is the *raison d'être* of the computer

Any of these operations when carried out correctly in a problem can be done by the computer more quickly, more accurately, more rationally (with no emotional stress) than by the human being.

These computer characteristics are impressive but they fall far short of competing with the human brain. The computer lacks two essential ingredients, originality and adaptability to unforeseen developments. The human brain can by the method of analogy, induction, or pure inspiration generate new ideas and predict the existence of phenomena as yet unknown. The computer follows a course completely prescribed by the programmer. The machine offers no criticism on the formulation of the problem and makes no suggestion concerning the use of alternative methods of solution, if they exist. The computer will solve a problem erroneously formulated as readily as one which is correct, unless the operator has programmed a self-check routine into the machine. Finally, the computer, on presenting its results, makes no attempt to evaluate their meaning. In short, even though it performs some simple thought processes, the computer is a slave which follows the bidding of the programmer.

The role of the EDA is not to supplant the human being who does the thinking about a problem, but rather to serve as a computational aid, providing data which make easier the task of the person solving the problem. We can better understand the role of the computer by examining how it is used to help solve problems.

1.7 Computer Analysis of a Problem

As we have already stated, the computer usually solves ordinary differential equations, which can either be abstract or represent some physical system. If they are abstract, the equations are of interest to us on their own merit. If they represent a physical system, the equations are a vehicle for the understanding of the system. In either case, we use the computer to extract information about the solution of the equations.

In a mathematical problem we strive for a solution which may be expressed in closed form[1] as a function of the problem parameters. If, for example, we ask the computer to integrate a sine, we would like as an answer

$$\begin{aligned} f(A, w, t) &= \int_0^t A \sin wx \, dx \\ &= \frac{A}{w} (\cos wt - 1) \end{aligned}$$

In this example the solution is a function of A, w, and t. Because we have the relationship in closed form we can perform mathematical operations on the parameters and compute the effect on the solution.

[1] *Closed form* is defined as any known tabulated function, e.g., sine, Bessel function, error function.

Unfortunately, the computer cannot present the solution in the form of abstract symbols. The EDA uses voltages in the problem to give quantitative data. When it is processed, we can obtain a functional relationship as a table or graph, *but never in closed form.* Unless we program the computer to process the data (techniques for doing this are described in Sec. 7.4 and Chap. 8), it will not give the functional relationship even in quantitative form.

1.8 Model Building

Before the computer can yield meaningful data, we must make certain that the equations correctly represent the physical system. The procedure by which the computer is used to make the equations conform to the physical system is called model building. It is one of the most important uses of the EDA. If the mathematical equations purport to describe a physical situation, they are useful only to the degree that they are accurate.

In all but the simplest systems, the equations are only an approximate representation of the true phenomena. Either we are unaware of all the factors or the correct equations are too unwieldy to program. A usual problem deals with only a few aspects of the behavior of the system. We therefore have a strong incentive to keep the mathematical model as simple as possible so that we can focus on the important parameters without being distracted by others. From a computer standpoint, this approach also appeals to us because a simpler problem reduces the likelihood of solution error.

Hence, as a first step in our study, we must make certain that the model is correct but as simple as it can be made. Suppose our problem is to study the effect of the sun's light pressure on the orbiting path of variously shaped earth satellites. In addition to the most obvious factors of the sun's light pressure and the earth's gravitational field, some others might include:

1. The gravitational field of the sun, moon, and other heavenly bodies
2. The oblateness of the earth
3. The drag of the earth's atmosphere

The sufficiency of the model will depend upon the nature of the results. A path which is to be specified with only a few feet of error will have a different model from one whose error can be several miles. We begin with some data on a satellite which has already been orbited, choosing the satellite whose path most closely approximates the one being studied. The simplest model is programmed, e.g., the solar pressure plus the earth's gravitational field. If it checks with the experimental data, we know that the model is adequate. If it does not check, we continue adding effects to the program until the calculated computer path corresponds to the satellite within the tolerated error.

As we move from the check case to a different shape and orbit of the satellite, we cannot be completely certain that the model is applicable. It is possible that some of the factors ignored in the model of the check case become important in the new system. If one cannot predetermine that an effect will be negligible, it should be programmed. Then if the computer study shows that the effect is small, it may be removed from the program of the problem. One must always be wary of simplified mathematical models, which should be continually checked by the latest available data.

Having formulated the mathematical model, we may use the computer to assist in designing and analyzing the system. The characteristics of satellites of several shapes and paths may now be observed in the computer room. The computer significantly reduces the time and cost of designing a satellite which is capable of flying a certain path.

Nothing dramatizes the value of the computer more than this type of application, where it serves as a computing aid in analysis and design work. In a short time the computer permits us to design and discard many systems before we find the system that meets our needs. The designer can make his mistakes where they cost him the least—on the computer.

1.9 Physical Simulation

If we are interested in the performance of some particular components which cannot be described mathematically,[1] we may combine them with a computer program to represent the entire system. This technique is called physical simulation. By reducing as much of the system as possible to a mathematical model, the study of the important components is easier to implement. The combined use of physical components and a mathematical model is feasible only because the phenomena in the mathematical model occur in precisely the same time sequence as their analogous counterparts in the physical system.[2] Unlike the mathematical model which may be time-scaled to run faster or slower than the actual system, a physical simulation must operate with the time scale of the hardware (i.e., real time). Since the computer operates with electrical signals, it may be connected directly to hardware elements whose inputs and outputs are electrical. Otherwise, converters must be used to make the hardware compatible with the computer. In a typical case, if the hardware output is a shaft rotation, it must be changed to an equivalent electrical signal before being processed by the computer.

[1] Human beings who are a part of the system, e.g., a pilot in a manned vehicle traveling in space, are "components" that usually cannot be represented mathematically. A physical simulation of which a human being is a part is sometimes called a human-factors problem.

[2] The computer is sometimes called a real-time computer because of its ability to imitate the physical system in this manner.

Converters can be elaborate devices, sometimes costing significantly more than the hardware and the computer. One example of this is a three-axis flight table which may rotate a platform along any of three orthogonal axes. This device has been used in the physical simulation of missile components when one is concerned with the effect of angular rotations on component performance.

Physical simulation appears to be a method of avoiding direct experimentation when the mathematical model is incomplete. However, there may be serious problems in making the physical simulation work because the converters cannot even be built. Suppose a radar antenna and receiver have been built for use with a vehicle homing on a radar beacon. A physical simulation would then be as shown in Fig. 1.2. The signal from the radar

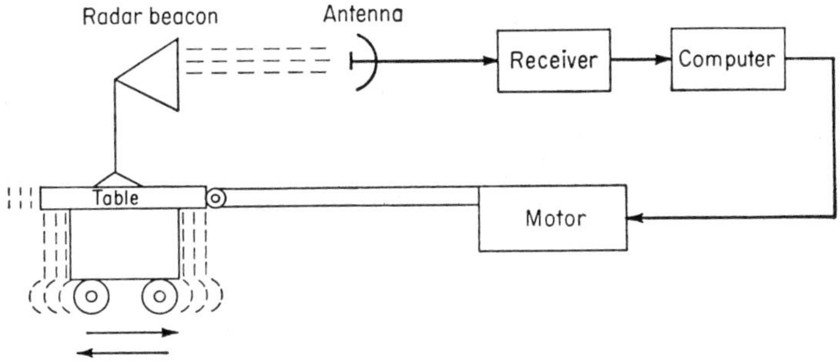

FIG. 1.2 Physical simulation of a radar problem.

beacon is picked up by the antenna and is processed by the receiver, whose output is fed to the computer. The computer's task is to make the necessary calculations which indicate how the receiver output is converted into a vehicle maneuver. Having done this, the computer determines the vehicle kinematics, which specify how the antenna is moving relative to the beacon. The computer output is used to move the beacon so that it maintains the proper position relative to the antenna. The scheme breaks down because we want to represent an actual separation of miles between the beacon and antenna by a more manageable equivalent distance of several feet. Unfortunately the radar beam characteristics are significantly different at ranges of miles and feet. Furthermore, the reflection from the walls of the chamber enclosing the beacon and antenna introduces additional errors. Hence, the antenna response for a separation of several feet may be different enough from the actual case to invalidate the results of the simulation.

The success or failure of a physical simulation depends strongly on whether the converters are available to do their required job. Since the computer itself causes relatively few problems, we shall not discuss physical simulation in this book in any detail.

2 Theory of Ordinary Differential Equations

2.1 Motivation

The programming and solution of ordinary differential equations is the easiest task performed on the computer. Indeed, the solution of these equations far overshadows in importance all other computer applications. So strong is the affinity of the analog computer for the differential equation that even problems such as polynomial root extraction and simultaneous algebraic equations are better solved when converted to equivalent differential equations. A thorough knowledge of these differential equations is indispensable to the computer user. Hence, in this chapter we cover the subject in some detail, stressing those properties which relate to the operation of the computer.

We begin the chapter with a discussion of characteristics common to the solution of all initial-value differential-equation problems. Initial-value problems, as distinguished from boundary-value problems, are the type programmed on an electronic differential analyzer and have conditions which must be satisfied to ensure that a unique solution exists. Obviously, if there is no unique solution, it is futile to program the problem on the computer.

Next we consider the available analytic methods which may be employed to solve these equations. Unfortunately, although we can make many observations about the general nature of the solution, we are able to get analytic solutions in only a few restricted cases. Nevertheless, even though their scope is limited, analytical techniques should be made familiar to the reader. An analytic solution, when it can be found, is far superior to a computer solution in providing insight into how the solution is a function of the problem variables. In our discussion the differential equations are divided into two categories, linear and nonlinear. There are striking differences in the analytical techniques which may be employed to solve these two classes of equations.[1]

[1] Unless otherwise qualified, the term *equation* will refer to the ordinary differential equation in the rest of this chapter.

2.2 General Properties of Ordinary Differential Equations

2.2.1 Definitions

Two sets of terms which may be unfamiliar to a person lacking a mathematical background are defined below. The reader should understand these definitions before proceeding to Sec. 2.2.2.

Existence and Uniqueness. If, for a given set of conditions, there is *one and only one* solution, the solution is said to exist and to be unique. As we shall show, there are situations where more than one solution results from a single set of conditions. Such a solution exists but is not unique. In the case of boundary-value problems, there are conditions for which the solution does not exist. Hence, differential equations fall into three categories:

1. The solution exists and is unique.
2. The solution exists but is not unique.
3. The solution does not exist.

The computer can solve only those problems which belong to category 1. Unless the solution is unique, there is no way of foreseeing how the computer will react to the problem. A problem that lacks a unique solution imposes requirements on computer performance which cannot be met. It is also true that because of limitations in individual computer elements, equations with unique solutions sometimes cannot be solved.

Necessary and Sufficient. In a mathematical proof one begins with a condition (hypothesis) and by a method of logical deduction reaches a conclusion. The condition and conclusion may each be a set of statements.

A *necessary* condition is a condition which is required to obtain the conclusion but which may lead to other conclusions as well.

A *sufficient* condition is a condition which follows logically from the conclusion. A *sufficient* condition is an overstatement of the conditions which are necessary to deduce the conclusion.

A *necessary and sufficient* condition is a condition which leads to *only one* conclusion. Conversely, given the conclusion, we know that it can be deduced from only one condition.

Summary. 1. When a necessary condition is satisfied, we do not know, without additional information, whether or not the conclusion holds. When a necessary condition is not satisfied, the conclusion does not hold.

2. When a sufficient condition is satisfied, the conclusion holds. When a sufficient conclusion is not satisfied, we do not know, without additional information, whether or not the conclusion holds.

3. When a necessary and sufficient condition is satisfied, the conclusion holds. When a necessary and sufficient condition is not satisfied, the conclusion does not hold. Only the necessary and sufficient condition by its presence or absence indicates the presence or absence of the associated con-

clusion. It is, therefore, coveted in mathematical theorems. Unfortunately, there are virtually no necessary and sufficient conditions applicable to ordinary differential equations. Almost all are sufficient conditions only.

Example 1. To conclude that a quadrilateral is a rectangle, it is *necessary but not sufficient* that two opposite sides are equal. It is *sufficient* but not necessary that all sides are equal and one angle is 90°. It is necessary and sufficient that two opposite sides are equal and parallel and one angle is 90°.

2.2.2 Equations of an nth-order System

An nth-order system may be expressed as an nth-order differential equation, as a set of n first-order differential equations, or as a set of equations some or all of which are higher than first-order. The nth-order differential equation has the form

$$a_n \frac{d^n y}{dx^n} + a_{n-1} \frac{d^{n-1} y}{dx^{n-1}} + \cdots + a_1 \frac{dy}{dx} + a_0 y = f\left(x; y, \frac{dy}{dx}, \ldots, \frac{d^{n-1}y}{dx^{n-1}}\right) \tag{2.1}$$

where x is called the independent variable, y is the dependent variable, and the a's are the coefficients. By common usage the solution of the equation is the y which, when expressed as a function of x only, say $y = g(x)$, will satisfy Eq. (2.1). However, the derivatives of the solution dy/dx, d^2y/dx^2, . . . , d^ny/dx^n may also legitimately be regarded as solutions of Eq. (2.1). The coefficients $a_0, a_1, a_2, \ldots, a_{n-1}$ may be constants or explicit functions of any combination of the variables x, y, and the derivatives of y. The coefficient a_n cannot be zero anywhere in the range of values of x for which the solution is obtained.[1] This restriction does not apply to any other coefficient.

A set of n first-order equations has the form

$$\frac{dy_i}{dx} = f_i(x; y_1, y_2, \ldots, y_n) \qquad i = 1, 2, \ldots, n \tag{2.2}$$

An nth-order equation (2.1) can be converted to a set of n first-order equations by the use of the relationship in the following equation set:

$$\begin{aligned} \frac{dy_1}{dx} &= y_2 \\ \frac{dy_2}{dx} &= y_3 \\ \frac{dy_{n-1}}{dx} &= y_n \\ \frac{dy_n}{dx} &= \frac{1}{a_n}\left[f(x; y_1, y_2, \ldots, y_n) - (a_{n-1}y_n + a_{n-2}y_{n-1} + \cdots + a_0 y_1)\right] \end{aligned} \tag{2.3}$$

[1] See Sec. 2.4.4 (method of Frobenius) for a discussion of how to solve the equation if a_n does go to zero.

where $$y_1 = y$$

and in general $$y_r = \frac{d^{(r-1)}y}{dx^{(r-1)}}$$

This transformation is important because we examine the set of first-order equations to see if the conditions for uniqueness are met. This test cannot be made on the nth-order equation. (See Sec. 2.2.4.)

Example 2. We wish to change the equation

$$e^{-y}\frac{d^3y}{dx^3} + 2\frac{d^2y}{dx^2} + 3t\left(\frac{dy}{dt}\right)^2 + \sin y = e^{-(x+d^2y/dx^2)} \tag{2.4}$$

to a set of three first-order equations. As prescribed by the procedure outlined in Eqs. (2.3), our final result is

$$\begin{aligned} \frac{dy_1}{dx} &= y_2 \\ \frac{dy_2}{dx} &= y_3 \\ \frac{dy_3}{dx} &= e^{y_1}\,[e^{-(x+y_3)} - 2y_3 - 3t(y_2)^2 - \sin y_1] \end{aligned} \tag{2.5}$$

The transformation from an nth-order equation to n first-order ones is routine because the procedure is clearly detailed in Eqs. (2.3) for all cases. No such general procedure exists for the inverse transformation from a set of n first-order equations to an nth-order equation. Hence, the inverse transformation is a more difficult operation and sometimes cannot be made at all. However, there is almost never a requirement for making the inverse transformation. The computer solves the problem with equal ease whether it is presented as nth-order or as a set of n first-order equations. There are also, on occasion, persuasive reasons for not even considering making the inverse transformation. The set of first-order equations often results from the formulation of a physical problem. The variables then represent quantities that have physical meaning and are therefore of some interest. If the inverse transformation is made, the resulting single equation can be solved for only one of the variables. The behavior of the remaining variables cannot be extracted from this equation.

Incidentally, the system may also be described by a set of equations which are higher than first-order in some of the variables. In this case the order of the system is equal to the sum of the orders of the highest-order derivative of each variable. For example, in the equation set

$$\begin{aligned} \frac{d^2\theta}{dt^2} + k_1\frac{d\theta}{dt} &= k_2\frac{d\gamma}{dt} \\ \frac{d^3\gamma}{dt^3} + k_3\frac{d^2\gamma}{dt^2} &= k_4\frac{d\theta}{dt} \end{aligned} \tag{2.6}$$

the highest order of θ is 2 and the highest order of γ is 3. Therefore, the order of the system is 5.

2.2.3 Initial Conditions

An initial-value problem is one in which the solution must satisfy a set of conditions for a single value of the independent variable x. For an nth-order system a *necessary* condition for a unique solution is that there are exactly n independent conditions.[1] In the nth-order equations these conditions are imposed on y and its derivatives. Similarly, for the set of first-order equations the conditions are specified for the variables $y_1, y_2, \ldots, y_n$.

If the problem has a unique solution, a set of initial conditions at one value of x completely specifies every system variable for all values of x. The particular value of x chosen for the initial conditions is dictated exclusively by the problem. From the standpoint of solving the problem, any value of x can be an initial value.

Suppose that for a set of first-order equations we take, at $x = a$, the initial values y_{1a}, y_{2a}, y_{na}. The problem is solved and, at $x = b$, we get the values of the variables $y_{1b}, y_{2b}, \ldots, y_{nb}$. Now suppose we solve the problem with the values $y_{1b}, y_{2b}, \ldots, y_{nb}$ chosen as initial conditions. Then at $x = a$ we shall find that the variables are $y_{1a}, y_{2a}, \ldots, y_{na}$. Furthermore, the solutions for initial conditions at $x = a$ and $x = b$ will be identical for all values of x. We shall use this property of differential equations in solving analytically a certain class of nonlinear equations (Sec. 2.5.3).

An nth-order differential equation is expressed in terms of $n + 1$ quantities, the dependent variable and its first n derivatives. The related set of first-order equations has $2n$ terms, each of the n variables and its first-order derivative. In other words, the total number of terms is equal to the number of initial conditions plus the number of equations being solved. For an nth-order system can we get a unique solution if we select initial conditions for any n of the terms? In one special case we do not have this freedom of choice.

Consider

$$\frac{d^2y}{dx^2} + \omega^2 y = 0 \tag{2.7}$$

or its equivalent form

$$\frac{dy_1}{dx} = y_2 \tag{2.8a}$$

$$\frac{dy_2}{dx} = -\omega^2 y_1 \tag{2.8b}$$

We observe that Eq. (2.7) is a single equation. It therefore has only one unknown. This means that one variable (either y or one of its derivatives)

[1] A condition is independent if it cannot be written as any linear combination of the other $n - 1$ conditions. The subject of independence of variables in linear algebraic equations can be found in mathematical texts, e.g., Ref. 23, pp. 3–5.

must be specified as a function of all the other variables at all values of x including the initial value. In Eqs. (2.8) two of the variables can be solved as a function of the others.

Suppose we attempted to solve Eq. (2.7) with initial conditions on y and d^2y/dx^2. If we substitute these initial values in Eq. (2.7), we cannot calculate the initial value of the remaining variable dy/dx. Because this term is undefined, the equation will not have a unique solution.

The reader may be perplexed by the necessity of having an initial condition for a term which is not even explicitly defined in the differential equation. In Eqs. (2.8) we see the reason why the initial condition of dy/dx cannot be ignored. The initial conditions on d^2y/dx^2 and y in Eq. (2.7) are the same as the initial conditions on dy_2/dx and y_1. Thus the initial value is not known for either term in Eq. (2.8a). Hence it cannot be solved and Eqs. (2.8) cannot have a unique solution.

The general rules to be followed for applying initial conditions are:

1. *In an nth-order differential equation every term from* y *to* $d^{n-1}y/dx^{n-1}$ *which is not in the equation must be one of the initial conditions of the problem.*

2. *In a set of first-order equations every term but one in each equation must have an initial value.*

If for an nth-order equation initial conditions are imposed on y and its first $n - 1$ derivatives and if for the set of first-order equations initial conditions are imposed on variables $y_1, y_2, \ldots, y_n$, the criteria of rules 1 and 2 will always be fulfilled. Incidentally, the computer programs initial conditions in this fashion because each integrator is equipped to apply an initial condition at its output.

2.2.4 Existence and Uniqueness of Solutions[1,2]

Before a problem can be solved on the computer one must show that a unique solution exists. If the equations are derived from a physical system, which has a solution for a given set of conditions, existence of a unique solution would seem obvious. However, we must remember that the equations may be only an approximate representation of the physical system. Hence, one should always make certain that the equations themselves present no difficulties no matter what the behavior of the physical system may be.

If, for example, we consider the first-order equation

$$\frac{dy}{dx} = f(x, y)$$

[1] See Ref. 24, pp. 62–91, for a thorough discussion of the existence and uniqueness of solutions. The solution to Example 3 is taken from this reference.

[2] We shall assume that the requirements on initial conditions, as outlined in Sec. 2.2.3, are realized.

a sufficient condition for a unique solution is that (1) $f(x, y)$ is continuous for all values of x and y in the solution, and (2) the Lipschitz condition is satisfied. The Lipschitz condition is satisfied only when the following inequality holds:

$$|f(x, y_a) - f(x, y_b)| < K|y_a - y_b| \tag{2.9}$$

where y_a and y_b are any two arbitrary values of y, and K is any positive number.

Example 3. For equations

$$\begin{aligned} \frac{dy}{dx} &= \sqrt{|y|} \\ f(x, y) &= \sqrt{|y|} \end{aligned} \tag{2.10}$$

$f(x, y)$ is continuous for all finite y. To test for the Lipschitz condition, assume that the inequality holds:

$$|\sqrt{|y_a|} - \sqrt{|y_b|}\,| < K|y_a - y_b| \tag{2.11}$$

However, for y_a and y_b both positive,

$$|\sqrt{|y_a|} - \sqrt{|y_b|}\,| = |\sqrt{y_a} - \sqrt{y_b}| \tag{2.12}$$

and

$$\begin{aligned} |y_a - y_b| &= |(\sqrt{y_a} - \sqrt{y_b})(\sqrt{y_a} + \sqrt{y_b})| \\ &= |(\sqrt{y_a} - \sqrt{y_b})|\,|\sqrt{y_a} + \sqrt{y_b}| \end{aligned} \tag{2.13}$$

Substituting in Eq. (2.11),

$$|\sqrt{y_a} - \sqrt{y_b}| < K|\sqrt{y_a} + \sqrt{y_b}|\,|\sqrt{y_a} - \sqrt{y_b}| \tag{2.14a}$$

$$1 < K|\sqrt{y_a} + \sqrt{y_b}| \tag{2.14b}$$

Next we choose an arbitrary finite value for K and let both y_a and y_b remain positive but become very small (in mathematical terminology, for $y_a > 0$, $y_b > 0$, let $y_a \to 0$, $y_b \to 0$). Then the value of the right side of Eq. (2.14*b*) will become less than 1. Hence, contrary to our original assumption, the Lipschitz condition is not satisfied at the point $y = 0$.† We can see from Eq. (2.14*b*) that the Lipschitz condition holds for all other values of y. In other words, if Eq. (2.10) is solved either in the range $-\infty < y < 0$ or $0 < y < \infty$, the solution is unique.

If the point $y = 0$ is included, we do not yet know whether or not the solution is unique. The Lipschitz condition is only a sufficient condition (see Sec. 2.2.1). We must test further to see if there is more than one solution, an almost impossible job in most problems higher than first-order. Our situation is beclouded by the fact that we can *never* show the solution is unique. If we find more than one solution, we have proved that the solution is not unique. However, if we find only one solution, we can never prove that no other solution exists.

In this example, for the initial condition $y = 0$ at $x = 0$, there are two solutions:

$$y = 0 \tag{2.15}$$

$$\begin{aligned} y &= \frac{x^2}{4} \qquad \text{for } x > 0 \\ y &= -\frac{x^2}{4} \qquad \text{for } x < 0 \end{aligned} \tag{2.16}$$

† To prove falsehood of an assertion, we need only show that the assertion fails in one case.

Parenthetically, we might add that, even without finding the two solutions, we would still know that the computer could not solve Eqs. (2.10). If it has to generate a function, the computer has difficulty when any of the first partial derivatives become large. In this case we have the worst of all possible circumstances because

$$\frac{df(x, y)}{dy} = \infty \qquad \text{for } y = 0$$

A further discussion of computer limitations will be found in Chaps. 6 and 7.

For an nth-order system of the form of Eqs. (2.2), a *sufficient* condition for a unique solution is:

1. Each $f_i(x, y_1, y_2, \ldots, y_n)$ is continuous in every variable in its argument.

2. The Lipschitz condition, shown by the inequality of Eq. (2.17), must hold for each $f_i(x, y, \ldots, y_n)$, where

$$|f_i(x; y_{1a}, y_{2a}, \ldots, y_{na}) - f_i(x; y_{1b}, y_{2b}, \ldots, y_{nb})| < K(|y_{1a} - y_{1b}| + |y_{2a} - y_{2b}| + \cdots + |y_{na} - y_{nb}|) \qquad (2.17)$$

The Lipschitz condition will always be satisfied if every first partial derivative of each $f_i(x, y_1, \ldots, y_n)$ with respect to each of the variables in its argument is finite:

$$\frac{\partial f_i}{\partial y_k} < \infty \qquad i = 1, 2, \ldots, n; k = 1, 2, \ldots, n$$

The partial derivatives need not be continuous. If the discontinuities are finite, the Lipschitz condition will be satisfied.

2.2.5 Variation of Initial Conditions and Parameters

In an nth-order system any constant coefficient will be designated as a parameter. By this definition the terms α and β are parameters of Eq. (2.18). The solution of

$$\frac{d^2y}{dx^2} + \alpha x \frac{dy}{dx} + \sin \beta y = e^{-x} \qquad (2.18)$$

is dependent upon the values of α and β as well as upon the initial conditions. In a typical problem we may want solutions for several sets of values of α, β, and the initial conditions. As we change these quantities from one case to another, what can we say about the nature of the solution?

Let us designate a parameter λ_i and the general initial condition $y^{(r)}(0)$, which is d^ry/dx^r evaluated at the initial value of x. Now let us modify the terms in equation set (2.2) to show the functional dependence on the m parameters, for example, $f_i(x; y_1, y_2, \ldots, y_n; \lambda_1, \lambda_2, \ldots, \lambda_m)$. Then if all the f_i terms are continuous in x, y, and λ and satisfy the Lipschitz condition of Eq. (2.17) the partial derivatives of the solution y, that is, $\partial y/\partial \lambda_i$ and $\partial y/\partial y^{(r)}(0)$, are continuous for all parameters and initial conditions.

For computer users this property has the following significance. Suppose we are getting a family of solutions for an nth-order system in which everything is held fixed except one λ_i. If we were to plot the result for any y_i at a particular value of x, say $x = x_a$, and if the conditions on the f_i's are satisfied, we know that the slope of the curves for any value of x will have no discontinuities anywhere. Without knowing the exact nature of the solution we expect the computer to be well-behaved as we vary either the initial conditions or the parameters. A large change in the value of the solution for a small change in the value of the parameter is a warning sign that the computer may be malfunctioning.

2.3 Linear and Nonlinear Equations Compared

2.3.1 Background

In Sec. 2.2 we discussed properties of differential equations which tell us something about the general nature of the solution, including whether or not the solution exists. However, this section, important as it may be to an understanding of differential equations, is merely a prologue to what we really want to know about the equations, namely, the exact solution of a specific equation. This solution may be obtained either analytically or on the computer. If we can find an analytic solution, we shall be able to express y as a function of x, that is, $y = f(x)$. The computer solution can give us a functional relationship but not in closed form. We shall now consider the analytical methods which may be used to solve these problems and in Chap. 3 shall take up the subject of computer solutions.

2.3.2 Definitions

It is customary in a discussion of analytic techniques to divide differential equations into two classes:

1. Linear equations
2. Nonlinear equations

There are marked differences between these two classes. Without exception a linear equation is much easier to solve than a nonlinear one.

An nth-order equation written as

$$a_n \frac{d^n y}{dx^n} + a_{n-1} \frac{d^{n-1} y}{dx^{n-1}} + \cdots + a_1 \frac{dy}{dx} + a_0 y = f(x) \tag{2.19}$$

is linear only if all the a's are either constants or functions of x alone. Some typical examples of linear and nonlinear equations are

Linear equations

$$\frac{d^2y}{dx^2} + \sin x \frac{dy}{dx} + e^{-x}y = x^2$$

$$\frac{dy}{dx} - y \ln x = 30$$

$$\frac{d^2y}{dx^2} + 3\frac{dy}{dx} - 2y = 0$$

Nonlinear equations

$$\frac{dy}{dx} + \sqrt{|y|} = 0$$

$$\frac{d^2y}{dx^2} + \left(\frac{dy}{dx}\right)^2 + y = \tan x$$

$$\frac{d^2y}{dx^2} - \frac{1}{x}\frac{dy}{dx} + \sin y = 0$$

2.3.3 Linear Equations and Superposition

The linear equation has an important property that is lacking in the nonlinear equation, which we shall discuss in terms of the nth-order equation. Suppose the solution to an equation is designated as $f(z; x)$, where z is a set of problem parameters. In this notation the z is used to indicate the functional form of the dependence on x [e.g., if $f(z; x) = \sin x$, then $f(z; \ ) = \sin (\ )$]. Then the linearity property states

$$f(\alpha_1 z_1 + \alpha_2 z_2; x) = \alpha_1 f(z_1; x) + \alpha_2 f(z_2; x) \tag{2.20}$$

where α_1 and α_2 are constants.

Equation (2.20) is sometimes called the superposition principle. Put into words, the equation says that, if we want the solution for a combination of two parameters, $\alpha_1 z_1 + \alpha_2 z_2$, we can solve by directly using $\alpha_1 z_1 + \alpha_2 z_2$. However, we also have the option of solving separately for the parameters z_1 and z_2 and then satisfying the right side of Eq. (2.20).

Note that there is nothing to prevent z_2 from containing some of the parameters of z_1. For example if $z_2 = z_1 + z_3$, then Eq. (2.20) would be modified:

$$f(\alpha_1 z_1 + \alpha_2 z_2; x) = (\alpha_1 + \alpha_2) f(z_1; x) + \alpha_2 f(z_3; x) \tag{2.21}$$

For any differential equation the parameters on which the solution depends are (1) initial conditions, (2) forcing function, and (3) differential equation coefficients. *Linearity properties hold only for initial conditions and forcing functions.* In Sec. 3.2 we show that the linearity property allows us to obtain a normalized solution by making $n + 1$ computer runs for an nth-order differential equation. With the normalized solution we can compute the solution for any forcing function and set of initial conditions.

2.4 Analytic Solutions of Linear Differential Equations[1]

2.4.1 Introduction

In the period 1675 to 1900 virtually every mathematician of consequence contributed to an understanding of differential equations. A partial list of these men is a roll call of the great mathematicians: Leibnitz, Newton, Bernoulli, Euler, Clairaut, Lagrange, Laplace, Fourier, Cauchy, Gauss, and Weierstrauss.[2] It may therefore surprise the reader that for all the prodigious achievements of these men very few differential equations can be solved analytically. Of all the differential equations which exist, only the first-order linear equation can always be solved (see Sec. 2.4.2).

However, there are techniques which can be applied to systems of higher order. We shall examine these techniques, stressing where they may be applied and also where they fail. We shall direct our attention to the single differential equation, yet we shall not slight any other formulation of the system equations. A set of first-order (or any other order) equations may always be transformed to a single differential equation when the system is linear [30].

A linear equation of the form

$$a_n \frac{d^n y}{dx^n} + a_{n-1} \frac{d^{n-1} y}{dx^{n-1}} + \cdots + a_1 \frac{dy}{dx} + a_0 y = f(x) \tag{2.22}$$

may, because of the property of superposition, be decomposed into two equations:

$$a_n \frac{d^n y_H}{dx^n} + a_{n-1} \frac{d^{n-1} y_H}{dx^{n-1}} + \cdots + a_1 \frac{dy_H}{dx} + a_0 y_H = 0 \tag{2.23}$$

$$a_n \frac{d^n y_p}{dx^n} + a_{n-1} \frac{d^{n-1} y_p}{dx^{n-1}} + \cdots + a_1 \frac{dy_p}{dx} + a_0 y_p = f(x) \tag{2.24}$$

which can then be solved separately.

Homogeneous Equation. Equation (2.23) is called the homogeneous or characteristic equation. The variable y_H and its derivatives have the same initial conditions as the variable y and its derivatives. The solutions of the homogeneous equation can always be written in the form

$$y_H(x) = c_1 u_1(x) + c_2 u_2(x) + \cdots + c_n u_n(x) \tag{2.25}$$

There are n different $u(x)$'s, the number being equal to the order of the differential equation. Every $u(x)$ separately, when substituted for y_H, is

[1] The reader can use any text on differential equations to supplement this brief treatment of the subject. My own personal preference inclines me to recommend Refs. 25 to 27.

[2] References 28 and 29 are two well-written histories of mathematics which place in proper perspective the work of these men and others.

a solution of Eq. (2.23). Note that the initial conditions play no part in determining whether or not $u(x)$ is a solution, but all n of the initial conditions must be used to calculate the constants $c_1, c_2, \ldots, c_n$. Although the n solutions are linearly independent,[1] they are not unique. For example, we could multiply each solution by a constant to form a new set of linearly independent solutions of the differential equations. In all, there exist an infinite number of linearly independent sets that satisfy the homogeneous equation. Nevertheless, no matter what the set of solutions, there is only one solution y_H for a given set of initial conditions. For each set of solutions the constants $c_1, c_2, \ldots, c_n$ assume values so that y_H is always the same.

A necessary and sufficient condition that the solutions of the homogeneous equation $u_1(x), \ldots, u_n(x)$ be linearly independent is that the Wronskian W does not go to zero anywhere for the range of the independent variable $(x_a < x < x_b)$. The Wronskian is the determinant

$$W = \begin{vmatrix} u_1^{(0)}(x) & u_2^{(0)}(x) & \cdots & u_n^{(0)}(x) \\ u_1^{(1)}(x) & u_2^{(0)}(x) & \cdots & u_n^{(1)}(x) \\ \cdots & \cdots & \cdots & \cdots \\ u_1^{(n-1)}(x) & u_2^{(n-1)}(x) & \cdots & u_n^{(n-1)}(x) \end{vmatrix} \tag{2.26}$$

where
$$u_K^{(r)}(x) = \frac{d^r u_K}{dx^r}$$

Example 4. Two solutions of the equation

$$\frac{d^2y}{dx^2} + y = 0 \tag{2.27}$$

are
$$u_1(x) = \sin x \tag{2.28a}$$
$$u_2(x) = \cos x \tag{2.28b}$$

The reader may verify this by substituting either Eq. (2.28*a*) or (2.28*b*) into Eq. (2.27).

We use the Wronskian to determine if the solutions are independent:

$$W = \begin{vmatrix} \sin x & \cos x \\ -\cos x & \sin x \end{vmatrix} = \sin^2 x + \cos^2 x = 1 \tag{2.29}$$

The solutions are linearly independent for all values of x. The general solution can then be written

$$y = c_1 \sin x + c_2 \cos x$$

where c_1 and c_2 are constants specified by the initial conditions.

Inhomogeneous Solution. Equation (2.24) is called the inhomogeneous equation, and its solution y_p the particular solution. The solution, obtained

[1] A set of functions $u_1(x), u_2(x), \ldots, u_n(x)$ is said to be linearly independent over a range $x_a < x < x_b$ if there does not exist a set of constants $\alpha_1, \alpha_2, \ldots, \alpha_n$, where $\alpha_1^2 + \alpha_2^2 + \cdots + \alpha_n^2 > 0$, such that $\alpha_1 u_1(x) + \alpha_2 u_2(x) + \cdots + \alpha_n u_n(x) = 0$ over the entire range.

with all initial conditions zero, is expressed as an integral, called the convolution integral,

$$y_p(x) = \int_{x_a}^{x} K(x, \alpha)f(\alpha)\, d\alpha \tag{2.30}$$

where x_a is the initial value of the independent variable.

$K(x, \alpha)$ is variously called the weighting function, the impulse response, or the system function. The weighting function is very important in the analysis of linear systems. Its properties are discussed further in Secs. 2.4.6 and 3.3 and Appendix 2.

The solution to Eq. (2.22) may be expressed in terms of the solutions to Eqs. (2.23) and (2.24) as

$$y = y_H + y_p$$

or $$y = c_1u_1(x) + c_2u_2(x) + \cdots + c_nu_n(x) + \int_{x_a}^{x} K(x, \alpha)f(\alpha)\, d\alpha \tag{2.31}$$

2.4.2 First-order Linear Differential Equations

The first-order linear differential equations can always be solved. An equation of the form

$$\frac{dy}{dx} + P(x)y = Q(x) \tag{2.32}$$

has as a solution

$$y(x) = \frac{1}{V(x)}\left[\int^{x} V(\omega)Q(\omega)\, d\omega + c\right] \tag{2.33}$$

where $$V(x) = \exp\left[\int^{x} P(z)\, dz\right]$$

and c is an arbitrary constant.

2.4.3 Constant-coefficient Homogeneous Equation

Polynomial-root Factoring Method. The problem of solving a constant-coefficient nth-order differential equation can be converted to a problem of solving for the roots of an nth-order polynomial.

Thus the equation

$$\frac{d^ny}{dx^n} + a_{n-1}\frac{d^{n-1}y}{dx^{n-1}} + \cdots + a_1\frac{dy}{dx} + a_0y = 0 \tag{2.34}$$

is written as

$$q^n + a_{n-1}q^{n-1} + \cdots + a_1q + a_0 = 0 \tag{2.35}$$

If Eq. (2.35) has n different roots, $q_1, q_2, \ldots, q_n$, then the solution to Eq. (2.34) is

$$y = c_1e^{q_1x} + c_2e^{q_2x} + \cdots + c_{n-1}e^{q_{n-1}x} + c_ne^{q_nx} \tag{2.36}$$

Where there is a multiple root, the form of Eq. (2.36) must be modified to obtain n independent solutions. Suppose that Eq. (2.35) has a multiple root q_1 of order m; that is,

$$q^n + a_{n-1}q^{n-1} + \cdots + a_1q_1 + a_0 = (q - q_1)^m(q^{n-m} + b_{n-m-1}q^{n-m-1} + \cdots + b_1q + b_0) \qquad (2.37)$$

Then the part of the solution resulting from the multiple root q_1 is

$$(c_1 + c_2x + c_3x^2 + \cdots + c_{n-1}x^{m-2} + c_mx^{m-1})e^{q_1x}$$

and the rest of the solution is the same as Eq. (2.36).

Integral Transforms, Laplace and Fourier.[1] An integral transform pair relates a function of a real variable with a function of a complex variable by means of two integrals. The most common integral transforms are the Laplace transform and the Fourier transform. Despite the fact that only the Laplace transform can be used for the constant-coefficient homogeneous equation, they will both be discussed in this section because they have so many characteristics in common.

The transform[2] pairs are shown in Eqs. (2.38) and (2.39):

Laplace transform

$$\begin{aligned} F(s) &= \int_0^\infty f(t)e^{-st}\,dt \\ f(t) &= \frac{1}{2\pi i}\int_{c-i\infty}^{c+i\infty} F(s)e^{st}\,ds \end{aligned} \qquad (2.38)$$

Fourier transform

$$\begin{aligned} G(\omega) &= \int_{-\infty}^{\infty} g(t)e^{-i\omega t}\,dt \\ g(t) &= \frac{1}{2\pi}\int_{-\infty}^{\infty} G(\omega)e^{i\omega t}\,d\omega \end{aligned} \qquad (2.39)$$

In these equations t is the real variable and s and $i\omega$ are complex. $F(s)$ is the transform of $f(t)$ and $G(\omega)$ is the transform of $g(t)$. The function $f(t)$ is the inverse transform of $F(s)$.

Every term in a differential equation is a function of the independent variable, which is real. Therefore, one can take the transform of each term of the differential equation and obtain a new equation in the complex variable of the transform. One then solves the transformed equation. The inverse transform of this solution is the solution of the differential equation.

In principle this procedure may be applied to any differential equation whose terms have transforms. However, with rare exceptions, one can solve the transformed equation only if the differential equation has constant

[1] For an additional discussion of the Laplace transform, see Chap. 4.

[2] See Refs. 31 and 32 for a more detailed discussion of the Laplace and Fourier transforms, respectively.

coefficients. In this case for an nth-order differential equation the transformed equation becomes an nth-order polynomial equation in powers of the complex variable ($i\omega$ or s). For almost all other differential equations the transformed equation becomes an integrodifferential equation, no easier to solve than the original one.

The Laplace transform method may be used to solve any constant-coefficient equation (homogeneous or inhomogeneous) for an arbitrary set of initial conditions. The Fourier transform method can be used only to solve a constant-coefficient inhomogeneous equation whose initial conditions are zero. Both transform methods have their main utility in obtaining the particular solution of the equation. As we shall see in Sec. 11.2.5, the Fourier transform may also be used to analyze the properties of noise.

A modification of the Laplace transform, called the Z transform, may be used to solve difference equations with constant coefficients. This method has had wide application in the analysis of sampled-data systems.[1]

2.4.4 Variable-coefficient Homogeneous Equations

The method of Frobenius may be used to obtain the solution of a homogeneous linear equation whose variable coefficients are polynomials in x. Another technique is the Picard iteration. Because of its applicability to nonlinear equations as well, we shall discuss the Picard iteration method in Sec. 2.5.2. Aside from these two methods there is very little one can do to solve the variable-coefficient equations[2] except guess at a solution and then see if it satisfies the equation.

Equidimensional Equation. An equation of the form

$$x^n \frac{d^n y}{dx^n} + b_{n-1}x^{n-1} \frac{d^{n-1}y}{dx^{n-1}} + \cdots + b_1 x \frac{dy}{dx} + b_0 y = 0 \qquad (2.40)$$

where the b's are constants, is called the *equidimensional* or *Euler equation.* It may be converted to a linear equation with constant coefficients by the change of independent variable:

$$x = e^t \qquad (2.41)$$

Singularities and the Method of Frobenius. As we have shown, an nth-order linear equation of the form

$$a_n \frac{d^n y}{dx^n} + a_{n-1}(x) \frac{d^{n-1}(y)}{dx^{n-1}} + \cdots + a_1(x) \frac{dy}{dx} + a_0(x)y = 0 \qquad (2.42)$$

[1] There is extensive literature on the subject of sampled-data systems. Suitable references are [33], [34], and [35, pp. 500–558].

[2] Reference 36, pp. 294–660, has a tabulation of the solutions of several linear and nonlinear equations.

may be equivalently represented as a system of n first-order equations:

$$\begin{aligned} \frac{dy_1}{dx} &= y_2 \\ &\cdots\cdots\cdots\cdots\cdots\cdots \\ \frac{dy_n}{dx} &= -\frac{1}{a_n}\left[a_{n-1}(x)y_n + a_{n-2}(x)y_{n-1} + \cdots + a_0(x)y_1\right] \end{aligned} \tag{2.43}$$

Any value of x for which $a_n(x)$ becomes zero is called a singular point. At this point the Lipschitz condition does not hold, and no unique solution exists. Ince [24, pp. 396–415] shows that the method of Frobenius will give a solution to Eq. (2.42) which is valid everywhere but at the singular point. If the singularity occurs at $x = d$, the method of Frobenius gives an explicit answer for Eq. (2.42) by assuming a solution in the form of the infinite series, expanded about the singular point

$$y(x - d, \sigma) = (x - d)^\sigma \sum_{j=0}^{\infty} c_j(x - d)^j \tag{2.44}$$

Equation (2.44) is substituted into Eq. (2.42), where all the coefficients a_i are represented by polynomials in $x - d$. After terms are combined properly, there will be n coefficients of the c's which are unspecified and are the c's associated with the lowest powers of x in Eq. (2.44) (that is, $c_0, \ldots, c_{n-1}$). All other c's can be expressed in terms of these. These unspecified c's have values determined by the initial conditions. There will also be an nth-order polynomial in σ of the form

$$\sigma(\sigma - 1) \cdots (\sigma - n + 1) + \sigma(\sigma - 1) \cdots (\sigma - n + 2)Q_{n-1}(d) + \cdots + \sigma Q_1(d) + Q_0(d) = 0 \tag{2.45}$$

giving n values which result in n independent solutions of Eq. (2.42). There is a special problem when one or more of the σ's is an integer. When this occurs, one must use special techniques to get the n independent solutions.

The method of Frobenius may also be used to solve n equations without singularities whose coefficients are polynomials in x. To date, its primary utility has been in solving several important second-order equations, some of which are

Bessel's equation

$$x^2 \frac{d^2y}{dx^2} + x\frac{dy}{dx} + (x^2 - a^2)y = 0 \qquad \text{singularity at } x = 0$$

Legendre's equation

$$(1 - x^2)\frac{d^2y}{dx^2} - 2x\frac{dy}{dx} + a(a + 1)y = 0 \qquad \text{singularities at } x = \pm 1$$

Gauss' equation

$$x(1 - x)\frac{d^2y}{dx^2} + [c - (a + b + 1)x]\frac{dy}{dx} - aby = 0$$

singularities at $x = 0, 1$

where a, b, and c are constants.

The major difficulty with the method of Frobenius is that the solution is in the form of an infinite series. Unless this series can be represented in some recognizable, closed form, it is virtually impossible to find out anything about the solution.

Suppose you were given the infinite series

$$f(x) = \sum_{k=1}^{\infty} (-1)^{k-1} \frac{x^{2k-1}}{(2k - 1)!} \tag{2.46}$$

and had to answer such questions as:

1. What are the zeros of $f(x)$ and where do they occur?
2. What are values of the maxima and where do they occur?
3. For $x = 45$, what is the value of $f(x)$?
4. What are the answers to the first three questions for d^2f/dx^2?

These questions are very difficult to answer if one must work with Eq. (2.46), but they become almost trivial when one can use Eq. (2.47) instead:

$$f(x) = \sin x \tag{2.47}$$

For this reason the method of Frobenius should be regarded as a last resort, to be used when there is no other method available.

Reduction of Order. If one knows one or more independent solutions of an nth-order equation, one may reduce the order of the equation by the number of solutions. Sometimes, but not often, this reduced equation can be solved.

To reduce the order one uses the following procedure. Suppose one knows r solutions: $u_k(x), \ldots, u_r(x)$. Then a substitution of the form

$$z = \begin{vmatrix} u_1^{(0)} & \cdots & u_r^{(0)} & \cdots & y^{(0)} \\ u_1^{(1)} & \cdots & u_r^{(1)} & \cdots & y^{(1)} \\ u_1^{(r)} & \cdots & u_r^{(r)} & \cdots & y^{(r)} \end{vmatrix} Q(x) \tag{2.48}$$

where $Q(x)$ is a convenient function and

$$u_s^{(r)} = \frac{d^r u_s}{dx^r}$$

will result in an $(n - r)$th-order equation in z.

When only one solution, $u_1(x)$, is known, the substitution

$$y = u_1(x)v(x) \tag{2.49}$$

into the original nth-order differential will result in a new equation of order $n - 1$ in the variable dv/dx [23, pp. 31–35].

2.4.5 Constant-coefficient Inhomogeneous Equations

Integral Transforms. In general, for an arbitrary forcing function whose transform can be evaluated, the integral-transform method is best for obtaining the particular solution of a constant-coefficient equation. Fourier and Laplace transforms are described in Sec. 2.4.3.

Method of Undetermined Coefficients. When the forcing function consists of a finite number of combinations of polynomials, sines and cosines, and exponentials, the method of undetermined coefficients is an even better method than the integral-transform method.

If x is a term in the forcing function, then those derivatives of $f(x)$ which with $f(x)$ form the largest set of linearly independent functions are called the family of $f(x)$. For example, the family of

$$f(x) = x \sin 2x \tag{2.50}$$

is

$$\{x \sin 2x,\ x \cos 2x,\ \sin 2x,\ \cos 2x\}$$

We determine the particular solution by first assuming that the solution consists of a sum of all members of a family where each term has an undetermined coefficient. The assumed solution is substituted into the original equation, and the coefficients are calculated.

When a member of the family is also a solution of the homogeneous equation, each member of the family must be multiplied by a high enough power of x so that no member is the same as the homogeneous solution.

2.4.6 Variable-coefficient Inhomogeneous Equations, and the Method of Variation of Parameters

If the equation has variable coefficients, the likelihood of our finding the particular solution is small, unless we have the complete set of solutions of the homogeneous equation. In this instance the particular solution can be assumed to be of the form

$$y_p = \sum_{k=1}^{n} c_k(x) u_k(x) \tag{2.51}$$

where $c_1(x), c_2(x), \ldots, c_n(x)$ are unknown coefficients which are functions of the independent variable, and $u_1(x), u_2(x), \ldots, u_n(x)$ are the set of solutions of the homogeneous equation.

If Eq. (2.51) is substituted into the original differential equation, the particular solution can be shown to be

$$y_p = \int_{x_a}^{x} \frac{\sum_{i=1}^{n} u_i(x) W_i(a)}{W[u_1(a), u_2(a), \ldots, u_n(a)]} f(a)\, da \tag{2.52}$$

where $f(a)$ is the forcing function, x_a is the initial value of x, $W(a) = W[u_1(a), u_2(a), \ldots, u_n(a)]$ is the Wronskian, and $W_i(a)$ is the Wronskian $W[u_1(a), \ldots, u_n(a)]$ with the ith column replaced as shown in Eq. (2.53):

$$W_i(a) = \begin{vmatrix} 1 & 2 & & i-1 & i & i+1 & & n \\ u_1^{(0)}(a) & u_2^{(0)}(a) & \cdots & u_{1-i}^{(0)}(a) & 0 & u_{1+i}^{(0)}(a) & \cdots & u_n^{(0)}(a) \\ u_1^{(1)}(a) & u_2^{(1)}(a) & & & 0 & & & \\ \cdot & & & & \cdot & & & \\ \cdot & & & & \cdot & & & \\ \cdot & & & & \cdot & & & \\ u_1^{(n-1)}(a) & & & & 1 & & & \end{vmatrix} \tag{2.53}$$

A comparison of Eqs. (2.30) and (2.52) shows that the weighting function $K(x, a)$ is

$$K(x, a) = \frac{\sum_{i=1}^{n} u_i(x) W_i(a)}{W(a)} \tag{2.54}$$

Since the solution is unique, there is only one weighting function associated with each linear differential equation. This function completely characterizes the inhomogeneous equation. Used in Eq. (2.52) in conjunction with an arbitrary forcing function $f(x)$, the weighting function will determine the particular solution.

Unfortunately, the weighting function can be evaluated analytically only if a linearly independent set of solutions of the homogeneous equation is known. The chances of finding such a set of solutions for a variable-coefficient equation are remote. However, with the computer we can obtain the next best thing to an analytic solution—a plot of the weighting function.

One must modify the differential equation in a particular fashion to solve for this function in one computer run. In Sec. 3.3 we discuss the weighting function and the computer technique for generating it.

Example 5. The equation

$$\frac{d^3y}{dt^3} - \frac{1}{t}\frac{d^2y}{dt^2} - \frac{dy}{dt} + \frac{1}{t}y = te^{2t} \tag{2.55}$$

has as a solution to its homogeneous equation

$$y_h = c_1e^t + c_2e^{-t} + c_3t \tag{2.56}$$

where

$$\begin{aligned} u_1(t) &= e^t \\ u_2(t) &= e^{-t} \\ u_3(t) &= t \end{aligned} \tag{2.57}$$

By Eq. (2.26)

$$W(x) = \begin{vmatrix} e^x & e^{-x} & x \\ e^x & -e^{-x} & 1 \\ e^x & e^{-x} & 0 \end{vmatrix} = 2x \tag{2.58}$$

Following Eq. (2.53)

$$W_1(x) = \begin{vmatrix} 0 & e^{-x} & x \\ 0 & -e^{-x} & 1 \\ 1 & e^{-x} & 0 \end{vmatrix} = e^{-x}(1 + x) \tag{2.59}$$

$$W_2(x) = \begin{vmatrix} e^{x} & 0 & x \\ e^{x} & 0 & 1 \\ e^{x} & 1 & 0 \end{vmatrix} = e^{x}(x - 1) \tag{2.60}$$

$$W_3(x) = \begin{vmatrix} e^{x} & e^{-x} & 0 \\ e^{x} & -e^{-x} & 0 \\ e^{x} & e^{-x} & 1 \end{vmatrix} = 2 \tag{2.61}$$

Equation (2.54) gives as the weighting function

$$K(t, x) = \frac{u_1(t)W_1(x) + u_2(t)W_2(x) + u_3(t)W_3(x)}{W(t)} \tag{2.62}$$

Substituting the pertinent quantities in Eq. (2.62), we obtain

$$K(t, x) = \frac{(x + 1)e^{t-x} + (x - 1)e^{t+x} - 2t}{2x} \tag{2.63}$$

The particular solution to Eq. (2.55) for a problem starting at $t = t_a$ is

$$y = \int_{t_a}^{t} K(t, x)e^{2x}\, dx \tag{2.64}$$

After some algebraic manipulation we get as a final result

$$y = e^t \left[\frac{3t - 4}{18}(e^{3t} + e^{3t_a}) - \frac{t}{2}e^t - \frac{t_a}{2}e^{t_a} - te^{2t_a - t}\right] \tag{2.65}$$

2.4.7 Summary

The form of the solution of the general nth-order inhomogeneous equation is always the sum of n independent solutions of the homogeneous equation plus an integral involving the weighting function and the forcing function. The solution can always be expressed in the form

$$y = c_1u_1(x) + c_2u_2(x) + \cdots + c_nu_n(x) + \int_{x_a}^{x} K(x, \alpha)f(\alpha)\, d\alpha \tag{2.66}$$

Because of linearity any u_i or combination of u_i's is a solution of the homogeneous equation. The weighting function $K(x, \alpha)$ is a function of the u_i's alone. The u_i's are determined by the order of the differential equation and the nature of the coefficients. They are independent of the forcing function and the initial conditions.

The set of u_i's, once found, can be used to calculate the weighting function. The u_i's and the weighting function are all that is needed to solve the linear differential equation for any combination of initial conditions and forcing function.

Table 2.1 lists the methods which may be employed to solve these equations. In each category the techniques are tabulated in the order of increasing difficulty to apply.

TABLE 2.1 METHODS FOR SOLVING HIGHER THAN FIRST-ORDER LINEAR DIFFERENTIAL EQUATIONS

I. *Homogeneous Equations*
  A. Constant coefficients
    1. Polynomial factoring
    2. Laplace transforms
    3. Picard iteration†
  B. Variable coefficients
    1. Method of Frobenius
    2. Picard iteration†

II. *Inhomogeneous Equations*
  A. Constant coefficients
    1. Undetermined coefficients
    2. Laplace or Fourier transform
    3. Variation of parameters
    4. Picard iteration†
  B. Variable coefficients
    1. Variation of parameters
    2. Picard iteration†

† The Picard iteration method may also be used to solve nonlinear equations. It is discussed in Sec. 2.5.2.

2.5 Analytic Solution of Nonlinear Equations [37–39]

2.5.1 Introduction

In Sec. 2.4 we showed that the structure of the solution is the same for all linear differential equations. Furthermore, by solving separately a homogeneous and an inhomogeneous equation and adding the two solutions, we have a systematic procedure for solving any linear equation. We have no such a priori knowledge about the nonlinear equation. The functional form of the solution is dependent not only upon the differential equation itself but also upon the forcing functions and the particular initial conditions chosen. For the nonlinear equation, every time we change the combination of initial conditions and forcing function we must solve the equation anew. In other words, we can get no general solution to a nonlinear differential equation from which we may derive a solution for a specific set of initial conditions and forcing function.

The linear equation, even with the property of superposition and the known form of the general solution, can almost never be solved for the variable-coefficient case. The nonlinear equation is virtually hopeless if one wants an exact analytic solution.

We shall discuss five techniques which may be used to solve nonlinear differential equations.[1] They are (1) Picard iteration, (2) piecewise linearization, (3) perturbation, (4) phase-plane analysis, and (5) describing func-

[1] For some additional techniques which may occasionally be used, see Ref. 40, pp. 335–353.

tion. All have severe restrictions, and only the Picard iteration method has general application to all nonlinear equations. Piecewise linearization and perturbation techniques convert the nonlinear equations to linear ones, which are easier to solve. Phase-plane analysis is a graphical technique which may be used for a special kind of second-order nonlinear equation. The describing-function method does not give the solution but instead determines whether or not a sinusoid is an approximate solution of the equation.

2.5.2 Picard Iteration

The Picard iteration method may be used for any differential equation, linear or nonlinear, which satisfies the condition for uniqueness. If we begin with the first-order equation

$$\frac{dy}{dx} = f(x, y) \tag{2.67}$$

and integrate,

$$\int_{x_0}^{x} \frac{dy}{dx}\, d\alpha = \int_{x_0}^{x} f(\alpha, y)\, d\alpha \tag{2.68}$$

or

$$y(x) = y_0 + \int_{x_0}^{x} f(\alpha, y)\, d\alpha \tag{2.69}$$

where y_0 is the initial value of y at $x = x_0$.

Equation (2.69) is an integral equation whose solution for y is the same as that of Eq. (2.67). The Picard iteration uses Eq. (2.69) by assuming any convenient y, designated y_1, to give y_2 shown in Eq. (2.70):

$$y_2 = y_0 + \int_{x_0}^{x} f(\alpha, y_1)\, d\alpha \tag{2.70}$$

Next, y_2 is used to obtain y_3:

$$y_3 = y_0 + \int_{x_0}^{x} f(\alpha, y_2)\, d\alpha \tag{2.71}$$

The process is repeated so that in general

$$y_n = y_0 + \int_{x_0}^{x} f(\alpha, y_{n-1})\, d\alpha \tag{2.72}$$

For large n, y_n will approach the y in Eq. (2.67), no matter what initial y_1 is chosen in the iteration.

Example 6. For the equation

$$\frac{dy}{dx} = x - y \qquad y = 1 \text{ at } x = 0 \tag{2.73}$$

the Picard iteration equation is

$$y_n(x) = 1 + \int_0^x [\alpha - y_{n-1}(\alpha)]\, d\alpha \tag{2.74}$$

As a starting value of y, we take $y = 1$. This is only for convenience. The y_n's will converge to the correct solution no matter what the starting point is. Then

$$y_1(x) = 1 + \int_0^x (\alpha - 1)\, d\alpha = 1 - x + \frac{x^2}{2} \tag{2.75}$$

$$y_2(x) = 1 + \int_0^x \alpha - \left(1 - \alpha + \frac{\alpha^2}{2}\right) d\alpha = 1 - x + x^2 - \frac{x^3}{3!} \tag{2.76}$$

$$y_3(x) = 1 + \int_0^x \alpha - \left(1 - \alpha + \alpha^2 - \frac{\alpha^3}{3!}\right) d\alpha = 1 - x + x^2 - \frac{x^3}{3!} + \frac{x^4}{4!} \tag{2.77}$$

This procedure will ultimately lead to the correct solution

$$y = 2e^{-x} - 1 - x \tag{2.78}$$

For a system of first-order equations,

$$\begin{aligned} \frac{dy_1}{dx} &= f_1(x, y_1, y_2, \ldots, y_n) \\ \frac{dy_2}{dx} &= f_2(x, y_1, y_2, \ldots, y_n) \\ &\cdots\cdots\cdots\cdots \\ \frac{dy_n}{dx} &= f_n(x, y_1, y_2, \ldots, y_n) \end{aligned} \tag{2.79}$$

The first step in the iteration is

$$\begin{aligned} y_{12} &= y_{10} + \int_{x_0}^x f_1(\alpha, y_{11}, y_{21}, \ldots, y_{n1})\, d\alpha \\ y_{22} &= y_{20} + \int_{x_0}^x f_2(\alpha, y_{11}, y_{21}, \ldots, y_{n1})\, d\alpha \\ &\cdots\cdots\cdots\cdots \\ y_{n2} &= y_{n0} + \int_{x_0}^x f_n(\alpha, y_{11}, y_{21}, \ldots, y_{n1})\, d\alpha \end{aligned} \tag{2.80}$$

where $y_{10}, y_{20}, \ldots, y_{n0}$ are the values of the respective y's at $x = x_0$
$y_{11}, y_{21}, \ldots, y_{n1}$ are the assumed initial functions of $y_1, y_2, \ldots, y_n$
$y_{21}, y_{22}, \ldots, y_{n2}$ are the first iteration of $y_1, y_2, \ldots, y_n$

The mth iteration has the form

$$\begin{aligned} y_{1,m+1} &= y_{10} + \int_{x_0}^x f_1(\alpha, y_{1,m}, y_{2,m}, \ldots, y_{n,m})\, d\alpha \\ &\cdots\cdots\cdots\cdots \\ y_{n,m+1} &= y_{n0} + \int_{x_0}^x f_n(\alpha, y_{1,m}, y_{2,m}, \ldots, y_{n,m})\, d\alpha \end{aligned} \tag{2.81}$$

The Picard iteration has three serious shortcomings:

1. We may not be able to integrate the necessary functions.

2. The rate of convergence to the correct solution may be very slow. How fast y_n approaches the correct solution y depends largely on how wisely we chose the original point y_0. The slower the rate of convergence the more steps are needed to get a good approximation to the solution.

3. The solution will almost always appear as an infinite series. Unless the series can be replaced by some known function, for example, e^x, $\sin x$, etc., we face the same problem as with the method of Frobenius: the solution is an infinite series with which we can do little.

For these reasons one may not often be able to use the Picard iteration.

2.5.3 Piecewise Linearization

As related in Sec. 2.2.3, a differential equation may be solved either over the entire range of the independent variable or in a set of contiguous segments which cover the whole range. When the equation is solved in segments, the final set of values of one segment becomes the initial set of values of the next. Sometimes an equation that is nonlinear over the entire range may be divided into segments, each of which is linear. Each segment is solved and pieced together to form the entire solution. This process of converting a nonlinear equation into a set of linear ones is called piecewise linearization.

Example 7. The equation

$$\frac{dy}{dt} = f(y) \tag{2.82}$$

$$\text{where } f(y) = \begin{cases} y & \text{for } -\infty < y < 10 \quad (2.83a) \\ 10 & \text{for } y \geq 10 \quad (2.83b) \end{cases}$$

is nonlinear if it is solved for the entire range of t. It is linear, however, in any segment which satisfies the condition for Eq. (2.83a) or (2.83b).

Assume that the initial condition is

$$y = 1 \qquad \text{at } t = 0$$

Solving Eq. (2.83a), we have

$$y = e^t \tag{2.84}$$

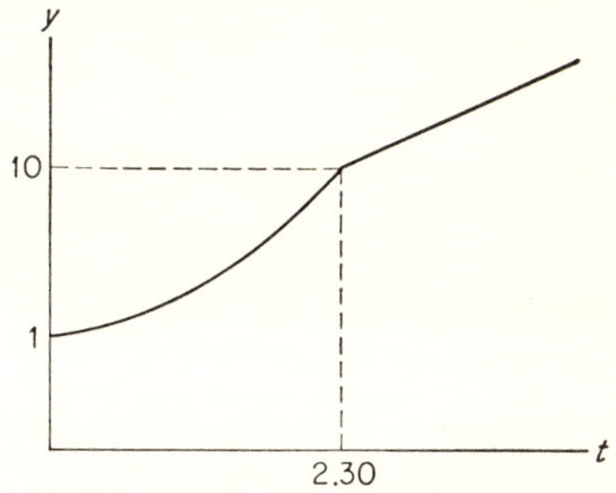

FIG. 2.1 Solution to a differential equation solved by piecewise linearization.

Equation (2.83a) holds until $y = 10$. We find the time at which this occurs:

$$\begin{aligned} 10 &= e^t \\ t &= \ln 10 = 2.30 \end{aligned} \tag{2.85}$$

At $t = 2.30$ we solve Eq. (2.83b) as follows:

$$y_2 = 10t + A \tag{2.86}$$

Substituting as an initial condition

$$\begin{aligned} y_2 &= 10 \qquad \text{at } t = 2.30 \\ A &= -220 \end{aligned}$$

Hence the solution to Eq. (2.82) is

$$\begin{aligned} y &= e^t & 0 &\leq t < 2.30 \\ y &= 10t - 220 & 2.30 &\leq t < \infty \end{aligned} \tag{2.87}$$

and is plotted in Fig. 2.1.

2.5.4 Perturbation Technique[1]

Although piecewise linearization is exact, much effort can be expended in determining the solution, especially if there are several segments involved.

If the variables in each nonlinear term deviate only a small amount about

[1] Perturbation is discussed in greater detail in Sec. 7.3.

some nominal value, we can make a Taylor series expansion and drop the higher-order terms. This perturbation technique will convert the nonlinear equation to a linear one.

The solution of the linear equation will approximate the original equation solution over a range of values of the variables in the truncated Taylor series. Since the perturbation method gives only an approximation to the correct solution, the user must employ the method with discretion. The amount of error one is willing to tolerate will determine the extent to which the method can be applied.

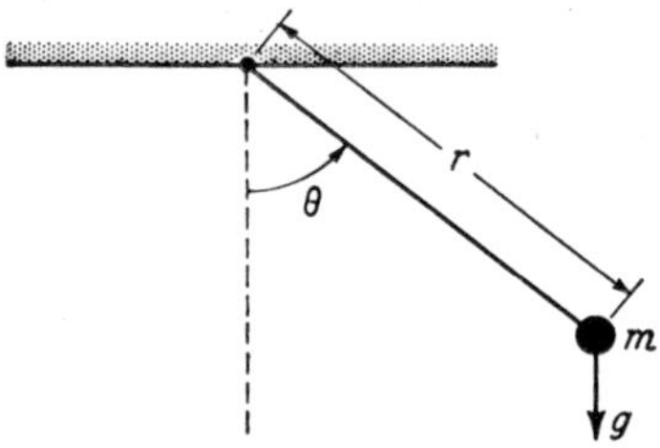

FIG. 2.2 A simple pendulum.

Example 8. For the pendulum of Fig. 2.2, where a mass m is attracted to a weightless rod, the equation of motion is

$$\frac{d^2\theta_T}{dt^2} + \frac{g}{r}\sin\theta_T = 0 \tag{2.88}$$

Equation (2.88) is a nonlinear equation which can be solved only in terms of elliptic integrals [41, pp. 115–130]. The solution is

$$t = \int_{\theta_0}^{\theta} \frac{d\theta}{\left(\frac{d\theta}{dt}\right)_{t=0}^2 + \frac{2g}{r}(\cos\theta - \cos\theta_0)} \tag{2.89}$$

where $\left(\frac{d\theta}{dt}\right)_{t=0}$ = initial value of $\frac{d\theta}{dt}$ at $t = 0$

θ_0 = value of θ at $t = 0$

However, if the angle is small we may make the approximation

$$\sin\theta \approx \theta$$

and Eq. (2.88) becomes

$$\frac{d^2\theta_A}{dt^2} + \frac{g}{r}\theta_A = 0 \tag{2.90}$$

The solution to Eq. (2.90) is

$$\theta_A = C_1 \sin kt + C_2 \cos kt \tag{2.91}$$

where $$k = \frac{g}{r}$$

The maximum value of the angle is its initial value for both θ_T and θ_A. The principal result of the approximation is that the period of the approximate angle θ_A remains the same regardless of the initial condition. The period of oscillation of the true angle θ_T increases as the angle increases. This situation has been known to the makers of pendulum clocks, to their despair, since the days of Christian Huygens (ca. 1659).

2.5.5 Describing-function Method[1]

The describing-function method may be used to calculate whether or not a sine wave is the approximate solution to an nth-order nonlinear homogeneous equation. If the sine is a solution, the describing-function method can determine the amplitude and frequency. If not, the method can ascertain certain special conditions which will make it a solution.

The describing-function method has had some application in the stability analysis of servomechanisms. A system is said to be stable if for any set of initial conditions the solution goes to a constant value. In a linear system there are three alternatives, none of which is dependent upon the initial conditions. All situations apply to an independent variable which is increasing.

1. The solution converges to zero and is called absolutely stable.
2. The solution oscillates and is called conditionally stable.
3. The solution diverges and is called unstable.

The nonlinear system has all the stability characteristics of the linear system plus an additional one called a limit cycle. (The solution of a nonlinear system may converge to a constant as well as zero.) A limit cycle is a periodic oscillation. However, it may be possible either to converge or to diverge to this oscillation depending upon the initial conditions chosen. Furthermore, there may be two or more different limit cycles. In this case the one excited will depend exclusively on the initial conditions.

The describing-function technique may be employed only if certain conditions are met.

For the set of equations

$$\begin{aligned} \frac{dy_1}{dx} &= f_1(y_1, \ldots, y_n, x) \\ &\cdots\cdots\cdots\cdots\cdots \\ \frac{dy_n}{dx} &= f_n(y_1, \ldots, y_n, x) \end{aligned} \tag{2.92}$$

all the $f_i(y_1, y_2, \ldots, y_n, x)$ except one must be linear combinations of the y's. The one exception must be of the form

$$\frac{dy_{n-2}}{dx} = f_j(y_{n-1}) \tag{2.93}$$

In addition to being a function of only one y_j, the function f_j has the following restrictions:

[1] Reference 35, chap. 10. This method has also been extended to the study of statistical problems by R. C. Booton. His work is described in Ref. 39, pp. 213–271.

1. If y_{n-1} is sinusoidal, then $f_j(y_{n-1})$ has the same period as y_{n-1}.
2. In a Fourier series expansion of $f_j(y_{n-1})$, only the fundamental is considered in the analysis. Therefore, $f_j(y_{n-1})$ is assumed to be a sine wave of the same period as y_{n-1}.

Using servomechanism terminology, these restrictions may be restated as follows:

1. In a given system there is only one nonlinearity.
2. The nonlinearity has only one input and its behavior is in no way determined by the independent variable; that is, the nonlinearity is a zero-memory device.
3. If the input to the nonlinearity is sinusoidal, the output is periodic and has the same period as the input.
4. Only the fundamental-frequency component of the nonlinear output is used for analysis.

The solution is most easily obtained graphically by means of a Nyquist plot. The principal advantage of the method is that it may be used for a system of any order.

Its major disadvantages are:

1. The method gives an approximate solution by assuming that only the fundamental of the output of the nonlinearity needs to be considered. There is no convenient method for determining the error in the approximate solution.
2. The behavior of the system is analyzed only in relation to the existence of an oscillation. Even when such an oscillation exists, the method cannot be used to determine the transient solution from the initial value to the point where it becomes periodic.
3. There is very little information about a system that does not have an oscillation. One cannot, in such case, get a solution for an arbitrary set of initial conditions and a forcing function.

2.5.6 Phase-plane Analysis[1]

The phase-plane method is a graphical technique which can be used to plot the dependent variable versus its first derivative. The plot is used to obtain the solution of a second-order system for any particular set of initial conditions.

The equation may be written as

$$\begin{aligned} \frac{dx}{dt} &= P(x, y) \\ \frac{dy}{dt} &= Q(x, y) \end{aligned} \tag{2.94}$$

[1] Reference 42 is the classic treatment of this subject. The text is clearly written and thorough. See also Ref. 35, chap. 11.

or
$$\frac{d^2x}{dt^2} = a\left(x, \frac{dx}{dt}\right)\frac{dx}{dt} + b\left(x, \frac{dx}{dt}\right)x = C \tag{2.95}$$

where C is a constant.

There are two restrictions on the differential equation indicated in these equations:

1. None of the quantities P, Q, a, b is a function of the independent variable.
2. There can be no forcing function other than a constant C.

The phase-plane analysis of Eq. (2.95) proceeds in the following steps:

1. With $y = dx/dt$, rewrite Eq. (2.95):

$$\frac{dy}{dt} + a(x, y)y + b(x, y)x = C \tag{2.96}$$

2. Divide by y:

$$\frac{dy/dt}{y} = \frac{dy}{dx} = -a(x, y) - b(x, y)\frac{x}{y} + \frac{C}{y} \tag{2.97}$$

since
$$\frac{\dfrac{dy}{dt}}{y} = \frac{\dfrac{dy}{dx}\dfrac{dx}{dt}}{\dfrac{dx}{dt}} = \frac{dy}{dx} \tag{2.98}$$

If a curve is drawn on the phase plane (y versus x) such that for each value of x and y the slope of the curve dy/dx satisfies Eq. (2.97), then x and y are solutions of Eqs. (2.94) or Eq. (2.95).

3. As an assistance in drawing the phase-plane plot, one first draws a set of isoclines, lines of constant slope, $dy/dx = k$. The isoclines are determined by setting $dy/dx =$ const and plotting the xy curve satisfying this condition. If, for example, $dy/dx = -1$, then the line marked -1 on the graph would indicate this isocline. Cross lines are drawn through the isocline all having the slope represented by the isocline, in this case -1. In a similar fashion other isoclines may be calculated and plotted. With these isoclines as aids we can get the phase portrait for any set of initial conditions.

Thus one obtains directly the relationship between the dependent variable x and its first derivative $y = dx/dt$. The information one is usually interested in, a plot of x versus the independent variable t, can also be obtained by a proper processing of the curve. The quantity t is determined by solution of the integral[1]

$$t = \int_{x_1}^{x_2} \frac{1}{y}\, dx \tag{2.99}$$

where $t =$ time for curve to pass from x_1 to x_2.

The usefulness of the phase-plane analysis may be summarized:

[1] The time can also be obtained graphically [35, p. 625].

Advantages

1. Once the isoclines have been determined, a graphical solution may be rapidly drawn for any set of initial conditions.

Disadvantages

1. The equation must be second-order, with the forcing function either zero or a constant. The terms of the equation cannot be functions of the independent variable.
2. The graphical solution is approximate. The error decreases as the number of isoclines used increases.

Parenthetically, we add that phase-plane plots are easy to obtain from an analog computer. As a result the graphical method of isoclines has been largely supplanted by the computer, which provides phase portraits more easily.

2.5.7 Summary

In a linear differential equation the solution can be obtained in two steps. First, a set of independent solutions to the homogeneous equations and a weighting function are found, and then the exact solution can be calculated for any initial conditions and forcing function. In the nonlinear equation we can get no part of the solution unless the exact initial conditions and forcing function are known. The various methods for solving nonlinear differential equations are listed with appropriate comments in Table 2.2.

TABLE 2.2

Method	*Comments*
1. Picard iteration method	The method works in principle for any nonlinear equation. However, quite often the necessary integrations cannot be carried out. The solution in the limit is exact
2. Piecewise linearization	The technique works in the one case where a nonlinear equation may be broken up into segments, each of which is linear over a range of the independent variable. The solution is exact
3. Perturbation	This method has wide application in the solution of nonlinear equations. The equations are linearized by assuming that certain quantities vary over a small range. The solution is approximate
4. Describing function	This procedure determines under what conditions a sinusoid is the approximate steady-state solution. The method works for any order equation which has only one nonlinearity of a particular kind
5. Phase plane	This is a graphical technique which gives an approximate solution for one type of second-order equation

3 Programming of Differential Equations

3.1 Computer Programming

3.1.1 Introduction

The importance of the computer is painfully obvious in our discussion of analytic methods for solving linear variable-coefficient and, to a greater degree, nonlinear differential equations. The analog or digital computer must be used to solve almost all differential equations. In this chapter we shall discuss the three primary methods of programming the equations on the computer. These techniques are (1) the general method, (2) canonical form, and (3) dummy variable. We shall consider the circumstances under which each method may be used.

We next show how the computer may be used to get a normalized set of solutions for a linear equation. This normalized set of solutions may be used to calculate the solution for any arbitrary set of initial conditions. One of the normalized solutions is the weighting function, the importance of which is discussed in Sec. 2.4, dealing with the properties of linear equations. Only a discrete approximation can be made to the weighting function, which is continuous, if the differential equation is programmed. However, the exact weighting function may be generated if the adjoint of the differential equation is formed. The adjoint technique has broad application in the solution of differential equations. We shall show how the adjoint of an equation is formed and give some simple rules for its programming.

We end the chapter with an examination of the linear boundary-value problem. Whereas the linear initial-value problem always has a unique solution, the boundary-value problem does not. Consequently, the initial-value problem is always easier to solve. We shall describe how the normalized equations may be used first to determine if a solution exists and then to find the initial conditions which will give this solution.

3.1.2 The General Method

Differential equations of the first degree may be programmed on the computer in a simple manner. Since the technique is the same for any such equation, linear or nonlinear, we shall consider only the simplest case, namely, the linear equation with constant coefficients. Suppose the equation to be programmed is

$$\frac{d^2y}{dt^2} + a_1\frac{dy}{dt} + a_0y = f(t) \tag{3.1}$$

As a first step we equate the *highest*-order derivative to the rest of the terms in the equation:

$$\frac{d^2y}{dt^2} = -a_1\frac{dy}{dt} - a_0y + f(t) \tag{3.2}$$

Next we assume that d^2y/dt^2 is available and place enough integrators in cascade to obtain the zeroth-order derivative, as shown in Fig. 3.1.[1]

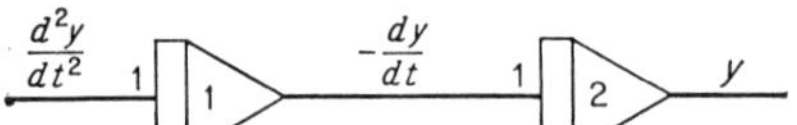

FIG. 3.1 The general method, the first step in programming Eq. (3.1).

Figure 3.1 contains all the derivatives of y which are required to satisfy the right side of Eq. (3.2). The next step is to multiply each of these derivatives by the proper coefficient; then sum and add the forcing function to obtain the result in Fig. 3.2.

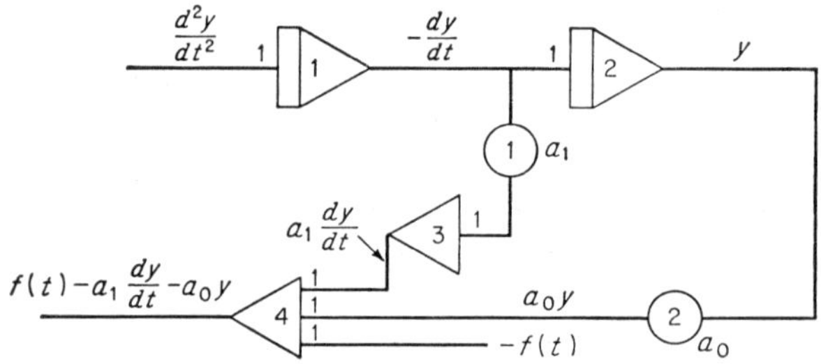

FIG. 3.2 The general method, an intermediate step in programming.

By Eq. (3.2) the output of amplifier 4 is equal to the input of integrator 1. Hence, when these points are connected, the loop will be closed and the computer will be constrained to solve Eq. (3.1). The addition of initial conditions, which appear at the outputs of integrators 1 and 2, completes the programming. The final computer diagram is shown in Fig. 3.3.[2]

Because our primary purpose was to illustrate the programming procedure with utmost clarity, our computer program has amplifier 3, which

[1] See Fig. 1.1 for a standard set of symbols.

[2] In the method described here as well as in the sections on canonical form and dummy variable, the computer is programmed directly from the original differential equation. Often, however, computer limitations require that the program be modified. This is called scaling the problem and is treated in detail in Chap. 4.

is superfluous. Amplifier 3 can be eliminated if we feed the output of potentiometer 1 into integrator 1 instead of amplifier 3. However, we can no longer observe d^2y/dt^2, because it is not the output of a single computer element.

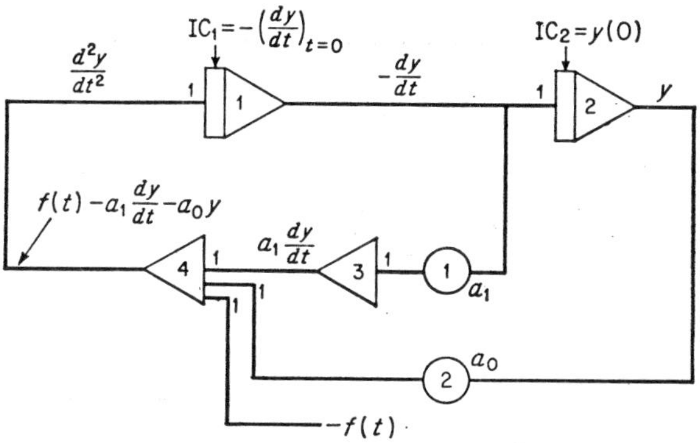

FIG. 3.3 The computer diagram of Eq. (3.1), programmed by the general method.

For the nth-order equation the computer is programmed in the same fashion. The steps to be followed are:

1. Set the highest-order derivative equal to all the other terms in the equation.
2. Draw in cascade a number of integrators equal to the order of the equation.
3. Assume that the highest-order derivative is available and assume that it is the input to the first integrator. Then the outputs of the integrators will be the dependent variable and its first $n - 1$ derivatives.
4. With these integrator outputs and the forcing function, generate the terms which are equal to the highest-order derivative.
5. Close the loop by feeding all these terms to the input of the first integrator.
6. Set up the problem initial conditions on the pertinent integrators. Each initial condition appears at the output of an integrator. Therefore, from our discussion in Sec. 2.2.3, if every integrator has an initial condition associated with it, there will be a unique solution.

The general method of programming was first developed by Lord Kelvin. We can recapture the moment of discovery as Lord Kelvin reported it to the Royal Society in 1876 [16]. Kelvin began his paper by stating that an equation of the form[1]

[1] Lord Kelvin's paper contains an error. Equation (3.4) holds when Eq. (3.3) reads

$$\frac{d}{dx}\left[\frac{1}{p(x)}\frac{dy}{dx}\right] = -y$$

$$\frac{d}{dx}\left[\frac{1}{p(x)}\frac{dy}{dx}\right] = y \tag{3.3}$$

can be solved by a method of successive approximations where

$$\begin{aligned} y_1 &= \int_0^x p(a)\left(c - \int_0^a y_0\,d\beta\right)da \\ &\cdots\cdots\cdots\cdots\cdots \\ y_n &= \int_0^x p(a)\left(c - \int_0^a y_{n-1}\,d\beta\right)da \end{aligned} \tag{3.4}$$

This is a modified Picard iteration method where y_0 is any convenient function and the initial condition is $y = 0$ at $x = 0$. He described how his brother's ball-and-disk integrator might be used to perform the iteration. Then he wrote:

> So far had I gone and was satisfied, feeling I had done what I wished to do for many years. But then came a pleasing surprise. Complete agreement [existed] between the function fed into the machine and that given out by it. . . . Thus I was led to a conclusion which was quite unexpected; and it seems to me very remarkable that the general differential equation of the second order with variable coefficients may be rigorously, continuously, and in a single process solved by a machine. . . .

In Fig. 3.3 we observe one of the useful characteristics of the analog computer. The dependent variable y, its derivatives, and also sums of these quantities appear simultaneously as voltages which can be readily measured.[1] In particular, we call attention to the fact that y and its derivatives always appear at the outputs of integrators. Note also that the quantities which appear at the input and output of each element are labeled in the diagrams and each computer element is identified by number. When programming, the reader should follow this labeling procedure. It will reduce programming errors and facilitate the operation of the computer.

The computer has three modes involved in problem solution: RESET, OPERATE, and HOLD [43, pp. 500–501, or 44, pp. 43–44]. At RESET the integrator outputs all have their initial values, and their inputs are disconnected. From RESET the computer may be switched to the OPERATE mode, where inputs to the integrators are closed and the computer solves the problem for an increasing independent variable, time.[2] From OPERATE the computer may be put into the HOLD mode. This stops the problem by disconnecting the inputs to the integrators and retaining their outputs at the moment of HOLD. This mode is particularly useful when we are interested in some of the problem quantities at a specific value of the independent variable. From HOLD we can go to either OPERATE or RESET. We can also

[1] The voltages are measured relative to some reference ground, which is usually not shown in the computer diagram.

[2] In Sec. 4.6 we discuss the method for solving a problem for a decreasing independent variable.

go from OPERATE to RESET. HOLD may also be used when we wish to stop a solution to change the scaling (see Chap. 7) or to alter some parameters (see Chap. 8).

Example 1. Given

$$\frac{d^2y}{dt^2} + 2\left(\frac{dy}{dt}\right)^2 + e^{-t}y = 1 \tag{3.5}$$

with initial values at $t = 0$,

$$y = 4$$
$$\frac{dy}{dt} = -3$$

we may program the computer by the general method.

Step 1. Set the highest-order derivative in Eq. (3.5) equal to all other terms:

$$\frac{d^2y}{dt^2} = 1 - 2\left(\frac{dy}{dt}\right)^2 - e^{-t}y \tag{3.6}$$

Step 2. Draw Fig. 3.4 having two integrators in cascade.

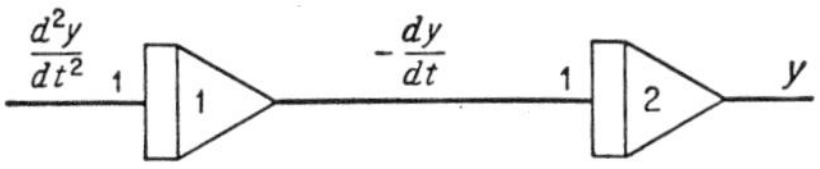

FIG. 3.4 The first step in the programming of Eq. (3.5) by the general method.

Step 3. Assume that the input to integrator 1 is d^2y/dt^2 and label integrator outputs accordingly.

Step 4. Generate the terms on the right side of Eq. (3.6).

Step 5. Feed all these terms to the input of integrator 1.

Step 6. Add the initial conditions.

Steps 4 to 6 are shown in Fig. 3.5.

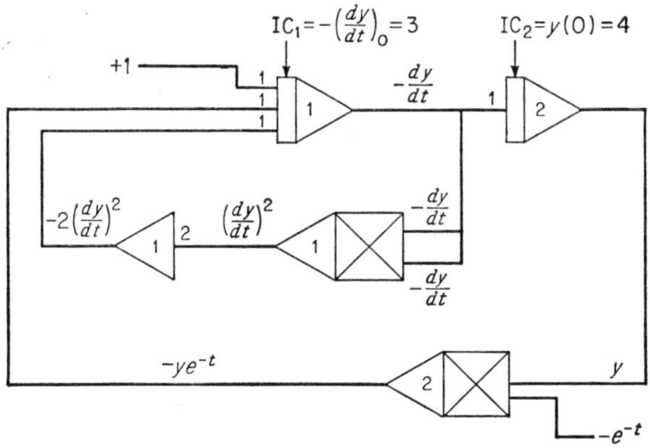

FIG. 3.5 The complete computer diagram of Eq. (3.5), programmed by the general method.

In constructing the diagram we have assumed that e^{-t} is being generated. Ways in which this function may be produced by the computer are discussed in Chap. 5.

Note that the sum of the three inputs to integrator 1 is d^2y/dt^2.

Alternative Use of the General Method. The general method is equally applicable when the problem to be programmed is in the form of a set of first- or higher-order equations. To program

$$\frac{dy_1}{dt} = f_1(y_1, y_2)$$
$$\frac{dy_2}{dt} = f_2(y_1, y_2) \tag{3.7}$$

we begin by assuming that the first-order derivatives dy_1/dt, dy_2/dt are available and integrating once.

Next, with the arguments y_1 and y_2 as inputs we generate $f_1(y_1, y_2)$ and $f_2(y_1, y_2)$. The outputs of these two function blocks, f_1 and f_2, are brought to the inputs of integrators, whose outputs are y_1 and y_2, to close the loop. The insertion of the initial conditions completes the setup of the problem.

To program the equation set

$$\frac{dy_1}{dt} + y_3\frac{dy_2}{dt} = 0$$
$$\frac{d^2y_2}{dt^2} + \frac{dy_2}{dt} + y_2 \sin t - y_1{}^2 = e^{-t} \tag{3.8}$$
$$\frac{dy_3}{dt} + y_1 + y_2 = 0$$

With initial conditions $y_1(0)$, $y_2(0)$, $\dot{y}_2(0)$, $y_3(0)$, we follow the procedure.

Step 1. Equate the highest-order term in each equation to all other terms:

$$\frac{dy_1}{dt} = -y_3\frac{dy_2}{dt} \tag{3.9a}$$
$$\frac{d^2y_2}{dt^2} = e^{-t} - \frac{dy_2}{dt} - y_2 \sin t + y_1{}^2 \tag{3.9b}$$
$$\frac{dy_3}{dt} = -y_1 - y_2 \tag{3.9c}$$

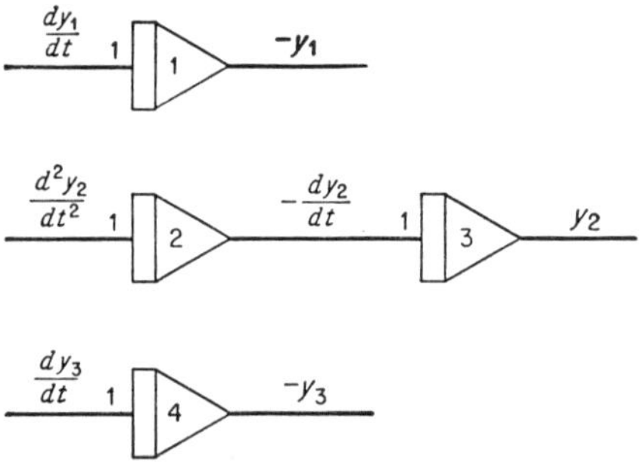

FIG. 3.6 The first step in the programming of Eqs. (3.8), a set of differential equations, by the general method.

Step 2. Assume that the highest-order derivative exists in Eqs. (3.9*a*), (3.9*b*), and (3.9*c*). Place enough integrators in cascade to generate the zeroth-order derivative in each case, as shown in Fig. 3.6.

Step 3. Generate the terms on the right side of Eq. (3.9*a*) and feed them into the input of integrator 1. The terms on the right side of Eq. (3.9*b*) are fed into integrator 2, and those in Eq. (3.9*c*) are fed into integrator 4. The complete diagram with initial conditions is shown in Fig. 3.7.

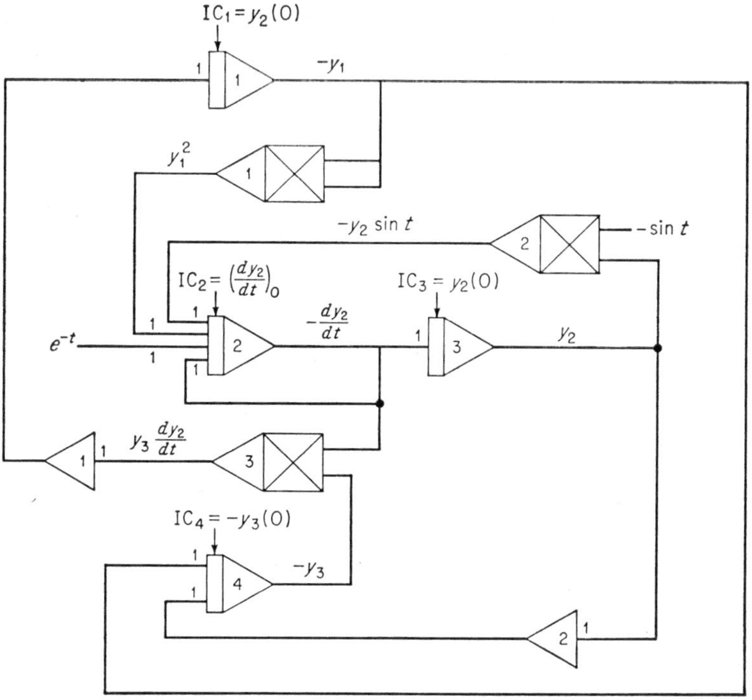

FIG. 3.7 The computer program of Eqs. (3.8).

Programming with Forcing-function Restrictions. The general method of programming works for all equations where the entire forcing function is available. On occasion, however, we must program the equation

$$\frac{d^n y}{dt^n} + a_{n-1}\frac{d^{n-1}y}{dt^{n-1}} + \cdots + a_1\frac{dy}{dt} + a_0 y$$
$$= b_m\frac{d^m x}{dt^m} + b_{m-1}\frac{d^{m-1}x}{dt^{m-1}} + \cdots + b_1\frac{dx}{dt} + b_0 x \quad (3.10)$$

where the forcing function consists of the variable x and its derivatives $dx/dt, \ldots, d^m x/dt^m$ but only the variable x is available. Such a situation arises often in servomechanism problems. In a system where x is the forcing function (input) and y is the dependent variable (output) and the a's and b's are constant, Eq. (3.10) represents a lead-lag network. When $n = 0$, the equation represents a lead network. When $m = 0$, the equation is called a lag network.

Because our computer has integrators but no differentiators[1] we have no way of generating the derivatives directly. The general method fails, and

[1] Electronic differentiators tend to increase computer errors by amplifying noise. They should be used only when there is no other way to obtain the derivative.

we must seek alternative ways of programming the problem. Two such methods are available, the *canonical* and *dummy-variable* methods. Both methods require that $m \leq n$. Furthermore, the a's and b's must be constants in the dummy-variable method.[1]

3.1.3 Canonical Form[2]

To facilitate our discussion we introduce the differential operator p defined as[3]

$$p = \frac{d}{dt}$$

Also
$$p^m = \frac{d^m}{dt^m}$$

where
$$p^m y = \frac{d^m y}{dt^m}$$

There is an associated integral operator

$$p^{-1} = \int_0^t [\quad]\, da$$

where
$$p^{-1}y = \int_0^t y(a)\, da$$

Thus p^{-m} is an m-fold integral, where

$$p^{-m}y = \underbrace{\int_0^x \int_0^{x_n} \cdots \int_0^{x_4} \int_0^{x_3} \int_0^{x_2}}_{m\text{-fold}} y(x_1)\, dx_1\, dx_2 \cdots dx_m$$

When the initial values of the function being operated on and certain of its derivatives are zero, then

$$\frac{p^n}{p^m} = p^{n-m}$$

We shall use the second-order equation

$$\frac{d^2y}{dt^2} + a_1 \frac{dy}{dt} + a_0 y = b_1 \frac{dx}{dt} + b_0 x \tag{3.11}$$

to illustrate the procedure. In operator notation, Eq. (3.11) becomes

$$p^2 y + a_1 p y + a_0 y = b_1 p x + b_0 x \tag{3.12}$$

As in the general method, we begin by equating the highest-order derivative

[1] The b's need not be constants in the canonical-form method if certain of their derivatives are known. See Prob. 4.12.

[2] The canonical-form method is discussed in Sec. 4.3 from a slightly different viewpoint.

[3] The important properties of this operator are discussed in Chap. 4.

to all the other terms:

$$p^2y = b_1px - a_1py + b_0x - a_0y \tag{3.13}$$

With constant coefficients Eq. (3.13) may be written

$$p^2y = p(b_1x - a_1y) + b_0x - a_0y \tag{3.14}$$

Next we integrate twice. In operator notation this is the same as multiplying both sides by p^{-2}. Thus

$$y = \frac{1}{p}(b_1x - a_1y) + \frac{1}{p^2}(b_0x - a_0y) \tag{3.15}$$

In setting up the computer diagram we start by cascading a number of integrators equal to the order of the differential equation, in this case two. The output of the last integrator is designated as y, and x is placed at the input as yet unconnected. This is shown in Fig. 3.8. Already we see a

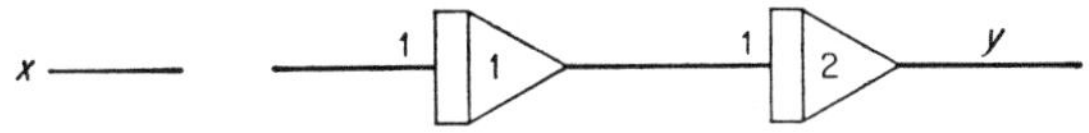

FIG. 3.8 The first step in programming Eq. (3.11) by the canonical-form method.

striking difference between the canonical form and the general method. In the canonical form we begin by assuming that the zeroth-order derivative exists at the output of the last integrator. In the general method we assume that the highest-order derivative exists at the input to the first integrator.

As a next step in the canonical form we must satisfy the assumption that y exists. This is done by causing each term on the right side of Eq. (3.15) to appear at the output of integrator 2.

Let us consider the terms in y first. To obtain $-(1/p)a_1y$ we must take the output y, multiply it by a coefficient a_1, and then operate on a_1y so that with one integration it appears at the output of integrator 2. This step is shown in Fig. 3.9.

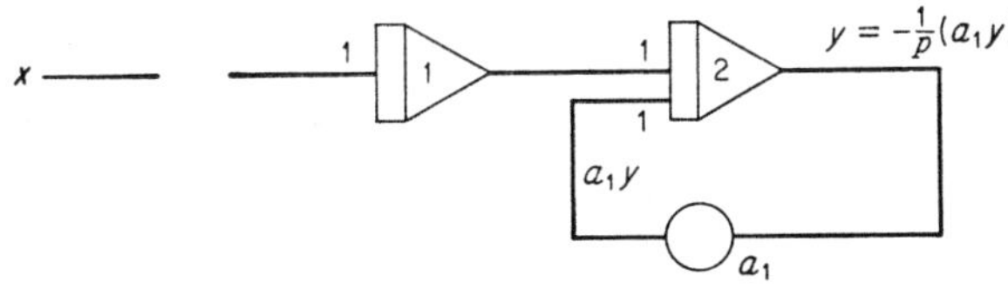

FIG. 3.9 An intermediate step in the programming of Eq. (3.11) by the canonical-form method.

Next, to get $-(1/p^2)a_0y$ at the output, we must take the assumed value y at the output of integrator 2 and integrate twice before the signal again

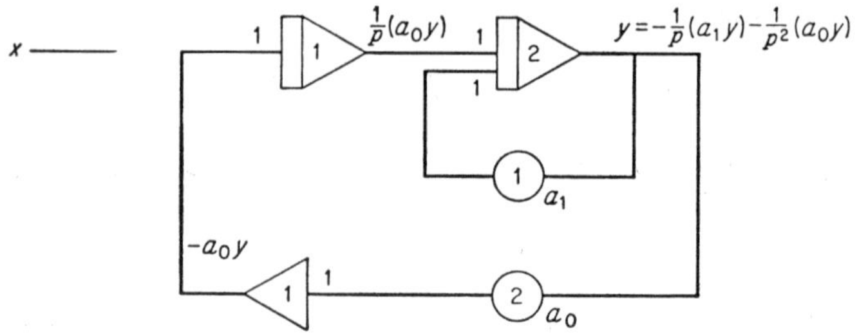

FIG. 3.10 An intermediate step in the programming of Eq. (3.11) by the canonical-form method.

reaches the output of integrator 2. The result is shown in Fig. 3.10. We repeat the process with the x's. The terms $(1/p)b_1(x)$ and $(1/p^2)b_0x$ are obtained by integrating x once and twice, respectively, by the time the signals reach the output of integrator 2. The final result is shown in Fig. 3.11.

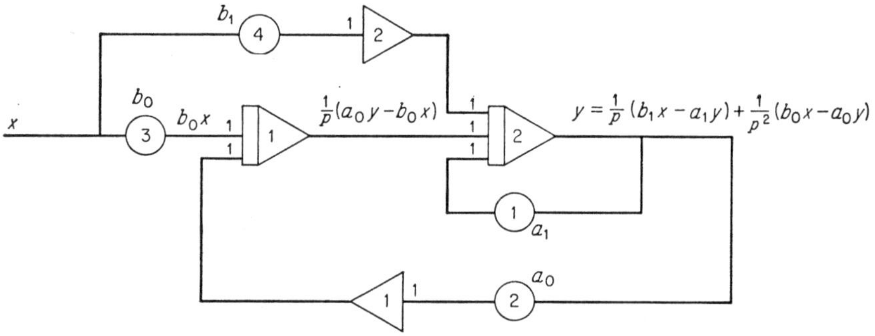

FIG. 3.11 Computer diagram of Eq. (3.11), programmed by the canonical-form method.

Canonical Form and Initial Conditions. The initial conditions of the problem are stated in terms of the dependent variable y and its derivatives. In programming the problem it is most convenient to apply these initial conditions at the outputs of the integrators. For the general method there are no difficulties because the outputs of the integrators are also y and its derivatives. However, this is not the case in the canonical form except for the integrator whose output is y.

Figure 3.11 shows that the output of integrator 2 is y, which initial value we know. However, the output of integrator 1 is $(1/p)(a_0y - b_0x)$, which initial value we must compute. We do this by integrating Eq. (3.14) once and rearranging terms, whereby

$$\frac{1}{p}(a_0y - b_0x) = (b_1x - a_1y) - py \tag{3.16}$$

The initial values of all terms on the right side of the equation are known.

Hence we can calculate the initial value of the output of integrator 1. For a higher-order differential equation we could follow a similar procedure for calculating the initial values applied to the other integrator outputs. There is, however, a simpler method, whose exposition we shall defer until Chap. 4, Example 3.

We shall also show in Sec. 4.3 that, because no derivatives of x are available, Fig. 3.11 contains all the terms which can give the first $n - m$ derivatives of y in Eq. (3.11). For the nth-order equation we may extract all the derivatives of the dependent variable in a canonical-form program.

Example 2. In the equation

$$\frac{d^2y}{dt^2} + 3\frac{dy}{dt} + 4y = 0.5\frac{dx}{dt} - 2x \tag{3.17}$$

with x available but not dx/dt, we wish to solve for y. This equation cannot be programmed with the general method.

First we write Eq. (3.17) in operator notation and equate the highest-order derivative to the remaining terms:

$$p^2y = 0.5px + 2x - 4y - 3py \tag{3.18}$$

Integrating twice,

$$y = \frac{1}{p}(0.5x - 3y) + \frac{1}{p}(2x - 4y) \tag{3.19}$$

Following the procedure outlined we assume that the variables x and y are already available with x unconnected and y at the output of integrator 2. Using these two quantities we generate at the output of integrator 2 all the terms on the right side of Eq. (3.18). The computer diagram is shown in Fig. 3.12.

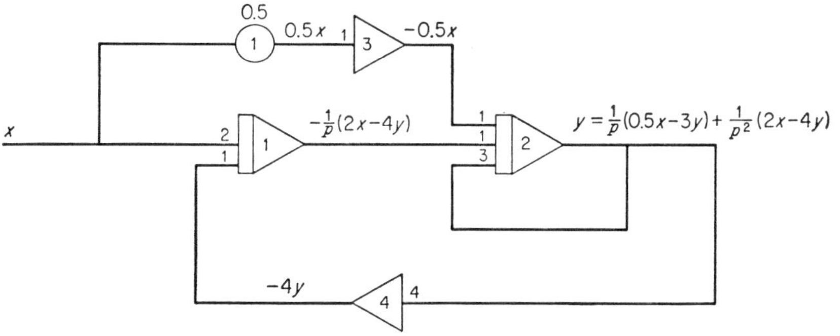

FIG. 3.12 Computer diagram of Eq. (3.17), programmed by the canonical-form method.

3.1.4 Dummy Variable

When compared with the canonical-form method, the dummy-variable method has one major restriction: both the a's and the b's must be constant in the differential equation. However, in problems where derivatives are required, the dummy-variable approach may be the better program to use.

These derivatives will be obtained with less effort than with a canonical-form program.

We illustrate the dummy-variable technique by programming the ubiquitous equation (3.11).

We begin with the dummy variable y_k, which is the solution to Eq. (3.20):

$$p^2y_k + a_1py_k + a_0y_k = x \tag{3.20}$$

First, multiply by b_0:

$$b_0p^2y_k + b_0a_1py_k + b_0a_0y_k = b_0x \tag{3.21}$$

Next, differentiate Eq. (3.20) and multiply by b_1. This is the same as multiplying by b_1p. Hence,

$$b_1p(p^2y_k) + b_1p(a_1py_k) + b_1p(a_0y_k) = b_1px \tag{3.22}$$

Now add Eqs. (3.21) and (3.22) and combine terms:

$$(b_1p + b_0)p^2y_k + (b_1p + b_0)(a_1py_k) + (b_1p + b_0)a_0y_k = b_1px + b_0x \tag{3.23}$$

Since the a's and b's are constant we may change the order of operation and rewrite Eq. (3.23):

$$p^2[(b_1p + b_0)y_k] + a_1p[(b_1p + b_0)y_k] + a_0[(b_1p + b_0)y_k] = b_1px + b_0x \tag{3.24}$$

For comparison we again write Eq. (3.11):

$$p^2y + a_1py + a_0y = b_1px + b_0x \tag{3.25}$$

We note that these equations are the same, term by term, if

$$y = (b_1p + b_0)y_k \tag{3.26}$$

In Eq. (3.26) we have a relationship between the dependent variable y, which we wish to find, and the dummy variable y_k, which we can easily find by using the general method to program Eq. (3.20). Furthermore, we can find the relationship between any derivative of y and derivatives of y_k simply by differentiating both sides of Eq. (3.26). In general, the relationship between variables is

$$p^ry = (b_1p + b_0)p^ry_k = b_1p^{r+1}y_k + b_0p^ry_k \tag{3.27}$$

Note that b_0 and b_1 are the coefficients of x and px in Eq. (3.11). In general, for the equation

$$\sum_{j=0}^{n} a_jp^jy = \sum_{i=0}^{m} b_ip^ix \tag{3.28}$$

where $m \leq n$, the dummy-variable equation is

$$\sum_{j=0}^{n} a_jp^jy_k = x \tag{3.29}$$

Therefore,
$$y = \sum_{i=0}^{m} b_i p^i y_k \tag{3.30}$$

and
$$p^r y = \sum_{i=0}^{m} b_i p^{i+r} y_k \tag{3.31}$$

To solve for y we first program Eq. (3.20) by using the general method. The resulting computer diagram is shown inside the block in Fig. 3.13. We now have y_k and its first two derivatives. To obtain y, we sum y_k and its first derivative to satisfy Eq. (3.26). This is shown in the dashed lines of Fig. 3.13. As indicated in the diagram, the output of amplifier 5 is $-y$.

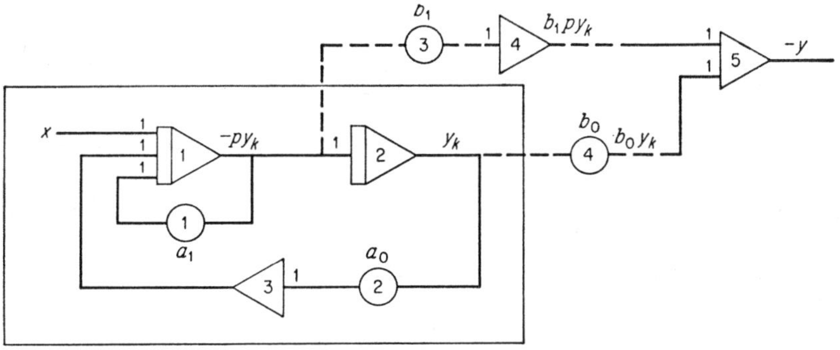

FIG. 3.13 Computer diagram obtained by the dummy-variable method. The portion enclosed by the solid-line rectangle is the program for Eq. (3.20). The dashed-line extensions are required to produce the solution of Eq. (3.13).

By satisfying Eq. (3.27) for $r = 1$ we can also get py. Note that the sum of the inputs to integrator 1 is $p^2 y_k$. If we wanted $p^2 y$ we would need $p^3 y_k$, which is not available. Thus py is the highest derivative of y we can obtain. In the general case the number of derivatives of y which the dummy-variable method can solve is restricted by the highest-order derivative of x in the forcing function. We may see this point by considering the nth-order Eq. (3.28) which has an mth-order derivative in x. As Eq. (3.29) shows, the highest-order derivative of y_k is also nth-order. Equation (3.31) relates a derivative of y with a sum of derivatives of y_k. The highest-order derivative of y_k in this equation is $m + r$. Since the highest-order derivative of y_k available on the computer is $p^n y_k$, it follows that we cannot get the rth derivative unless

$$m + r \leq n$$

or

$$r \leq n - m$$

Thus we pay a price for not having to differentiate x. For every differentiation of x we avoid, we have one less derivative of y we can obtain.

Initial Conditions. As in the canonical-form method we face the problem of having initial conditions stated for y and its derivatives and the integrator outputs in terms of the variable y_k and its derivatives. For the computer diagram 3.13 we can obtain the initial conditions for y_k and py_k by solving the simultaneous set of equations

$$\begin{aligned} b_1 py_k + b_0 y_k &= y \\ b_1 p^2 y_k + b_0 y_k &= py \\ p^2 y_k + a_1 py_k + a_0 y_k &= x \end{aligned} \tag{3.32}$$

where, of course, the initial values of y_k and py_k are found by inserting the initial values for y, py, and x.

In the nth-order equation of the form of (3.28) the initial conditions are determined by solving the simultaneous set of equations

$$\begin{aligned} \sum_{i=0}^{m} b_i p^{i+k} y_k &= p^k y \qquad k = 0, 1, \ldots, n = 1; n \text{ equations} \\ \sum_{j=0}^{m} a_j p^{i+r} y_k &= p^r x \qquad r = 0, 1, \ldots, m - 1; m \text{ equations} \end{aligned} \tag{3.33}$$

for y_k and its first $n - 1$ derivatives. The number of simultaneous equations we must use to get these initial values is

$$\begin{aligned} \text{No. of simultaneous equations} = {} & \text{order of differential equation } (n) \\ & + \text{highest derivative in forcing function } (m) \end{aligned}$$

For a higher-order system one must expend considerable effort to determine the numerical values of the initial conditions.

Example 3. The equation

$$\frac{d^3y}{dt^3} - 7\frac{d^2y}{dt^2} + 5\frac{dy}{dt} + 2y = 1.5\frac{d^2x}{dt^2} - 3.2\frac{dx}{dt} - 4.8x \tag{3.34}$$

has the initial values at $t = 0$:

$$\begin{aligned} y &= 1 & x &= 3.3 \\ \frac{dy}{dt} &= 2.4 & \frac{dx}{dt} &= 6.5 \\ \frac{d^2y}{dt^2} &= 8 & \frac{d^2x}{dt^2} &= 16 \end{aligned}$$

The equation has been programmed by the dummy-variable method, and we wish to find the initial conditions on y_k and its derivatives.

In Eq. (3.34) we observe

$$n = 3 \qquad m = 2$$

Therefore, three initial conditions are required on y_k and we need five equations to find them. By equation set (3.33), these equations are

$$\begin{aligned} 1.5p^2y_k - 3.2py_k - 4.8y_k &= 1 \\ 1.5p^3y_k - 3.2p^2y_k - 4.8py_k &= 2.4 \\ 1.5p^4y_k - 3.2p^3y_k - 4.8p^2y_k &= 8 \\ p^3y_k - 7p^2y_k + 5py_k + 2y_k &= 3.3 \\ p^4y_k - 7p^3y_k + 5p^2y_k + 2py_k &= 6.5 \end{aligned} \tag{3.35}$$

Following the rules for solving simultaneous linear algebraic equations, we get as our initial conditions

$$\begin{aligned} y_k &= 0.4383 \\ py_k &= 0.4976 \\ p^2y_k &= 1.0077 \end{aligned} \tag{3.36}$$

3.2 Normalized Equations for Linear Systems

We mentioned in Sec. 2.4 that a linear equation of the form

$$\sum_{i=1}^{n} a_i(t)p^iy = g(t) \tag{3.37}$$

has a solution

$$y(t) = \sum_{j=1}^{n} c_ju_j(t) + \int_{t_0}^{t} K(t, \alpha)g(\alpha)\, d\alpha \tag{3.38}$$

where t_0 is the initial value of t and the c's are constants whose values are fixed by the initial conditions.

We showed that, if a linearly independent set of solutions

$$u_j(t) \qquad j = 1, 2, \ldots, n$$

of the homogeneous equation is known, then the weighting function $K(t, \alpha)$ may be calculated. Unfortunately, when one or more of the a_1's in Eq. (3.37) are not constant, it is almost impossible to find all the u_j's. Without them we cannot calculate the weighting function.

The computer cannot provide us with analytic expressions. However, it can give us graphical plots of the n linearly independent solutions and the weighting function. All these quantities may be obtained in exactly $n + 1$ computer runs. With these $n + 1$ plots we can construct the solution to Eq. (3.38) for any set of initial conditions and arbitrary forcing function. For this reason we call these plots the normalized solution of the differential equation.

Because it presents special problems we defer our discussion of the weighting function until the next section. We shall now concern ourselves with the method for obtaining the graphical solutions to the homogeneous equation

$$\sum_{i=0}^{n} a_ip^iy = 0 \tag{3.39}$$

To simplify our notation we shall use the following symbols:

1. $\alpha_ny^{(n)}$ represents an initial condition of magnitude α_n on the nth derivative of y.

2. $f_{(0)}{}^{(n)}(t)$ is the solution y to Eq. (3.39) for the initial condition $y^{(n)}$, that is, $\alpha_n = 1$. In general, the symbol $f_{(q)}{}^{(n)}(t)$ is the qth derivative of

the solution when there is a unity initial condition on the nth derivative and all other initial conditions are zero.

Let us suppose that initial conditions $a_0 y^{(0)}, a_1 y^{(1)}, \ldots, a_{n-1} y^{(n-1)}$ are applied to Eq. (3.39). Then the solution as a function of these initial conditions is

$$y = f_0(a_0 y^{(0)} + a_1 y^{(1)} + \cdots + a_{n-1} y^{(n-1)}, t) \tag{3.40}$$

As indicated in Sec. 2.3.3, we may separate the effects of the initial conditions:

$$y = a_0 f_0^{(0)}(t) + a_1 f_0^{(1)}(t) + \cdots + a_{n-1} f_0^{(n-1)}(t) \tag{3.41}$$

Equation (3.41) *reveals that a set of n solutions, each of which is a function of only one nonzero initial condition, constitutes a complete set of linearly independent solutions of Eq.* (3.39).

These solutions are easily obtained on the computer. Equation (3.39) is programmed. We then solve the equation where all initial conditions except one are zero. The nonzero initial condition is unity. The waveform of the particular integrator output, which is y, will be plotted to give the solution associated with that initial condition. If, for example, the initial condition were $y^{(n)}$, the solution would be

$$y = f_0^{(n)}(t) \tag{3.42}$$

After n such runs, each with a different nonzero initial condition, we would have the entire set of independent solutions. Then, given any combination of initial conditions, unity or otherwise, we can solve Eq. (3.41) for y.

This technique may be extended to secure normalized equations for the derivatives of y. If we want a normalized equation for $d^r y/dt^r$, then by differentiating Eq. (3.41),

$$\frac{d^r y}{dt^r} = a_0 f_{(r)}^{(0)}(t) + a_1 f_{(r)}^{(1)}(t) + \cdots + a_{n-1} f_{(r)}^{(n-1)}(t) \tag{3.43}$$

The normalized quantities in Eq. (3.43) are obtained in the same fashion as those of Eq. (3.41). The only difference is that for any initial condition the variable $d^r y/dt^r$ is recorded instead of y.

3.3 The Weighting Function[1]

The weighting function $K(t, \alpha)$ is the solution to the homogeneous differential equation [45, p. 144]

$$\sum_{r=0}^{n} a_r p^r y = \delta(t - \alpha) \tag{3.44}$$

where all initial conditions are zero and $\delta(t - \alpha)$ is a delta function.[2]

[1] See Appendix 2 for an additional discussion of the properties of the weighting function.
[2] See Appendix 1 for a discussion of the delta function.

A delta function cannot be simulated exactly. However, the homogeneous equation

$$\sum_{r=0}^{n} a_r p^r y = 0 \tag{3.45}$$

whose initial conditions $\alpha_n y^n$ ($n = 0, 1, \ldots, n-1$) are

$$\alpha_r = \begin{cases} 0 & \text{for } r \neq n-1 \\ \dfrac{1}{a_n(\alpha)} & \text{for } r = n-1 \end{cases}$$

has the same solution as Eq. (3.44) and can be solved exactly.

The weighting function $K(t, \alpha)$ is a function of two variables, t and α. The variable α is the time[1] at which the nonzero initial condition is applied to the $(n-1)$st derivative, and t is the time at which y is measured. A typical plot of the weighting function might be the one shown in Fig. 3.14.

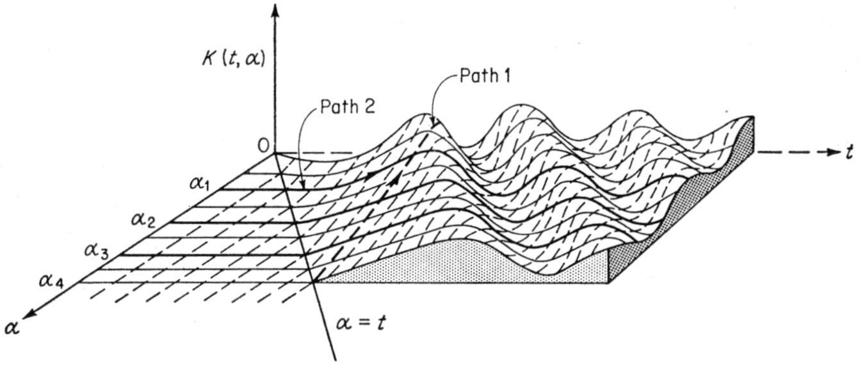

FIG. 3.14 A plot of a typical weighting function. Path 2 is a computer solution to a differential equation which begins at $t = \alpha_1$. Path 1 is the computer solution of the adjoint of the differential equation.

This figure shows an important property of this function:

$$K(t, \alpha) = 0 \qquad \text{for } \alpha > t \tag{3.46}$$

If we want the solution at any particular time, say t_2, we must evaluate the convolution integral

$$y(t_2) = \int_{t_0}^{t_2} K(t_2, \alpha) f(\alpha)\, d\alpha \tag{3.47}$$

For this calculation we need the weighting function whose t is fixed at t_2 and whose α is varied from t_0 to t_2. In Fig. 3.14, $K(t_2, \alpha)$ is shown as path 1.

[1] It is common practice in computer terminology to call the independent variable *time*. I shall adhere to that custom. Thus the phrase *time at which* should be understood by the reader to mean *the value of the independent variable at which.*

Suppose we wish to generate $K(t_2, \alpha)$ using Eq. (3.45). While in RESET the computer is set up for some initial time of α, say α_1. All coefficients are set for this value, and the $(n - 1)$st derivative has an initial condition $1/a_n(\alpha_1)$. The computer is switched to OPERATE and starting at α_1 generates a solution in the direction of increasing time. The solution is

$$y(t) = K(t, \alpha_1) \tag{3.48}$$

and is shown by path 2 in Fig. 3.14. The arrow indicates the direction of the solution. At $t = t_2$, Eq. (3.48) becomes

$$y(t_2) = K(t_2, \alpha_1) \tag{3.49}$$

which is one discrete point of the weighting function. Similarly, for starting values $\alpha_2, \alpha_3, \ldots, \alpha_m$, we can get other points $K(t_2, \alpha_2)$, $K(t_2, \alpha_3)$, . . . , $K(t_2, \alpha_m)$ of the weighting function. We want to be able to draw path 1 in Fig. 3.14. No matter how many runs we make, we can get only a discrete approximation to the curve. The number of runs required to plot the weighting function will depend upon the behavior of the curve and the desired accuracy.

In one special case, when the coefficients of the equation are constant, the continuous weighting function can be plotted by solving Eq. (3.45). Because the coefficients are constant they remain the same for any initial value α_1 and readout t_2. The weighting function is a function of the difference of these variables. Hence,

$$K(t_2, \alpha_1) = K(t_2 - \alpha_1) \tag{3.50}$$

Thus, one computer run can give the solution, which is plotted as a function of the difference of the initial and readout values. In Fig. 3.14, starting with $t = \alpha$ all curves with increasing t, for example, path 2, will be the same as those with decreasing α, for example, path 1.

For the variable-coefficient case the computer will plot the weighting function in one run if it solves an equation related to Eq. (3.45). The related form is called the adjoint equation.

3.4 The Adjoint Equation[1]

The term *adjoint* is used in conjunction with linear operators. The adjoint operator is obtained from the original operator by a transformation that satisfies certain conditions. Just as the original operator, the adjoint

[1] For an excellent discussion of linear operators and their adjoints see Ref. 45, particularly pp. 42–50, 173–174. See also Ref. 44, pp. 233–246, and Ref. 46, pp. 1052–1083, for a discussion of the adjoint equation and its analog-computer applications.

is also linear. Two of the most common linear operators which have adjoints are:

1. The matrix of coefficients for a set of simultaneous algebraic equations
2. The differential-equation operator

Beginning with Lagrange, mathematicians for more than two centuries have studied adjoints. Their interest has probably been aroused because the nature of the solutions of the homogeneous adjoint equations can give much information about the existence of a solution of the inhomogeneous equations from which the adjoint came. One by-product of their work has been to find a property of the adjoint equation which enables us to plot the weighting function in one computer run. On that by-product we focus our attention.

If we have the equation

$$\sum_{r=0}^{n} a_r p^r y = Ly = 0 \tag{3.51}$$

where

$$L = \sum_{r=0}^{n} a_r p^r$$

is a linear differential operator, then the adjoint of Eq. (3.51) is

$$\sum_{r=0}^{n} (-1)^r p^r (a_r u) = Mu = 0 \tag{3.52}$$

M is called the adjoint operator. When

$$L = M$$

Eq. (3.51) is said to be self-adjoint. Equation (3.53), which is called the *Sturm-Liouville equation*, is self-adjoint:

$$p(x)\frac{d^2y}{dx^2} + \frac{dp}{dx}\frac{dy}{dx} + q(x)y = 0 \tag{3.53}$$

By comparison, when we run the problem backward (see Chap. 4), we make the transformation of independent variable $t' = -t$. Thus we get the transformation

$$L'y = \sum_{r=0}^{n} (-1)^r a_r p^r y \tag{3.54}$$

If and only if all the a_r's are constant, the adjoint operator M and the operator L' are equal, that is, $L'y = My$.

For equation

$$e^{-t}\frac{dy^3}{dt^3} + t^2\frac{d^2y}{dt^2} + 3\frac{dy}{dt} + y \sin t = 0 \tag{3.55}$$

the adjoint equation is

$$\frac{d^3}{dt^3}(e^{-t}u) + \frac{d^2}{dt^2}(t^2u) - \frac{d}{dt}(3u) + u \sin t = 0 \tag{3.56}$$

For the equations

$$\begin{aligned} Ly &= \delta(t - \alpha) \\ Mu &= \delta(t - t') \end{aligned} \tag{3.57}$$

the associated weighting functions are

$$\begin{aligned} y &= K(t', \alpha) \\ u &= H(\alpha, t') \end{aligned} \tag{3.58}$$

Because L and M are adjoint operators,

$$K(t', \alpha) = H(\alpha, t') \tag{3.59}$$ †

In particular, for $t' = t_2$,

$$K(t_2, \alpha) = H(\alpha, t_2) \tag{3.60}$$

In words, Eq. (3.59) states that, if we want to get the weighting function $K(t_2, \alpha)$, we can obtain it by getting $H(\alpha, t_2)$, the weighting function for the adjoint equation, where the initial condition is applied at time t_2 and the output is read at time α. Since t_2 is a constant and α is a variable, the solution of the adjoint equation will give the weighting function for all α. Thus it is possible to obtain the weighting function in one computer run.

At first glance Eq. (3.52) seems to indicate that there may be a formidable effort in converting from the original differential equation to its adjoint and then programming. However, all this effort can be bypassed by working with the original equation.

We shall demonstrate the technique by showing how one programs the computer to plot the weighting function of Eq. (3.55). To illustrate a point we shall make a minor modification by changing the coefficient of the dy/dt term from 3 to a_0. The procedure involved in programming the adjoint differential equation consists of the following steps:

Step 1.[1] Program the original differential equation, using the inhomogeneous-equation representation [Eq. (3.44)] with a delta function as the forcing function. Figure 3.15 is the computer diagram for Eq. (3.44) with a delta function added.

† Reference 24, pp. 256–259.

[1] This procedure was developed by Laning and Battin and reported in Refs. 47 and 48, pp. 239–247, 398–403.

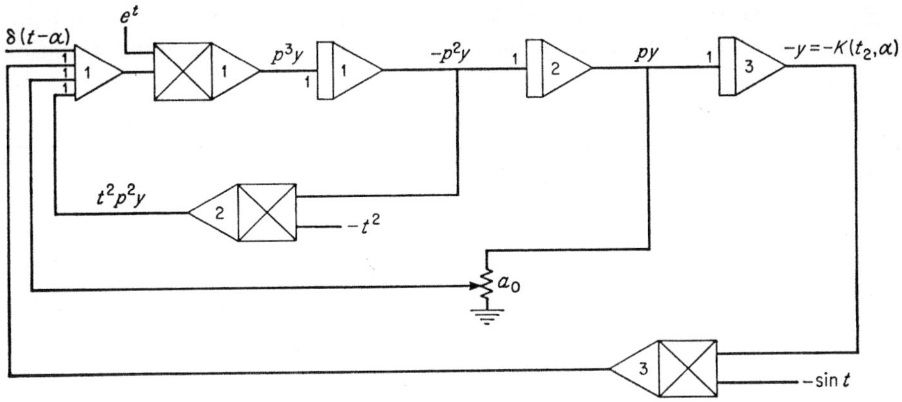

FIG. 3.15 The computer program of Eqs. (3.58).

Step 2. Disconnect all elements in Fig. 3.15 and turn them around in place, as in Fig. 3.16.

Step 3. Connect the turned-around elements so that an *input to* an element in Fig. 3.15 becomes an *output from* that element in the adjoint diagram. Similarly an *output from* an element in Fig. 3.15 becomes an *input to* that element. Omit from this step the delta function and all variable-coefficient terms (for example, e^t, t^2, $\sin t$).

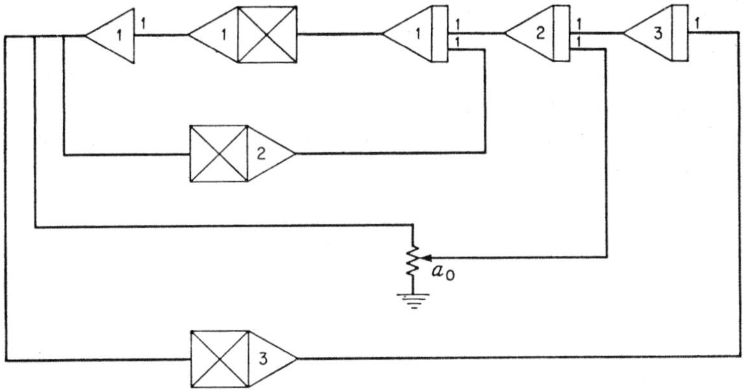

FIG. 3.16 A step in the programming of the adjoint of Eqs. (3.58).

For example, in Fig. 3.15 the *output* of integrator 1 goes to the *input* of integrator 2 and multiplier 2. In the adjoint diagram, integrator 1 has two inputs. One input comes from the output of integrator 2, and the other comes from the output of multiplier 2. The interconnection of the elements by the rule of steps 2 and 3 is shown in Fig. 3.16.

Step 4. Make a change of variable on all variable coefficients from t to $t_2 - t$ (for example, e^t becomes e^{t_2-t}), and put these at the inputs to the same multipliers as in Fig. 3.15.

Step 5. At the input to the element (integrator 3) from whose output the Green's function was obtained in Fig. 3.15, insert the delta function $\delta(t - 0)$. From the output of the element (amplifier 1) at whose input the delta function was applied in Fig. 3.15, plot the Green's function $K(t_2, t_2 - t)$. The complete computer diagram for the adjoint equation, incorporating steps 4 and 5, is shown in Fig. 3.17.

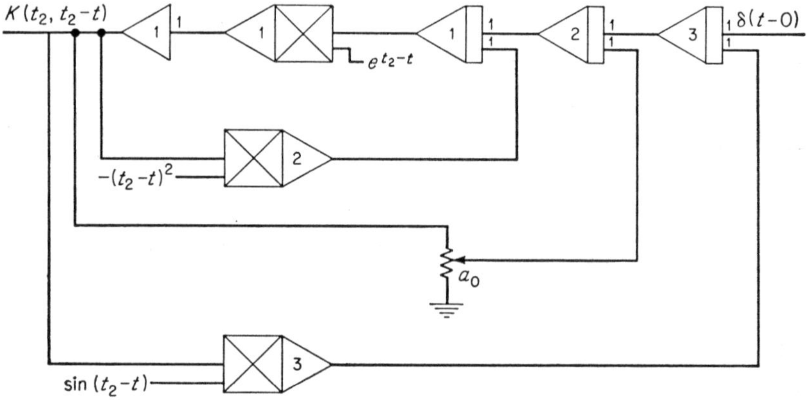

FIG. 3.17 The computer program of the adjoint of Eqs. (3.58).

We used the potentiometer a_0 to demonstrate how to handle gains of other than unity at the inputs of amplifiers and integrators. In Fig. 3.15 it can represent a gain on amplifier 1. In the adjoint simulation, shown in Fig. 3.17, the gain is at the input of integrator 2. In both cases the gain appears between the same two elements, amplifier 1 and integrator 2. So will it be for any gain factor.

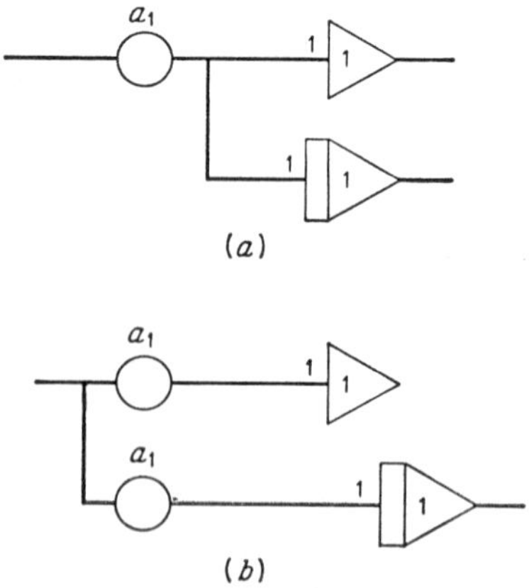

FIG. 3.18 Two ways to draw a segment of a computer diagram. If the diagram contains potentiometers with two or more outputs, like (*a*), they should be redrawn in the form of (*b*) before being converted to the computer diagram for the adjoint equation.

The potentiometer is a device which can have any number of outputs but only one input. If in step 1 we have a potentiometer that has more than one output, then in the adjoint simulation the same potentiometer would have more than one input. We can avoid this difficulty by changing any potentiometer having more than one output to a set of parallel potentiometers having only one output each, as indicated in Fig. 3.18*a* and *b*. Figure 3.18*b* is then used in the conversion to the adjoint diagram.

In one circumstance we may simplify the construction of the adjoint equation. Suppose Fig. 3.19*a* is a computer block diagram for the generation of a weighting function and Fig. 3.19*b* is the diagram for the adjoint equation. The terms $G(p)$, $H(p)$, $Q(p)$ are linear differential operators. $\bar{G}(p)$ is the term we obtain after making the transformation of variable from t to $t_2 - t$ on the adjoint of $G(p)$. Strictly speaking, we cannot call $\bar{G}(p)$ the adjoint of $G(p)$. The same relationship holds for $H(p)$ with $\bar{H}(p)$ as well as $Q(p)$ with $\bar{Q}(p)$. If, for example, $H(p)$ has only constant coefficients, then $\bar{H}(p)$ is precisely the same as $H(p)$.†

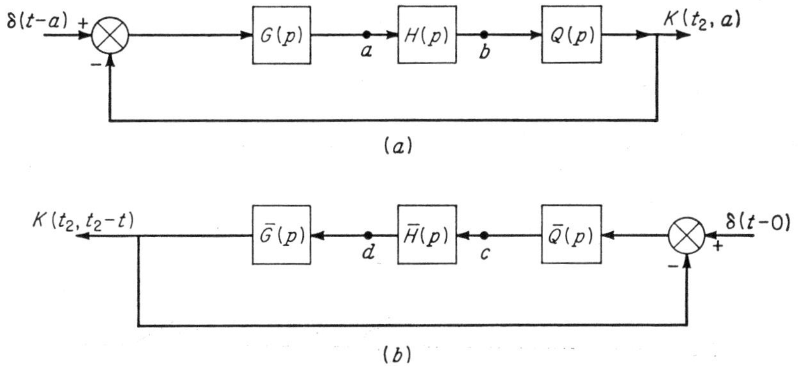

FIG. 3.19 Block diagrams of (*a*) a differential equation and (*b*) its adjoint.

Because $H(p) = \bar{H}(p)$ we do not have to reverse each element in $H(p)$ to get $\bar{H}(p)$. Instead we reverse the block as a unit. Thus point a in Fig. 3.19*a* corresponds to point c in Fig. 3.19*b* with a similar correspondence between points b and d. This block reversal may be made for any-order constant-coefficient operator, provided that neither the forcing function nor the output is included inside the block [44, pp. 237–239].

If Fig. 3.20*a* is the computer diagram of a differential equation, then Fig. 3.20*b* is its adjoint program.

To complete the computer setup in Fig. 3.17 we must change the delta function $\delta(t - 0)$ into an equivalent initial condition. This conversion may be performed if we use the following rule.

† Despite the fact that $H(p) = \bar{H}(p)$, it is incorrect to say that $H(p)$ is self-adjoint. $H(p)$ would be self-adjoint only if $H(p) = \bar{H}(p)$ without a change of variable.

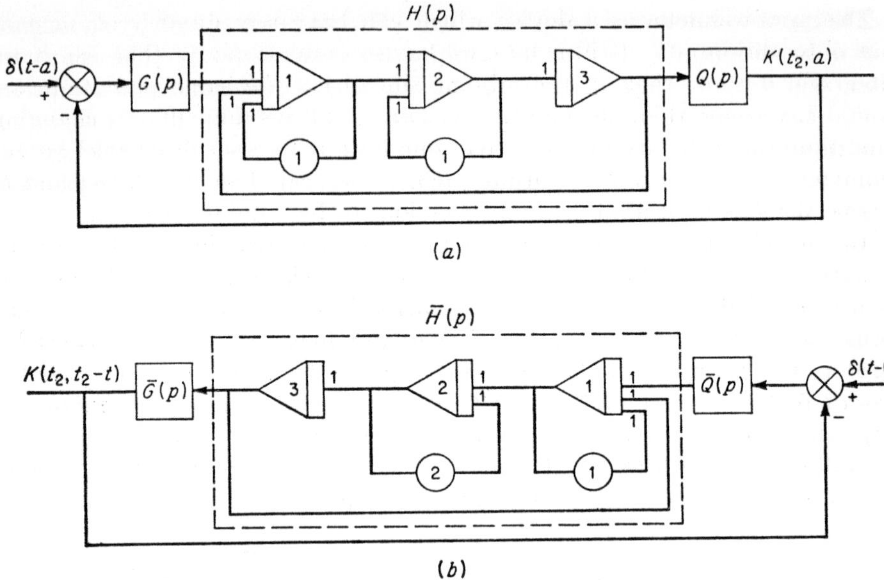

FIG. 3.20 Block diagrams of (*a*) a differential equation and (*b*) its adjoint when one element $H(p)$ contains only constant coefficients.

Pass the delta function through all computer elements[1] except integrators, and perform the operation indicated in each case. At the output of the first integrator through which the delta function passes, apply an initial condition equal to the value of the coefficient of the delta function at the time when the delta function is not zero.

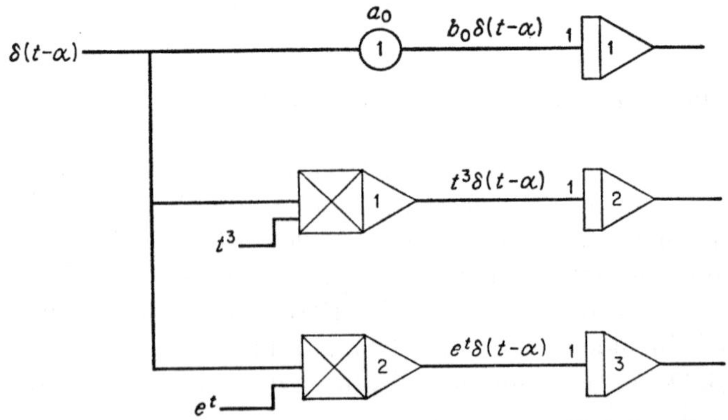

FIG. 3.21 The effect of computer elements on a delta-function input.

In Fig. 3.21 we eliminate the delta function, which is nonzero at $t = a$, in the following manner.

[1] The permissible elements are potentiometers, amplifiers, and multipliers.

The input to integrator 1 is $b_0\delta(t - a)$. This can be replaced by an output initial condition $-b_0$. (The integrator also inverts sign.) The input to integrator 2 is $t^3\delta(t - a)$. It can be replaced by an output initial condition $-a^3$. Likewise the input to integrator 3 can be replaced by an output initial condition $-e^a$. To summarize, the delta function in Fig. 3.21 may be removed and replaced by the appropriate initial conditions. The solution will be the same in both cases.

As we have already stated, the computer solves the problem in the direction of increasing time t. However, because of the transformation of variable from t to $t_2 - t$, the adjoint equation is solved by running backward in time compared with the solution of the original equation. Thus, in the original equation to get $K(t_2, \alpha_1)$, the problem starts at α_1 and runs to t_2. In the adjoint the problem begins at t_2 $(t = 0)$ and runs for increasing time, which is the same as decreasing α, until $t_2 - t = \alpha_1$. The nature of the plot of the weighting function obtained by the adjoint equation is shown in path 1 of Fig. 3.14. The plot begins at $K(t_2, \alpha = t_2)$ and moves in the direction of decreasing α.

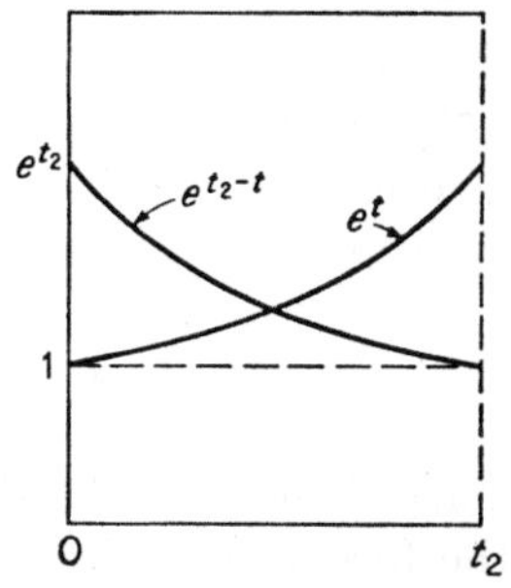

FIG. 3.22 A plot of the response of a time-varying function e^t and its adjoint equivalent e^{t_2-t}.

Figure 3.22 compares the variation of the coefficient e^t and its adjoint counterpart e^{t_2-t} as a function of t.

In the one case of a system which has several inputs and one output, the adjoint equation can generate all the weighting functions in one computer run. Consider Fig. 3.23 as a typical example. The quantities $f_1(t)$, $f_2(t)$,

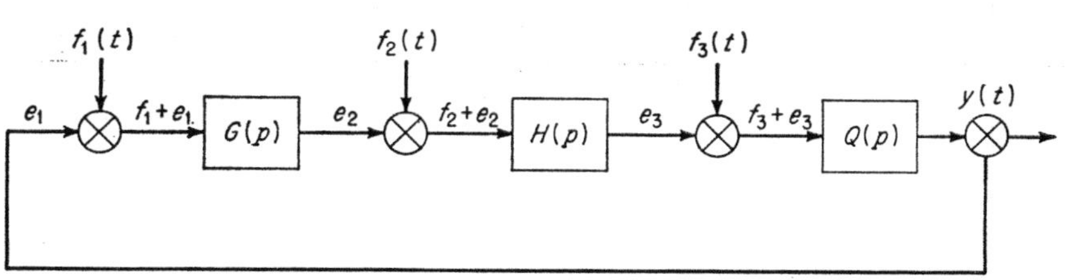

FIG. 3.23 A block diagram with three inputs and one output.

and $f_3(t)$ are the inputs, and $y(t)$ is the output. The terms $G(p)$, $H(p)$, and $Q(p)$ are linear differential operators. For the moment let us assume there is only one input, say $f_1(t)$. To get the weighting function associated with a forcing function at this point, we replace $f_1(t)$ by a delta function. By the procedure outlined in Sec. 3.3 we can find a discrete approximation to the weighting function, which we designate as $K(t, \alpha)$. In a similar fashion, if we apply the delta function at the $f_2(t)$ input alone, we would find its weighting function $K_2(t, \alpha)$, and finally at the $f_3(t)$ input the weighting function is $K_3(t, \alpha)$. Because of linearity we can add the effects of the forcing functions, so that the output due to these forcing functions is

$$y(t_2) = \int_0^{t_2} [K_1(t_2, \alpha)f_1(\alpha) + K_2(t_2, \alpha)f_2(\alpha) + K_3(t_2, \alpha)f(\alpha)]\, d\alpha \quad (3.61)$$

where $t = 0$ is the initial value of t.

Suppose that instead of the way just discussed we applied delta functions simultaneously at all the inputs. Would we be able to obtain the discrete weighting functions in less computer runs than before? This method would give us a composite weighting function $G(t, \alpha)$, which would be the sum of the other weighting functions, i.e.,

$$G(t, \alpha) = K_1(t, \alpha) + K_2(t, \alpha) + K_3(t, \alpha) \quad (3.62)$$

There is no way the single function $G(t, \alpha)$ can in general be combined with the forcing functions to produce the integral of Eq. (3.61). Therefore, to get an output with more than one input delta function is a waste of time that gives no fruitful information about the weighting function, unless

$$f_1(t) = f_2(t) = f_3(t)$$

Now let us set up the adjoint equation for Fig. 3.23. The line above the original operator indicates that the operator has been converted to its adjoint. The diagram for the adjoint equation is Fig. 3.24.

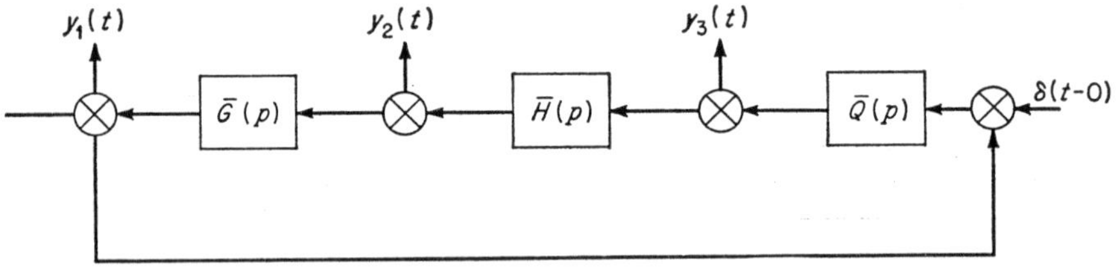

FIG. 3.24 The adjoint of a system which has three inputs and one output. Note that the adjoint has one input and three outputs.

The adjoint equation has only one input and three outputs, where

$$\begin{aligned} y_1(t) &= K_1(t_2, t_2 - t) \\ y_2(t) &= K_2(t_2, t_2 - t) \\ y_3(t) &= K_3(t_2, t_2 - t) \end{aligned} \quad (3.63)$$

All three outputs may be recorded simultaneously. Therefore, the adjoint computer program can record every weighting function associated with a single output.

3.5 Boundary-value Problems[1]

In Chap. 2 we deal exclusively with the initial-value problem. However, the differential equation can also be solved when conditions on the dependent variable and its derivatives are imposed at several values of the independent variable. A problem having these conditions is called a boundary-value problem. The most common boundary-value problem, called the two-point boundary-value problem, is one which has conditions imposed at two values of the independent variable. We shall confine our discussion to this case although the general boundary-value problem presents no added conceptual difficulties.

The boundary-value problem is much harder to solve than the initial-value problem. There are no convenient tests which we can make to determine if the solution exists or is unique. We have only a necessary condition which states that the number of independent boundary conditions must be equal to the order of the differential equation.

For Eq. (3.64), which presents no difficulty as an initial-value problem,

$$\frac{d^2y}{dt^2} + y = 0 \tag{3.64}$$

the solution is

$$y = c_1 \sin t + c_2 \cos t \tag{3.65}$$

If we choose as boundary conditions

$$y(0) = 1 \qquad y(\pi) = 0$$

then we must be able to find a c_1 and c_2 which will satisfy the following equation set:

$$\begin{aligned} y(0) &= 1 = c_1 \sin 0 + c_2 \cos 0 \\ y(\pi) &= 0 = c_1 \sin \pi + c_2 \cos \pi \end{aligned} \tag{3.66}$$

In trying to solve Eqs. (3.66) we find that we get two different values for c_2 and that c_1 is indeterminate (can assume any value). Hence there is no solution to Eq. (3.64) for the boundary conditions chosen.

If for Eq. (3.64) we choose the boundary conditions

$$y(0) = 0 \qquad y(\pi) = 0$$

then the solution of Eq. (3.66) would yield

$$c_1 = \text{indeterminate} \qquad c_2 = 0 \tag{3.67}$$

This boundary-value problem also has no solution.

The computer can be programmed for initial conditions but not boundary-value conditions. It can solve a boundary-value problem only if we impose initial conditions which will generate a solution that passes through the

[1] See Ref. 18, pp. 18–22, 137–149, for further discussion.

boundary values. The linear boundary-value problem must be solved in two steps.[1] First we must determine if a unique solution exists and then we must calculate the initial conditions which must be inserted to give the solution. Both of these steps can be carried out if we have a normalized set of equation solutions.

Now suppose we must solve a third-order equation with boundary conditions at $t = 0, 1$:

$$\frac{d^3y}{dt^3} + a_2\frac{d^2y}{dt^2} + a_1\frac{dy}{dt} + a_0y = 0 \tag{3.68}$$

$$y(0) = 0 \qquad \left(\frac{dy}{dt}\right)_{t=1} = \dot{y}(1) = 1 \qquad y(1) = 2$$

We first set up the equations to determine the initial conditions. Using the normalized solutions of Eq. (3.43),

$$y(0) = 0 = \alpha_0 f_0^{(0)}(0) + \alpha_1 f_0^{(1)}(0) + \alpha_2 f_0^{(2)}(0) \tag{3.69a}$$
$$\dot{y}(1) = 1 = \alpha_0 f_1^{(0)}(1) + \alpha_1 f_1^{(1)}(1) + \alpha_2 f_1^{(2)}(1) \tag{3.69b}$$
$$y(1) = 2 = \alpha_0 f_0^{(0)}(1) + \alpha_1 f_0^{(1)}(1) + \alpha_2 f_0^{(2)}(1) \tag{3.69c}$$

For any normalized equation which is evaluated at the initial value of the independent variable or one of its derivatives (in this case at $t = 0$)

$$f_n^{(s)}(0) = 0 \qquad \text{for } s \neq n \tag{3.70}$$

Hence we can solve Eq. (3.69*a*) for α_0 and get

$$\alpha_0 = 0$$

Equations (3.69*b*) and (3.69*c*) can be solved for α_1 and α_2. The result is

$$\alpha_1 = \frac{\begin{vmatrix} 1 & f_1^{(2)}(1) \\ 2 & f_0^{(2)}(1) \end{vmatrix}}{\begin{vmatrix} f_1^{(1)}(1) & f_1^{(2)}(1) \\ f_0^{(1)}(1) & f_0^{(2)}(1) \end{vmatrix}} \qquad \alpha_2 = \frac{\begin{vmatrix} f_1^{(1)}(1) & 1 \\ f_0^{(1)}(1) & 2 \end{vmatrix}}{\begin{vmatrix} f_1^{(1)}(1) & f_1^{(2)}(1) \\ f_0^{(1)}(1) & f_0^{(2)}(1) \end{vmatrix}} \tag{3.71}$$

If the denominator of Eq. (3.71) is not zero, then Eqs. (3.69) give the initial conditions that will solve Eq. (3.68), satisfying the boundary values.

[1] A high-speed repetitive computer can be used by trial and error to determine rapidly the initial conditions which must satisfy the specified boundary conditions. However, the difficulty in applying this method increases sharply with the number of such initial conditions which must be found.

4 Block-diagram Operations, Scaling, and Operational-amplifier Applications[1]

4.1 Introduction

We have already seen in Chap. 3 that a computer diagram, which consists of symbolic representations of operations in proper sequence, indicates how a problem is programmed. This record of the variables appearing at the input and output of each element is essential for the orderly and accurate operations of the computer. However, the computer diagram may be used as more than a picture of the manner in which elements are connected.

Often a program derived directly from the problem equations may lack certain variables we wish to observe, or may tax the computer beyond its capacity to perform accurately. In either case, we need not return to the original equations to remedy the fault. With considerably less effort we can manipulate the computer diagram itself to eliminate any defects in the program.

In large problems the computer will contain several elements interconnected with one another, sometimes to the point of obscuring the major constituent parts of the problem. A block diagram, by combining several computer elements into an equivalent one, compresses the problem representations into fewer operations. Were the block diagram merely a shorthand notation for the computer diagram, it would be a convenience and nothing more.

However, in removing the distraction of detail found in the computer diagram, the block diagram often permits us to analyze the problem more easily. Some properties of the solution which may be determined from the block diagram include (1) stability, (2) asymptotic behavior, and on occasion (3) parameter sensitivity. In addition, we may use the block diagram to rearrange, add, or remove a sequence of operations. This also may be

[1] Signal-flow graphs, originated by S. J. Mason, provide an alternative means for the manipulation of block diagrams.

done on the computer diagram. Because there are fewer elements involved, block-diagram manipulation will tend to reduce the likelihood of error. When large numbers of elements are involved, it may be best to alter the block diagram first and then change the computer diagram accordingly.

In this chapter we shall examine some of the properties and manipulation rules of the block diagram and its more detailed equivalent, the computer diagram. We shall deal only with those parts of the problem which can be represented by linear operations. Nonlinear operations, whose properties differ markedly from linear operations, are considered separately in Chap. 5.

The manipulation of computer diagrams is equivalent to time or amplitude scaling (or a combination of both) of the computer program. These two techniques allow us to extend the range of operation of the computer and, hence, the accuracy of the solution. They are, therefore, two of the most important subjects in computer technology and as a result are discussed in some detail.

We close the chapter with a discussion of the operational amplifier. We shall show how from this element come the summer and integrator, and also how in conjunction with resistors and capacitors it can form a large class of transfer functions.

4.2 Operator Characteristics[1]

The block-diagram concept stems from that of the mathematical operator. If L is an operator, then in the equation

$$Lx = y \tag{4.1}$$

we may say that L operates on (or transforms) x to form y. For the operators L_1 and L_2, Eq. (4.2),

$$L_1L_2y = z \tag{4.2}$$

means that L_2 operates on y, and L_1 operates on the resulting function. If

$$L_2y = w \tag{4.3}$$

then

$$L_1L_2y = L_1w \tag{4.4}$$

It makes no sense to say that L_1 operates on L_2 and the resulting operator operates on y. *An operator can operate only on a variable and not on another operator.*

If Eq. (4.2) is a differential equation, L_1 and L_2 are called differential operators. A linear operator L which is associated with a linear equation has the superposition property (see Sec. 2.3.3)

$$L(\alpha x + \beta y) = \alpha Lx + \beta Ly \tag{4.5}$$

where α and β are constants.

[1] For further reading on this subject see the section by W. Kaplan in Ref. 49, pp. 8-01 to 8-20, as well as Ref. 50. The more mathematically inclined may consult Ref. 51.

The commutative property for two different operators is defined by

$$L_1L_2y = L_2L_1y \tag{4.6}$$

In general, differential operators are commutative only when they are linear and have constant coefficients. However, if any linear operator is multiplied by a constant, that constant may be commuted. Using Eq. (4.5) we may write

$$\begin{aligned} L_1L_2(\alpha\beta y) &= L_1[\alpha L_2(\beta y)] = L_1[\beta L_2(\alpha y)] = \alpha L_1(\beta L_2 y) = \beta L_1(\alpha L_2 y) \\ &= \alpha L_1 L_2(\beta y) = \beta L_1 L_2(\alpha y) = \alpha\beta L_1 L_2 y \end{aligned} \tag{4.7}$$

The basic differential operator p is

$$p = \frac{d}{dt} \tag{4.8a}$$

and

$$py = \frac{dy}{dt} \tag{4.8b}$$

Then by convention

$$p^n = \frac{d^n}{dt^n} \tag{4.9a}$$

and

$$p^n y = \frac{d^n y}{dt^n} \tag{4.9b}$$

The inverse, or integral, operator is

$$p^{-1} = \int^t [\quad]\, d\alpha \tag{4.10a}$$

and

$$p^{-1}f(t) = \int^t f(\alpha)\, d\alpha \tag{4.10b}$$

Extending the integral operator to nth order,

$$p^{-n} = \underbrace{\int^t \int^{x_{n-1}} \cdots \int^{x_2} \int^{x_1}}_{n\text{-fold}} [\quad]\, d\alpha\, dx_1\, dx_2 \cdots dx_{n-2}\, dx_{n-1} \tag{4.11a}$$

and

$$p^{-n}f(t) = \int_0^t \int_0^{x_{n-1}} \int^{x_2} \int^{x_1} f(\alpha)\, d\alpha\, dx_1\, dx_2 \cdots dx_{n-2}\, dx_{n-1} \tag{4.11b}$$

A differential equation of the form

$$\begin{aligned} &\frac{d^n y}{dt^n} + a_{n-1}\frac{d^{n-1}y}{dt^{n-1}} + \cdots + a_1\frac{dy}{dt} + a_0 y \\ &\qquad = \frac{d^m x}{dt^m} + b_{m-1}\frac{d^{m-1}x}{dt^{m-1}} + \cdots + b_1\frac{dx}{dt} + b_0 x \end{aligned} \tag{4.12}$$

may be written in operator notation as

$$\begin{aligned} &p^n y + a_{n-1}p^{n-1}y + \cdots + a_1 py + a_0 y \\ &\qquad = p^m x + b_{m-1}p^{m-1}x + \cdots + b_1 px + b_0 x \end{aligned} \tag{4.13}$$

or

$$\begin{aligned} &(p^n + a_{n-1}p^{n-1} + \cdots + a_1 p + a_0)y \\ &\qquad = (p^m + b_{m-1}p^{m-1} + \cdots + b_1 p + b_0)x \end{aligned} \tag{4.14}$$

An equivalent form for Eq. (4.14) is

$$\frac{y}{x} = \frac{p^m + \cdots + b_1 p + b_0}{p^n + \cdots + a_1 p + a_0} \tag{4.15}$$

In going from Eq. (4.14) to Eq. (4.15) we have performed what seems to be an algebraic operation. The reader is warned to remember that in no way are the left and right sides of Eq. (4.15) equal in the usual sense. Unlike an algebraic equation we cannot substitute the right side of Eq. (4.15) for the left side. A ratio of operators by itself has no functional value whatever. Equation (4.15) has meaning only as an alternative representation for Eq. (4.14).

Any nth-order linear operator may also be written as a product of first-order operators or, for that matter, higher-order operators. Thus Eq. (4.14) may be factored as

$$[(p + c_1)(p + c_2) \cdots (p + c_n)]y = [(p + d_1)(p + d_2) \cdots (p + d_m)]x \tag{4.16}$$

where the c's and d's are either constants or functions of the independent variable. The c's and d's can be regarded as coefficients of the zeroth-order differential operator p^0. If the c's and d's are all constants, i.e., if all coefficients in Eq. (4.14) are constant, all the first-order operators are commutative. If they are not constant, then the order of operation cannot be commuted. Any adjoining sequence of operations containing only constant zeroth-order operator coefficients may be commuted. The equation

$$[(p + e^{-t})(p + 3)(p - 5)(p + 2)]y = x \tag{4.17}$$

may be written as

$$x = [(p + e^{-t})(p + 3)(p + 2)(p - 5)] \tag{4.18}$$

or any of the four other permutations of the operators $(p + 3)$, $(p + 2)$, $(p - 5)$. However, under no condition may the position of operator $p + e^{-t}$ be changed with any of the other three.

Equation (4.16) looks very much like a factored polynomial. Suppose some of the c's and d's are equal. Can we then cancel these factors from both sides of the equation and thereby reduce the order of the equation? We know that operators are not algebraic quantities and yet there is enough of a similarity of form to tempt us.

Some sufficient conditions which permit us to make such cancellation are:

1. All initial conditions are zero.

2. The operators to be canceled are on the extreme left. Note that an operator on the extreme left is the last in the sequence to operate on the variable.

Since constant-coefficient-equation operators can be commuted, any operator can be placed on the extreme left. Therefore, in this case like factors may be canceled from both sides of the equation no matter what their order.

The equation

$$(p + \sin t) \sum_{L=0}^{n} a_i p^i y = (p + \sin t) \sum_{r=0}^{m} b_r p^r x \tag{4.19}$$

may be reduced by one order to

$$\sum_{L=0}^{n} a_i p^i y = \sum_{r=0}^{m} b_r p^r x \tag{4.20}$$

if the initial conditions are zero. However, the equation

$$[(p + \sin t)(p + 1)]y = [(p + 1)(p + \sin t)]x \tag{4.21}$$

cannot be reduced. Observe that

$$p^2 y = p^2 x \tag{4.22}$$

implies

$$y = x$$

only if also $dx/dt = dy/dt$ initially.

The block-diagram representation of Eq. (4.15) is shown in Fig. 4.1. The arrows indicate that the signal x is operated on by the terms in the block to produce y. The function x is called the *input*, and y is called the

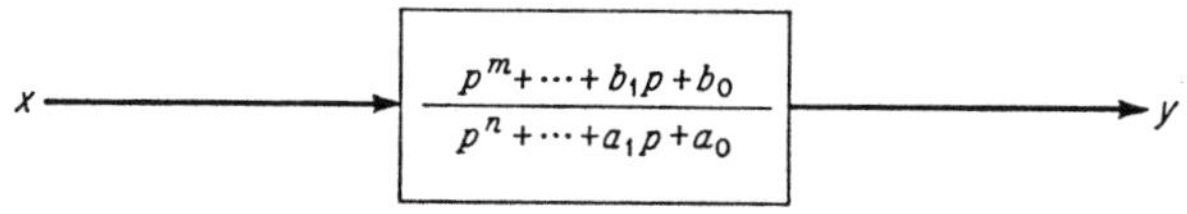

FIG. 4.1 Block diagram of a linear operator.

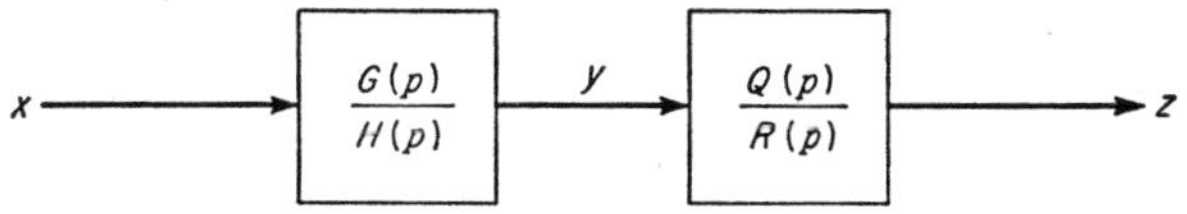

FIG. 4.2 Two operators in cascade.

output. In a block diagram the input is always processed to yield the output. Because the signal flows only in one direction the block diagram is said to be a unilateral representation. However, bilateral problems, such as the determination of current in branches of a four-terminal network, have a unilateral form.

We can cascade the blocks as shown in Fig. 4.2, from which we may write the following relationships:

$$\frac{z}{x} = \frac{G(p)}{H(p)} \frac{Q(p)}{R(p)} \tag{4.23}$$

where $G(p)$, $H(p)$, Q_p, and R_p are differential operators which are ratios of polynomials in p. Differential operators can always be put in this form.

When we draw operators in cascade, we imply that no operator in any way loads down, i.e., alters, the characteristics of any other operator. Were we to change the order of the operators in Fig. 4.2, their form would not change. Should these operators arise from physical systems, loading effects may seriously change the nature of the operator.

The computer block diagram has properties which are strikingly similar to those of the block diagram used by control systems engineers.[1] For constant-coefficient equations both the computer operator and control systems engineer represent problems in substantially the same topological form, a reason for the ease with which both groups can communicate with each other. In particular, the large body of knowledge accumulated in the field of servomechanisms may be appropriated directly for use by computer personnel.

If we take the Laplace transform of both sides of Eq. (4.12), the resulting equation has the form

$$(s^n + a_{n-1}s^{n-1} + \cdots + a_1 s + a_0)Y(s) = (s^m + b_{m-1}s^{m-1} + \cdots b_1 s + b_0)X(s) \qquad (4.24)$$

provided all initial conditions are zero.

The block diagram derived from Eq. (4.24) is shown in Fig. 4.3. The terms inside the block diagram are called the transfer function. If we compare Figs. 4.1 and 4.3, we can see that both blocks are the same except that

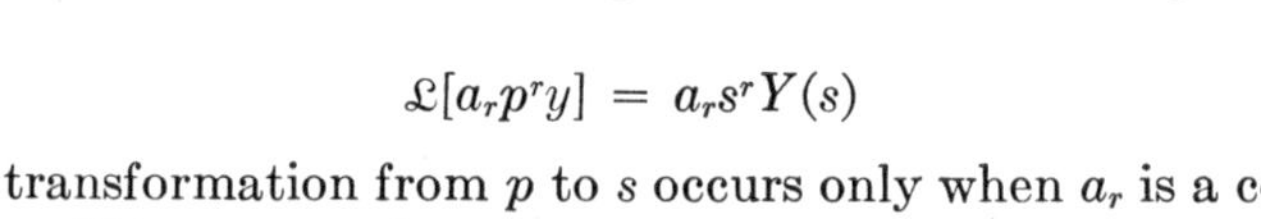

FIG. 4.3 A linear operator using the Laplace transform operator *s*.

in one case we use the operator p and in the other case the operator s. This equivalence results because the Laplace transform of any term in Eq. (4.12) is

$$\mathcal{L}[a_r p^r y] = a_r s^r Y(s) \qquad (4.25)$$

This direct transformation from p to s occurs only when a_r is a constant and the initial condition on $p^r y$ is zero.

In Sec. 3.3 we discussed the weighting function for an equation like (4.12), where $m = 0$. We can now extend this concept to apply to any arbitrary m. Such a modified weighting function[2] would satisfy the equation

$$y(t) = \int_0^t K(t, \alpha)x(\alpha)\, d\alpha \qquad (4.26)$$

From Eq. (4.15) we may write

$$y(t) = \frac{p^m + b_{m-1}p^{m-1} + \cdots + b_1 p + b_0}{p^n + a_{n-1}p^{n-1} + \cdots + a_1 p + a_0} x(t) \qquad (4.27)$$

[1] There are many books on the subject of control systems, sometimes called servomechanisms. One good book worth the reader's perusal is Ref. 52.

[2] For more about the modified weighting function, see Appendix 2.

Equating the right sides of Eqs. (4.26) and (4.27),

$$\int_0^t K(t, \alpha)x(\alpha)\, d\alpha = \frac{p^m + b_{m-1}p^{m-1} + \cdots + b_1p + b_0}{p^n + a_{n-1}p^{n-1} + \cdots + a_1p + a_0} x(t) \qquad (4.28)$$

Because both the integral and the differential operators transform a function $x(t)$ into the same function $y(t)$, they are equal and can be written as

$$\int_0^t K(t, \alpha)[\quad]\, d\alpha = \frac{p^m + b_{m-1}p^{m-1} + \cdots + b_1p + b_0}{p^n + a_{n-1}p^{n-1} + \cdots + a_1p + a_0} [\quad] \qquad (4.29)$$

The Laplace transform of $K(t, \alpha)$ is the transfer function $H(s)$. Thus

$$\mathcal{L}[K(t, \alpha)] = H(s) \qquad (4.30)$$

Incidentally, all our remarks about the modified weighting function involving the Laplace transform apply to all linear differential equations, with variable as well as constant coefficients.

4.3 Block-diagram Manipulation[1]

In this section we enumerate the rules which may be used to alter a block-diagram configuration. We shall use Fig. 4.4 to define the terminology used in these rules.

Definition of Terminology

1. *Element or block.* An operator which can be represented as a ratio of polynomials in p and is designated as a single unit in the diagram, for example, $F(p)$.
2. *Input.* This term has two meanings. It applies to the function being operated on by a block as well as to the point where the function feeds into the block. Therefore, input is an abbreviated form either for the input point or for the input signal to the block.
3. *Output.* The term applies either to the result of an operation on the input to a block or to the point where this function comes from the block.
4. *Summing junction.* A point at which two or more signals are added or subtracted, e.g., point A. The positive and negative signs at the input to the summing junction indicate whether the signal is added or subtracted. Thus

$$\epsilon(t) = f(t) + r(t) - g(t) \qquad (4.31)$$

Two or more inputs to either an amplifier or an integrator constitute a summing junction.

[1] Some of these rules for block-diagram alteration were taken from Ref. 53, pp. 92–98. See also the section written by S. Jennings and A. A. Winkeljohann in Ref. 49, pp. 20-61 to 20-66.

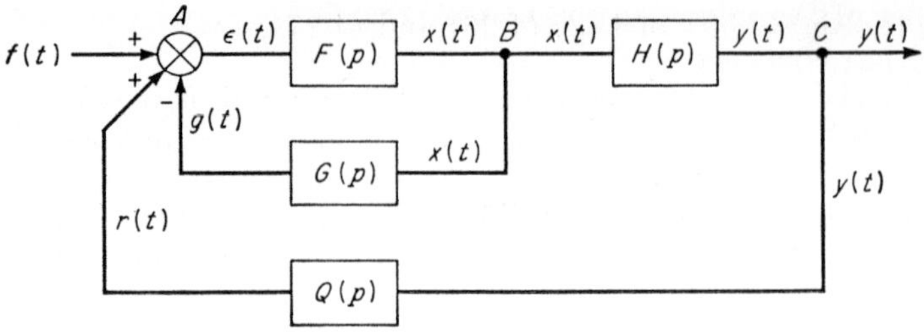

FIG. 4.4 A typical system block diagram.

5. *Node.* A point at which a signal branches off into two or more directions, e.g., points B and C. Two or more outputs from any computer element constitute a node.

6. *Forward path.* The path that the signal takes from a summing junction to a node or from one node to another.

7. *Forward branch.* All the elements in a forward path.

8. *Feedback path.* The path that a signal takes from a node to a summing junction, e.g., the path from B to A through operator $G(p)$.

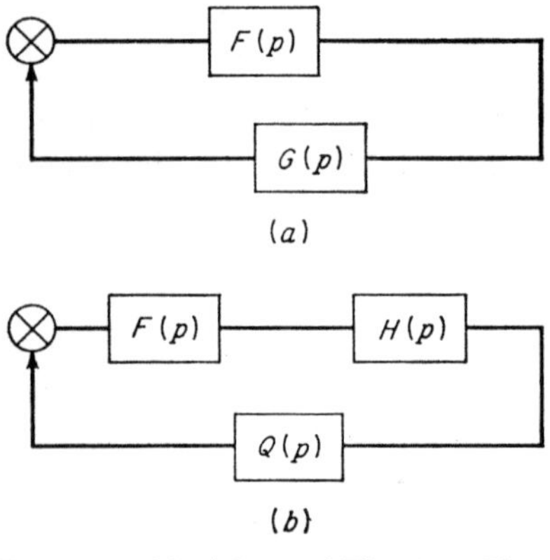

FIG. 4.5 (*a*) A loop of Fig. 4.4; (*b*) a loop of Fig. 4.4.

9. *Feedback branch.* All the elements in a feedback path.

10. *Loop.* Any closed path around which the signal can pass. Because all blocks are unilateral, signal can flow in only one direction. The two loops in Fig. 4.4 are shown in Fig. 4.5. CAUTION: This definition of a loop differs from the one used in servomechanism applications.

11. *Forcing function.* Any input to a summing junction which has not been operated on by any element in the block diagram. There may be more than one forcing function.

12. *Readout.* Any function in the block diagram whose behavior is being recorded. This function is a solution of the problem represented by the block diagram. There may be more than one function being read out.

Manipulation Rules

Rule 1. For an operator $G(p)$ we define the reciprocal as $1/G(p)$ and the inverse as $G^{-1}(p)$, where

$$G(p)G^{-1}(p) = G^{-1}(p)G(p) = 1 \tag{4.32}$$

Then

$$G^{-1}(p) = \frac{1}{G(p)}$$

when either of the following conditions is met:

a. $G(p)$ has constant coefficients.

b. Either the numerator or denominator of $G(p)$ is unity.

Rule 2. There are four possible ways in which a forcing function may be moved to a new summing junction. These are illustrated in Fig. 4.6:

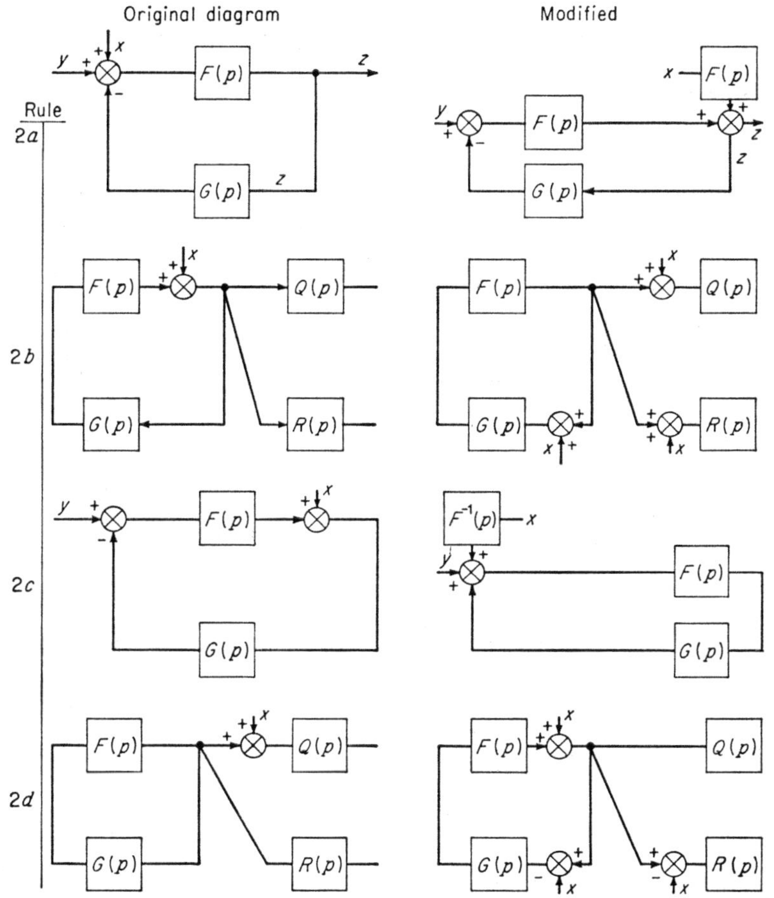

FIG. 4.6 Illustration of rules for manipulating the position of the forcing function

a. When moved forward through a block, it must be added at the new point after being operated on by a block identical to the one through which it passed.

b. When moved forward past a node, the forcing function must be added in all branches of the node except the one whence it came.

c. When moved backward through a block, the forcing function must be multiplied by the inverse of the block.

d. When moved backward through a node, the forcing function is subtracted from all branches of the node except the one whence it came.

If the forcing function is moved past a readout point, the readout point must be treated as a branch of a node.

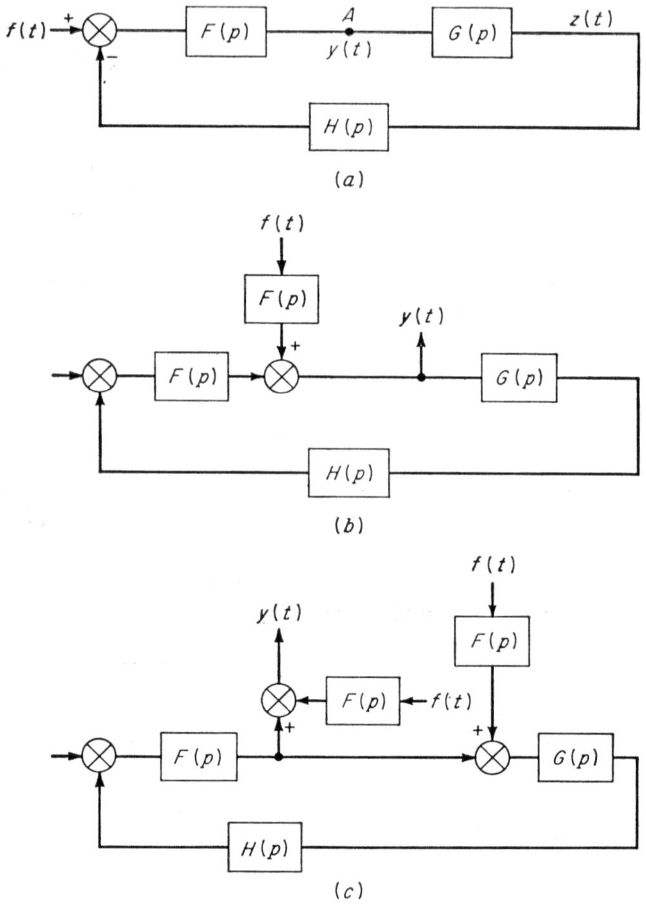

FIG. 4.7 (*a*) A system block diagram; (*b*) first step in moving the forcing function in (*a*) past a readout point; (*c*) altered block diagram of (*a*) after the forcing function has been moved past a readout point.

Example 1. In the block diagram of Fig. 4.7*a* there is one readout point A, from which the variable y is obtained. To move the forcing function to the input of the operator $G(p)$, we proceed through the following series of steps. These steps are illustrated in Fig. 4.7*b* and *c*.

Step 1. By rule 2*a* pass $f(t)$ through operator $F(p)$.

Step 2. By rule 2*b* move the forcing function past nodal point A.

Example 2. We wish to program the equation

$$p^2y + a_1py + a_0y = px + b_0x \tag{4.33}$$

where only the function x is available. We begin by setting up Eq. (4.33) by the general method (Fig. 4.8*a*). The inputs to integrator 1 are a summing junction, and its output is a nodal point. By rule 2*a* we pass *px* through integrator 1 (Fig. 4.8*b*). Then by rule 2*b* we move the forcing function past nodal point *A* (Fig. 4.8*c*). After eliminating the summing junctions, we obtain Fig. 4.8*d* as our computer diagram.

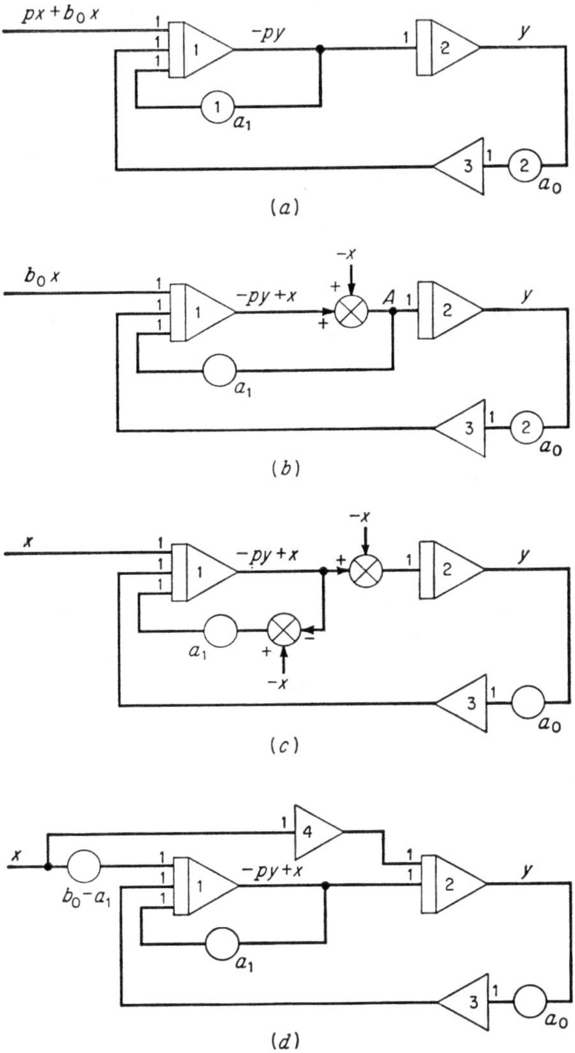

FIG. 4.8 Steps in programming a forcing function to avoid generating derivatives (see Example 2).

The output of integrator 1 has changed as a result of the manipulation of the forcing function. The initial condition on this integrator must, therefore, be modified accordingly so that it has the value of $-py + x$ at the initial time.

There are two points worth emphasizing in this example:

1. In programming the problem we did not use either the dummy-variable or canonical-form method. However, for derivatives of x higher than first-order the procedure in this

example is cumbersome. The canonical-form technique is preferable, especially when all initial conditions are zero.

2. The technique in this example may be used in block diagrams containing nonlinearities and variable coefficients provided we do not need to pass the function through an integrator after it has passed through either a nonlinear or variable-coefficient operator.

Rule 3. The order of operators may be changed only when the operators are commutative. Constants multiplying linear operators may always be

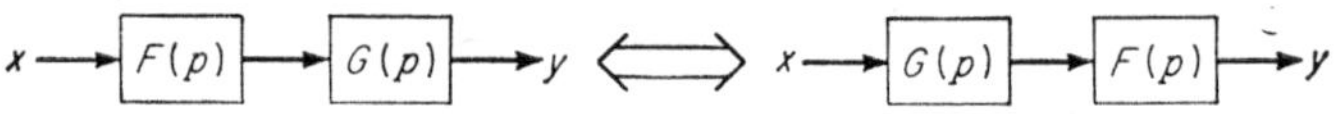

FIG. 4.9 Commutation of operators.

commuted. Otherwise, except for special cases, only operators having constant coefficients are commutative. Commutative operators may be manipulated as shown in Fig. 4.9.

Rule 4. The order of summing junctions may be interchanged, as illustrated in Fig. 4.10.

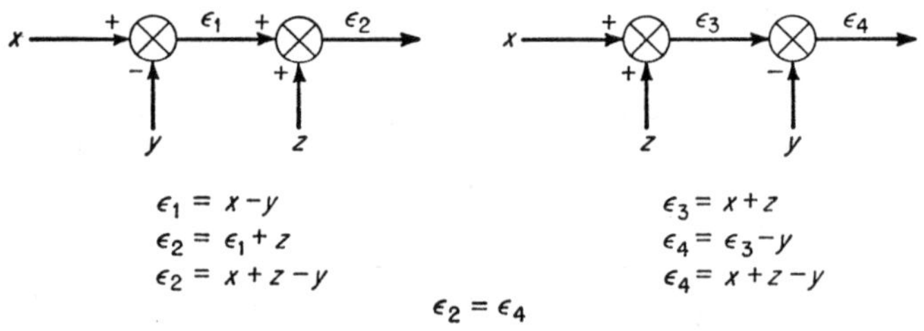

FIG. 4.10 Interchangeability of summing points.

Rule 5. *The Constant-loop-gain Rule. A multiple-loop block diagram, containing only linear constant-coefficient differential operators, may have the positions of any number of elements altered provided the gain of each loop after manipulation is the same as the gain before manipulation.* The gain of a loop is the product of all the operators in the loop. Later in this chapter we shall use this rule to amplitude-scale in an almost effortless manner.

Let us consider by example what this rule means. After the basic ideas have been explained, we shall show how the rule may be extended to block diagrams with nonlinearities and variable coefficients.

In Fig. 4.11 there are two loops shown by the arrows. The gains of the two loops are

$$\begin{aligned} \text{Loop 1} &= L_1 = -K_1G(p) \\ \text{Loop 2} &= L_2 = -K_1K_2G(p)H(p) \end{aligned} \tag{4.34}$$

An alternative way of setting up the computer for the same problem is shown in Fig. 4.12. A comparison of Fig. 4.11 with Fig. 4.12 will show that the loop gains are the same in both cases. Hence, Figs. 4.11 and 4.12 are equivalent computer setups for a particular problem. However, e_2 and e_3 in Fig. 4.12 are not the same as their counterparts e_a and e_b in Fig. 4.12.

Manipulation of the position of a block changes the variable which appears at its output. Further, every integrator associated with the block has its output variable changed too. Thus, when we move a block we must calculate new initial conditions for the integrator involved.

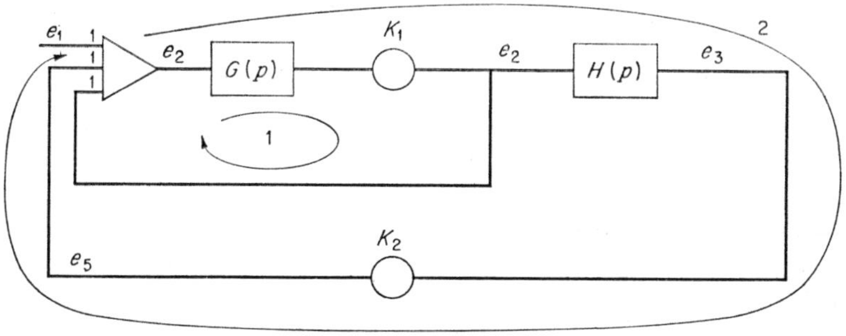

FIG. 4.11 A block diagram (see Fig. 4.12).

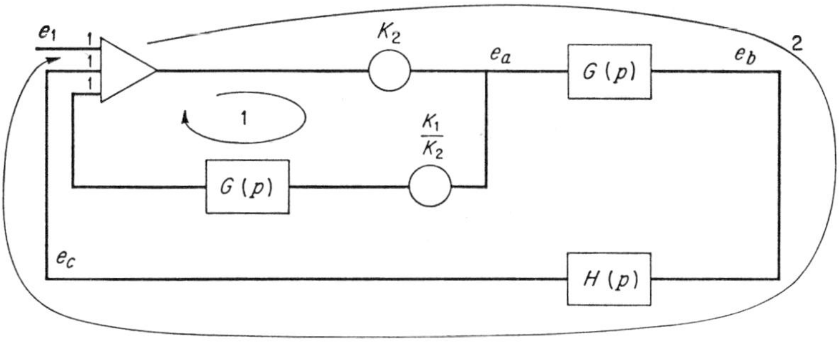

FIG. 4.12 An alternative way of drawing the block diagram of Fig. 4.11. The two figures are equivalent by the constant-loop-gain rule.

We shall now show heuristically that the manipulation rule holds and may be applied to change the computer program from Fig. 4.11 to Fig. 4.12. This will also show how computer variables are scaled after loop rearrangement, and also how new initial conditions can be determined. Although we are considering a two-loop system, we shall demonstrate why the manipulative rule is valid for multiple-loop systems of any order.

We shall first calculate the transfer function of e_3/e_1 in Fig. 4.11 and then show that this transfer function is invariant under manipulation if the loop gains remain constant. This is most easily done by determining first the innermost-loop transfer function. The innermost loop is that loop whose gain can be ascertained without opening any feedback path in the computer.

The transfer function so calculated becomes one operator in the next loop, and the process is repeated until the overall transfer function is secured.

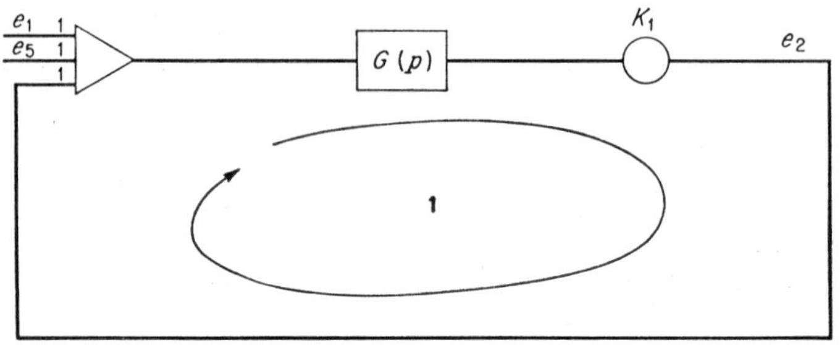

FIG. 4.13 Loop 1 of Fig. 4.11.

Following this outlined procedure we start with loop 1, which is shown in detail in Fig. 4.13. Using basic feedback theory we find that the transfer function is

$$\frac{e_2}{e_1 + e_5} = \frac{-K_1G(p)}{1 + K_1G(p)} = \frac{\text{forward gain}}{1 + L_1} \tag{4.35}$$

Figure 4.11 may now be represented in the equivalent form shown in Fig. 4.14, from which we obtain

$$\frac{e_3}{e_1} = \frac{-\dfrac{K_1G(p)H(p)}{1 + K_1G(p)}}{1 + \dfrac{K_1K_2G(p)H(p)}{1 + K_1G(p)}} = \frac{\text{forward gain}}{1 + \dfrac{L_2}{1 + L_1}} \tag{4.36}$$

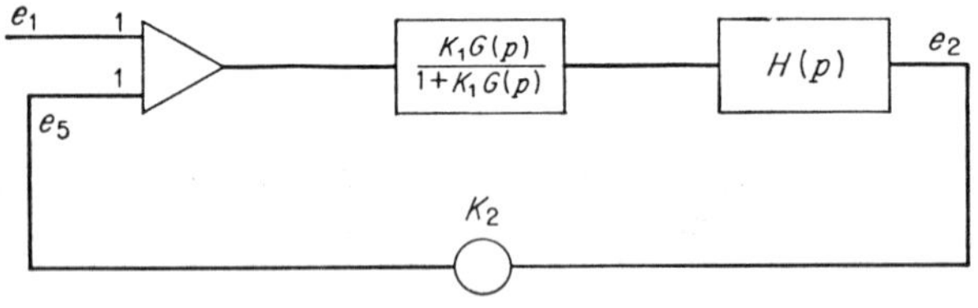

FIG. 4.14 A representation of Fig. 4.11 after loop 1 is replaced by its equivalent.

In Eqs. (4.35) and (4.36) we have the beginnings of a recursive relationship which would show us the general result for a system of any number of loops. This important general equation states that for a multiple system with loop gains $L_1, L_2, \ldots, L_n$ the transfer function between any two points e_i and e_j is

$$\frac{e_j}{e_i} = \frac{\text{forward gain}}{1 + f(L_1, L_2, \ldots, L_n)} \tag{4.37}$$

The forward gain is the transfer function from e_i to e_j with the feedback path from e_j to e_i open. If all the loop gains are the same before and after manipulation, Eq. (4.37) shows that the denominator of the transfer function remains constant and that the ratio of the output e_j to the input e_i is proportional to the forward gain only. Thus, a particular variable will be available before and after manipulation if the forward gain between it and a given input e_i does not change.

We may now relate the variables appearing in Fig. 4.11 with those in Fig. 4.12. From Fig. 4.11

$$\frac{e_2}{e_1} = \frac{\dfrac{K_1G(p)}{1 + K_1G(p)}}{1 + f(L_1, L_2)} \tag{4.38}$$

and

$$\frac{e_3}{e_1} = \frac{\dfrac{K_1G(p)H(p)}{1 + K_1G(p)}}{1 + f(L_1, L_2)} \tag{4.39}$$

From Fig. 4.12

$$\frac{e_a}{e_1} = \frac{\dfrac{K_2}{1 + K_1G(p)}}{1 + f(L_1, L_2)} \tag{4.40}$$

and

$$\frac{e_b}{e_1} = \frac{\dfrac{K_2G(p)}{1 + K_1G(p)}}{1 + f(L_1, L_2)} \tag{4.41}$$

From Eqs. (4.38) and (4.40)

$$e_a = \frac{K_2}{K_1G(p)}\, e_2 \tag{4.42}$$

Also from Eqs. (4.39) and (4.41)

$$e_b = \frac{K_2G(p)}{K_1H(p)}\, e_3 \tag{4.43}$$

If we rewrite Eq. (4.43) as

$$K_1H(p)e_b = K_2G(p)e_3 \tag{4.44}$$

we see in this case that e_b and e_3 are related through a differential equation. From the initial conditions of the problem we either shall have or can calculate all the initial values of e_3 and its derivatives for the right side of Eq. (4.44). However, we cannot get the initial value of e_b, because the left side of the equation contains not only e_b but some of its derivatives. There will, however, be enough other equations like (4.44) relating known to transformed initial values so that we shall be able to solve for the initial values on all integrators in the transformed diagram.

Sometimes a computer diagram may be modified using the constant-loop gain rule so that derivatives of a variable may be obtained without additional elements. Suppose Fig. 4.15 represents a computer diagram as it is

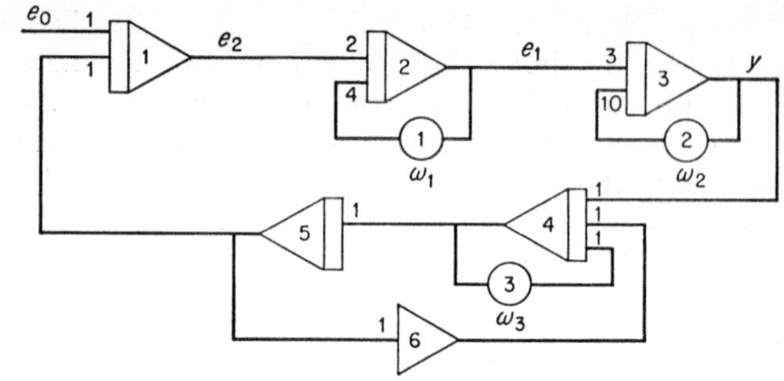

FIG. 4.15 A block diagram (see Fig. 4.16).

originally drawn. In addition to y, we want the derivatives dy/dt and d^2y/dt^2. The derivative dy/dt is readily available at the input to integrator 3 and is

$$py = -(3e_1 + 10\omega_2 y) \tag{4.45}$$

However, d^2y/dt^2 can be secured only by a proper summation of the variables in the circuit. Relating e_1 and e_2,

$$pe_1 = -(2e_2 + 4\omega_1 e_1) \tag{4.46}$$

Differentiating Eq. (4.45),

$$p^2y = -(3pe_1 + 10\omega_2 py) \tag{4.47}$$

Substituting Eqs. (4.46) and (4.47) in (4.45),

$$p^2y = 6e_2 + 12e_1 + 30\omega_2 e_1 + 100\omega_2{}^2 y = 6e_2 + 12e_1 - 10\omega_2 py \tag{4.48}$$

A simpler way of obtaining the two derivatives is shown in the modified circuit of Fig. 4.16. Note that the loop gains in Figs. 4.15 and 4.16 are the

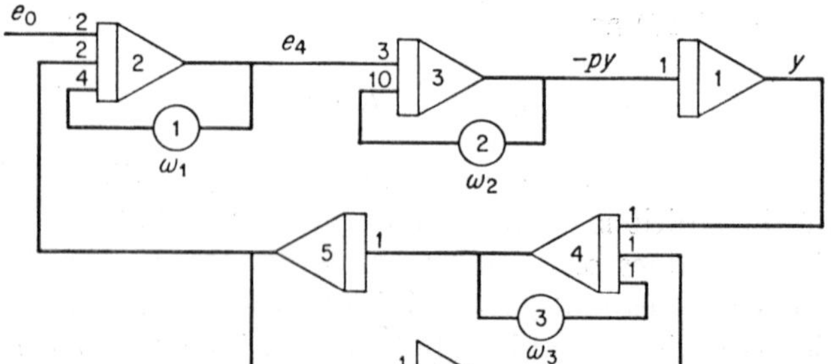

FIG. 4.16 A block diagram which is equivalent to Fig. 4.15. This figure has the advantage of giving dy/dt more easily than Fig. 4.15.

same. Now $-py$ is the voltage at the input to integrator 1 (or output of integrator 3). Also p^2y is available at the input summing point of integrator 3 and is

$$p^2y = -3e_4 + 10\omega_2 py \tag{4.49}$$

Finally, it should be pointed out that the loop-gain criterion is a sufficient but not necessary condition for permissible circuit modification. A critically damped or overdamped second-order differential equation is

$$\frac{d^2y}{dt^2} + 2a\frac{dy}{dt} + b^2 = 0 \tag{4.50}$$

where $$a > |b|$$

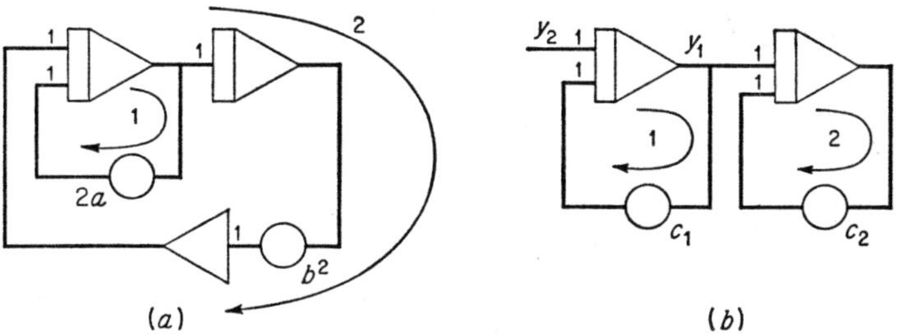

FIG. 4.17 (*a*) A block diagram—see (*b*); (*b*) a block diagram which has the same loop gains as (*a*). This demonstrates that the loop-gain rule applies only if the structure of the loop remains the same.

For such an equation Fig. 4.17*a* and *b* are correct computer diagrams if

$$c_1 + c_2 = 2a \qquad c_1c_2 = b^2 \tag{4.51}$$

The two loops in Fig. 4.17*a* have different gains from those of Fig. 4.17*b*.

Example 3. In this example we shall show how we can program a problem by the general method and then manipulate it into the canonical form. For equation

$$\frac{d^3y}{dt^3} + a_2\frac{d^2y}{dt^2} + a_1\frac{dy}{dt} + a_0y = b_2\frac{d^2x}{dt^2} + b_1\frac{dx}{dt} + b_0x \tag{4.52}$$

Fig. 4.18 is the computer diagram when the general method is used. First we move integrator 3 in the manner shown in Fig. 4.19, and then we move integrator 2 as shown in Fig. 4.20. Figures 4.18 and 4.20 have the same loop gains. However, we must deter-

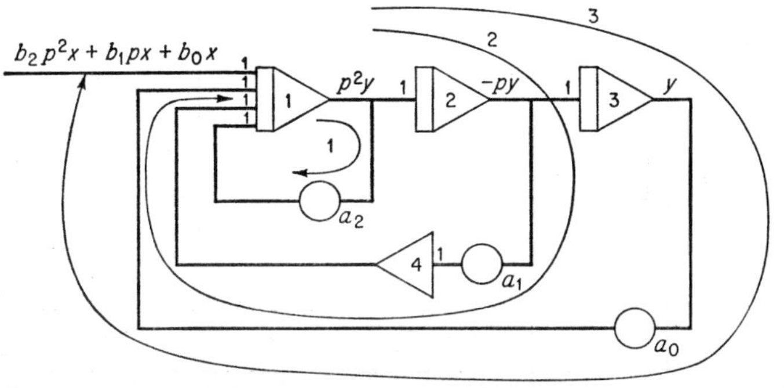

FIG. 4.18 A computer diagram programmed by the general method.

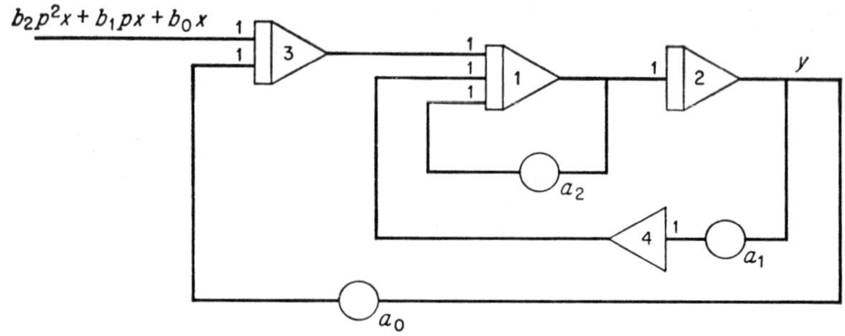

FIG. 4.19 A step in the conversion of Fig. 4.18 to a canonical-form program.

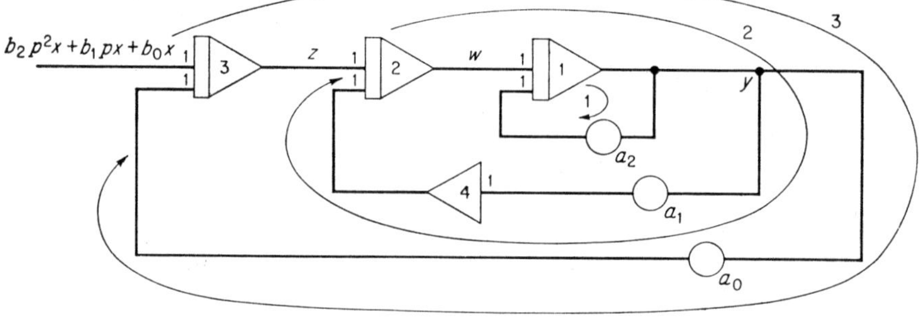

FIG. 4.20 A step in the conversion of Fig. 4.18 to a canonical-form program.

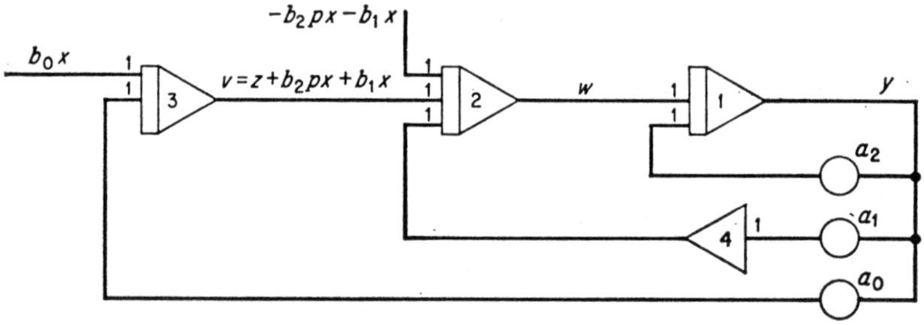

FIG. 4.21 A step in the conversion of Fig. 4.18 to a canonical-form program.

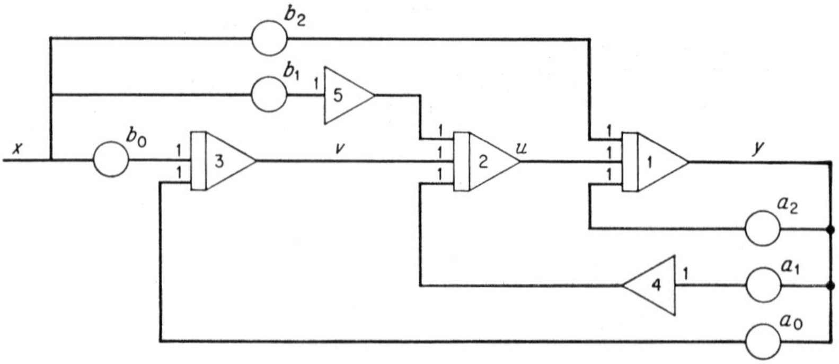

FIG. 4.22 A step in the conversion of Fig. 4.18 to a canonical-form program.

mine the outputs of integrators 1 and 2 in Fig. 4.20. Using the method outlined in Eqs. (4.44) to (4.49), we find

$$w = -(p + a_2)y \qquad z = (p^2 + a_2p + a_1)y \tag{4.53}$$

Next, as shown in Fig. 4.21, we pass the derivatives of x through integrator 3 (see Example 2). Finally, passing b_1px through integrator 2 and connecting all terms in x, we have Fig. 4.22, the canonical form for Eq. (4.52).

The variable at the output of integrator 3 is

$$v = (p^2 + a_2p + a_1)y + (b_2p + b_1)x \tag{4.54}$$

and at the input

$$u = -(p + a_2)y - b_2x \tag{4.55}$$

From these equations we can calculate the initial conditions on the integrators.

For the reader who dislikes the effort of this type of calculation, we point out that the initial conditions of many engineering problems of this sort are zero, where the only variable of interest is y.

Constant-coefficient operators may be manipulated in a diagram which contains nonlinear and variable-coefficient linear operators. Now the order of operation must be maintained except for constant-coefficient operators.

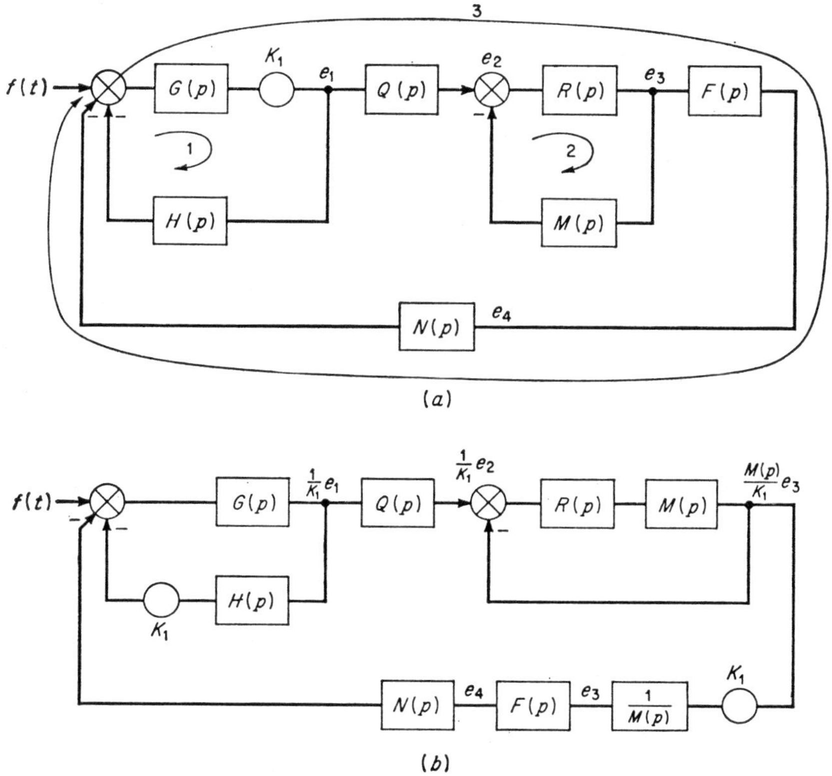

FIG. 4.23 (*a*) A block diagram where $Q(p)$ and $F(p)$ are not constant-coefficient operators—see (*b*); (*b*) a block diagram which is equivalent to (*a*).

The constant-loop-gain rule will continue to hold if the variable at the input to each nonlinear operator is the same after manipulation as before and if the variable at the input to a variable-coefficient linear operator after manipulation differs from the one before manipulation by no more than a constant.

Example 4. In Fig. 4.23*a*, $Q(p)$ is a variable-coefficient operator and $F(p)$ is a nonlinear operator. Now suppose we manipulate loop 1 by putting K_1 in the feedback path and loop 2 by putting $M(p)$ in the forward path. Then Fig. 4.23*b* represents the same system as Fig. 4.23*a*.

4.4 Background of Time and Amplitude Scaling[1,2]

We have seen in Chap. 3 that the computer voltage and solution time are analogous to the dependent and independent problem variables, respectively. Furthermore, when the computer is programmed directly from the problem equations, there is a one-to-one relationship between the computer and problem variables; i.e., one volt $\equiv$ one unit of problem dependent variable, and one unit of computer time $\equiv$ one unit of problem independent variable. However, the computer has inherent limitations, which bound the range of errorless operation of each of the components. To ensure that the computer does not attempt to solve the problem outside the proper operating range, we must often modify the simple one-to-one scaling between the computer and problem variables.

Amplitude and time scaling are the two means at our disposal for combating computer shortcomings. Amplitude scaling changes the voltage level of the components; time scaling alters the length of time required by the computer to solve the problem. A technique of scaling, simple to implement and relatively immune to error, is described in this chapter. The programmer initially sets up the computer directly from the original set of equations describing the problem. Scaling from the one-to-one relationship to any other may be carried out directly on the computer diagram by the application of a few rules of block-diagram manipulation.

In this chapter we consider scaling by a factor which remains constant during the computer run. The subject of variable scaling is covered in Sec. 7.2.

4.5 Amplitude Scaling

The computer has elements that cannot perform the mathematical operations for which they were designed if the problem voltage magnitudes are

[1] For an excellent discussion of this subject written from a somewhat different viewpoint, see Ref. 43, pp. 72–115. For additional reading consult Ref. 44, pp. 127–128, and Ref. 46, pp. 98–100, 379–382.

[2] Scaling that involves nonlinear elements will be treated in Sec. 5.2.2. Noise scaling is discussed in Chap. 11.

either too high or too low. For the operational amplifier, an upper limit on the maximum voltage is reached because the output current can range only between zero and a value determined by its design. The servomultiplier cannot have an input voltage exceeding that on its follow-up potentiometer. And so the list can be compiled for all the computer elements.

Nor can the voltage remain below a certain level for too long an interval of computer time during the problem run. Amplifier noise and integrator voltage drift become significant compared with the voltage generated by the solution. Electromechanical devices, like the relay and servomultiplier, have threshold voltages below which they do not perform their operations accurately. The permissible lower level of voltage in a computer element is determined largely by the nature of the problem. The question that must be answered is: How sensitive is the problem solution to the voltage corrupted by noise and drift? A lack of solution repeatability from one computer run to another could indicate that the voltage level is too low somewhere in the problem. If the programmer does not have other indications to the contrary, a good empirical criterion for existing precision computers is that errors can result in a computer element if the maximum voltage does not exceed 1 volt during the problem run.

Amplitude scaling may be carried out either through a modification of the problem equations or by a direct alteration of the computer diagram. Of the two methods the alteration of the computer diagram is much preferred. Changing the mathematical equations requires so much more effort that it is recommended only for those who enjoy work for its own sake. However, we shall discuss the mathematical technique because it may shed some light on the process. We shall show how to scale so that the problem voltage does not exceed the capability of a computer element. The presumption is that in so doing we also prevent the voltage from falling below its minimum operating level. In Chap. 7 we discuss in some detail how to scale when this is not the case.

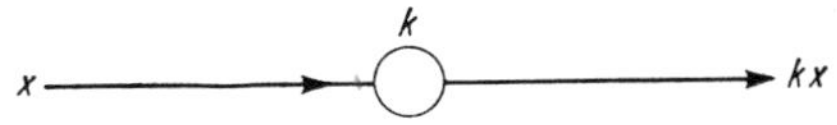

FIG. 4.24 Effect of a potentiometer on scaling.

First we must clarify the notation we shall use. As Fig. 4.24 shows, the input to the potentiometer is a voltage corresponding to the variable x, and its output is a voltage corresponding to kx. Let us assume that x represents distance and that the scale at the potentiometer input is

$$1 \text{ ft} \equiv 1 \text{ volt} \tag{4.56}$$

Thus the computer voltage and the problem variable may be said to be in one-to-one scaling, because

$$1 \text{ unit of problem variable} \equiv 1 \text{ unit of voltage}$$

For 1 volt at the input, there will be k volts at the output. However, the distance x remains 1 ft. The voltage level has changed, but no operation has been performed on the problem variable. Thus, at the output, k volts represents 1 ft in x. The scale now becomes

$$\frac{1}{k}\text{ ft} \equiv 1\text{ volt} \tag{4.57}$$

The coefficient k has the dimension of feet per volt or, in general, problem-variable dimension per volt.

This scaling procedure is applicable to all computer elements. Thus, in general, the scaling on the physical quantity changes as

$$\frac{1}{\text{Gain of the element}}$$

The advantages of this method of amplitude scaling are twofold. First, the scaling is performed as the computer is being programmed. Second, the scale appears on the computer diagram at the output of each element.

The factor k must be chosen so that the voltage does not exceed its maximum. If the problem-variable maximum is x_{max} and the voltage maximum is V_{max}, then

$$k \leq \left|\frac{V_{max}}{x_{max}}\right| \tag{4.58}$$

If we work with the mathematical equation, we can ensure that proper factors will appear on each of the computer elements in the following manner. Suppose we must program the equation

$$\frac{d^2y}{dt^2} + a\frac{dy}{dt} + by = f(t) \tag{4.59}$$

where a and b are both constants. The maximum values for the problem variables are

$$\left|\frac{d^2y}{dt^2}\right|_{max} \qquad \left|\frac{dy}{dt}\right|_{max} \qquad |y|_{max}$$

The voltage maximum in all cases is V_{max}. Then our factors are

$$k_2 = \frac{|V_{max}|}{\left|\dfrac{d^2y}{dt^2}\right|_{max}} \qquad k_1 = \frac{|V_{max}|}{\left|\dfrac{dy}{dt}\right|_{max}} \qquad k_0 = \frac{|V_{max}|}{|y|_{max}} \tag{4.60}$$

We then rewrite Eq. (4.59):

$$\frac{1}{k_2}\left(k_2\frac{d^2y}{dt^2}\right) + \frac{a}{k_1}\left(k_1\frac{dy}{dt}\right) + \frac{b}{k_0}(k_0y) = f(t) \tag{4.61}$$

The quantities in the parentheses are the variables we want to program. Using the standard programming techniques we obtain Fig. 4.25. In this diagram we see that y and its derivatives are scaled differently. Only the

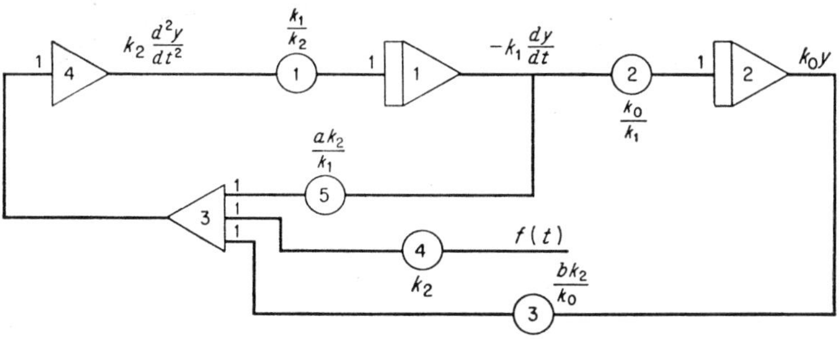

FIG. 4.25 Computer diagrams of Eq. (4.61).

addition of initial conditions remains to complete the program. If the problem initial conditions are

$$\left(\frac{dy}{dt}\right)_0 = m \qquad (y)_0 = n$$

then

$$-k_1\left(\frac{dy}{dt}\right)_0 = -k_1 m \quad \text{volts} \tag{4.62}$$

and

$$k_0 y_0 = k_0 n \quad \text{volts}$$

The same scaling can be done on the computer diagram. We begin by programming the computer directly from the equations on a one-to-one scale (the scale factor is unity everywhere). Then we change the scale factor at each point to whatever value we want. Finally, using the constant-loop-gain rule (Sec. 4.3), we ensure that each loop gain is the same after the scale-factor change as it was before. Incidentally, amplitude scaling refers only to changing the variable level by a constant. Therefore, we use the constant-loop-gain rule as it applies to the manipulation of constant gains and not operators.

Example 5. In the equation

$$\frac{d^3y}{dt^3} + a\frac{d^2y}{dt^2} + b\frac{dy}{dt} + cy = 0 \tag{4.63}$$

with initial conditions

$$y = 5 \qquad \frac{dy}{dt} = -3 \qquad \frac{d^2y}{dt^2} = 2.5$$

we are to program the computer so that we have the quantity 10 dy/dt.

We begin by programming Eq. (4.63) on a one-to-one scale, with Fig. 4.26 as the resulting computer diagram. As shown by the arrows, there are three loops having the gains

$$\begin{aligned} \text{Loop 1} &= -\frac{a}{p} \\ \text{Loop 2} &= -\frac{b}{p^2} \\ \text{Loop 3} &= -\frac{c}{p^3} \end{aligned} \tag{4.64}$$

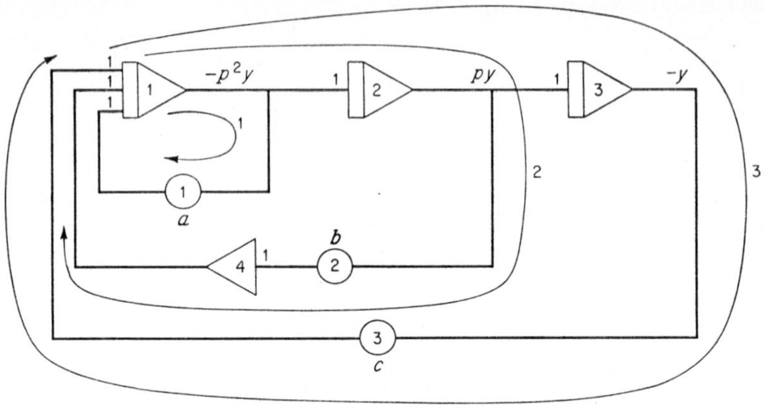

FIG. 4.26 Computer diagram programmed directly from Eq. (4.63).

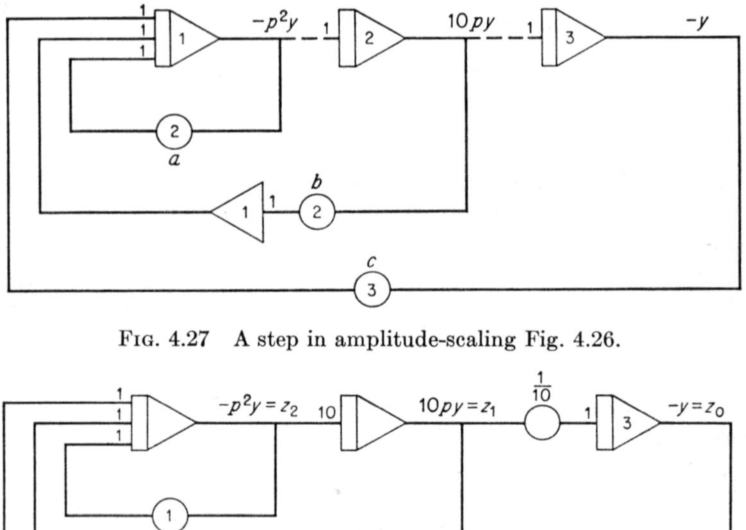

FIG. 4.27 A step in amplitude-scaling Fig. 4.26.

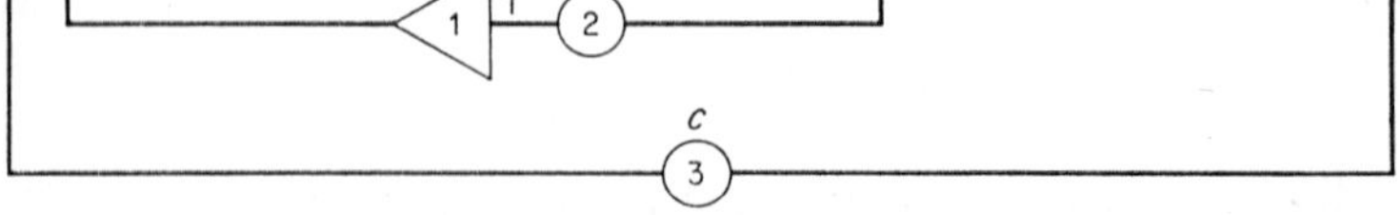

FIG. 4.28 An amplitude-scaled diagram equivalent to Fig. 4.26.

As an interim step, shown in Fig. 4.27, write $10py$ at the output of integrator 2. We have $-p^2y$ at the output of integrator 1 and $-y$ at the output of integrator 3. Therefore, the diagram can be consistent only if there is a gain of 10 between the outputs of integrators 1 and 2 and a gain of $\frac{1}{10}$ between the outputs of integrators 2 and 3. After inserting these gains, we find that loop 3 has the same gain as in Fig. 4.26. We must, however, include a gain factor in loop 2. When this is done, as shown in Fig. 4.28, we have the correct circuit with the amplitude scaling on the output of integrator 2. For

Fig. 4.28 our initial conditions are

$$\begin{aligned} z_2 &= -(p^2y)_0 = -2.5 \text{ volts} \\ z_1 &= 10(py)_0 = -30 \text{ volts} \\ z_0 &= -(py)_0 = -5 \text{ volts} \end{aligned} \tag{4.65}$$

The primary and virtually insurmountable difficulty one encounters in scaling a new, unfamiliar problem is that there is almost no way of determining the maximum values of the variables. In his book,[1] Jackson cites an ingenious empirical method which he has developed and used with some success.

For the constant-coefficient equation

$$\frac{d^ny}{dt^n} + a_{n-1}\frac{d^{n-1}y}{dt^{n-1}} + \cdots + a_1\frac{dy}{dt} + a_0y = Bu(t) \tag{4.66}$$

where all initial conditions are zero and $Bu(t)$ is a step forcing function of magnitude B,† the maxima for the variables can be approximated by the relationships

$$\begin{aligned} \left|\frac{d^ny}{dt^n}\right|_{\max} &= B \\ \left|\frac{d^ry}{dt^r}\right|_{\max} &= \frac{B}{a_n} \qquad 1 \le r \le n-1 \\ |y|_{\max} &= \frac{2B}{a_0} \end{aligned} \tag{4.67}$$

There are no analytical methods that will indicate the maximum values of variables in large-scale nonlinear or linear variable-coefficient problems. Here we must be guided by the results of a previously run similar (we hope) problem. In a physical problem we may know the range of some variables without knowing much about the others. If, for example, the problem is to consider the amount of gasoline consumed by an automobile under certain road conditions, we know that the quantity of gasoline used cannot exceed the amount in the fuel tank at the start of the test.

What if there is no analytic method, no previously run similar problem, and no physical knowledge to define the range of variable? In a complex problem this will all too often be the case. When so confronted all we can do is make an educated guess about the range of each variable and then scale it accordingly. The computer is programmed on this basis, and the problem is run. With a probability approaching unity, this first computer run will find some voltages too large and others too small. (We assume that the computer contains safeguards, e.g., fuses on the potentiometers, to protect components from being damaged by excessive voltages.) The voltages

[1] Reference 43, pp. 101–103. Jackson calls his method the *equal-coefficients rule*.
† See Appendix 1.

in the computer circuit may be checked rapidly with a voltmeter and waveform recorder. As an added aid each operational amplifier has an overload monitor, which records when the voltage at the summing junction departs significantly from zero. A lighted bulb on the front panel identifies the overloaded amplifier, which cannot behave like a linear element because the required output voltage is too great.

When all the troublesome voltages have been identified, they can be scaled on the computer diagram by the technique illustrated in Example 5. No recourse to the problem equations is necessary. When we scale some of the variables, we also change voltage levels in other parts of the circuit. After scaling, a voltage which was previously within specifications may now place an element outside accurate operation. We must scale by trial and error by proceeding in the following manner:

1. Make a computer run and observe all the voltages.
2. Scale the variables where the voltages fall outside the accurate operating range.
3. Make a computer run with the new scale factors.
4. Rescale the variables as indicated by the results of step 3.

In this way we continue alternating between making computer runs and scaling. Ultimately we find a scale that will permit the computer to solve the problem accurately. At first reading one may think that this method of scaling involves a lot of work. In practice this is not the case. With a small amount of experience one can change the scale on any element very rapidly, because the diagram shows which other elements must also be changed to keep the loop gain constant. Furthermore, in most problems there will be only a few variables that must be changed after the first computer run. Most important, this preliminary work is not wasted on the computer operator; he learns his computer diagram and a great deal about the nature of the problem solution as well.

4.6 Time Scaling

In the entire spectrum of computer programming problems, no concept is more difficult for the novice and the experienced operator alike to grasp than time scaling. The situation is even more tragic because the difficulty is mainly semantic, with the confusion stemming from use of the word *time*. The computer user's vocabulary embraces the phrases *problem time, computer time,* and *real time.* Problem time refers to the problem independent variable. The term arose because in the early days of computing most problems had time as their independent variable. Computer time is the independent variable of the computer solution. Real time is the time base of our physical world. It has no place at all in our discussion of time scaling except when we must connect to the computer devices whose behavior can be measured only in real time. Such devices would include equipment used in physical simulation and recorders driven by motors whose speeds are in terms of real time.

Time scaling is merely an alteration of computer running time. *Such an alteration affects the relation between the problem independent variable, hereafter called problem time, and computer time.* One should always scale the computer to run as rapidly as possible for correct solutions. The longer each individual run takes, the longer the time required to accumulate the necessary data. In some long problems involving statistical evaluations, a factor of 2 in problem running time can mean a difference of several weeks in total time on the computer. However, there is an upper speed with which the computer can solve a problem. Computer elements only approximate the mathematical operations for which they were designed. Servomotors are velocity-limited and can handle accurately only those input signals which change at a rate less than the velocity limit of the motor. Amplifiers and integrators introduce extraneous time lags which can convert a stable problem to an unstable computer setup. Such a system can be made stable only by changing loop gains.[1] Since amplitude scaling keeps the loop gain fixed, this is clearly not the way to stabilize a circuit. One can, however, change the gain of any loop that has at least one integrator by time scaling. Usually one must lower the loop gain to stabilize the circuit, a condition equivalent to slowing down the problem by time scaling. It is also possible to run the problem too slowly, in which case drifts in the amplifiers and integrators may produce errors in the solution.

In most complex problems it is virtually impossible to ascertain beforehand whether a given time scaling is within the bounds of correct computer operation. In such a situation, as in amplitude scaling, one takes the most educated guess on the time scaling and programs the computer accordingly. The problem is run first with this time scale and then with a scale that slows down the computer by a factor of 2.† If both solutions are the same, the faster time scale is not too fast.

As we saw in Chap. 3, the independent variable of the problem corresponds to computer time. If we start, for example, with a scale of

$$1 \text{ unit of independent problem variable} \equiv 1 \text{ unit of computer time}$$

we shall speed up or slow down the problem depending on how we change this scale. IMPORTANT: *The computer has its own time reference, and all problem time is measured relative to it.*

The integrator, which is the only component whose operation has the dimension of time, is the computer time reference.[2] One unit of computer

[1] See Ref. 49, pp. 21-01 to 21-81, by W. E. Sollecito and S. G. Neque, for an extensive summary of stability in feedback systems.

† This feature is built into some computers.

[2] If we use *RC* networks in conjunction with the operational amplifier to get higher-order transfer functions, each capacitor has the dimension of time associated with it.

time is defined as the length of time it takes the output of a unit gain integrator to build up to one unit of voltage where the input is one unit of voltage.[1] In the case of general-purpose computers, one unit of time is 1 sec, but it must be emphasized that a computer can be designed with any other time base.[2] We may also define the time reference in terms of the period of a standard sine-wave oscillation. The definition we shall use is somewhat better because it relates directly the time and voltage references. The voltage reference is almost universally 1 volt. However, in principle it can be anything else that suits our convenience.

The relation of the integrator to the time base may be illustrated by the following simple example. Suppose a body is moving with a constant velocity $V = 1$ fps. The distance the body will travel is

$$d = \int_0^{t_1} V\, dt = Vt_1 \tag{4.68}$$

Let us assume that there is the relationship

$$1 \text{ volt} = 1 \text{ fps} \tag{4.69}$$

To find the distance traveled by the body, the computer is set up as shown in Fig. 4.29. If $V = 1$ volt, then at 1 sec there will be 1 volt at the

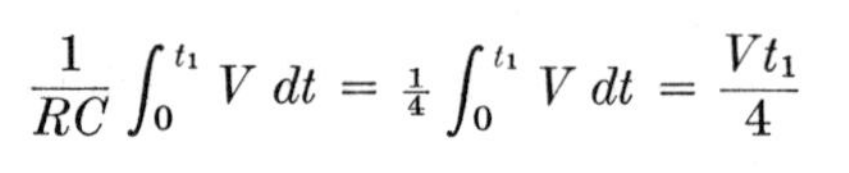

FIG. 4.29 An integration.

output of the integrator. However, consider an integrator where $RC = 4$. In this case the output of the integrator is[3]

$$\frac{1}{RC}\int_0^{t_1} V\, dt = \tfrac{1}{4}\int_0^{t_1} V\, dt = \frac{Vt_1}{4} \tag{4.70}$$

For a computer where all $RC = 4$, the time base is 4 sec. If $t_1 = 4$, then $Vt_1/4 = d$. The problem is set up on such a computer in precisely the same manner as on a computer where $RC = 1$. If the operator were aware only of the computer time reference, the solutions of the two computers would be indistinguishable. Actually the computer with $RC = 1$ will solve the problem four times as fast as the computer which has $RC = 4$.

We shall now show the mathematical basis of time scaling. Next we shall indicate how such scaling may be performed directly on the computer diagram without recourse to the original system equations. The procedure involved in making a time-scale change is quite straightforward. If the

[1] The gain of each integrator is calibrated against one integrator whose gain is specified as unity.

[2] One high-speed repetitive computer has a time base of $\frac{1}{3{,}000}$ sec.

[3] See Sec. 4.9 and, in particular, Eqs. (4.84).

independent variable of the problem is x, a linear transformation is made such that

$$x = \frac{t}{\alpha} \tag{4.71}$$

where t = computer independent variable

When $\alpha > 1$, the problem is slowed down by α; when $\alpha < 1$, the problem is sped up by α. Consider Eq. (4.72),

$$\frac{d^2y}{dx^2} + b_1 \frac{dy}{dx} + b_2 y = f(x) \tag{4.72}$$

If we make the transformation

$$x = \frac{t}{\alpha}$$

$$\frac{1}{1/\alpha^2} \frac{d^2y}{dt^2} + b_1 \frac{1}{1/\alpha} \frac{dy}{dt} + b_2 y = f\left(\frac{t}{\alpha}\right) \tag{4.73a}$$

or

$$\frac{d^2y}{dt^2} + \frac{b_1}{\alpha} \frac{dy}{dt} + \frac{b_2}{\alpha^2} y = \frac{1}{\alpha^2} f\left(\frac{t}{\alpha}\right) \tag{4.73b}$$

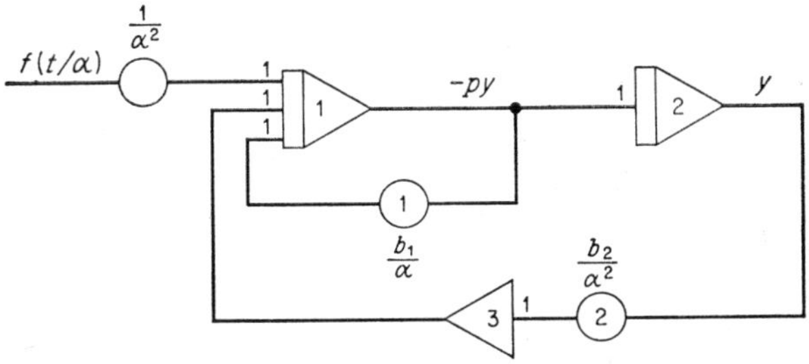

FIG. 4.30 A time-scaled equation programmed directly from Eq. (4.73*b*).

Figure 4.30 is the computer diagram of Eq. (4.73*b*), where

$$py = \frac{dy}{dt} \qquad y' = \frac{dy}{dx} \tag{4.74}$$

and

$$py = \frac{y'}{\alpha} \qquad p^2y = \frac{y''}{\alpha^2} \tag{4.75}$$

By application of the rules of amplitude scaling for loop-gain manipulation, Fig. 4.31 may be obtained from Fig. 4.30. From observation of Fig. 4.31 we make the following important statement about time scaling on a computer diagram: *To time-scale by α, multiply the gain of each integrator by a factor $1/\alpha$.* When this is done, the variable at the output of each integrator remains the same as before scaling (in plotting the outputs of the integrators we must represent each unit of the independent variable x by $1/\alpha$ units of the computer time t). With this type of scaling, initial conditions at the output of each integration are the same voltage before and after scaling. Note that the input $f(t/\alpha)$ must also be time-scaled. If this input is generated by integration operations, a correct change of these integrator input gains automatically time-scales the input. However, if the input is some external source, it must first be recorded and then rescaled. Where electronic or electromechanical gear is used in conjunction with a computer in a physical simulation, the gear cannot usually be modified. Under these conditions the computer cannot be time-scaled.

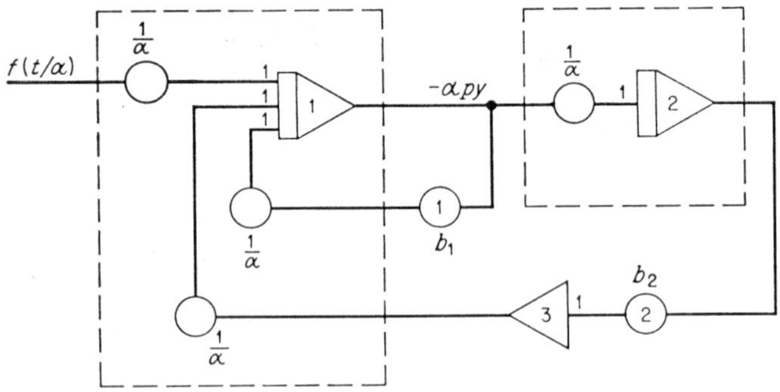

FIG. 4.31 A computer diagram equivalent to Fig. 4.30. Elements enclosed by dashed lines represent integrators with all input gains changed to allow for time scaling.

It is more than coincidence that the time-scaling problem can be set up with the $1/\alpha$ gain factor appearing at each integrator input. Let us consider a computer whose integrators have a gain $1/\alpha$ times the gain of the integrators used in Fig. 4.31. Thus the new integrator would be represented by the unit in Fig. 4.31 enclosed by the dashed lines. Such an integrator could be built by increasing the value of the capacitor of the original integrator by a factor α. Using this new integrator we could, if we wished, program Eq. (4.72) on a one-to-one relationship between x and the new time base of the computer which is now the original time base multiplied by α. (Remember that the time reference is determined by the integrator.) Again we stress that, if the input is generated by the computer, it will automatically be time-scaled properly. If, however, the input comes from an external source, its time scale must be made to conform to that of the computer.

4.6.1 Running the Problem "Backward"

The computer solves a problem in the direction of increasing computer independent variable, time. Thus the problem initial conditions must be given at the minimum value of this independent variable. If we are interested in the solution in the range of independent variable $x_a < x < x_b$ with the initial conditions given at x_1, where

$$x_a < x_1 < x_b$$

we get the solution in the range

$$x_1 < x < x_b$$

by applying the initial conditions and running the problem in the direction of increasing computer time. However, getting the solution in the range

$$x_a < x < x_1$$

is another matter. We can achieve this objective only by running the problem backward from x_1. To run the problem backward we make the time-scale change

$$t = -x \tag{4.76}$$

In Eq. (4.76) we see that an increase in computer time results in a decrease in problem variable x. Comparing this equation with Eq. (4.71), we observe that

$$\alpha = -1 \tag{4.77}$$

Hence, the computer is programmed to run backward by multiplying each integrator input by the factor -1.

Suppose we must solve a problem in the range $x_a < x < x_b$ and have two sets of initial conditions at x_a and x_b. Do we run the problem forward or backward? We run in the direction in which the solution converges. In so doing we shall get a more accurate computer solution.[1] If the solution converges for increasing independent variable, it will diverge for decreasing independent variable, and vice versa.

We are, however, at liberty to run the problem forward or backward at our discretion only if the problem is reversible. To be reversible a problem must meet the following criterion. We begin the problem at one value of independent variable x_a and plot the solution to a second value x_b. The problem is stopped at x_b, and with the problem variables retained on the integrator outputs, the computer operation is reversed. If the plot from x_b to x_a retraces the plot from x_a to x_b, the problem is reversible.

A problem will be reversible if all the computer elements perform single-valued operations. We define a single-valued operation as one where for each value of the input there is only one relationship between the input and the output. Figure 4.32 is an example of an operation which is not single-valued. The variable v is the input and y is the output, with the arrows designating the value of y as a function of the direction of v. In the two ranges $v > \alpha$ and $v < -\alpha$ this operation is single-valued. If the problem is confined to either of these two ranges, it is reversible. In the intermediate range $-\alpha < v < \alpha$ the value of y is dependent not only upon the

[1] In Sec. 7.4 we justify this statement.

input v but also upon its direction. Hence, a problem encompassing this range is not reversible. In irreversible problems the computer will obtain

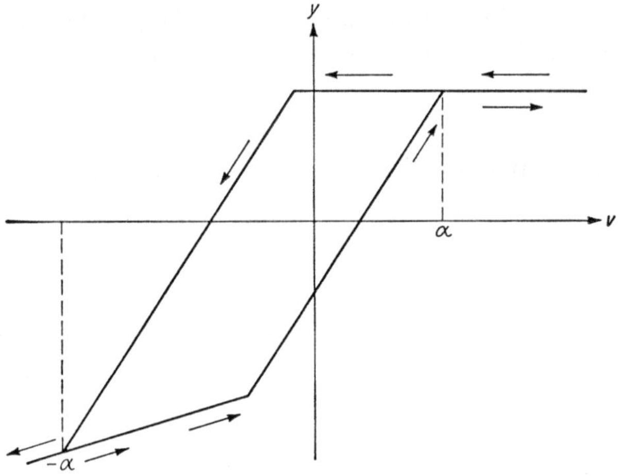

FIG. 4.32 A double-valued function.

the correct solution only by changing its independent variable in precisely the same direction as that specified for the problem.

4.7 Scaling the Adjoint Equation

The rules for setting up the adjoint computer diagram, discussed in Sec. 3.4, apply whether the diagram of the original differential equation is on a one-to-one basis or is scaled. When the weighting function in the original diagram is scaled, the weighting function in the adjoint equation is scaled the same way.

Assume that Fig. 4.33*a* is the computer diagram of the original system. In Fig. 4.33*b* we introduce scale factor $a_1a_2a_3$ having the effect on the weighting function shown. If we draw the adjoint equation using Fig. 4.33*b* as our model, the outcome is Fig. 4.33*c*, with the weighting function having the same scale factors as before. Because the adjoint equation does not have the same computer voltage as the original one, optimum scaling of the two computer diagrams will be different.

A delta function can be replaced by a corresponding initial condition at the output of the appropriate integrator (see Sec. 3.4). Because an initial condition is unaffected by time scaling, it follows that the delta function is replaced by the same initial condition after time scaling as before scaling. Thus, the gain of the integrator through which the delta function must pass changes with time scaling, and the delta-function amplitude changes a

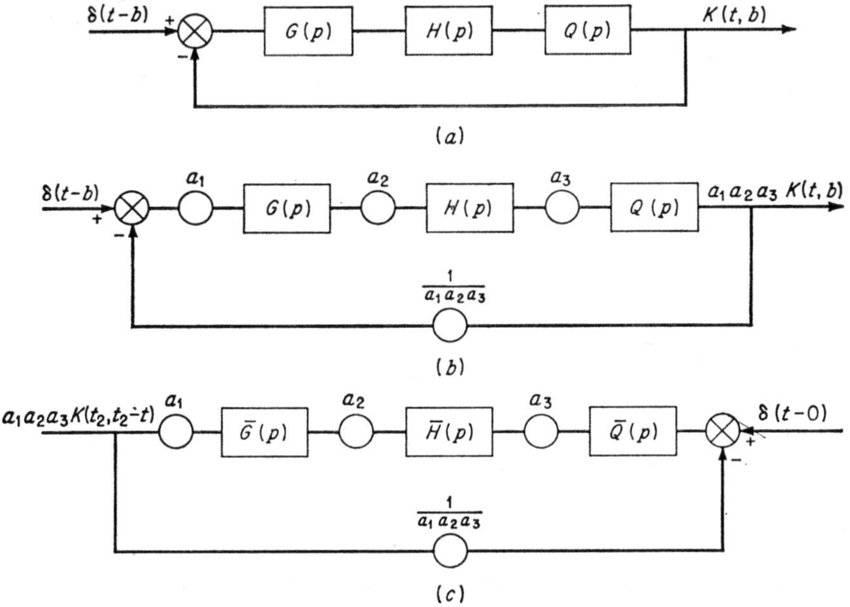

FIG. 4.33 (*a*) A block diagram of a computer program to evaluate a weighting function; (*b*) (*a*) amplitude-scaled; (*c*) block diagram of the adjoint of (*b*).

compensating amount to leave the integrator output the same.[1] Only amplitude scaling can therefore affect the integrator initial condition.

4.8 Summary

Amplitude and time scaling may be done directly on the computer diagram. First set up the computer diagram directly from the problem equation on a one-to-one amplitude and time scale. Then to change the scale observe the following rules.

4.8.1 Amplitude Scaling

1. Change the scale factor of the pertinent elements.
2. Insert gains which make computer variables consistent with the scale factors in the diagram.
3. Add gains where needed so that all loop gains are the same as they were in the diagram with a one-to-one amplitude scale.
4. Relabel all computer diagram variables which have been changed as a result of the scaling. Two variables are related as the gain between them.
5. The ordinate of Fig. 4.34 shows how scaling by K affects the voltage of a computer variable.

[1] See Appendix 1.

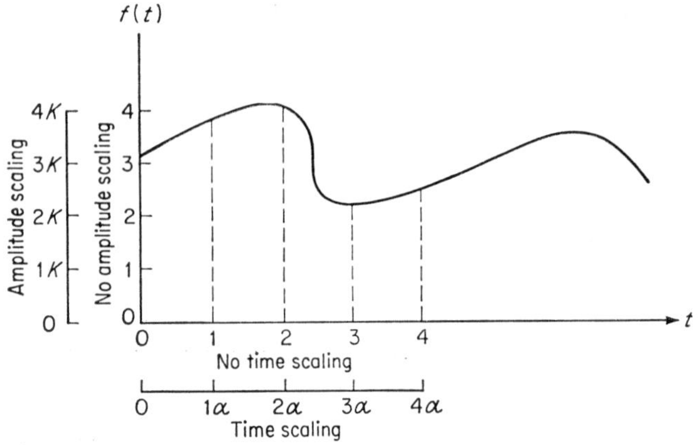

FIG. 4.34 A graph showing the effects of amplitude and time scaling. Note that we vary the ordinate with amplitude scaling and the abscissa with time scaling.

4.8.2 Time Scaling

1. To scale by a factor of α, change all the inputs to each of the integrators by a factor $1/\alpha$.

2. Initial conditions on the integrators remain unchanged. The variable at the output of each integrator is the same as before scaling. However, α units of computer time t are equal to one unit of independent variable x.

3. If the problem has an input generated by a source external to the computer, this input must have the same time scale as the computer.

4. The abscissa of Fig. 4.34 shows the effect on the computer generation of a variable after scaling by α.

4.9 Operational Amplifier[1]

In programming a differential equation we must use a number of integrators at least equal to the order of the differential equation. We may use even more integrators if they are needed to generate either the forcing function or some of the coefficients. Thus, where only constant coefficients are involved, the simulation of a transfer function requires a number of integrators equal to the order of its denominator. What do we do if the problem requirement exceeds our computer integrator capacity by one or more? If the excess is about 100, we may mumble a few words about the impracticality of the problem and forget it. But what if the excess is a few or, worse luck, only one? For such a circumstance we are succored by the operational amplifier, which we may program to do the work of more than one integrator.

[1] See Refs. 43, pp. 37–48, 514–550; 46, pp. 35–67, 458–493; and 54.

The operational amplifier, shown symbolically in Fig. 4.35, is the heart of the computer. From it come the amplifier, integrator, and inverter. An operational amplifier has the following distinguishing characteristics:

1. A high negative-gain voltage amplifier with several inputs and a single output.
2. A high input impedance. For any input voltage the current flowing into the amplifier is negligible.

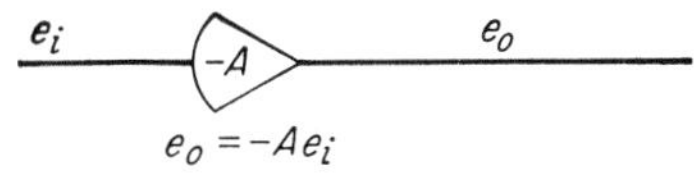

FIG. 4.35 Operational amplifier.

3. A low output impedance. The output is not loaded down by anything attached to it.

When two-terminal-pair networks which consist of linear passive elements (invariably resistors and capacitors) are connected to the operational amplifier as shown in Fig. 4.36*a*, we can obtain first-order and higher transfer

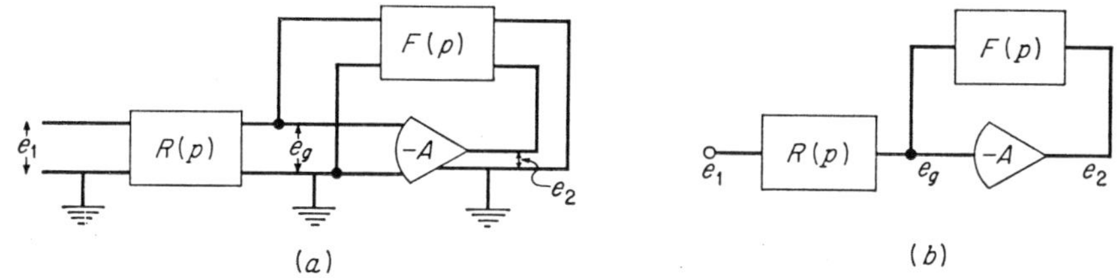

FIG. 4.36 (*a*) An operational amplifier showing all connections of the input and feedback elements; (*b*) an equivalent diagram to (*a*) where all ground connections are removed.

functions. Because two of the four terminals are grounded, the circuit is always drawn in the simpler form of Fig. 4.36*b*. A two-terminal-pair network, say $R(p)$ in Fig. 4.37, has two different sets of characteristics, each of which uniquely defines the network.

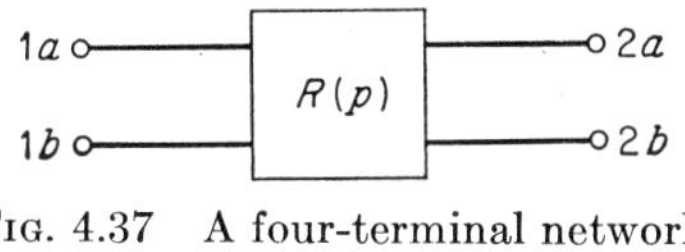

FIG. 4.37 A four-terminal network.

The first set embraces the open-circuit impedances $z_{11}(p)$, $z_{22}(p)$, $z_{12}(p)$, $z_{21}(p)$. With a current source I_1 across terminals 1*a*-1*b* and a voltage E measured across the same terminals with terminals 2*a*-2*b* open, we have the driving-point impedance $z_{11} = E_1/I_1$. Similarly with the current applied and the voltage measured across 2*a*-2*b* and terminals 1*a*-1*b* open, the other driving-point impedance is $z_{22} = E_2/I_2$. When current I_1 is applied across 1*a*-1*b* and the voltage E_2 is measured across the open terminals 2*a*-2*b*, the resulting ratio $z_{12} = E_2/I_1$ is called the transfer impedance. Because we deal with linear passive elements in the network, the relationship $z_{12} = z_{21}$ always holds.

The second set of functions which uniquely describe the network are the admittances y_{11}, y_{22}, y_{12}, y_{21}. These are defined in terms of a voltage source across one set of terminals with the other set of terminals shorted. The ensuing current of interest then flows through the terminal pair and into the network block. With the voltage and current associated with the same terminal pair as their subscripts, we have the relationships

$$\begin{aligned} y_{11} &= \frac{I_1}{E_1} \\ y_{22} &= \frac{I_2}{E_2} \\ y_{12} &= y_{21} = \frac{I_2}{E_1} = \frac{I_1}{E_2} \end{aligned} \tag{4.78}$$

The reciprocals of the short-circuit admittances are the short-circuit impedances.

The open-circuit impedances and short-circuit admittances are related by the equation set

$$\begin{array}{llll} z_{11} = \dfrac{y_{22}}{Y} & z_{22} = \dfrac{y_{11}}{Y} & z_{12} = \dfrac{-y_{12}}{Y} & Y = y_{11}y_{22} - y_{12}^2 \\ y_{11} = \dfrac{z_{22}}{Z} & y_{22} = \dfrac{z_{11}}{Z} & y_{12} = \dfrac{-z_{12}}{Z} & Z = z_{11}z_{22} - z_{12}^2 \end{array} \tag{4.79}$$

Because e_g is almost zero, the terminal pair connected to the amplifier input in Fig. 4.36*a* is essentially shorted. Under these circumstances, if

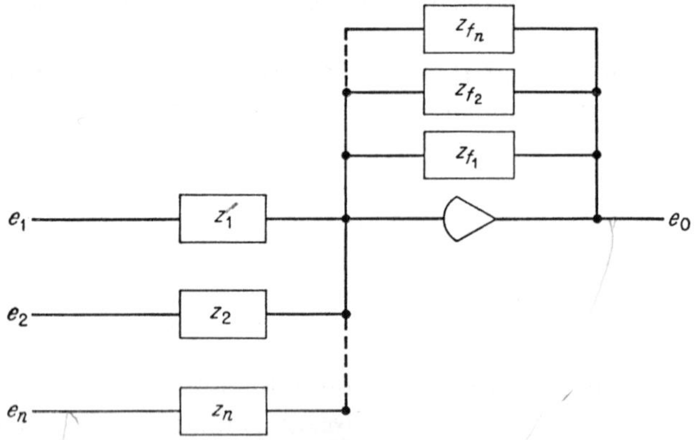

FIG. 4.38 An operational amplifier with multiple input and feedback elements.

z_i and z_f are the short-circuit transfer impedances of $R(p)$ and $F(p)$, respectively, then our transfer function e_o/e_1 is

$$\frac{e_0}{e_1} = -\frac{z_f}{z_i} \tag{4.80}$$

Note that the *transfer function* is a ratio of *transfer impedances*.[1]

For a more general connection of the operational amplifier shown in Fig. 4.38, the relationship between the inputs and output is to a close

[1] Henceforth, in this book, transfer impedance will refer to the short-circuit transfer impedance.

approximation

$$e_0\left(\frac{1}{Z_{f_1}} + \frac{1}{Z_{f_2}} + \cdots + \frac{1}{Z_{f_n}}\right) = -\left(\frac{e_1}{Z_1} + \frac{e_2}{Z_2} + \cdots + \frac{e_n}{Z_n}\right) \quad (4.81)$$

Incidentally, by convention the Z's enclosed in the blocks are the transfer impedances of the respective networks.

Equation (4.81) shows that the effective transfer admittance of networks in parallel is the sum of their individual admittances. However, with some exceptions we do not merely add the transfer impedances when we cascade networks. For the networks shown in Fig. 4.39 the overall transfer impedance Z_{ab} satisfies the condition

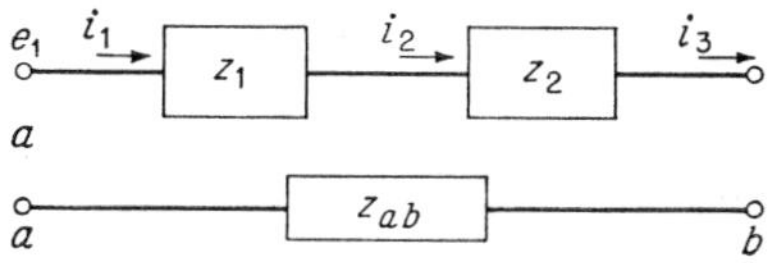

FIG. 4.39 Two transfer impedances in cascade.

$$Z_{ab} = Z_1 + Z_2$$

when

$$i_1 = i_2 = i_3$$

This condition is met when each network in cascade has its elements attached to two terminals, one input and one output, with no internal elements connected to ground.[1] In such a network all driving-point and transfer impedances are equal.

A potentiometer used in cascade with a transfer impedance will alter it by a constant factor (the scale factor of the potentiometer is a transfer function). The transfer impedance from point a to point b in Fig. 4.40

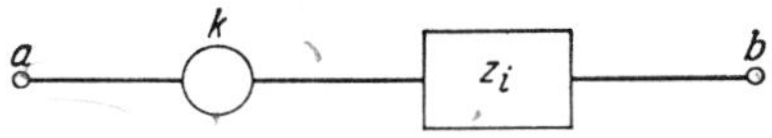

FIG. 4.40 A transfer impedance in cascade with a potentiometer.

is Z_i/k. A cascaded potentiometer changes the transfer impedance by the reciprocal of the potentiometer scale factor.

Equation (4.80) shows how the operational amplifier may be used to provide a wide range of transfer functions, determined by the transfer impedances we can select for Z_i and Z_f. For example, if Z_i and Z_f are the same except for a constant multiplying factor, Fig. 4.38 becomes an amplifier. A computer amplifier uses resistors for Z_i and Z_f, in which case

$$\frac{e_o}{e_i} = -\frac{R_f}{R_i} \quad (4.82)$$

Standard, high-precision computers use

$$R_f = 10^6$$

and R_i assumes the values for whatever gains are chosen. These gains are usually 1, 4, 5, and 10.

[1] See table 3.1 in Ref. 35, p. 168, which gives the transfer functions for several network configurations.

We may construct an integrator if

$$Z_f = M(p) \qquad Z_i = pM(p) \tag{4.83}$$

The standard computer integrator consists of a capacitor C_f as the feedback element and a resistor as the input element. Then

$$Z_f = \frac{1}{C_f p} \qquad Z_i = R_i \tag{4.84}$$

and

$$\frac{e_o}{e_i} = -\frac{1}{C_f R_i p}$$

The gain of the integrator is the product $1/C_f R_i$. The standard capacitor value is 10^{-6} farad.

By using more than one capacitor in the input and feedback transfer impedances we may generate transfer functions of higher than first order and replace a corresponding number of integrators. A table of various transfer impedances is given in Appendix 3. The table uses only resistors and capacitors. We can use Appendix 3 to calculate transfer impedances with inductors. We may replace a resistor R or capacitor C by an inductor L and make the transformation in the transfer impedance:

$$R \to pL \qquad pC \to \frac{1}{pL} \tag{4.85}$$

A transfer impedance has two sets of terminals. In Fig. 4.41, Z_i and Z_f are input and feedback elements, respectively, whose terminals are distinguished by letters. There are four different ways (all shown in Fig. 4.41)

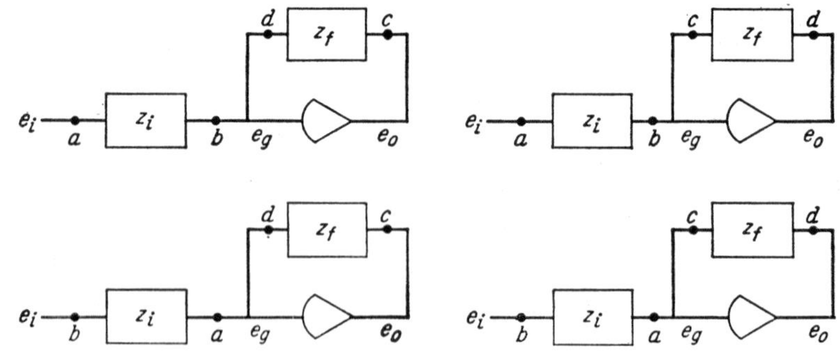

FIG. 4.41 Four equivalent circuits.

whereby these transfer impedances may be connected with an operational amplifier. *All these computer diagrams have the same transfer function* because the output impedance at e_g is almost zero. As we have already indicated, the transfer impedance measured between two points is the same no matter which point is the input and which is the output when all elements are bilateral. We can turn the blocks around because the input and output of each block have approximately a zero-impedance load.

5 *Explicit-function Generation*

5.1 Introduction

In this and the next chapter we shall examine the problem of function generation, treating explicit functions in this chapter and implicit functions in the following one. The variable z as a function of the two variables x and y may be expressed *explicitly* as

$$z = f(x, y) \tag{5.1}$$

or *implicitly* as

$$g(x, y, z) = 0 \tag{5.2}$$

The distinction between these two forms will be discussed in the chapter dealing with implicit functions.

With the termination of Chap. 6 we shall have completed our discussion of all the essential facets of computer programming. The rest of the book will be devoted to techniques for solving problems and improving computer accuracy.

An explicit function may be generated either as the solution of a differential equation or in a memory device storing the functional relationship. Solving differential equations is the computer's forte, and storage is its weakest attribute. Where possible, we should use the former method of function generation.

We shall begin the chapter with a discussion of the differential-equation method, which is applicable only for functions of one variable. This variable on an unmodified integrator is time,[1] but the integrator may be altered so that functions of other problem variables may be generated. We call this operation generalized integration. In the text we shall show how generalized integration can be implemented and how the equation may be constructed to generate the desired function as a solution.

We shall next discuss the utilization of storage devices first for functions of one variable and next for functions of more than one variable. This division is deliberate, based on the relative difficulty of generating functions.

[1] We continue to use the term *time* synonymously with *problem independent variable*.

A function of one variable may be generated exactly or with arbitrarily small errors. Although in a few special cases a function of two or more variables may be reproduced exactly, usually it can only be approximated and then with a good amount of effort and engineering judgment. We shall consider the various ways in which we can generate functions. For functions of more than one variable our emphasis will be on simplifying and approximating techniques.

The generation of functions represents nonlinear operations. Amplitude scaling therefore presents problems not encountered in the linear operations described in Chap. 4. Scaling of nonlinear elements will be considered in all its important ramifications.

We close the chapter with a consideration of time delays, sometimes called transport lags. These operations appear in physical problems like those concerned with chemical processing plants. We shall be most interested in a method called the Padé approximation, which from the standpoint of the computer is easiest to program.

5.2 Functions of One Variable

5.2.1 Differential-equation Function Generation

The function of a single variable may be represented either graphically or as an equation like Eq. (5.1). In the latter case we may generate the function as one solution of a homogeneous differential equation of order n† if:

1. The function and its first n derivatives are continuous.
2. The derivatives can be calculated.

Our procedure is to differentiate a number of times until with a proper selection of coefficients the homogeneous equation can be formed.[1] The equation thus constructed can be linear or nonlinear. In either event the solution will be the function we seek when we impose the correct initial conditions.

Example 1. To generate the function

$$y = e^{-at} \tag{5.3}$$

we differentiate once:

$$\frac{dy}{dt} = ae^{-at} \tag{5.4}$$

Combining Eqs. (5.3) and (5.4),

$$\frac{dy}{dt} + ay = 0 \tag{5.5}$$

† The minimum order of the equation depends upon the nature of the function.

[1] An extensive list of differential equations and their solutions can be found in Ref. 36, pp. 293–660.

Computer diagram 5.1 is the program for Eq. (5.5). From Eq. (5.3) we get the initial condition

$$y = 1 \qquad \text{at } t = 0$$

Example 2. For the function

$$y = \frac{1}{1+t} \qquad -1 < t < \infty \tag{5.6}$$

the first derivative is

$$\frac{dy}{dt} = -\frac{1}{(1+t)^2} \tag{5.7}$$

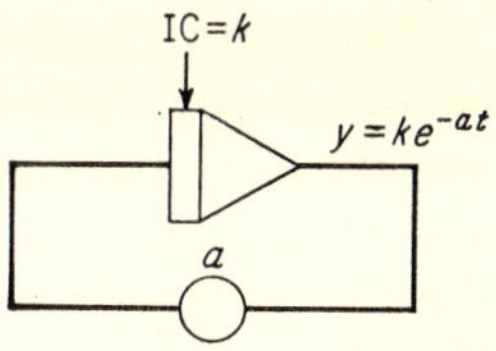

FIG. 5.1 A circuit for generating an exponential.

From Eqs. (5.6) and (5.7) we obtain the nonlinear equation

$$\frac{dy}{dx} + y^2 = 0 \tag{5.8}$$

with initial condition

$$y = 1 \qquad \text{at } t = 0$$

Each function is the solution of a minimum-order differential equation. That is, there is no equation of lesser order for which the function can be a solution. This minimum-order equation is unique when the highest-order derivative has a unity coefficient.

However, we can construct a limitless number of equations beyond the minimum order with two or more equations for each order.

Let $u_1(t)$ be a solution of the differential equation

$$\sum_{r=0}^{n} a_r p^r y = 0 \tag{5.9}$$

Then, if we multiply the differential equation by any linear operator, such that

$$\left(\sum_{m=0}^{l} b_m p^m\right)\left(\sum_{r=0}^{n} a_r p^r y\right) = 0 \tag{5.10}$$

$u_1(t)$ will be a solution of the resulting differential equation.

Incidentally, if the differential equation is second- or higher-order, we need not know the other solutions to establish the correct initial conditions. For an nth-order equation with the desired solution

$$y = f(t) \tag{5.11}$$

the initial conditions are y and its first $n - 1$ derivatives evaluated at the initial point $t = a$.

Powers of t. When generating integer powers of t, we can use multipliers. A much more convenient method involves integrators in cascade. The result is shown in Fig. 5.2. The initial condition on every integrator is zero at $t = 0$.

Sines and Cosines. Sines and cosines of dependent variables are used so frequently that a special device called a resolver is available to generate them.[1] These trigonometric functions are also solutions of differential

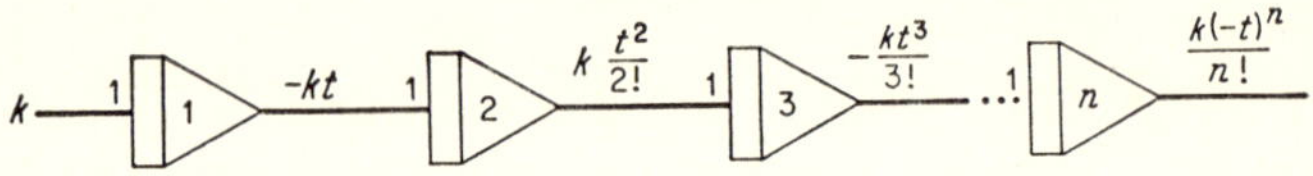

FIG. 5.2 A circuit for generating powers of t.

equations. We shall construct the equation, although presently we shall have difficulties which can be avoided only by recourse to generalized integration. The solution we want is

$$y = \sin g(t) \qquad \text{or} \qquad y = \cos g(t) \tag{5.12}$$

Since the same differential equation will produce both the sine and cosine, we shall find the equation which generates $\sin g(t)$. Thus

$$\begin{aligned} \frac{dy}{dt} &= \frac{dg}{dt}\cos g(t) \\ \frac{d^2y}{dt^2} &= \frac{d^2g}{dt^2}\cos g(t) - \left(\frac{dg}{dt}\right)^2 \sin g(t) \end{aligned} \tag{5.13}$$

Combining equations,

$$\begin{aligned} \frac{d^2y}{dt^2} - \frac{d^2g/dt^2}{dg/dt}\frac{dy}{dt} + \left(\frac{dg}{dt}\right)^2 y &= 0 \\ \text{or} \qquad \frac{d^2y}{dt^2} - \frac{\ddot{g}}{\dot{g}}\frac{dy}{dt} + (\dot{g})^2 y &= 0 \end{aligned} \tag{5.14}$$

In the special case where

$$y = \sin \omega t \tag{5.15}$$

we have

$$g(t) = \omega t \qquad \frac{dg}{dt} = \omega \qquad \frac{d^2g}{dt^2} = 0 \tag{5.16}$$

Equations (5.14) reduce to

$$\frac{d^2y}{dt^2} + \omega^2 y = 0 \tag{5.17}$$

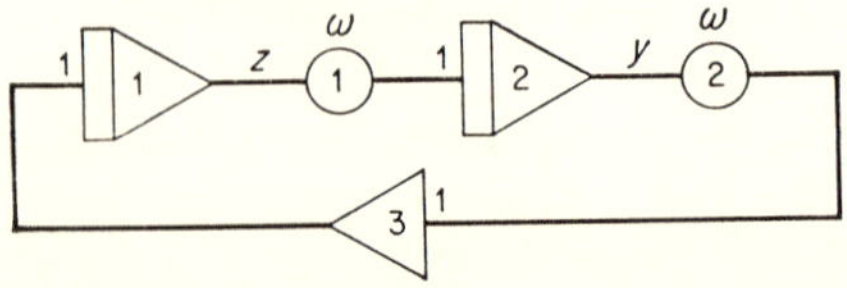

FIG. 5.3 A circuit for generating a sine wave.

The computer diagram is shown in Fig. 5.3. In this computer program y can be either $\sin \omega t$ or $\cos \omega t$ depending upon the integrator initial values.

[1] The application of the resolver is discussed in Sec. 6.6.

With the distribution of loop gain in Fig. 5.3, if

then
$$\begin{aligned} y &= \sin \omega t \\ z &= \cos \omega t \end{aligned} \tag{5.18}$$

An alternative program would eliminate potentiometer 1 and set potentiometer 2 at ω^2. Then if $y = \sin \omega t$, the output of integrator 1 is $(\cos \omega t)/\omega$.

This special case is by far the most common of all sine functions, and it is one that the computer generates with ease. The situation becomes more complicated when we try to simulate Eqs. (5.14). The problem resides in the term $\ddot{g}/\dot{g}$. If $\dot{g}$ approaches zero, even though $\ddot{g}$ also approaches zero at the same time, the computer will be unable to perform the division $\ddot{g}/\dot{g}$ accurately. Therefore, if the computer is to generate the function, we must require that $\dot{g}$ always be the same sign and reasonably far from zero. This is an unfortunate restriction which would bar the use of this method of function generation in the solution of several important problems, e.g., frequency-modulation systems. We bypass these difficulties with a different computer program which uses generalized integration and which requires only that $\dot{g}$ be continuous.

Generalized Integration [56, 133]. The integrators in the computer integrate with respect to time. Therefore, we can program only differential equations whose solutions are functions of this variable. If we are to generate functions of any other variable, we must convert the integrator to integrate with respect to that variable. This we can do, modifying the integrators to generate functions of any problem *dependent* variable. We shall designate this type of integration *generalized integration*. The name is clearly a misnomer, for true generalized integration would be integration with respect to any variable, dependent or independent, which would allow us to generate functions of more than one variable (e.g., to solve partial differential equations). Because we can find no better name to express the operation, we shall use this one.

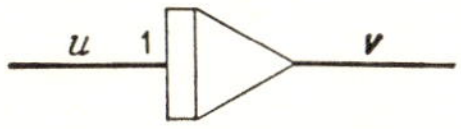

FIG. 5.4 An integrator.

To see how generalized integration is performed, let us begin with the single integrator in Fig. 5.4. If the input u is

$$u = \frac{d}{dt}\left(\frac{d^{n-1}y}{dx^{n-1}}\right) \tag{5.19}$$

then the output v is

$$v = -\int^t \frac{d}{dt}\left(\frac{d^{n-1}y}{dx^{n-1}}\right) dt = -\frac{d^{n-1}y}{dx^{n-1}} \tag{5.20}$$

But we may also write

$$\frac{d}{dt}\left(\frac{d^{n-1}y}{dx^{n-1}}\right) = \frac{d}{dx}\left(\frac{d^{n-1}y}{dx^{n-1}}\right)\frac{dx}{dt} = \frac{d^n y}{dx^n}\frac{dx}{dt} \tag{5.21}$$

and

$$v = -\int^t \frac{d^n y}{dx^n}\frac{dx}{dt}\,dt = -\int^x \frac{d^n y}{dx^n}\,dx \tag{5.22}$$

Equation (5.22) may be programmed as in Fig. 5.5, whereby the portion

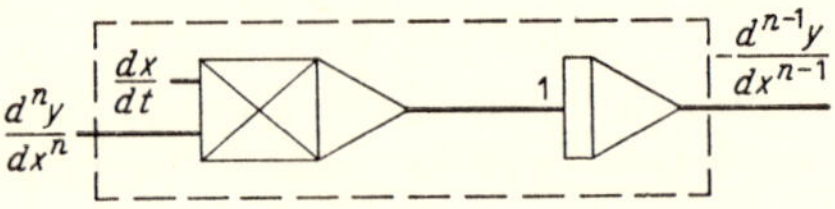

FIG. 5.5 A generalized integrator.

enclosed by the dashed lines is an integrator performing its integration with respect to the variable x.

If we want to generate the function

$$y = \sin x \tag{5.23}$$

we solve the equation

$$\frac{d^2y}{dx^2} + y = 0 \tag{5.24}$$

For Eq. (5.24) the computer diagram is Fig. 5.6. Taking

$$x = g(t)$$

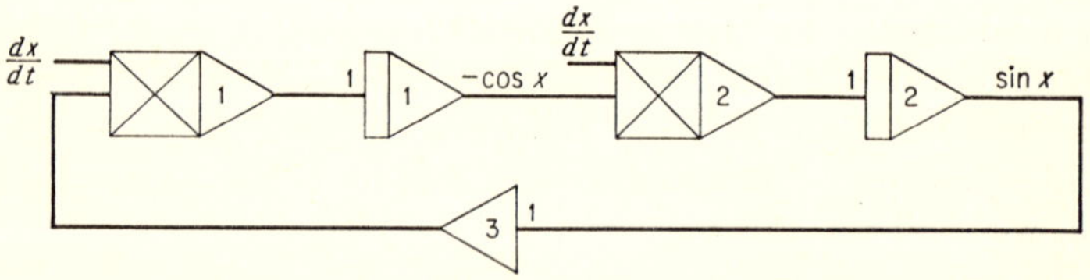

FIG. 5.6 Generation of sin ωt by the method of generalized integration.

Fig. 5.6 will generate the function of Eq. (5.12). Note that Figs. 5.3 and 5.6 are the same when

$$x = \omega t \tag{5.25}$$

Example 3. The equation for a frequency-modulated wave has the form

$$y = \sin(\omega_0 t + a \sin \omega_1 t) \tag{5.26}$$

The coefficients ω_0, ω_1, and a are constants.

To generate this function we use the method of generalized integration where

$$x = \omega_0 t + a \sin \omega_1 t \tag{5.27}$$

$$\frac{dx}{dt} = \omega_0 + a\omega_1 \cos \omega_1 t$$

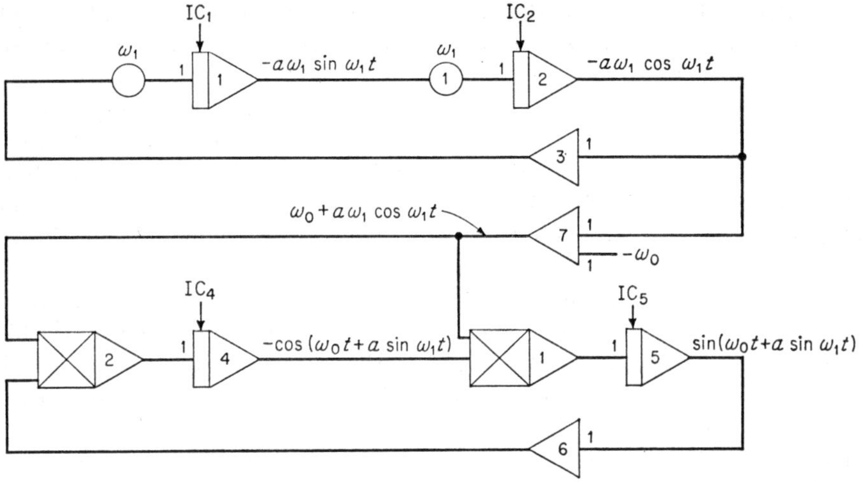

FIG. 5.7 Generation of a frequency-modulated wave.

The computer diagram is shown in Fig. 5.7. If the problem starts at $t = 0$, the initial conditions on the integrators are

$$\begin{aligned} IC_1 &= 0 \\ IC_2 &= -a\omega_1 \\ IC_4 &= -1 \\ IC_5 &= 0 \end{aligned} \tag{5.28}$$

Scaling. The rules for scaling outlined in Chap. 4 are directly applicable for differential-equation function generation. Amplitude scaling alters the magnitude of the function. Time scaling affects the argument of the function, and is used here when the rest of the problem is time-scaled.

5.2.2 Storage Elements[1,2]

When the functional relationship is known only graphically or when the differential-equation method consumes equipment needed elsewhere, we program the function on storage elements called function generators. Until a change is made, each storage element retains the programmed function. The function generators fall into two principal categories:

1. Curve followers, which represent the function exactly
2. Generators, which approximate the function by straight-line segments

Curve Followers. In the plot of a function $y = f(x)$ the abscissa represents x and the ordinate represents y. A curve follower is designed so that

[1] See Ref. 46, pp. 331–357, for a detailed description of some of these devices.

[2] Noise generators which produce functions having statistical properties are discussed in Chap. 11.

a point driven in the x direction will seek out and record the y value. The two principal curve followers are the photoelectric device and the xy plotter.

The photoelectric device consists of a cathode-ray tube whose face is divided into an opaque and a transparent region. The boundary between these regions is the desired function. The electron beam of the tube is biased to move toward the transparent region. Once in this region a photocell sees the beam and drives it toward the opaque region. With the opposing effects of the bias and photocell signals the spot reaches equilibrium at the boundary, which is the beam deflecting voltage y.

The xy plotter uses a line of high electrical conductivity, which is drawn on a plane surface. Through the line passes an alternating current which produces a magnetic field. A coil above the line senses the magnetic field. As it is driven in the x direction, the coil moves in the y direction so that it remains directly above the curve.

The curve follower will generate the desired function if the rate of movement in the x direction is slow enough to permit the nulling in the y direction without error. If the rate of x is too large, we can amplitude-scale the curve to reduce its slope. This can be done by scaling either x or y. The technique is discussed in the latter part of this section. We can also time-scale the problem, slowing it down to fit the limitations of the curve follower.

Straight-line-segment Generators. We can, if we choose, approximate a function by a curve consisting of straight-line segments. As we show in Sec. 2.2.4, such a straight-line curve satisfies the Lipschitz condition, a necessary prerequisite for the existence of unique solutions. Three devices for generating straight-line curves are (1) the diode function generator, (2) the tapped potentiometer, and (3), especially for discontinuous functions, the relay. Of these the diode function generator is most widely used. It is preferred over curve followers because it is easier to program and can operate at higher frequencies in x, thereby allowing the problem to be run on a faster time scale.

Curve Fitting. Any bounded curve may be approximated with straight-line segments with an arbitrarily small error if we use a sufficient number of segments.[1] Alas, for practical considerations we are limited in the number of segments we can use. No matter how we place the segments, the curve will differ from the correct curve. How we draw the approximate curve depends upon the nature of the error we specify. One would want to choose that type of error which produces the least error in the problem solution. There are several error criteria, and undoubtedly one can find problems for which each is best.

Before examining some of these criteria let us consider some of the general aspects of curve fitting, i.e., approximating one curve with another. By this process we construct the approximate curve to pass through a number of points either on or close to the correct curve. The term *close* means that

[1] This statement is proved in the Weierstrauss-Stone approximation theorem.

the points are chosen to minimize some kind of error. The approximate curve will have a number of parameters whose values can be adjusted.

In our discussion we assume that the functional form of the approximate curve is known. The problem then reduces to finding the coefficients, designated here as parameters, which make the approximate curve best conform to the correct curve. For example, in a straight-line-segment approximation the parameters are the end points and slopes of the individual segments, whereas in a polynomial approximation, that is,

$$g(x) = a_0 + a_1x^2 + \cdots + a_nx^n$$

the parameters are the coefficients of the powers of x.

There are three possible situations:

1. The number of parameters exceeds the number of points. The problem of fitting is said to be *underspecified* because we cannot get unique values for all the parameters. The problem can be solved only if arbitrary values are chosen for enough of the parameters so that the remaining number of parameters equals the number of points.

2. The number of parameters equals the number of points. The problem is *precisely specified* because we can get an algebraic set of equations whose solution yields a unique set of parameter values.

3. The number of parameters is less than the number of points. The problem is *overspecified.* It may be solved either by using a number of points equal to the number of parameters or by using all the points and calculating the parameter values so that the error function, which relates the points and the approximate curve, is minimized. If we have some compelling reason to have the approximate curve pass through certain points, the problem should be precisely specified by choosing the desired points and enough other points to equal the number of parameters. Otherwise one should use as many of the points as one can.

When we discard points, we are deliberately not using some information that we have, a profligacy we cannot always afford. Therefore we minimize some error function, which permits us to solve for the parameters in terms of as many of the points as we wish to use. The most common of all the error-function minimization techniques is the method of least squares.[1]

The Method of Least Squares.[2] From ancient times astronomers were beset by their inability to know how to process surplus data. Spurred by this need to solve overspecified problems, Legendre (in 1806) and Gauss (in 1809) independently discovered the method of least squares. An argument, fomented almost entirely by Legendre, broke out between the two

[1] In Chap. 8 we discuss computer methods for finding parameters which minimize error functions. The term *distance function* is used in that chapter instead of error function.

[2] I am indebted to Ref. 57, pp. 212–214, 305–306, for some of the material in this section.

men concerning whose was the prior work. The historical records show that both men share equally the credit for this remarkable discovery.

Briefly this is the method of least squares. Suppose a function $y = f(x)$ is specified at the n points $x_1, x_2, \ldots, x_n$. To this we want to fit the function $f_a(x)$, which has three parameters α, β, and γ. Then in the method of least squares we minimize the function ϵ with respect to the parameters where

$$\epsilon = \sum_{r=1}^{n} [f(x_r) - f_a(x_r)]^2 \tag{5.29}$$

We take the first partial derivatives of ϵ with respect to all the parameters and set each partial derivative equal to zero. Thus

$$\begin{aligned} \frac{\partial \epsilon}{\partial \alpha} &= 0 = -\sum_{r=1}^{n} \left\{ 2[f(x_r) - f_a(x_r)] \frac{\partial f_a(x_r)}{\partial \alpha} \right\} \\ \frac{\partial \epsilon}{\partial \beta} &= 0 = -\sum_{r=1}^{n} \left\{ 2[f(x_r) - f_a(x_r)] \frac{\partial f_a(x_r)}{\partial \beta} \right\} \\ \frac{\partial \epsilon}{\partial \gamma} &= 0 = -\sum_{r=1}^{n} \left\{ 2[f(x_r) - f_a(x_r)] \frac{\partial f_a(x_r)}{\partial \gamma} \right\} \end{aligned} \tag{5.30}$$

In equation set (5.30) the number of equations is equal to the number of unknown parameters. The method of least squares changes an overspecified problem to a precisely specified one with a unique set of values for the parameters.

We may want to approximate a function over a region $a \leq x \leq b$ instead of at a number of discrete points. In that case the error function becomes an integral,

$$\epsilon_1 = \int_a^b [f(x) - f_{a_1}(x)]^2 \, dx \tag{5.31}$$

and the equations to be solved for the parameters are

$$\begin{aligned} \frac{\partial \epsilon_1}{\partial \alpha} &= 0 = -\int_a^b 2 \left\{ [f(x) - f_{a_1}(x)] \frac{\partial f_{a_1}}{\partial \alpha} \right\} dx \\ \frac{\partial \epsilon_1}{\partial \beta} &= 0 = -\int_a^b 2 \left\{ [f(x) - f_{a_1}(x)] \frac{\partial f_{a_1}}{\partial \beta} \right\} dx \\ \frac{\partial \epsilon_1}{\partial \gamma} &= 0 = -\int_a^b 2 \left\{ [f(x) - f_{a_1}(x)] \frac{\partial f_{a_1}}{\partial \gamma} \right\} dx \end{aligned} \tag{5.32}$$

Least squares is not the only error-minimization method we can use. Another error criterion is

$$\epsilon_2 = \int_a^b |f(x) - f_{a_2}(x)| \, dx \tag{5.33}$$

To obtain the parameter equations we differentiate as in equation set (5.32).

If we let

$$|f(x) - f_{a_2}(x)| = |g(x)| \tag{5.34}$$

then

$$\frac{\partial \epsilon_2}{\partial \alpha} = 0 = -\int_a^b \frac{\partial |g|}{\partial \alpha}\,dx \tag{5.35}$$

but by the chain rule

$$\frac{\partial |g|}{\partial \alpha} = \frac{\partial |g|}{\partial g}\frac{\partial g}{\partial \alpha} \tag{5.36}$$

where

$$\frac{\partial |g|}{\partial g} = \begin{cases} 1 & \text{when } g > 0 \\ -1 & \text{when } g < 0 \end{cases} \tag{5.37}$$

$$\frac{\partial g}{\partial \alpha} = -\frac{\partial f_{a_2}}{\partial \alpha}$$

Hence Eq. (5.35) becomes

$$\frac{\partial \epsilon_2}{\partial \alpha} = -\int_a^b \pm \frac{\partial f_{a_2}}{\partial \alpha}\,dx = 0 \tag{5.38}$$

and the equations for the other parameters are

$$\begin{aligned} \frac{\partial \epsilon_2}{\partial \beta} &= -\int_a^b \pm \frac{\partial f_{a_2}}{\partial \beta}\,dx = 0 \\ \frac{\partial \epsilon_2}{\partial \gamma} &= -\int_a^b \pm \frac{\partial f_{a_2}}{\partial \gamma}\,dx = 0 \end{aligned} \tag{5.39}$$

The effect of minimizing the criterion ϵ_1 or ϵ_2 may be seen in an examination of the error between the true and approximate curves, $f(x) - f_a(x)$. In the method of least squares the error term is squared, thereby weighting the larger errors more heavily than the smaller ones. When the absolute value of the error is used, the larger and smaller errors are weighted equally. Hence if we were to plot the error term as a function of x resulting from the minimization of ϵ_1 and ϵ_2, the curve associated with ϵ_1 would have a lower maximum. However, for a major portion of the range $a < x < b$, the other curve will have a smaller error. We would, therefore, be inclined to use ϵ_1 when the larger the error the worse the solution and ϵ_2 when we want to keep the overall error as small as possible.

Example 4. Using error criteria ϵ_1 and ϵ_2, we want to approximate $\sin x$ by the straight line $a + bx$ in the range $0 < x < \pi/2$. Considering first ϵ_1,

$$\epsilon_1 = \int_0^{\pi/2} (\sin x - a_1 - b_1 x)^2\,dx \tag{5.40}$$

Then

$$\begin{aligned} \frac{\partial \epsilon_1}{\partial a_1} &= -2\int_0^{\pi/2} (\sin x - a_1 - b_1 x)\,dx = 0 \\ \frac{\partial \epsilon_1}{\partial b_1} &= -2\int_0^{\pi/2} x(\sin x - a_1 - b_1 x)\,dx = 0 \end{aligned} \tag{5.41}$$

Performing the integration called for and then solving algebraically for a_1 and b_1, we get

$$a_1 = \frac{8}{\pi^2}(\pi - 3) \qquad b_1 = \frac{24}{\pi^3}(4 - \pi) \tag{5.42}$$

Next, for ϵ_2,

$$\epsilon_2 = \int_0^{\pi/2} |\sin x - a_2 - b_2 x| \, dx \tag{5.43}$$

and

$$\frac{\partial \epsilon_2}{\partial a_2} = \int_0^{\pi/2} \pm dx = 0 \tag{5.44a}$$

$$\frac{\partial \epsilon_2}{\partial b_2} = \int_0^{\pi/2} \pm x \, dx = 0 \tag{5.44b}$$

We may write Eq. (5.44*b*) as

$$\int_0^{\alpha} x \, dx + \int_{\beta}^{\pi/2} x \, dx = \int_{\alpha}^{\beta} x \, dx \tag{5.45}$$

where α and β represent the condition $\sin x - a_2 - b_2 x = 0$. From Eq. (5.45)

$$\beta - \alpha = \frac{\pi}{4} \qquad \beta + \alpha = \frac{\pi}{2} \tag{5.46}$$

Performing the necessary algebra, we get

$$a_2 = 0.109 \qquad b_2 = \frac{2.172}{\pi} \tag{5.47}$$

Figure 5.8 compares the two error curves resulting from a minimization of ϵ_1 and ϵ_2.

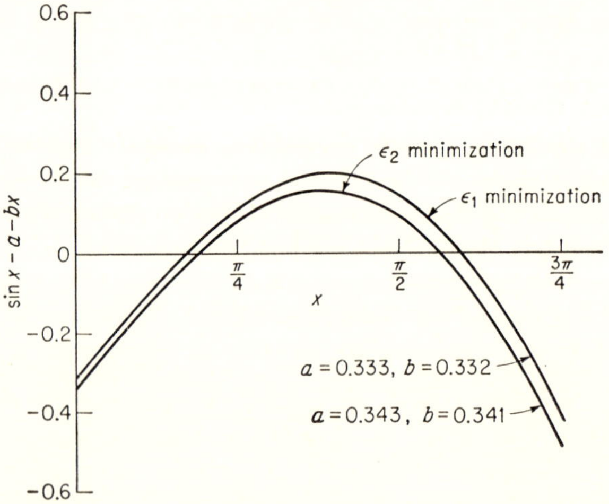

FIG. 5.8 A comparison of two minimization methods. [See Eqs. (5.40) and (5.43).]

Although we can devise many error criteria, each with its special advantages, the method of least squares stands alone for computational reasons. Not only are its mathematical operations easier to perform analytically,[1]

[1] These minimization operations can be performed on a computer, as we show in Chap. 8. When the computer is used, the least-squares method has no particular advantage.

but it has a unique property when the approximate function is comprised of orthonormal functions.

For a class of functions $\phi_i(x)$, $i = 1, 2, \ldots, n$, if

$$\int_a^b \phi_r(x)\phi_s(x)\,dx = \begin{cases} k & \text{when } r = s \\ 0 & \text{when } r \neq s \end{cases} \tag{5.48}$$

the functions comprise an orthonormal set when $k = 1$ and an orthogonal set otherwise.

Suppose the approximate function $f_a(x)$ is made up as a sum of functions drawn from a class of functions $f_i(x)$, $i = 1, 2, \ldots, n$. Thus

$$f_a(x) = \sum_{r=1}^{m} \alpha_r f_r(x) \qquad m \leq n \tag{5.49}$$

where α_r is a constant.

When we are constructing the approximate function, we would like to use as few of the functions $f_i(x)$ as possible. For example, assume that we approximate the correct function $f(x)$ by two $f_i(x)$ so that the error term is

$$\delta_1 = f(x) - \alpha_1^{(1)}f_1(x) - \alpha_2^{(1)}f_2(x) \tag{5.50}$$

When we minimize some error criterion, we shall obtain values for the parameters $\alpha_1^{(1)}$ and $\alpha_2^{(1)}$. If, after this is done, we find that the error term δ_1 is too large, we may add another term and get a new error term,

$$\delta_2 = f(x) - \alpha_1^{(2)}f_1(x) - \alpha_2^{(2)}f_2(x) - \alpha_3^{(2)}f_3(x) \tag{5.51}$$

Carrying out the minimization procedure, we again obtain values of $\alpha_1^{(2)}$ and $\alpha_2^{(2)}$ in addition to $\alpha_3^{(2)}$. What if any relation do these values have to $\alpha_1^{(1)}$ and $\alpha_2^{(1)}$? If the $f_i(x)$ constitute an orthonormal set and if the minimization criterion is the method of least squares, then

$$\alpha_1^{(1)} = \alpha_1^{(2)} \qquad \alpha_2^{(1)} = \alpha_2^{(2)} \tag{5.52}$$

Thus, under these conditions, if we are building up an approximate function by adding one term at a time, we need only calculate the coefficient of the last term added. All other coefficients remain the same. This property holds only for orthonormal functions. Otherwise when a new term is added, all the terms must be recalculated. However, it is possible by the Gram-Schmidt orthogonalization process [58, p. 427] to change any set of functions into an orthonormal set.

Straight-line approximations to curves may be made for several reasons other than the minimization-of-error criterion. We show a sample of some two-segment approximations in Fig. 5.9. The straight-line segments are drawn, meeting the following conditions:

Figure 5.9*a*: The end points of the straight-line segments lie on the curve. This construction guarantees that the approximate curve is correct at certain values of x and also in this case that the error is always the same sign. It also happens to be the easiest straight-line approximation to draw.

Figure 5.9*b*: The straight-line segments are tangent to the curve at $x = 0$ and $x = 1$. If in the problem the value of x tends to vary a relatively small amount around one of the tangent points, the straight-line segment will be a close representation of the curve.

Figure 5.9*c*: The two curves intersect at points $x = 1$ and $x = 2$. This could be to minimize some error criterion or to ensure that the error between the true and approximate curves is the same sign in the range $1 \leq x \leq 2$.

These three cases are but a smattering of the almost endless number of criteria by which the straight-line segmented curve may be drawn.

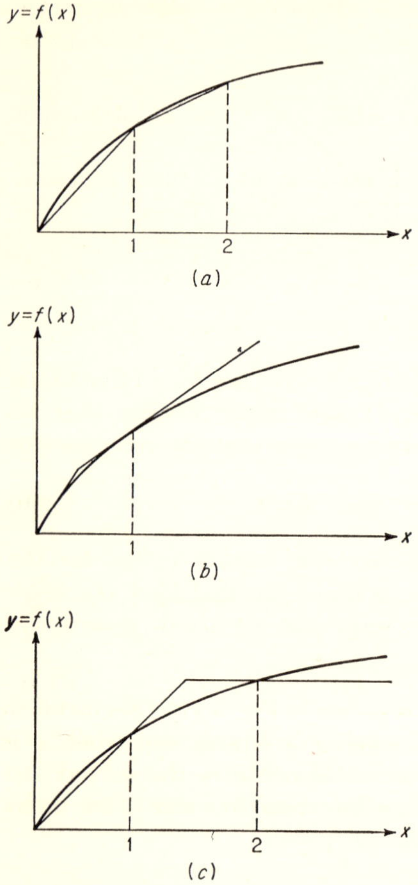

FIG. 5.9 Straight-line approximations based on different criteria.

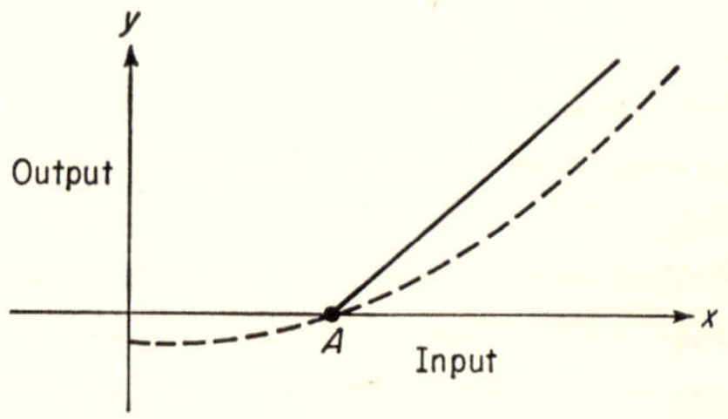

FIG. 5.10 A comparison of ideal and actual diode response. The solid line is ideal, and the dashed line is actual.

Diode Function Generator.[1] The diode function generator consists of individual diode elements and auxiliary equipment which are combined to produce the straight-line segment. Each idealized diode has a zero output for one range of input values x and a constant slope for all other values of x. Figure 5.10 shows the response of an idealized diode in the function generator with the dashed line representing actual diode behavior. The region of constant slope is called the *conducting region* of the diode, and point A, the

[1] See Refs. 59 to 61 for some early work on this device.

transition between zero output and conduction, is called the *break point*. The diode, by itself, always has a break point at $x = 0$. In the function generator this break point is shifted by biasing equipment. The position of the break point and the sign of the slope identify the diode segment as

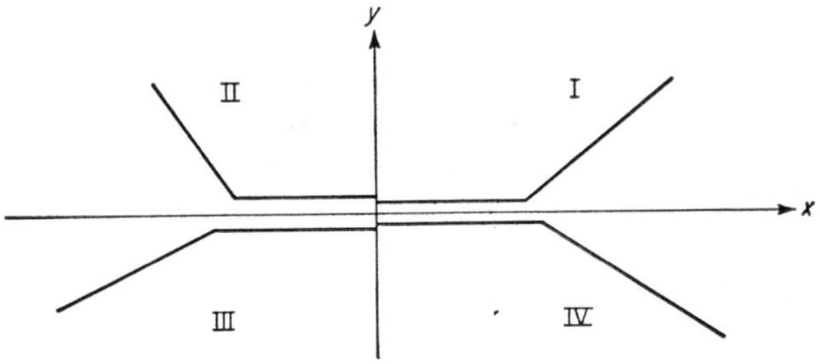

FIG. 5.11 The four quadrants of diode response.

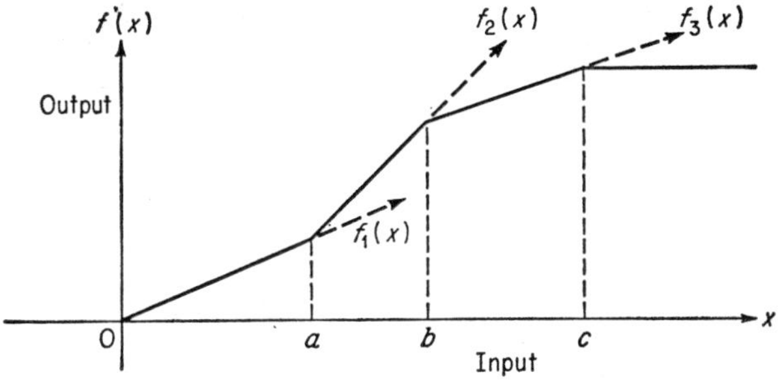

FIG. 5.12 A straight-line function.

being in one of the four quadrants. This is shown in Fig. 5.11. Segments in the first and fourth quadrants conduct when $x \geq 0$ and also when x is greater than the value of the break point. Elements in the second and third quadrants conduct when $x \leq 0$ and x is less than the value of the break point. The values of break point and slope of an element in any quadrant may be changed continuously by potentiometers.

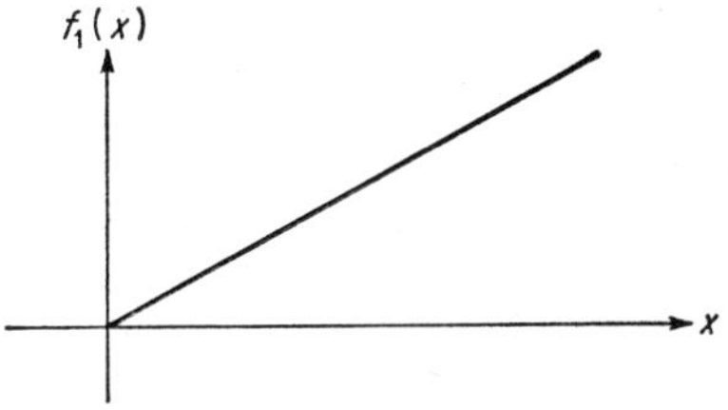

FIG. 5.13 An individual diode segment of Fig. 5.12.

The segments are combined so that for a common input the output is the sum of the outputs of all the elements. How are the break points and slopes adjusted to result in a straight-line-segment curve, shown in Fig. 5.12? As a first step we take a diode with a break point at $x = 0$. The diode slope is then adjusted so that the function $f(x)$ reads the correct value at the next line-segment end point $x = a$. This element, $f_1(x)$, is shown in Fig. 5.13 and is obviously in the first quadrant. At $x = a$ we set the break point of the second diode and adjust its slope so that we get $f(x)$ at the next break point, $x = b$.

The first diode alone produces a function which is the dashed-line extension of Fig. 5.12 at $x = a$. We note that $f(x) > f_1(x)$ for $a < x < b$. Hence the second diode must have a positive slope, which places it also in the first quadrant. At $x = b$ we place the break point of the third diode. The adjustment of its slope so that the three diodes generate $f(x)$ at $x = c$ means that the third diode is in the fourth quadrant [because $f(x) < f_2(x)$ for $b < x < c$]. The last diode with break point at $x = c$ is also in the fourth quadrant.

By an analogous procedure we can generate functions for $x < 0$. We begin at $x = 0$ and construct the function in the direction of decreasing x.

The quadrant of the diode segment which must be added at each break point is indicated in Table 5.1.

TABLE 5.1 DIODE SEGMENT BEHAVIOR

Value of input	Role of diode segment	Quadrant of segment
$x \geq 0$	Slope increase	I
	Slope decrease	II
$x \leq 0$	Slope increase	IV
	Slope decrease	III

We have already seen that the break points of the diode segments correspond with the end points of the straight-line segments. All that remains is a determination of what the slope of each element should be. It is a simple derivation to show that the sum of the slopes of conducting diodes must be the slope of the straight-line segment.[1]

Tapped Potentiometer [62, 63]. Another method, less flexible than the diode function generator in ease of changing from one function to another, is the tapped potentiometer. This device modifies the resistance characteristics of a linear potentiometer, which is one whose resistance from any point to the arm varies linearly with the displacement of the arm. Taps are placed at equal increments of resistance. The end points of the linear segments must also be placed at these tap points; this puts some restriction on the types of functions that can be generated. The arm of the potentiometer is positioned by a servomotor. In its linear (untapped) state the potentiometer is used to form products. Thus the advantage of this method of function generation is that the potentiometer may also perform the dual role of multiplication as well. A schematic for this potentiometer and its servomotor is shown in Fig. 5.14.

As a first step in programming we select all the extreme values of the curve, including its two end points. Next, assuming that these extrema exist, we determine the shunt resistor across successive tie points. We are

[1] See Eq. (5.78) for a mathematical representation of the diode function generator.

guided by the fact that the same current flows through all tie points between adjacent extrema. Therefore, the segment which has the largest slope must have the largest resistance and hence remains unshunted. The other, lesser voltage drops are obtained by shunting resistors across the tie points.

Finally, to obtain the proper voltages at the extrema, we assume that we have the three reference voltages ± 100 volts and zero (ground). We then

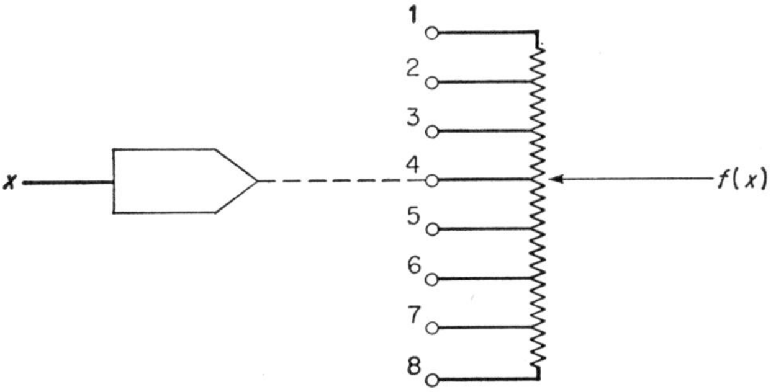

FIG. 5.14 A symbolic diagram of a tapped potentiometer.

connect these voltages through series resistors to the extrema subject to the conditions that $+100$ volts goes to positive maxima and -100 volts goes to negative minima. The ground reference is connected to positive minima and negative maxima. The current which must flow in or out of the extrema has already been calculated. We therefore use series resistors such that the voltage drop across each of them will give the correct voltage at the extrema.

A more expensive way from the standpoint of equipment is to place a voltage at each tap point to conform to the correct value at that point of the function being generated.

The Relay.[1] The relay is a switch whose position is determined by the relative magnitudes of two input variables. A schematic drawing of the relay is shown in Fig. 5.15. The inputs x and y, which may be either fixed or variable, are the inputs of a differential amplifier, called the relay amplifier. The output of this amplifier controls a double-pole double-throw relay with arms A_1 and A_2. For the condition $x - y \geq 0$, the arm A_1 connects with contact point $+1$, and the arm A_2 connects with contact point 2. Points -1 and -2 are both open. Symbolically we may represent this and the alternate action

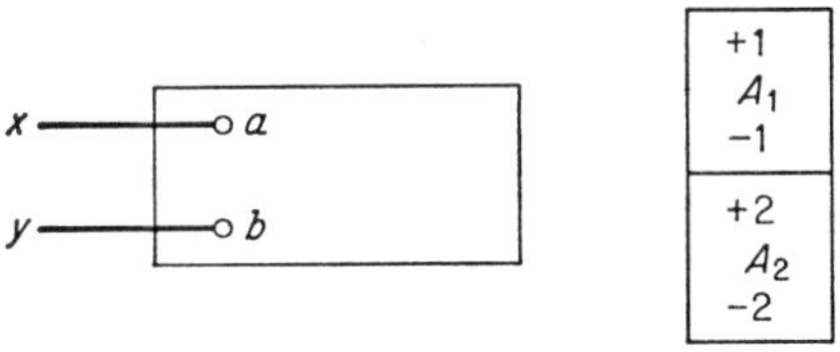

FIG. 5.15 A symbolic diagram of a relay.

[1] For further reading on the relay and its applications, see Refs. 46, pp. 261–290, and 44, pp. 428–434.

as follows:

$$\begin{aligned} \text{If } x - y \geq 0, \text{ then } A_1 &\leftrightarrow +1 \\ A_2 &\leftrightarrow +2 \\ \text{If } x - y < 0, \text{ then } A_1 &\leftrightarrow -1 \\ A_2 &\leftrightarrow -2 \end{aligned} \tag{5.53}$$

Note that the positions a and b at the input of the relay are not the same. If x at input a and y at input b put the relay in position $A_1 \leftrightarrow +1$, then y at input a and x at input b put the relay in the opposite position, $A_1 \leftrightarrow -1$.

A continuous or discontinuous function composed of straight-line segments can be generated by a group of relays, each of which switches at a different straight-line end point. However, the relay has an even more important role. Because it can choose between two alternatives, the relay

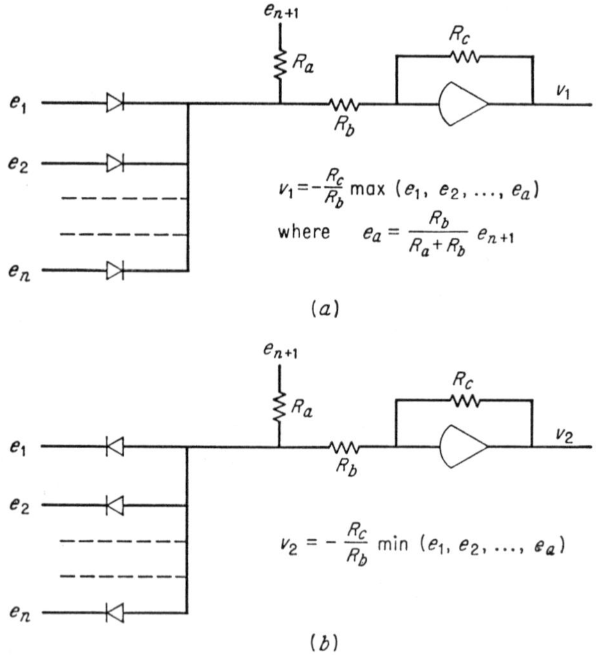

FIG. 5.16 (*a*) A circuit which selects an instantaneous maximum; (*b*) a circuit which selects the instantaneous minimum.

makes decisions. In so doing the relay comes closest of all computer elements to simulating thought processes. With the relay, parameters or even the type of equation to be solved may be changed during the problem; forcing functions may be added or removed; a set of sequential operations may be programmed. The relay is one of the most valuable components of the computer, imparting to it a flexibility not otherwise attained. In the following pages we shall list and describe some of the more important relay applications.

1. *Decision making.* We have already seen that the relay may make a choice based upon which of two variables is larger. Two diode circuits, one of which selects the maximum and the other of which selects the minimum of several variables (see Fig. 5.16*a* and *b*), widen our decision-making capability. Instead of being restricted to using a single variable (or constants) at the relay inputs, we may use a maximum or minimum of a set of variables as well.

2. *Generation of triangular and rectangular waves.* Figure 5.17*a* is a relay circuit which will generate triangular and rectangular waves. The outputs

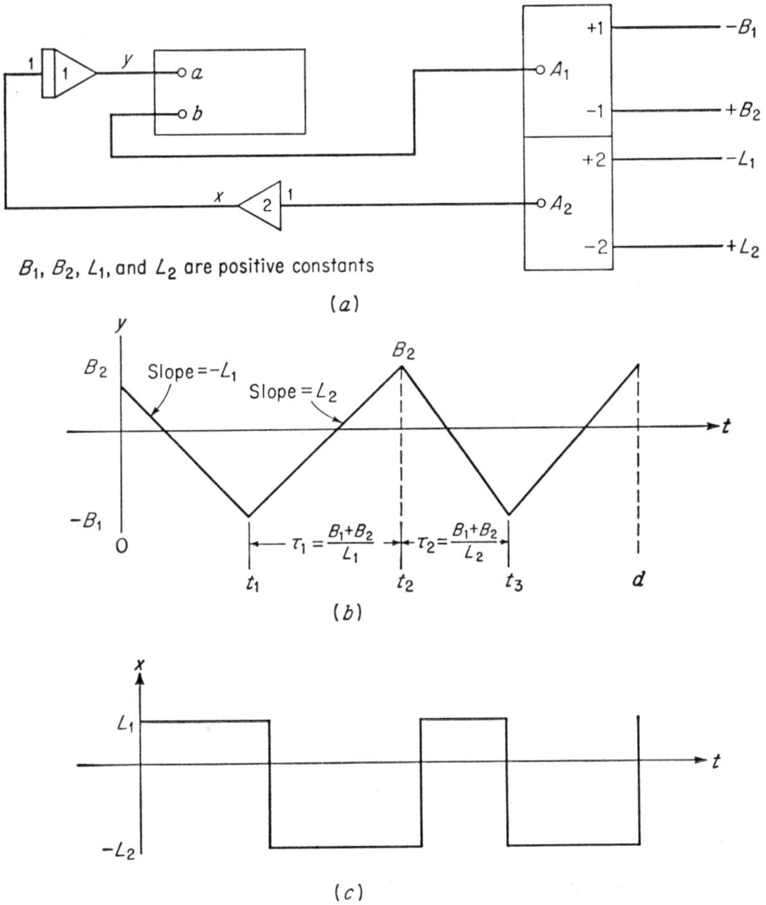

FIG. 5.17 (*a*) A circuit which generates rectangular and triangular waves; (*b*) a triangular wave generated by (*a*); (*c*) a rectangular wave generated by (*a*).

of integrator 1 and amplifier 2 are plotted in Fig. 5.17*b* and *c*. To describe the operation we take as a starting point at $t = 0$ the relay position $A_1 \leftrightarrow +1$, $A_2 \leftrightarrow +2$. Thus we begin with $y - (-B_1) > 0$. The input to integrator

1 is a positive voltage L_1, which is integrated with a reversal in sign so that y is driven negative. At $t = t_1$ the relay switches because $y - (-B_1) < 0$. The twofold effect is to put the voltage B_2 at the relay input and the voltage $-L_2$ at the integrator input. The quantity y begins increasing until $y > B_2$. At this point, $t = t_2$, the relay switches and the cycle repeats itself.

Figure 5.17*b* shows that the slopes of the triangular wave are $-L_1$ and L_2 and the peak amplitudes are $-B_1$ and B_2. The rectangular wave has as its peaks L and $-L_2$. Both waveforms have the same period of oscillation:

$$\tau = (B_1 + B_2)\left(\frac{1}{L_1} + \frac{1}{L_2}\right) \tag{5.54}$$

Although in the interest of clarifying the discussion we presumed that all relay voltages were constant, they may be variables. Thus in the rectangular wave we may modulate the width and height of the pulses.

3. *Sampled-data system simulation.* There are problems in which some of the variables may be a sequence of numbers occurring at discrete intervals of time. One such example is a digital-computer control of a continuous process. The discrete variables are called pulses, and the system in which they appear is a sampled-data system.[1]

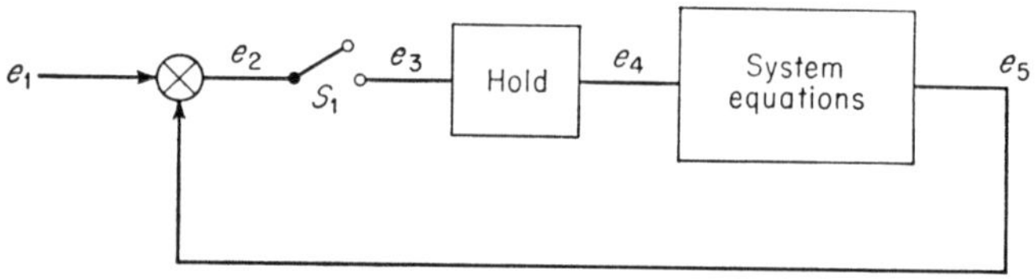

FIG. 5.18 A sampled-data system.

A system where only one variable is a sequence of numbers is shown in Fig. 5.18. Here e_3 is the only variable which is discrete. The operation from e_2 to e_4 is called sample and hold, with the switch S_1 designated as the sampler. S_1 is alternately opened and closed with

$$\begin{aligned} e_3 &= e_2 && \text{when } S_1 \text{ is closed} \\ e_3 &= 0 && \text{when } S_1 \text{ is open} \end{aligned}$$

The hold circuit processes e_3 so that e_4 has nonzero values at points other than when S_1 is closed.

The sampler is easily simulated by a relay in which the arm is alternately switched from one pole to another. One relay amplifier input may be a pulse generated by another relay circuit or by any other suitable waveform like a sine wave. The other input is a constant voltage.

[1] There is extensive literature on the subject of linear sampled-data systems. Three books on the subject are Refs. 33 to 35. For a discussion of analog-computer simulation of these systems, see Ref. 46, pp. 1259–1264, and Ref. 66.

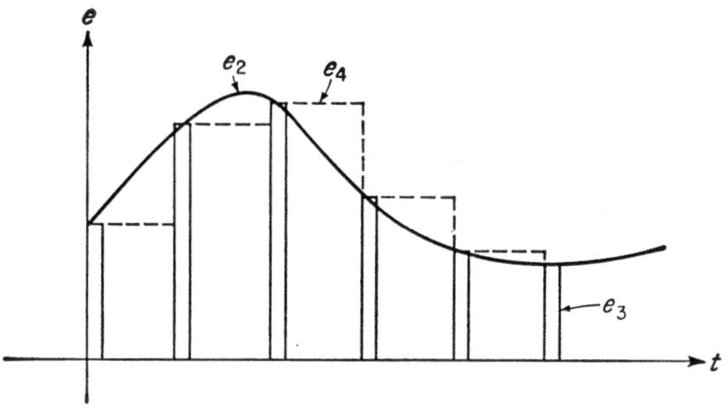

FIG. 5.19 Waveform illustrating a zero-order hold.

There are several types of data hold. We shall demonstrate only the zero-order data hold, whose properties are illustrated in Fig. 5.19. The solid line is e_2, and the pulses represent e_3. In this illustration the pulses are assumed to be of infinitesimal width. The zero-order data hold, whose output is e_4, holds e_3 at its value from the time S_1 opens until it closes again.

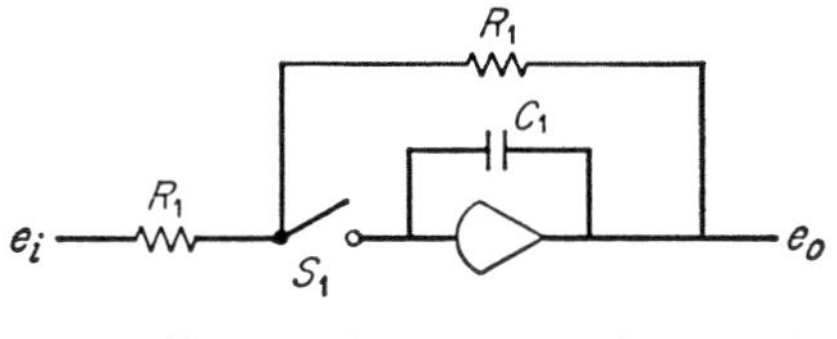

FIG. 5.20 A computer diagram for a zero-order sample-and-hold circuit.

A zero-order hold is shown in Fig. 5.20.

When S_1 is closed, the circuit is a first-order lag having the characteristics

$$\frac{e_o}{e_i} = -\frac{1}{C_1R_1p + 1} \tag{5.55}$$

Otherwise e_o is the value of the output at the time S_1 opens. This value is only approximately e_i because, as Eq. (5.55) shows, e_i and e_o are not the same when S_1 is closed. However, if the time constant R_1C_1 is small enough, the error will be small. If, for example, e_i is a rectangular pulse applied at $t = 0$, then at $t = 4.6R_1C_1$,

$$e_o = 0.99e_i \tag{5.56}$$

For errors of less than $0.01e_i$ the pulse width should be about five times the time constant. Note that the circuit of Fig. 5.20 combines the sample-and-hold operation.

Incidentally this circuit may be used to store the value of a computer variable. If e_i is the variable, then S_1 will remain closed during the problem and, provided the time constant is small enough, e_o will faithfully follow e_i. At the specified time the switch is opened and e_o will remain the value of e_i at that time.

4. *Maximum-value readout.* A problem variable $x(t)$ has a maximum value $x(\tau)$ in a time interval $\alpha \leq \tau \leq t$. Although α and t can both be variables, we shall consider the case where the instantaneous maximum value is recorded at all times from the beginning of the computer run. Thus, $\alpha = 0$ and t is the problem independent variable. For notational purposes

$$x_m(t) = x(\tau) \tag{5.57}$$

There are two tasks that our circuit must perform. It must store $x_m(t)$ when

$$x(t) < x_m(t) \tag{5.58}$$

and it must record $x(t)$ when

$$x(t) = x_m(t) \tag{5.59}$$

and

$$\frac{dx}{dt} > 0$$

An exact method for satisfying conditions in Eqs. (5.58) and (5.59) requires the availability of dx/dt and is shown in Fig. 5.21.

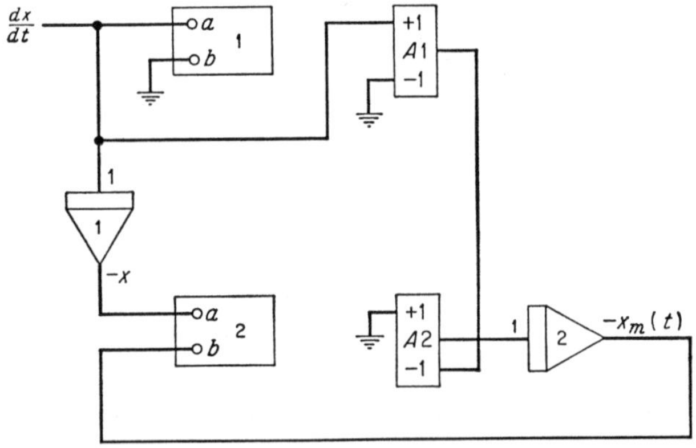

FIG. 5.21 A circuit for recording a maximum value in a time interval.

Some Special Diode Circuits. There are several special straight-line functions which arise frequently in physical situations. Because it has interesting ramifications, we shall consider only the limiter in detail. Absolute-value, dead-space, and bang-bang limiter circuits are shown in Fig. 5.22. Circuits for other straight-line functions can be found in Refs. 44 and 46.

The Limiter.[1] Suppose we confine a variable y to the region $-a \leq y \leq b$, where a and b are positive constants. If dy/dt is positive for a sufficiently

[1] The circuit diagrams used in our discussion of the limiter employ diodes. There is an equivalent circuit with relays replacing the diodes. When the diode is conducting, the substitute relay switch is closed. Otherwise the switch is open.

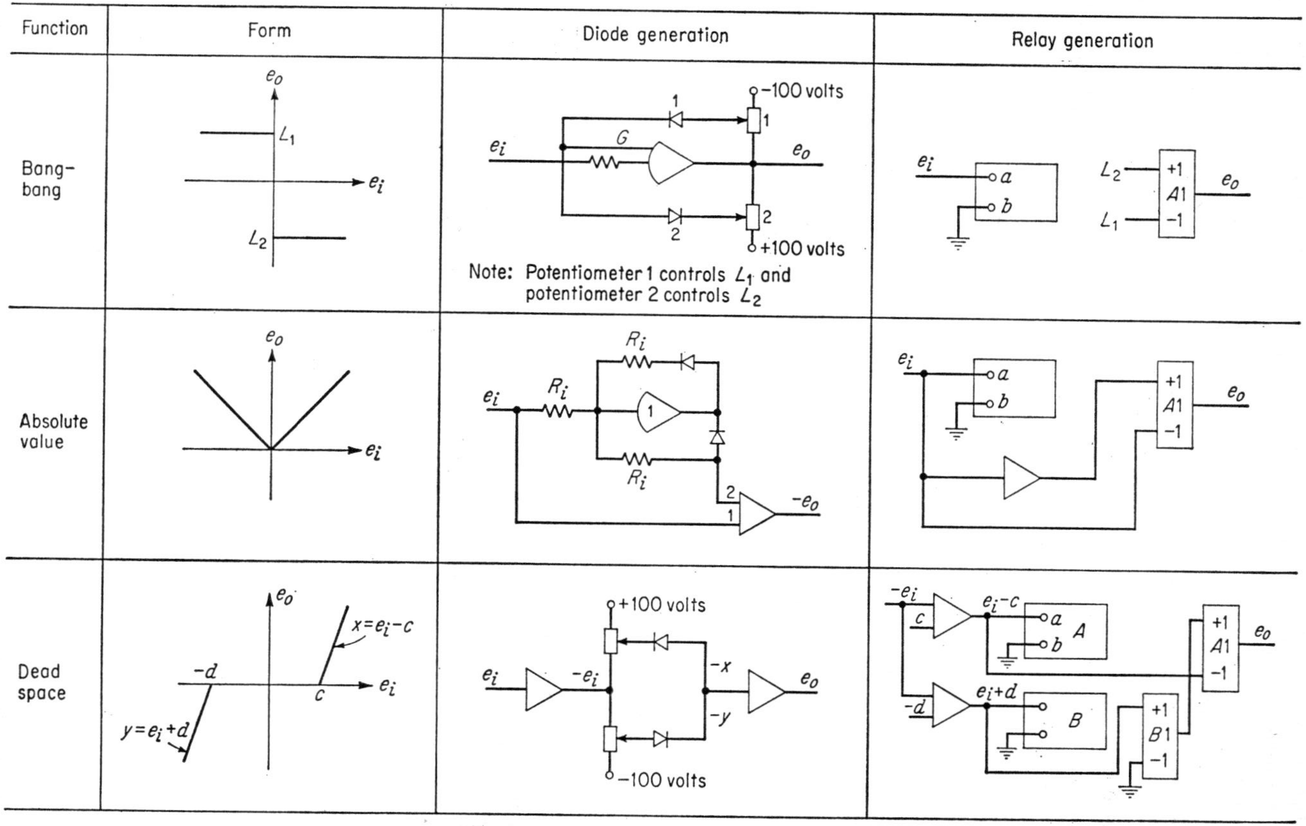

FIG. 5.22 Three special straight-line functions generated by diodes and relays.

long time, y will reach its upper bound. However, when $y = b$, $dy/dt = 0$. Furthermore, y will decrease from its limit at the first time when $dy/dt < 0$. We can use similar arguments for the behavior of y at its lower bound, $-a$.

Let us now consider two examples which program this limiting differently on the computer. First, suppose we are applying positive or negative thrust of constant magnitude to a mass being propelled in air. Eventually, because of air resistance, the mass will reach a terminal velocity V_T. Furthermore, it will remain at V_T until the thrust reverses sign, at which point the velocity will be reduced.

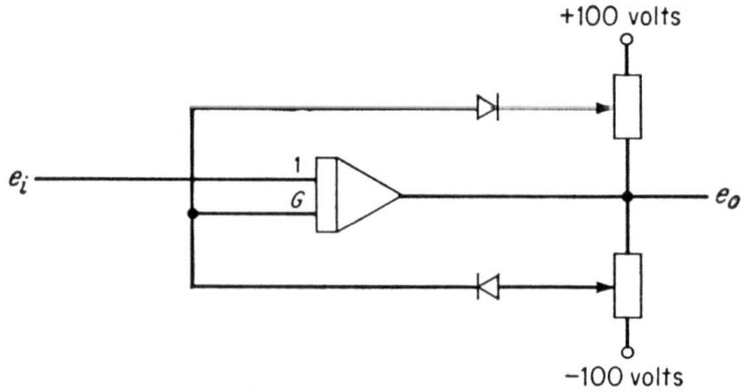

FIG. 5.23 A limiter.

Figure 5.23 is the computer circuit for representing this phenomenon. The circuit is an integrator until one of the diodes conducts. At this point the feedback capacitor is shunted by a very low impedance and the output voltage remains at the limit level. When e_i changes sign, the magnitude of e_o decreases, the diode stops conducting, and the circuit becomes an integrator again.

If we also want the derivative of the limited output, we obtain it by the addition of two amplifiers, as shown in Fig. 5.24. This circuit also allows

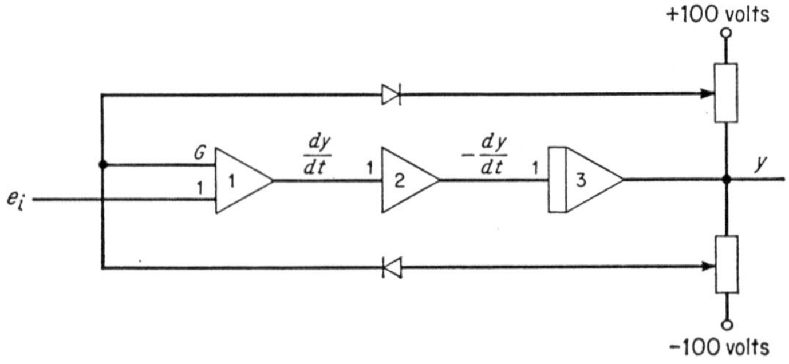

FIG. 5.24 A limiter circuit in which the output derivative is available.

us to limit dy/dt separately.

As a second example, suppose we are applying thrust to the same mass and making a velocity measurement, designated V_m. Our measuring device has a maximum value M, which is less than terminal velocity. Whenever $V > M$, $V_m = M$ and $dV_m/dt = 0$. Note, however, that for $V < M$, we have the condition $dV/dt = dV_m/dt$. Therefore, if V_m is at the limit, it will begin decreasing when V falls below M. This phenomenon is shown in Fig. 5.25*a* as hard limiting, which can be achieved by a relay

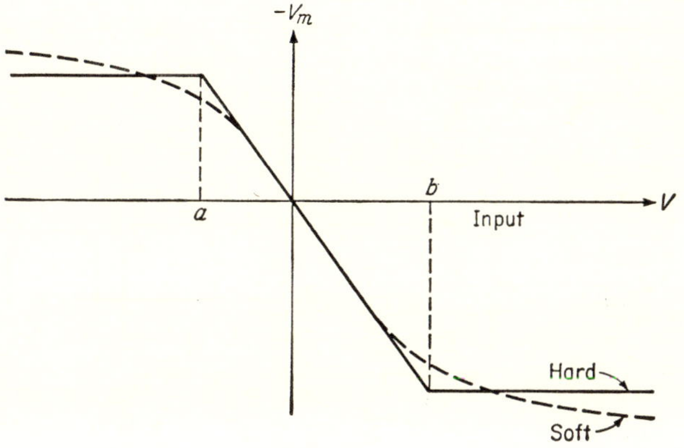

(*a*)

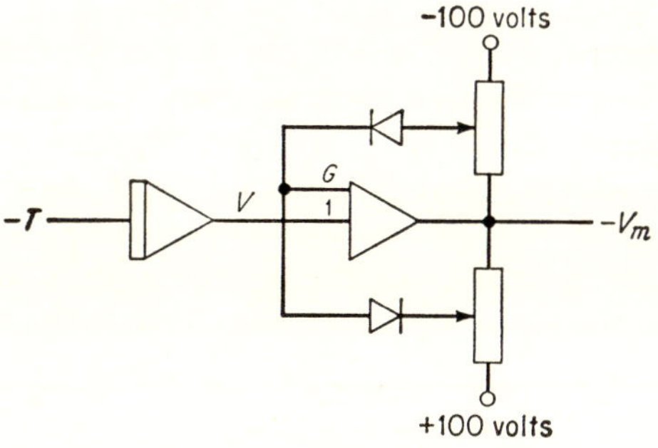

(*b*)

FIG. 5.25 (*a*) A limiter. The dashed lines represent soft limiting, and the solid lines represent hard limiting. (*b*) A limiter.

circuit. The dashed line (soft limiting) represents the performance of the diode circuit shown in Fig. 5.25*b*.

Composite Functions. Sometimes by combining relatively simple nonlinearities in cascade or parallel we may construct more complex nonlinearities. Two such examples are shown in Fig. 5.26. With some ingenuity the reader can devise many more.

Scaling. In Chap. 4 we discussed thoroughly the time and amplitude scaling of linear operators. Time scaling of nonlinearities presents no difficulties. As in the linear case, they can be scaled by changing the gain

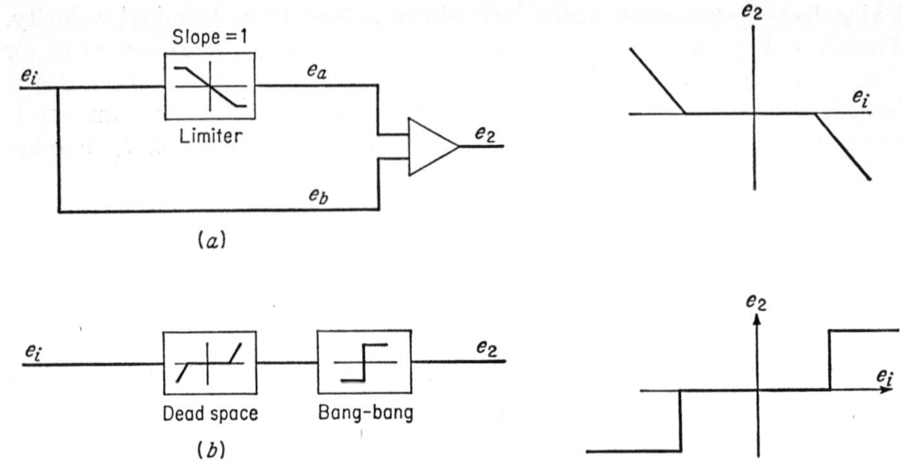

FIG. 5.26 Two examples of composite functions.

of integrator inputs. However, nonlinear elements are directly affected by amplitude scaling of their inputs. Care must be taken in such scaling or serious errors can be the consequence.

To review some salient aspects of scaling, suppose we wish to scale both the input and output of the linear operator shown in Fig. 5.27*a* to the one

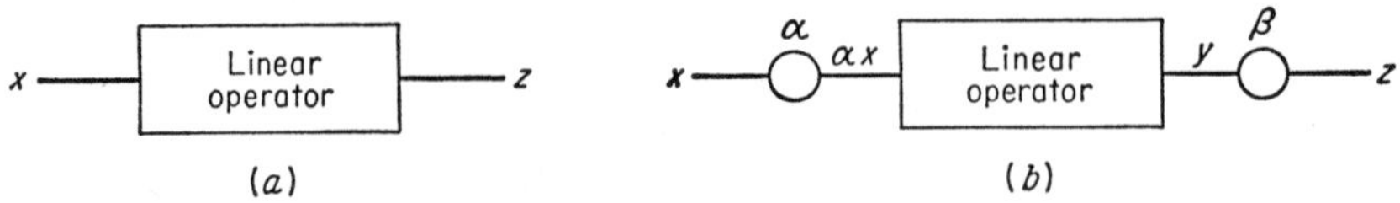

FIG. 5.27 Scaling on a linear operator.

in Fig. 5.27*b*. If x is scaled by the factor α, then, because the operator is linear,

$$y = \alpha z \tag{5.60}$$

and $$\beta = \frac{1}{\alpha} \qquad \text{or} \qquad \beta\alpha = 1$$

Supposing, instead, that the operator is nonlinear, can we scale x and get a constant scale factor on z? If we can, we are able to scale without altering the operator. The answer is a qualified yes. Under certain special conditions this can be done. If Fig. 5.28*a* represents the nonlinear operation, we shall now examine when the scaling shown in Fig. 5.28*b* is valid.

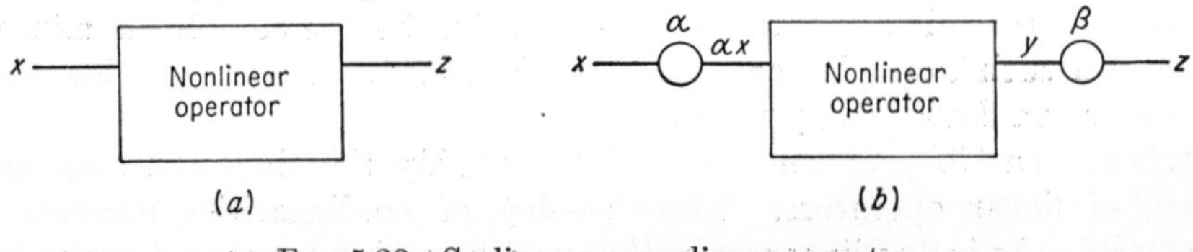

FIG. 5.28 Scaling on a nonlinear operator.

Figure 5.28*a* implies that the relationship

$$z = f(x) \tag{5.61}$$

exists. From Fig. 5.28*b* we see that

$$y = f(\alpha x) \tag{5.62}$$

If β, a constant scale factor, is to establish the relationship

$$\beta y = z \tag{5.63}$$

$$\beta = \frac{f(x)}{f(\alpha x)} = \text{const} \tag{5.64}$$

When the condition of Eq. (5.64) is met, we can scale separately the input and output of a nonlinear operator. We can also, in loops containing such nonlinearities, apply the constant-loop-gain rule of Chap. 4 for amplitude scale.

Note that, in general, for nonlinear operators

$$\alpha\beta \neq 1 \tag{5.65}$$

Example 5. We examine the operators

(*a*) $z = x^{\frac{3}{2}}$

(*b*) $z = uv$

(*c*) $z = e^{-x}$

to see which can be scaled by the circuit in Fig. 5.28.

(*a*) $z = x^{\frac{3}{2}}$. When the input is scaled by α,

$$y = (\alpha)^{\frac{3}{2}} x^{\frac{3}{2}} \tag{5.66}$$

Continuing our algebraic manipulation,

$$z = \beta y = \beta \alpha^{\frac{3}{2}} x^{\frac{3}{2}} \tag{5.67}$$

Thus

$$\beta = \frac{1}{\alpha^{\frac{3}{2}}} = \text{const} \tag{5.68}$$

and the nonlinearity may be scaled externally.

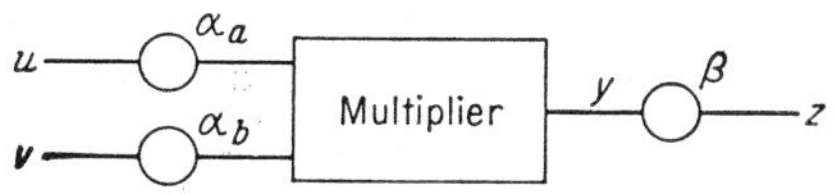

FIG. 5.29 Scaling on a multiplier.

(*b*) $z = uv$. For this operation we have the circuit of Fig. 5.29. For $\alpha_a = \alpha_b = \beta = 1$, the operation yields

$$z = uv \tag{5.69}$$

with scaling

$$y = \alpha_a \alpha_b uv \tag{5.70}$$

and

$$z = \beta y = \beta \alpha_a \alpha_b uv \tag{5.71}$$

Thus

$$\beta = \frac{1}{\alpha_a \alpha_b} = \text{const} \tag{5.72}$$

and the circuit may be scaled externally.

(c) $z = e^{-x}$. With scaling,

$$y = e^{-\alpha x}$$

and
$$z = \beta y = \beta e^{-\alpha x} \tag{5.73}$$

Therefore
$$\beta = e^{x(\alpha - 1)} \tag{5.74}$$

In Eq. (5.74) we see that β is a function of x and is, therefore, not a constant. Hence β cannot be scaled by a fixed potentiometer setting.

When the scale factor does not pass through the nonlinearity as a constant, the scaling compels us to alter the nonlinearity. If the nonlinearity is represented by the curve in Fig. 5.30*a*, it can be obtained from a function

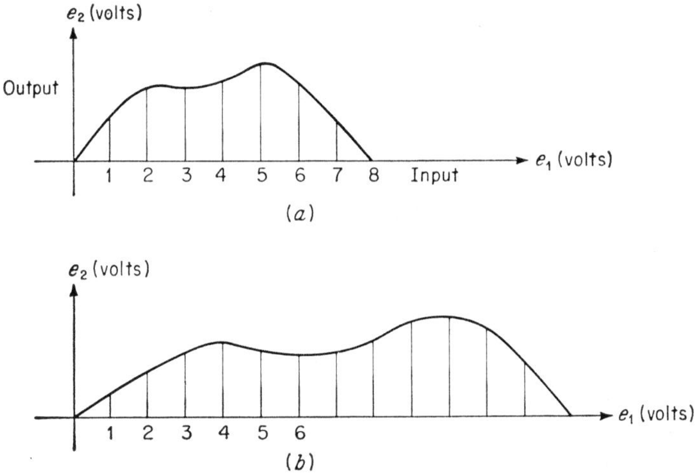

FIG. 5.30 (*a*) An arbitrary function; (*b*) (*a*) input-scaled.

generator. However, we can scale either the input or output simply by plotting a new curve to conform to this scaling. If the input is to be αe_1 instead of e_1, we generate a new curve, multiplying the input (horizontal) axis by α. If we do nothing to the vertical axis, the output axis will still have unity scaling. Figure 5.30*b* is such a curve with the input scaled by a factor of 2 and the output unscaled. We can also scale only the output or we can scale input and output—each by a different amount if we want.

Of particular interest to us is the scaling of straight-line-segment curves. In this case the break points of the diodes are changed when the input is scaled. The reader can verify that the following statements are pertinent:

1. If the input to a diode function generator is scaled by a factor k, the break points of all the diodes must be multiplied by k. Then (*a*) if the slopes of all the diode segments remain the same, the output is scaled by a factor k; (*b*) if all the slopes are multiplied by a factor m, the output is scaled by the product km.

2. If the input has unity scaling and the slopes of the elements are changed by a factor m, then the output is scaled by m.

5.3 Functions of Two Variables

As we have already stated, the analog computer performs very well its arithmetic operations. However, its storage elements (memory devices) are relatively cumbersome to use and not too flexible. Nowhere is this more clearly illustrated than in the generation of functions of two or more variables. Functions of two variables may be generated either by devices designed for the purpose or by decomposition into combinations of functions of one variable. Except in special cases, functions of two variables cannot be generated exactly. Therefore, approximations must be made to the functions of interest. We shall describe the various methods available for generating functions of two variables and leave to the reader the responsibility of determining which one best suits his needs.

No device exists beyond the experimental stage which can generate functions of three or more variables. These functions are beyond the present capability of computer simulation unless they can be approximated by combinations of functions of one or two variables.

Our discussion will first cover the equipment which can generate these functions and then the mathematical methods whereby they are simplified as combinations of functions of one variable.

5.3.1 Equipment

Several devices have been devised which can be used to generate functions of two variables. We shall consider only three: (1) the conducting sheet, (2) the tapped potentiometer, and (3) the diode function generator.[1]

The Conducting Sheet. On the conducting sheet, voltages are applied along lines of silver paint. Because of their low resistance compared with that of the material of the conducting sheet, these are equipotential lines. A probe which can be moved anywhere on the sheet as a function of two variables, x and y, taps off the voltage at that point. When the point is between equipotential lines, the voltage is an interpolation based upon Laplace's equation.[2] If equipotential lines representing lines of constant contour of $f(x, y)$ are plotted as shown in Fig. 5.31, the probe reads directly. A modified xy plotter in which x and y coordinates are fed externally can be used to position the probe. In the standard xy plotter used for functions of one variable only the x coordinate is driven externally. The y coordinate is positioned by a nulling process internal to the plotter.

Quite often the data are presented in the form of Fig. 5.32. Either one can replot them in the form of Fig. 5.31 (sometimes a laborious job), or the

[1] See Ref. 46, pp. 347–357, for a discussion of some other function generators of two variables.

[2] The Laplace equation in two dimensions is $\partial^2 V/\partial X^2 + \partial^2 V/\partial Y^2 = 0$. In this equation the two equipotential lines between which the voltage V is interpolated are the boundary conditions.

function may be solved implicitly.[1] Figure 5.33 shows how this may be done on the xy plotter. The variable u is represented along the x axis, and the function $f(u, v)$ along the y axis. On the conducting sheet are equipotential lines of the variable v.

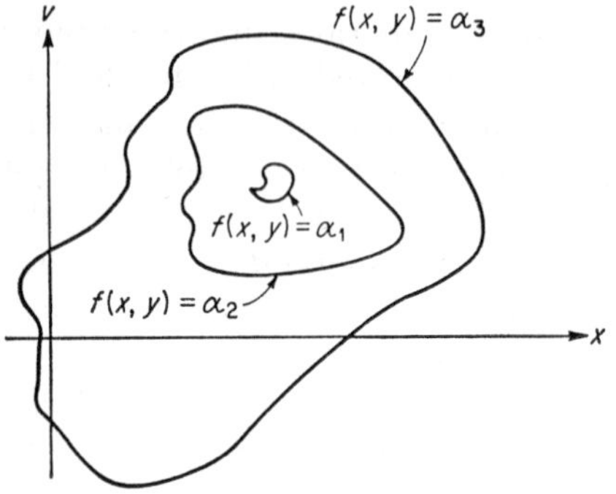

FIG. 5.31 A function of two variables represented by a contour plot.

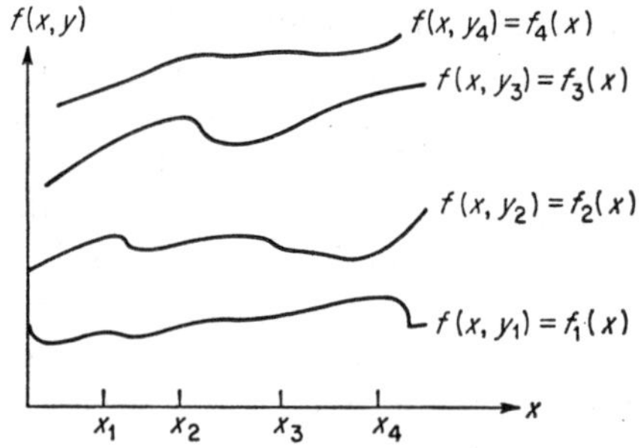

FIG. 5.32 A function of two variables which is plotted for discrete values of one variable.

The probe is positioned in the x direction by the variable u_i. The voltage that the probe senses is the variable v_p. This voltage is differenced with an input voltage v_i at the input of the high-gain amplifier. The output of the high-gain amplifier is fed into the servomotor which positions the probe

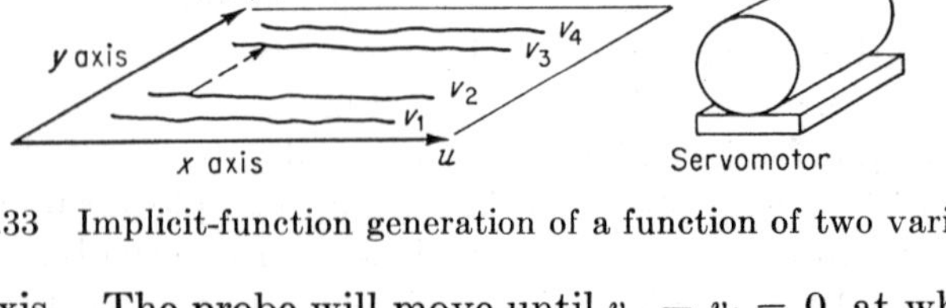

FIG. 5.33 Implicit-function generation of a function of two variables.

along the y axis. The probe will move until $v_p - v_i = 0$, at which time the output of amplifier 1 is $f(u_i, v_i)$.

In the implicit-function solution we must have the voltage on the sheet either increase or decrease monotonically to ensure that the setup of Fig. 5.33 will null to the proper solution. Where this is not the case, we may some-

[1] This technique was originally proposed by H. F. Meissinger. See Chap. 6 for a discussion of the solution of implicit functions.

times bias the function such that

$$f'(u, v) = f(u, v) + kv \tag{5.75}$$

In this case we would subtract kv from the output to get $f(u, v)$. Biasing may also be used to keep equipotential lines from intersecting, because in that instance the technique will not work. We do not have this problem in other methods for generating functions of two variables.

We may alter the interpolation on the conducting sheet by drawing lines of silver paint without any applied voltage between those lines which have voltages applied. These additional lines will then also be equipotential lines.

Tapped Potentiometer. The tapped potentiometer is precisely the same in construction as the one used to generate functions of a single variable. The difference is that variable instead of fixed voltages are at each tap point. The schematic is shown in Fig. 5.34. We note that at each of the tap points we generate a function of a single variable x, whereas the other variable y is held fixed at y_i. The variable y is positioned by the servomotor, and the arm of the potentiometer reads out the function $f(x, y)$. Since the potentiometer is tapped at equal increments of resistance, the functions at the taps are generated for equal increments of y. Between taps there is a linear interpolation in y if we neglect output loading. For $y_i < y < y_{i+1}$,

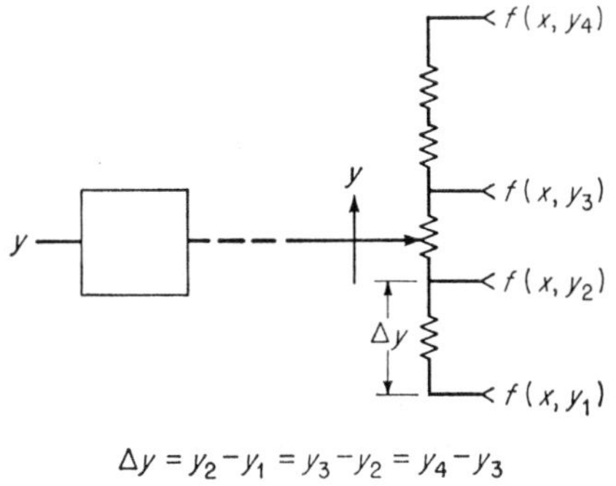

FIG. 5.34 Generation of a function of two variables with a servomotor and tapped potentiometer.

$$f(x, y) = f(x, y_i)[1 - \lambda_i(y)] + f(x, y_{i+1})\lambda_i(y) \tag{5.76}$$

where
$$\lambda_i(y) = \frac{y - y_i}{y_{i+1} - y_i}$$

We have at our disposal an alternative circuit where the tapped potentiometer is replaced.[1] In this method the function $f(x, y)$ is generated by the summation

$$f(x, y) = \sum_{i=1}^{n} \alpha_i(y) f(x, y_i) \tag{5.77}$$

The characteristics of $\alpha_i(y)$ are shown in Fig. 5.35. A cursory examination of Eqs. (5.76) and (5.77) will show that they are the same for all values of y. The summation operation in Eq. (5.77) involves relays, function generators,

[1] See Ref. 18, pp. 58–62, for a more detailed discussion.

and multipliers. Overall, if electronic multipliers are used, this circuit can handle faster variations in y than the tapped potentiometer.

Variable-break-point Diode Function Generator. As we show in our discussion of the diode function generator (Sec. 5.2.2), the slope at the output of the function generator is equal to the sum of the slopes of all the conducting diodes. When the break points of the diodes are made variable, this principle can be used.[1]

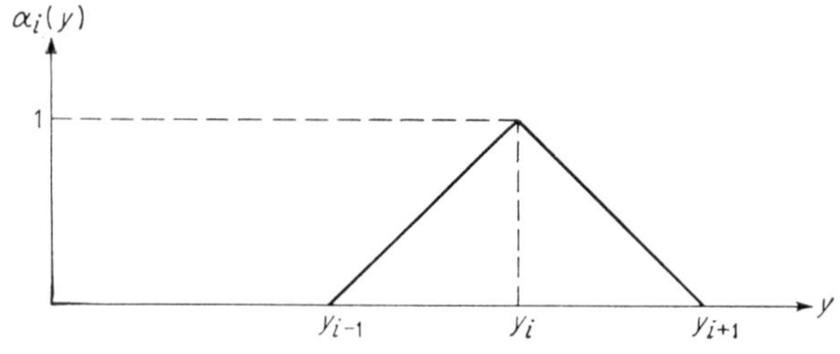

FIG. 5.35 A plot of $\alpha_i(y)$ in Eq. (5.77).

Suppose that, instead of being fixed, the diode break points become a function of a variable y and x is the input to the function generator. Then, for $x > 0$, the equation for the generator output is

$$f(x, y) = \sum_{i=1}^{n} k_i(x - x_i)u[x - h_i(y)] \tag{5.78}$$

where $h_i(y)$ $(i = 1, 2, \ldots, n)$ are the diode break points
k_i are the diode slopes

and

$$u[x - h_i(y)] = \begin{cases} 1 & x \geq h_i(y) \\ 0 & x < h_i(y) \end{cases}$$

Now, provided

$$h_1(y) < h_2(y) < \cdots < h_n(y)$$

no matter how y changes, the diodes will have their break points occurring in the same order for increasing x. A typical function that might be generated in this fashion is shown in Fig. 5.36. Here the lines h_i are the loci of the individual diodes as well as the end points of the straight-line segments. Between any two consecutive loci the lines of all the curves are parallel.

A function of two variables must usually be altered from its original form to be generated by the diode function generator. The steps to follow are:

[1] This technique was originated by Meissinger and is described by him in Refs. 67 and 68.

1. Redraw each curve $f(x, y_1)$, $f(x, y_2)$, . . . , $f(x, y_n)$ in the family so that it consists of a set of line segments, each one of which is parallel to a similar segment of the other curves.

2. Determine graphically the curves for the diode break points as a function of y.

3. Program the proper slope on each diode. Generate the diode break-point curves and connect to the correct diodes.

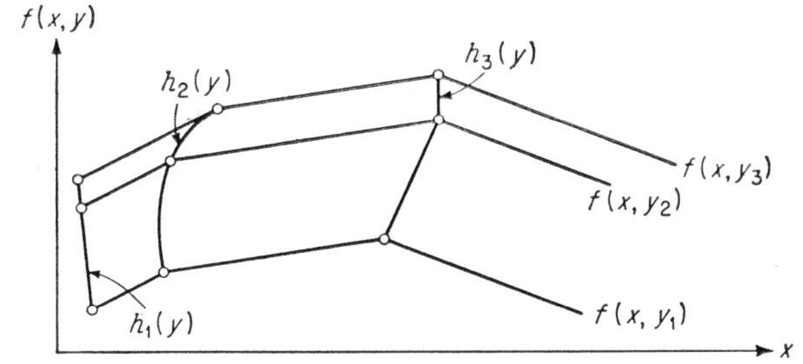

FIG. 5.36 The Meissinger technique for the generation of functions of two variables with a diode function generator.

5.3.2 *Mathematical Methods of Simplifying Functions*

The three devices we discussed can all be used to generate functions of two variables. Each has defects which cannot often be ignored. The conducting sheet is a cumbersome device which is hard to change from one function to another. Also, because a servomotor moves the probe, it is a relatively slow device which limits the speed of the problem time scale. The tapped potentiometer and diode function generator with some exceptions both require an extra function generator for each tap or diode break point. The tapped potentiometer like the conducting sheet uses a servomotor. Our attention in this section will be on methods for simplifying functions of two or more variables into combinations of functions of one variable. Some possible combinations for functions of two variables are tabulated in Eqs. (5.79) to (5.83):

$$f(x, y) = g(x) + h(y) \tag{5.79}$$
$$f(x, y) = g(x)h(y) \tag{5.80}$$
$$f(x, y) = A + g(x)h(y) \tag{5.81}$$
$$f(x, y) = r(x) + g(x)h(y) \tag{5.82}$$
$$f(x, y) = g_1(x)h_1(y) + g_2(x)h_2(y) \tag{5.83}$$

The ways of nature are not so benign that all functions of two variables will satisfy one of these equations. However, on occasion, there may be a close approximation. In that case we choose the best representation and fit to the correct function by some minimization procedure, usually least squares. What are the clues that indicate which representation is the most suitable?

We begin by assuming that the function is in the form of Fig. 5.32. If the curves tend to be parallel, Eq. (5.79) applies. If there is a proportionality between the curves, then Eq. (5.80) expresses the relationship. Thus, for two arbitrary values of f_x at x_i and x_j, if

$$f_1(x_i):f_2(x_i):f_3(x_i):f_4(x_i)::f_1(x_j):f_2(x_j):f_3(x_j):f_4(x_j)$$

we have the proportionality relationship which states that the ratio of any two terms on the left side is equal to the ratio of any two terms on the right side with the same subscripts.

If the proportionality exists after the subtraction of a constant, Eq. (5.81) is the one to use. Incidentally, in Eqs. (5.79) to (5.81) we may take

$$g(x) = f(x, y_i) \tag{5.84}$$

where $i = 1, 2, 3, 4$. Once $g(x)$ is so chosen with a y_i, the $h(y)$ can be calculated for the equation of interest.

If there is proportionality after the subtraction of a function of one of the variables, Eq. (5.82) is satisfied. Unfortunately, it is much easier to write about such a function $r(x)$ than to find it.

When all else fails, try Eq. (5.83). In its most general form this equation is written

$$f(x, y) = \sum_{i=1}^{n} g_i(x)h_i(y) \tag{5.85}$$

This is much too general to be useful, for there is no way of determining the functional form of the individual $g_i(x)$ and $h_i(x)$. One form which has worked in many cases (e.g., aerodynamic coefficients) is

$$f(x, y) = \sum_{k=1}^{n} a_k(x)y^{k-1} \tag{5.86}$$

For many functions an n of 3 is sufficient, which means that only three functions of x—a_1, a_2, a_3—must be determined. These can be evaluated by any curve-fitting methods, including least squares.

Levitt [69] proposes a different formulation of Eq. (5.85) which has the advantage of generating exactly the family of continuous curves in Fig. 5.32. For a family of n curves the approximate function has the form

$$r(x, y) = g_1(x)h_1(y) + g_2(x)h_2(y) + \cdots + g_n(x)h_n(x) \tag{5.87}$$

We start by choosing $h_j(x)$ as

$$h_j(y) = \begin{cases} \prod_{k=1}^{j-1} (y - y_k) & \text{for } j \neq 1 \\ 1 & \text{for } j = 1 \end{cases} \tag{5.88}$$

For a family of three curves

$$r(x, y) = g_1(x) + (y - y_1)g_2(x) + (y - y_1)(y - y_2)g_3(x) \tag{5.89}$$

By setting

$$r(x, y_i) = f(x, y_i) \qquad \text{for } i = 1, 2, 3 \tag{5.90}$$

we can solve for $g_1(x)$, $g_2(x)$, and $g_3(x)$ and get

$$\begin{aligned} g_1(x) &= f(x, y_1) \\ g_2(x) &= \frac{f(x, y_2) - g_1(x)}{y_2 - y_1} \\ g_3(x) &= \frac{f(x, y_3) - g_1(x)}{(y_3 - y_1)(y_3 - y_2)} - \frac{g_2(x)}{y_2 - y_1} \end{aligned} \tag{5.91}$$

This method can be extended to larger families of curves. In this case it is advantageous to generate more than one approximation function, each of which is fitted to three consecutive curves. This produces a parabolic interpolation between curves and reduces the number of multiplication operations needed.

Thus for a five-curve family where $f_3(x)$ is the middle curve, choose, for $y \geq y_3$,

$$r(x, y) = g_3(x) + (y - y_3)g_4(x) + (y - y_3)(y - y_4)g_5(x) \tag{5.92}$$

and, for $y \leq y_3$,

$$r(x, y) = g_3(x) + (y - y_3)g_2(x) + (y - y_3)(y - y_2)g_1(x) \tag{5.93}$$

Levitt's technique requires only that the function being approximated be known graphically, and as he shows in his paper, the method may be extended to functions of three variables.

When an analytic expression of the function is known, we may use either a Taylor series expansion or Eterman's technique. The Taylor series expansion may in principle be used without restriction but is of interest only when the series converges rapidly. This is true only if the higher-order derivatives are zero or if we are interested only in a small region around the nominal value of each variable. If we can neglect the higher-order terms, the Taylor series expansion for a function of two variables becomes

$$\begin{aligned} f(x, y) = f(x_0, y_0) &+ \left(\frac{\partial f}{\partial x}\right)_{x_0,y_0} (x - x_0) + \left(\frac{\partial f}{\partial y}\right)_{x_0 y_0} (y - y_0) \\ &+ \left(\frac{\partial^2 f}{\partial x^2}\right)_{x_0,y_0} (x - x_0)^2 + 2\left(\frac{\partial^2 f}{\partial x\, \partial y}\right)_{x_0,y_0} (x - x_0)(y - y_0) \\ &+ \left(\frac{\partial^2 f}{\partial y^2}\right)_{x_0,y_0} (y - y_0)^2 \end{aligned} \tag{5.94}$$

In his book Eterman[1] derives a closed-form expression for the approximation of the function $f(x_1, x_2, \ldots, x_n) - M$:

$$f(x_1, x_2, \ldots, x_n) - M = g_1(x_1) + g_2(x_2) + \cdots + g_n(x_n) \tag{5.95}$$

[1] In describing this method and in Example 6 we follow closely Ref. 18, pp. 62–82. This book contains a very good discussion of methods of simplifying functions of more than one variable.

The functions $g_i(x_i)$ are determined by the method of least squares. The term M is a constant which is the mean value of $f(x_1, x_2, \ldots, x_n)$, which we want to approximate, and is by definition

$$M = \frac{1}{\prod_{k=1}^{n}(b_k - a_k)} \int_{a_1}^{b_1} \int_{a_2}^{b_2} \cdots \int_{a_{n-1}}^{b_{n-1}} \int_{a_n}^{b_n} f(x_1, x_2, \ldots, x_n)\, dx_1\, dx_2 \cdots dx_n \tag{5.96}$$

The expression for $g_i(x_i)$ is

$$g_i(x_i) = \frac{1}{\prod_{\substack{k=1\\k\neq i}}^{n}(b_k - a_k)} \int_{a_1}^{b_1} \cdots \int_{a_{i-1}}^{b_{i-1}} \int_{a_{i+1}}^{b_{i+1}} \cdots \int_{a_n}^{b_n} [f(x_1, x_2, \ldots, x_n) - M]\, dx_1 \cdots dx_{i-1}\, dx_{i+1} \cdots dx_n \tag{5.97}$$

Example 6. Let $f(x_1, x_2) = e^{-x_1x_2}$ in the region $0 \leq x_1 \leq 1$, $0 \leq x_2 \leq 1$. Then

$$M = \int_0^1 \int_0^1 e^{-x_1x_2}\, dx_1\, dx_2 = \int_0^1 \frac{1}{x_1}(1 - e^{-x_1})\, dx_1 = 0.799 \tag{5.98}$$

$$g_1(x_1) = \int_0^1 (e^{-x_1x_2} - 0.799)\, dx_2 = \frac{1 - e^{-x_1}}{x_1} \tag{5.99}$$

$$g_2(x_2) = \int_0^1 (e^{-x_1x_2} - 0.799)\, dx_1 = \frac{1 - e^{-x_2}}{x_2} \tag{5.100}$$

Hence

$$f(x_1, x_2) \cong g_1(x_1) + g_2(x_2) + M \tag{5.101}$$

Eterman's method may also be used when the function is known only at a discrete number of points. In that instance Eq. (5.96) and (5.97) are changed to equivalent summations. For the case of a function of two variables $f(x_1, x_2)$ they become, with r points in x_1 and s points in x_2,

$$M = \frac{1}{rs} \sum_{j=1}^{r} \sum_{k=1}^{s} f(x_{1j}, x_{2k}) \tag{5.102}$$

and

$$g_i(x_{1j}) = \frac{1}{s} \sum_{k=1}^{s} f(x_{1j}, x_{2k}) \tag{5.103}$$

$$g_2(x_{2k}) = \frac{1}{r} \sum_{j=1}^{r} f(x_{1j}, x_{2k}) \tag{5.104}$$

and

$$f(x_1, x_2) = \sum_{j=1}^{n} g_1(x_{1j}) + \sum_{k=1}^{m} g_2(x_{2k})$$

5.4 Time Delay

A time delay, sometimes called a transport lag, arises when an event is measured at some time after its occurrence.

The problems of interest for computer simulation are those where the time-delayed measurement is used to control some variable. Very often this quantity is the one whose measurement has been delayed. One future example is a vehicle traversing the surface of the moon whose movements are controlled from an earth station. The vehicle senses its environment and transmits its information to earth. The information is processed there, and the command is then sent to the moon. Two time delays, each somewhat more than one second, appear in this sequence of operations. The first delay is in the signal from moon to earth, and the other in the return signal from earth to moon.

A more down-to-earth example arises in a chemical processing plant. Here the chemicals are mixed continuously in a certain proportion. The yield is measured at some later time. On the basis of the measurement the proportion of each chemical may be changed to improve the yield.

To these examples may be added control of nuclear reactors, quality control of production-line processes, and petroleum distillation control. Whatever the example, a time delay has a destabilizing effect on feedback systems. The time delay is therefore an important factor in system performance, and it must be simulated accurately.

One likely device[1] is the magnetic-tape recorder. In this device the variable is recorded on the tape by one magnetic head and is read out at some later time by a second head displaced from the first. The delay time is a function of the tape speed and distance between heads.[2] The magnetic recorder has one major restriction: the time delay cannot easily be varied over a large range of values.

For the greatest flexibility in simulating time delays the elements of the computer (potentiometers, integrators, and amplifiers) can be programmed to approximate the phenomenon to a high degree of accuracy for conditions to be specified.

The function $g(t)$ is delayed by time T, relative to the function $f(t)$, if

$$g(t) = f(t - T) \tag{5.105}$$

We can, perhaps, best understand the properties of the time delay which must be simulated by the computer by observing its characteristics in the frequency domain.

[1] Another device is a delay line comprised of passive electrical elements. This device has too many limitations to merit consideration for general problem use.

[2] The formula $\tau = D/v$, where τ = delay time, D = distance between heads, and v = velocity of time, holds only when the problem has not been time-scaled.

Taking the Laplace transform of Eq. (5.105),

$$\mathcal{L}[g(t)] = \mathcal{L}[f(t - T)] \tag{5.106a}$$

$$G(s) = e^{-Ts}F(s) \tag{5.106b}$$

where

$$\mathcal{L}[f(t)] = F(s)$$

Letting $s = j\omega$,

$$G(j\omega) = e^{-jT\omega}F(j\omega) \tag{5.107}$$

Equation (5.107) states that, when a signal of any frequency is delayed, its magnitude is unchanged and its phase is shifted by a negative amount equal to the product of the frequency and delay, ωT.

Thus if

$$f(t) = A \sin \omega_1 t \tag{5.108}$$

$$g(t) = A \sin (\omega_1 t - \omega_1 T) = A \sin (\omega_1 t - \phi_1)$$

Our first inclination in attempting to find a computer simulation is to expand e^{-Ts} in a Maclaurin series:

$$e^{-Ts} = 1 - Ts + \frac{(Ts)^2}{2} + \cdots + \frac{(-Ts)^n}{n!} \tag{5.109}$$

The power of s multiplying $F(s)$ represents a differentiation of $f(t)$ of the same order [see Eq. (4.25)], so that the inverse Laplace transform of Eq. (5.106b) is

$$g(t) = \left[1 - T\frac{d}{dt} + T^2\frac{d^2}{dt^2} + \cdots + (-T)^n\frac{d^n}{dt^n}\right]f(t) \tag{5.110}$$

This expansion has several undesirable features. Because the series in Eq. (5.109) does not converge rapidly for large s, both the amplitude ratio $g(t)/f(t)$ and the phase characteristics do not conform to that required of a time delay either for a large frequency or for the time in the neighborhood of $t = 0$.† As Eq. (5.110) shows, the expansion requires higher-order derivations of $f(t)$. Because differentiation is the *bête noire* of analog computing, we avoid this abomination unless we have no alternative. Fortunately, we have recourse to another method, called the Padé approximation.[1] This results in a linear transfer function having the same order numerator and denominator[2] where the delay of Eq. (5.110) is approximated as

$$\frac{h(t)}{f(t)} = \frac{P_n(p)}{Q_n(p)} \tag{5.111}$$

† These variables have the relationship that, as we let $s \to \infty$, then $\omega \to \infty$ and $t \to 0$.

[1] For a thorough discussion of how the Padé approximation may be used to represent a function by an equivalent one which is a ratio of rational polynomials, see Ref. 70, pp. 760–782.

[2] The Padé expansion can also have a different order numerator and denominator. However, under these conditions the magnitude ratio of output and input of a sine wave will not be unity.

where

$$
\begin{aligned}
h(t) &\approx g(t) \\
P_n(p) &= 1 - \frac{Tp}{2} + \frac{n(n-1)}{(2n)(2n-1)}\frac{(Tp)^2}{2!} + \cdots + \frac{(-1)^n n!}{(2n)!}(Tp)^n \\
Q_n(p) &= 1 + \frac{Tp}{2} + \frac{n(n-1)}{(2n)(2n-1)}\frac{(Tp)^2}{2!} + \cdots + \frac{n!}{2n!}(Tp)^n
\end{aligned}
\tag{5.112}
$$

and in general

$$Q_n(p) = P_n(-p)$$

Table 5.2 lists the term $P_n(p)$ in the Padé approximation for transfer functions for orders up to $n = 6$.

TABLE 5.2 PADÉ APPROXIMATION FOR TIME-DELAY SIMULATION

Order	$P_n(p)$
$n = 0$	1
$n = 1$	$1 - \frac{Tp}{2}$
$n = 2$	$1 - \frac{Tp}{2} + \frac{(Tp)^2}{12}$
$n = 3$	$1 - \frac{Tp}{2} + \frac{(Tp)^2}{10} - \frac{(Tp)^3}{120}$
$n = 4$	$1 - \frac{Tp}{2} + \frac{3(Tp)^2}{28} - \frac{(Tp)^3}{84} + \frac{(Tp)^4}{1{,}680}$
$n = 5$	$1 - \frac{Tp}{2} + \frac{(Tp)^2}{9} - \frac{(Tp)^3}{72} + \frac{(Tp)^4}{1{,}680} - \frac{(Tp)^5}{7{,}560}$
$n = 6$	$1 - \frac{Tp}{2} + \frac{5(Tp)^2}{44} - \frac{(Tp)^3}{132} + \frac{(Tp)^4}{1{,}584} - \frac{(Tp)^5}{5{,}080} + \frac{(Tp)^6}{665{,}280}$

As seen in Figs. 5.37 to 5.39, the Padé approximation is very good in the range $0 \le \omega T \le 2$ for $n = 2$ and in the range $0 \le \omega T \le 6$ for $n = 4$. Notwithstanding the extra complexity, the sixth-order approximation is not so accurate as the other two. Morrill [71] has suggested the "mop-up" equalizer as a means of extending the accuracy of the approximation beyond fourth order. In his procedure the error between the phase shift of the true time delay and the fourth-order Padé expansion is evaluated. A mop-up equalizer transfer function is calculated to minimize the error. The equalizer and fourth-order Padé expansion in cascade become the approximation to the time delay. The accuracy of this method is shown in Fig. 5.39, where for a second-order mop-up the error is less than 2 per cent for $0 \le \omega T \le 12$.

Whereas with standard computer equipment the delay of a continuous signal can only be approximated, the delay of a sampled signal can be generated exactly. All that is needed is a correct sequence of switching operations which transfer voltages from one part of the circuit to another with the prescribed delay.

We shall consider the simulation of a first-order data hold which will show the reader how such delays are accomplished. The behavior of the first-order hold is shown in Fig. 5.40. The input to the device is a sequence of

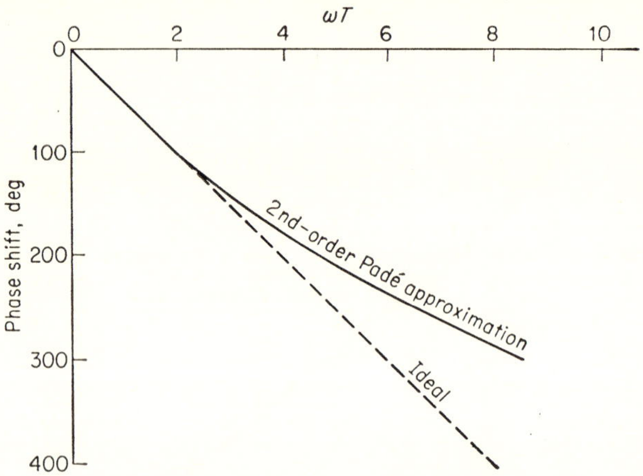

FIG. 5.37 A second-order Padé approximation compared with an ideal time delay.

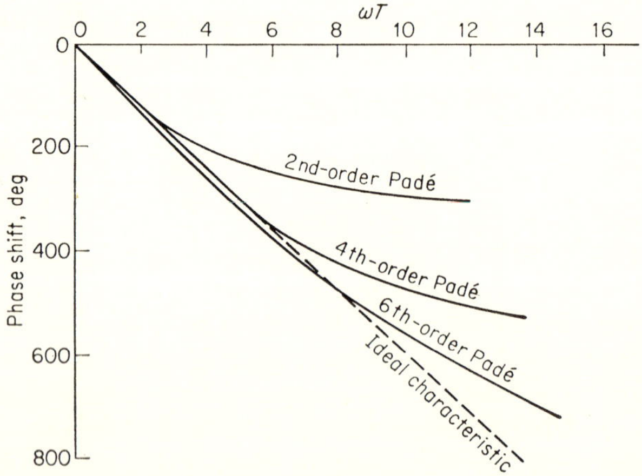

FIG. 5.38 Several Padé approximations compared with an ideal time delay.

samples at equally spaced intervals of time, whose magnitudes are represented by the height of the arrows. The first-order hold extrapolates from one sample point to the next by using the value of the latest point and the slope between it and the previous point. The solid line constitutes the response of the output, with the dashed-line extension showing how the previous point is used.

For $NT < t < (N + 1)T$ the equation of the first-order hold is

$$f(t) = f(NT) + \{f(NT) - f[(N - 1)T]\} \frac{t - NT}{T} \tag{5.113}$$

Figure 5.41*a* is the computer mechanization of Eq. (5.113). To explain the sequence of operations we begin at $t = NT$, with all switches open and

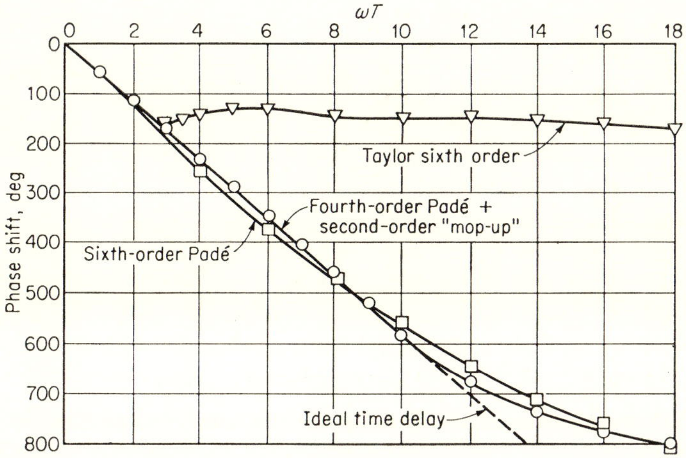

Fig. 5.39 A comparison of some approximations with an ideal time delay.

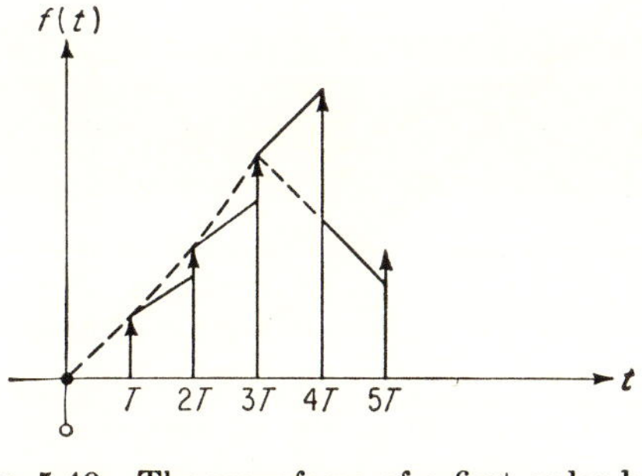

Fig. 5.40 The waveform of a first-order hold.

the values in the circuit as marked in the diagram. The first-order hold is being generated at the output of amplifier 6. The term $t - NT$ is the sawtooth wave $g(t)$ shown in Fig. 5.41*b* and may be approximated by the circuit of Fig. 5.17*a* or produced by a special sawtooth-wave generator. At time t_1, where $NT < t_1 \leq (N + 1)T$, switch 2 is closed and the output of amplifier 2 becomes $f(NT)$, and the output of amplifier 4 is $f(t) - f(NT)$. At time t_2, where $t_1 < t_2 < (N + 1)T$, switch 2 is opened. At time

$$t_3 = (N + 1)T$$

both switch 1 and switch 3 are closed and then reopened rapidly. The output of amplifier 1 will then be $-f(N + 1)T$, and the output of amplifier 3 will be $f[(N + 1)T] - f(NT)$. The output of amplifier 6 will now be a first-order hold in the interval $(N + 1)T \leq t \leq (N + 2)T$. Thus, if we repeat the sequence of switches opening and closing, we shall always be able

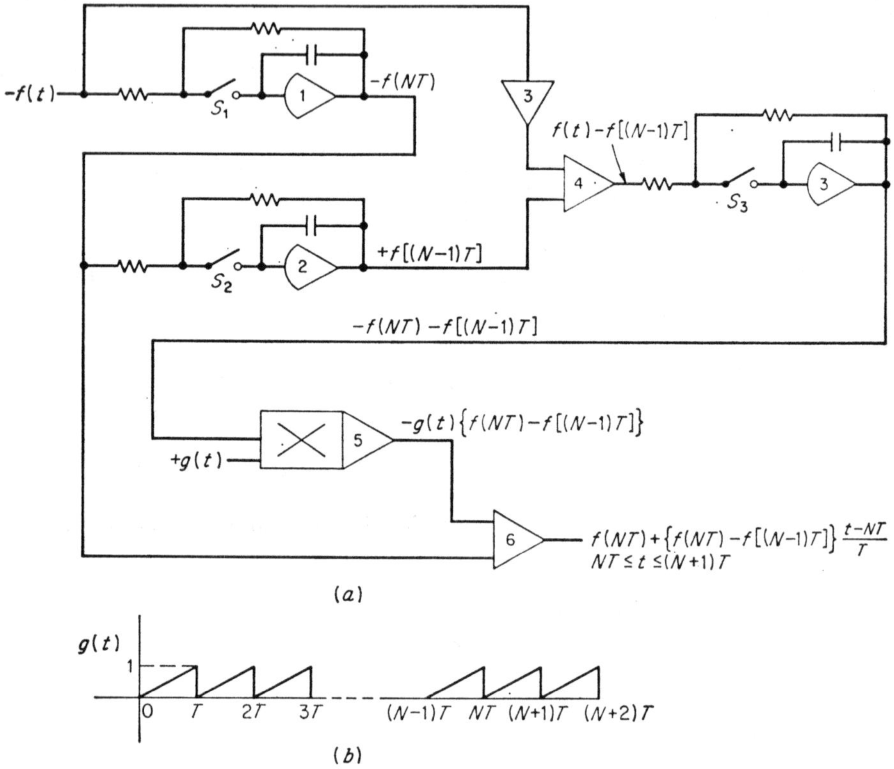

FIG. 5.41 (*a*) A circuit which generates a first-order hold; (*b*) a sawtooth waveform.

to store, at the output of amplifier 2, the sample delayed by one period. If we wanted to store the signal delayed by two or more sampling periods, we need only cascade other storage elements after amplifier 2. The output of each element would store a signal delayed by one sampling period from the element which supplied its input (e.g., as the output of amplifier 2 is delayed relative to the output of amplifier 1).

6 Implicit Functions[1]

6.1 Introduction

A functional relationship between variables may be expressed in either the implicit or the explicit form. For the three variables x, y, z, with z the dependent variable, the relationship is expressed implicitly as

$$f(x, y, z) = 0 \tag{6.1}$$

and explicitly as

$$z = g(x, y) \tag{6.2}$$

We have thus far considered only the solution of equations where the dependent variable is given explicitly. As we recall, a differential equation such as Eq. (6.3),

$$\sum_{k=0}^{n} a_k p^k y = \sum_{j=1}^{m} b_j p^j x \tag{6.3}$$

is solved for $p^n y$ as a function of the other variables. We thus convert Eq. (6.3) to the explicit form

$$p^n y = g(y, py, \ldots, p^{n-1}y; x, px, \ldots, p^m x) \tag{6.4}$$

and then program. Equation (6.3) may also be written in implicit form as

$$f(y, py, \ldots, p^n y; x, px, \ldots, p^m x) = 0 \tag{6.5}$$

Equation (6.5) will be programmed differently from Eq. (6.4). However, not too surprisingly, both completed computer diagrams will be the same.

The example of the differential equation notwithstanding, we do not always have an explicit form which can be programmed by the computer. For one reason, equations are solved in implicit form because they can only be formulated that way. Such a case is Eq. (6.6),

$$x + \sin y + e^{-y} = 0 \tag{6.6}$$

where y cannot be written as an explicit function of x. Another and perhaps more common reason is that the computer lacks the components to solve the problem explicitly.

[1] In this chapter we shall discuss only functions of real variables.

The computer has elements which are designed to perform special tasks, e.g., multipliers and sine-wave generators. If these operations must be performed, they can be done explicitly. However, if the inverse[1] operation is required, the computer handles this by solving an equation in implicit form.

For example, to perform the division

$$z = \frac{y}{x} \tag{6.7}$$

the computer solves the implicit-function equation

$$zx - y = 0 \tag{6.8}$$

which requires only a multiplication but no division.

As another illustration, a computer which has only a sine-wave generator obtains an arc sine,

$$y = \sin^{-1} x \tag{6.9}$$

by solving the equation

$$x - \sin y = 0 \tag{6.10}$$

The chapter begins with a discussion of the conditions under which a solution exists in implicit form. Next we outline the procedure for programming the computer—a procedure quite different from the programming of explicit functions. We also derive an approximate equation for the error in the solution.

As our next topic we discuss implicit-function equations which have multiple solutions. Only one of these solutions is the correct one, and the computer can obtain any of them. How the computer can be made to single out the one correct solution from the many is the purpose of our discussion about multiple solutions.

Then we show how the inverse of any computer operation may be generated by the implicit-function method already described. We can, of course, build a computer element which is the inverse operation of some other element. However, at the small additional cost of a high-gain amplifier any computer operation may be programmed to generate its own inverse. In terms of lower cost and added flexibility, the implicit-function technique is preferred for generating inverse operations.

We next examine how the computer can solve simultaneous implicit-function equations for two or more variables. When trying to solve these equations, we face problems of computer stability which dwarf in complexity similar problems encountered in the programming of a single equation. There are procedures which can mitigate these problems. We shall discuss these procedures after first enumerating the conditions under which a solution exists and then describing the programming technique.

As we show at the close of the chapter, one of the many applications of

[1] If $y = f(x)$, the inverse operation is $x = f^{-1}(y)$.

the implicit-function technique is in the transformation from rectangular to polar coordinates. This will be a convenient point for us to introduce the resolver, a device which by generating sines and cosines can handle the rotation of coordinate systems. These rotations arise in problems involving the flight dynamics of rigid bodies. Because this subject is one of the most important problem areas treated by the computer, we shall examine rotations in some detail—even though they do not usually require the solution of implicit-function equations.

6.2 Mathematical Considerations[1]

If Eq. (6.1) has a solution, it must be expressible in the explicit form of Eq. 6.2, where $g(x, y)$ is a continuous single-valued function. The term *single-valued* is used in a restricted sense which merits some comment. In its usual connotation the term means that for each value of x and y there is only one value of z. As we use the term here, $g(x, y)$ is single-valued if, for the entire range of values of x and y, one continuous function can describe the relationship between z and the variables x and y.

To illustrate this point, take

$$y = \sin x$$

For any given y, there will be several values of x which will satisfy the equation. However, for any arbitrary small range of x, there will be only one value of x for each value of y (except near $y = \pm 1$). We call this function single-valued in the *small*. By comparison, if there is only one value of x for each value of y we have the common single-valued function, also called single-valued in the *large*.

As another example consider

$$F(x, z) = z^2 - x = 0 \qquad (6.11)$$

Then z is multiple-valued[2] by the usual definition because

$$z = \pm \sqrt{x} \qquad (6.12)$$

However, if the point $x = 0$ is excluded, z may be regarded by us as a single-valued continuous function of x.

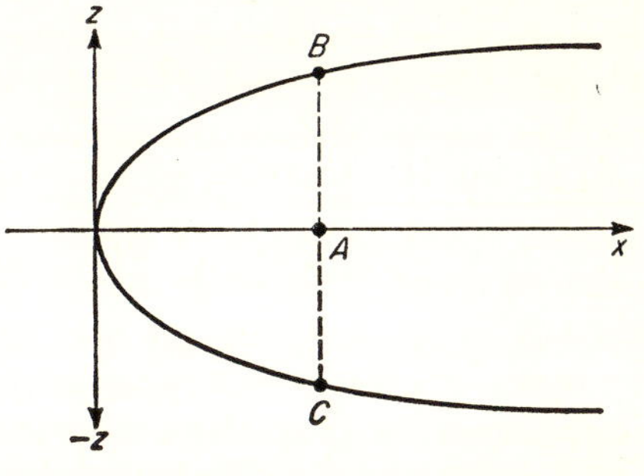

FIG. 6.1 A plot of the equation $z^2 = x$.

Figure 6.1 is a plot of Eq. (6.12). With it we shall use some heuristic arguments to explain why $x = 0$ is a troublesome point.

[1] This discussion is based upon material found in Ref. 72, pp. 238–251. See also Ref. 73, vol. 2, pp. 111–121, which explains implicit functions from a geometric viewpoint.

[2] In practical problems where multiple-valued functions arise, the nature of the problem ordinarily gives some clue about the branch of the functions which should be used.

Suppose we begin with x at point A. From here we can project to the curve at either point B or point C. We pick the point by designating which solution of the equation we want. Thus, let us choose $z = \sqrt{x}$, and with it point B. Incidentally, as we describe in Sec. 6.4, the computer itself is programmed to make such a choice.

As x is varied from its starting value, z will move on the curve from its initial point B. While x remains larger than zero, z will always be obtained from the single-valued continuous function $\sqrt{x}$.

However, if x is allowed to go to zero our problems begin. For at $x = 0$ the functional relationship may be either $z = \sqrt{x}$ or $z = -\sqrt{x}$. Furthermore, after being zero, if x is increased, z can become either positive or negative. Once we get into the neighborhood of $x = 0$, we can no longer represent z as a single-valued function of x. Theorem 6.1 gives a formal statement of *sufficient* conditions for the existence of a single-valued continuous solution to an implicit-function equation. It also shows in precise mathematical terms why $x = 0$ must be avoided in this instance.

THEOREM 6.1.[1] Let $x = x_0$, $y = y_0$, $z = z_0$ be a set of values which satisfy the equation

$$F(x, y, z) = 0 \tag{6.13}$$

and let us suppose that the function F and its first partial derivatives are continuous in a region,

$$|x - x_0| < a \qquad |y - y_0| < b \qquad |z - z_0| < c$$

where a, b, c are positive constants. Then, if the derivative F_z ($F_z = \partial F/\partial z$) is not zero at $x = x_0$, $y = y_0$, $z = z_0$, there is one and only one continuous function of the independent variables x and y, for example, Eq. (6.2), which satisfies Eq. (6.13) and which assumes the value $z = z_0$ when $x = x_0$ and $y = y_0$.

Applying this theorem to Eq. (6.11), we see

$$\frac{\partial F}{\partial z} = 2z \tag{6.14}$$

$$\frac{\partial F}{\partial z} \begin{cases} \neq 0 & \text{when } x \neq 0 \qquad (6.15a) \\ = 0 & \text{when } x = 0 \qquad (6.15b) \end{cases}$$

The conditions for a unique solution are met when Eq. (6.15*a*) is satisfied. However, we should note that $F_z \neq 0$ is only a sufficient condition. In the equation

$$F(z, x) = (z - x)^3 = 0 \tag{6.16a}$$

$$\frac{\partial F}{\partial z} = 3(z - x)^2 \tag{6.16b}$$

$$\frac{\partial F}{\partial z} = 0 \qquad \text{at } x = 0$$

[1] This statement of the theorem is taken from Ref. 27, vol. I, p. 35.

Yet the explicit solution is

$$z = x \tag{6.17}$$

everywhere, including $x = 0$.

6.3 Computer Programming

Equations in implicit form are programmed in a somewhat different manner from equations in explicit form. *An important rule to remember is: All equations in implicit form are solved on the analog computer using a high-gain amplifier in a closed loop.*

Conversely, if the programmed equations of the computer are written in terms of all the inputs of any operational amplifier, they will be in implicit form.

The procedure for programming the computer is as follows:

1. The dependent variable is assumed to be available.
2. All the terms in the equation which add to zero are summed at the input of a high-gain amplifier.
3. The output of the high-gain amplifier is the dependent variable which is fed back to close the loop.

Example 1. Our task is to program a computer, which has only multiplier elements, to perform the division

$$z = \frac{x}{y} \tag{6.18}$$

Because the computer cannot perform the division directly,[1] the equation must be solved in implicit form:

$$F(x, y, z) = zy - x = 0 \tag{6.19}$$

Both x and y are available inputs, and z is assumed to exist. We begin then by putting the necessary terms at the input to the high-gain amplifier, as shown in Fig. 6.2.

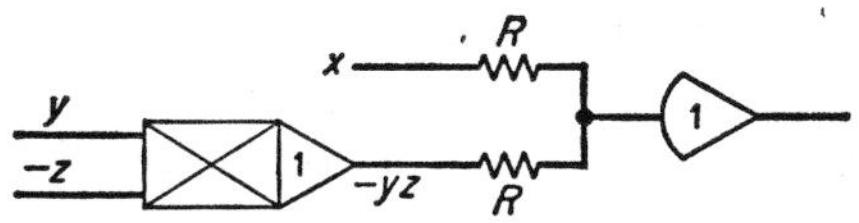

FIG. 6.2 An initial step in programming Eq. (6.19).

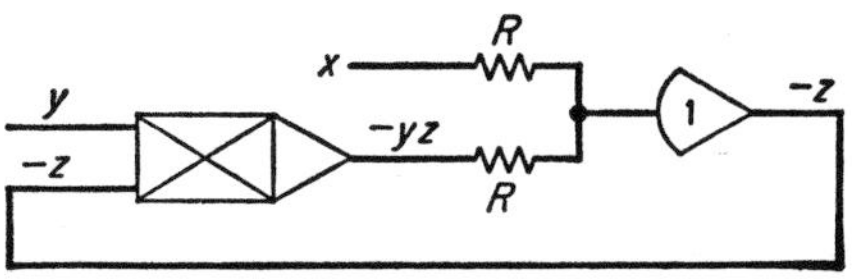

FIG. 6.3 A computer diagram for Eq. (6.20).

All that remains is to determine the output of amplifier 1 and close the loop. When y is positive, the output is $-z$ and the complete computer diagram is Fig. 6.3. Using this diagram we can easily verify that the output of amplifier 1 is correct. We merely let y be a positive constant value. Then the circuit is that of a conventional feedback amplifier whose transfer function is known (see Sec. 4.9). Should y be negative, Fig. 6.3 will

[1] We can also perform division by first forming the term $1/y$ on a function generator and then multiplying by x to get the resultant product x/y.

not give the solution to Eq. (6.18). This point is discussed in Sec. 6.4, which deals with multiple solutions.

Differentiating Eq. (6.19),

$$F_z = y \qquad (6.20)$$
$$F_z = 0 \qquad \text{when } y = 0$$

In Fig. 6.4 we have made a plot of Eq. (6.18) for $x = a$. The figure demonstrates that in the neighborhood of $y = 0$, z cannot be a single-valued continuous function of x and y.

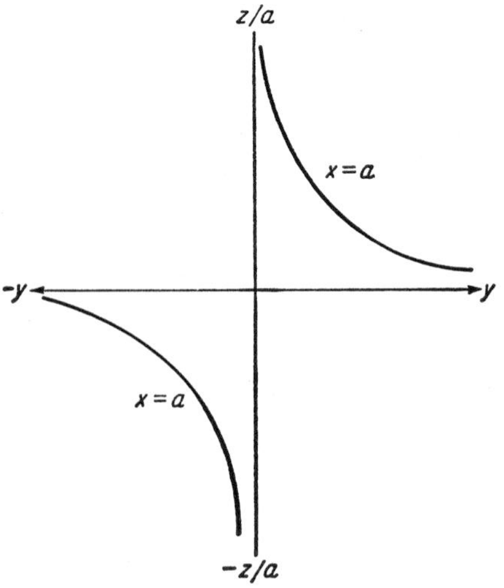

FIG. 6.4 A plot of Eq. (6.18) for $x = a$.

6.3.1 *Error Analysis*[1]

The computer technique for solving implicit-function equations produces errors even if all the components are assumed to be ideal, i.e., errorless. We shall now obtain an approximate equation for the solution error. Because component shortcomings will invariably make the error larger, the equation will represent the minimum error possible in the solution of the implicit equation.

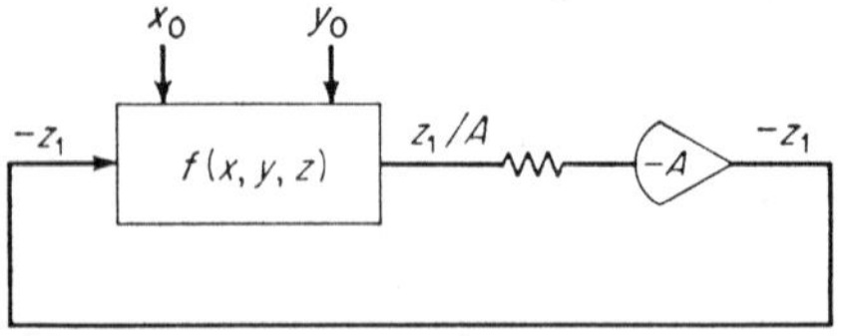

FIG. 6.5 A general circuit for the solution of an implicit function.

The computer, in its ideal state, has an inexact solution because the gain of the operational amplifier is finite. In Fig. 6.5 is a block diagram of the computer diagram which attempts to solve the equation

$$F(x, y, z) = 0 \qquad (6.21)$$

for z. The block is used to represent the operation that for inputs of x, y,

[1] The discussion follows the approach used in Ref. 75, pp. 15–21, 160–162.

and z generates the output $F(x, y, z)$. If the amplifier gain is $-A$ and the output is $-z_1$, then the input is z_1/A instead of the correct value, zero.

For the values $x = x_0$, $y = y_0$, Fig. 6.5 solves the equation

$$F(x_0, y_0, z_1) = \frac{z_1}{A} \tag{6.22}$$

instead of

$$F(x_0, y_0, z_0) = 0 \tag{6.23}$$

There will invariably be some difference between the correct value z_0 and the computer solution z_1. We get an estimate of this error by expanding $F(x_0, y_0, z_1)$ in a Taylor series around z_0:

$$\frac{z_1}{A} = F(x_0, y_0, z_1) \tag{6.24}$$

$$\frac{z_1}{A} = F(x_0, y_0, z_0) + \left(\frac{dF}{dz}\right)_{x_0,y_0,z_0} (z_1 - z_0) + \frac{1}{2}\left[\left(\frac{d^2F}{dz^2}\right)_{x_0,y_0,z_0} (z_1 - z_0)^2\right]$$

If we assume that the error $\Delta z_e = z_1 - z_0$ is small, then all terms containing higher powers of Δz_e are negligible compared with $z_1 - z_0$. Under these conditions, good approximation to Eq. (6.24) becomes

$$\frac{z_1}{A} = F(x_0, y_0, z_0) + (z_1 - z_0)\left(\frac{\partial F}{\partial z}\right)_{x_0,y_0,z_0} \tag{6.25}$$

Using Eq. (6.23), we get

$$\Delta z_e \approx \frac{z_1}{A\left(\frac{\partial F}{\partial z}\right)_{x_0,y_0,z_0}} \tag{6.26}$$

Since the error is small,

$$z_1 \approx z_0 \tag{6.27}$$

and we may also write Eq. (6.26) as

$$\Delta z_e \approx \frac{z_0}{A\left(\frac{\partial F}{\partial z}\right)_{x_0,y_0,z_0}} \tag{6.28}$$

Either Eq. (6.26) or (6.28) shows that the larger the amplifier gain A, the smaller the error. Where $\partial F/\partial z$ is finite and $A \to \infty$, then $\Delta z \to 0$.

Although Eq. (6.27) applies only when the error is small, it may also be used to give a qualitative indication of the condition under which the error becomes excessive. For a finite A the error Δz_e ceases to be small when $\partial F/\partial z \to 0$. As we have already seen in Theorem 6.1, this is the same condition that indicates a solution may not exist. If the solution is not single-valued, the computer error will become large near $\partial F/\partial z = 0$.

If in the neighborhood of $\partial F/\partial z = 0$ the solution is single-valued, as, for instance, Eq. (6.16*a*), the error may still be small. In making the Taylor series expansion we keep the lowest-order error term whose coefficient is not zero. If this coefficient is $\partial^n F/\partial z^n$, Eq. (6.24) becomes

$$\frac{z_1}{A} \approx \frac{1}{n!}\left(\frac{\partial^n F}{\partial z^n}\right)_{x_0,y_0,z_0}(z_1 - z_0)^n \tag{6.29}$$

or

$$\Delta z_e \approx \left[\frac{z_1 n!}{A\left(\frac{\partial F}{\partial z^n}\right)_{x_0,y_0,z_0}}\right]^{1/n} \tag{6.30}$$[1]

6.3.2 Functions and Their Inverses

Given that a functional relationship between two variables is

$$y = f(x) \tag{6.31}$$

then the inverse relationship is defined as

$$x = f^{-1}(y) \tag{6.32}$$

A computer which can generate a particular functional relationship can also secure the inverse by solving the implicit-function equation

$$f(x) - y = 0 \tag{6.33}$$

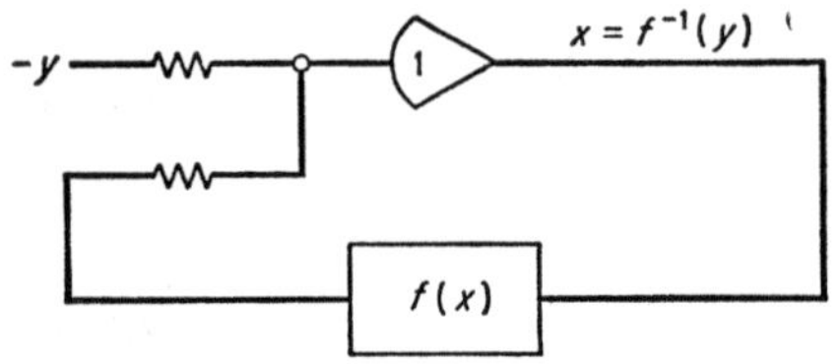

FIG. 6.6 A circuit which generates the inverse of a function.

Figure 6.6 is the outcome. This diagram illustrates the rule for generating inverse functions:

RULE: *If the computer can generate a given function $f(x)$, it can generate the inverse by placing the function generator in the feedback loop of an operational amplifier. Then if a variable $-y$ is applied at the input of the amplifier, the output will be $f^{-1}(y)$.*

One illustration of this rule is shown in Example 1. Division was performed when a multiplier was put in the feedback loop.

Example 2. A computer problem may call for the generation of a variable raised to a fractional power. If we want the function

$$z = x^{a/b} \tag{6.34}$$

where a and b are both integers and

$$b > a$$

then what we must do is set up the computer to generate

$$z = x^{1/b} \tag{6.35}$$

or

$$z^b = x$$

In so doing we shall also obtain all the terms

$$x^{2/b}, x^{3/b}, \ldots, x^{(b-1)/b}$$

[1] See Eqs. (6.58) to (6.61) for an example of the application of this error analysis.

The computer diagram is slightly different when b is an even or an odd integer, as can be seen in Fig. 6.7. As a specific illustration we show in Fig. 6.8 the circuit for

$$z = x^{\frac{1}{3}} \tag{6.36}$$

When, in Eq. (6.34), $a > b$ we again proceed as we did for $a < b$ by programming

$$z = x^{1/b}$$

To get $x^{a/b}$ we perform multiplications in addition to those in the feedback loop.

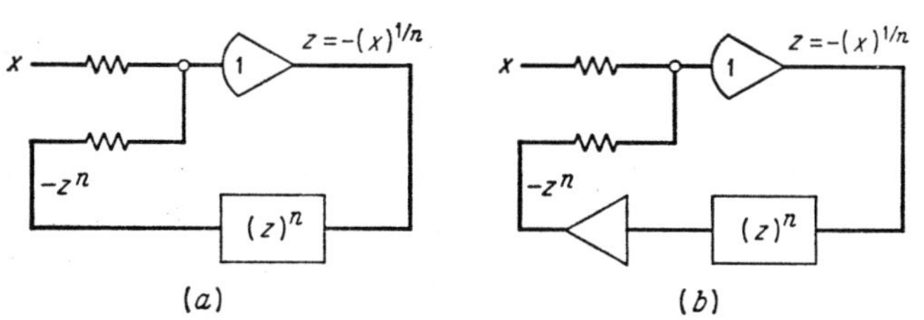

FIG. 6.7 (*a*) Generation of fractional exponents when n is odd; (*b*) generation of a fractional exponent when n is even.

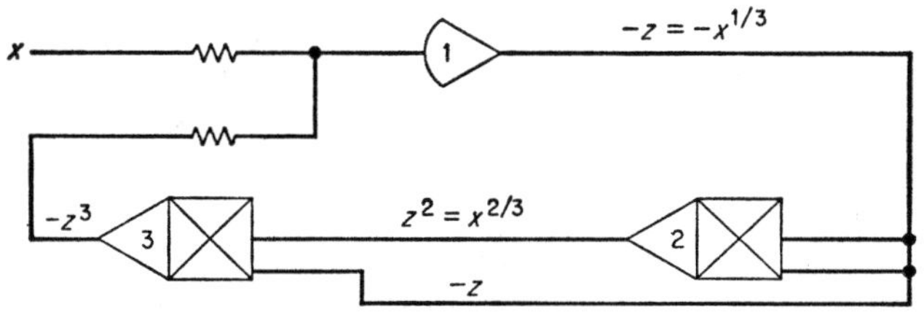

FIG. 6.8 Generation of $x^{\frac{1}{3}}$ and $x^{\frac{2}{3}}$.

6.4 Multiple Solutions

An implicit-function equation which has multiple solutions is disquieting to the novice programmer. Of all the solutions which will satisfy the equation he is presumably interested in only one. How is he to program the computer so that it will single out the solution of interest and reject the others? Before answering this question let us consider an example of an equation having a multiple solution.

To solve for the angle θ in the equation

$$\theta = \tan^{-1} \frac{y_1}{x_1} \tag{6.37}$$

we program an implicit-function equation in terms of available sine and cosine generators.[1]

[1] The programming of the computer to solve Eq. (6.37) is discussed in Sec. 6.6.

Thus we rewrite Eq. (6.37) as

$$x_1 \sin \theta_1 = y_1 \cos \theta_1 \tag{6.38}$$

In implicit form this equation is

$$x_1 \sin \theta_1 - y_1 \cos \theta_1 = 0 \tag{6.39}$$

If we restrict ourselves to the region $-\pi < \theta < \pi$, we can verify that not only is θ_1 a solution but also

$$\theta_2 = \theta_1 - \pi \tag{6.40}$$

The problems involved in setting up the computer where multiple solutions exist can best be illustrated by considering an arbitrary function $F(x, y, z)$, such as the one plotted for $x = x_0$, $y = y_\theta$ in Fig. 6.9. Points

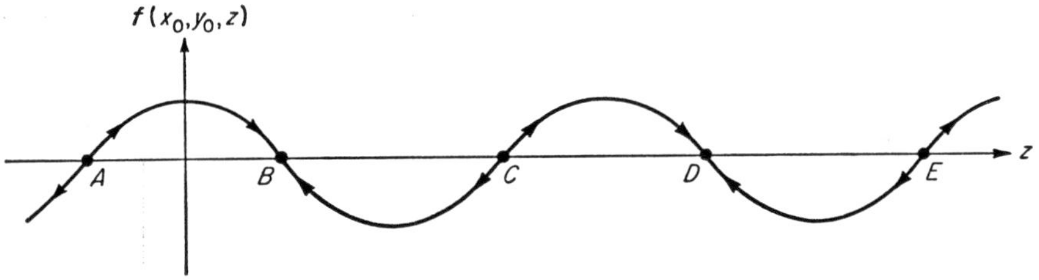

FIG. 6.9 A waveform with arrows which show the direction in which the computer varies z. Multiple solutions exist at $z = B, D$.

A, B, C, D, and E represent solutions for z in implicit form for the equation

$$F(x, y, z) = 0 \tag{6.41}$$

However, any computer program will be able to obtain only alternative values of z (either A, C, and E or B and D). The reason for this is that the closed-loop simulation will null (converge to) on certain solutions of z but will diverge from others.

Referring to Fig. 6.9, let us suppose that the computer has obtained one of the multiple solutions, say $z = B$. A small change in F would result in a large one for z, were it not for the fact that the loop is closed. We shall refer to this behavior as the tendency of z to change with F.

The specific computer program will determine what this tendency is. In Fig. 6.9 we have assumed, as shown by the direction of the arrows, that z tends to increase when F is positive. Note that an inverter directly before or after the high-gain amplifier will reverse this tendency (for F positive, z tends to decrease).

The arrows show quite clearly why certain solutions are possible and why others are not. Only those cases for F positive and negative where the arrows move toward $F = 0$ are possible computer solutions. In Fig. 6.9 these are B and D only. If the problem begins with an initial value of $A < z < C$, the computer will converge to $z = B$. For $C < z < E$ the computer will converge to $z = D$. Under no conditions can we solve for any of the solutions $z = A$, $z = C$, $z = E$ with this program.[1]

If we want any of the unobtainable solutions, we must, as we have already indicated, add an inverter either directly before or after the operational amplifier. Then it is possible to get A, C, or E but not B or D. Where multiple solutions exist, we program to obtain some but in the process exclude others. This partial loaf is not so unpalatable as it appears because in a problem we are programming to get one particular solution.

The following criteria may be used to test if a particular computer program can give a desired implicit-equation solution.[2]

1. If $F(x_0, y_0, z)$ is the input and z is the output of the operational amplifier, then the computer can null on a solution z_1, where $\partial F/\partial z > 0$.
2. If $-F(x_0, y_0, z)$ is the input and z is the output of the operational amplifier, then the computer can null on a solution z_1, where $\partial F/\partial z < 0$.

Consider Fig. 6.5, where $\partial F/\partial z > 0$. We begin with the output of the amplifier z_1, which is a solution to the equation

$$F(x_0, y_0, z_1) = \frac{z_1}{A}$$

Thus

$$F(x_0, y_0, z_1) \approx 0$$

Let us assume that the output of amplifier 1 is perturbed and becomes $z_1 + \Delta z$. Because $\partial F/\partial z > 0$, z_1/A becomes positive. Consequently, $z_1 + \Delta z$ will be decreased. By similar arguments we can show that, if z_1 changed to $z_1 - \Delta z$, the closed-loop operation would drive the output of the amplifier in a positive direction. Hence, z_1 is a stable solution. In a like manner we can show that criterion 2 will also produce stable solutions.

Apart from the satisfaction of performing the calculations, does one derive any benefit in determining the criterion before the computer is programmed? A simpler procedure is to draw the computer diagram on the assumption that the correct solution will result. If the assumption is wrong, we merely add an inverter in the loop. No other course of action is open to us. This

[1] It is true that, if the computer begins with $z = A$, $z = C$, or $z = E$, it will, under ideal conditions, remain at any of these initial values. However, these are points of unstable equilibrium. Any small error in F will cause the computer to move to a stable point ($z = B$, $z = D$). Since noise in the computer elements will guarantee that such an error will occur, we may assume without qualification that the solutions at unstable equilibria are unobtainable.

[2] These criteria were first developed by Mengel and Melahn [76] when they found the conditions under which the computer solution of an implicit-function equation is stable.

procedure may not use the intellectual capacity of the computer operator, but it cannot be surpassed in its simplicity.[1]

As may be seen from Fig. 6.9, if F is continuous then the obtainable and unobtainable solutions alternate. Hence if n is the total number of solutions, a given computer program will be able to get $n/2 \pm 1$ solutions for n odd and $n/2$ solutions for n even. Obviously for a particular program the obtained solutions may be used to define the region of the unobtained ones. Between any two values of z obtained, there must be at least one solution to the implicit-function equations which the computer cannot get. (There can be more than one because we may have missed one or more of the obtainable solutions.)

6.5 Simultaneous Equations

The relationship between several variables is expressible in a set of simultaneous equations in implicit form. The functional relation between the dependent variables x and y may be written implicitly in terms of the independent variables u and v:

$$F(u, v, x, y) = 0 \qquad G(u, v, x, y) = 0 \tag{6.42a}$$

Under what condition does the solution exist in the explicit form

$$x = r(u, v) \qquad y = s(u, v) \tag{6.42b}$$

Theorem 6.1 states the sufficient conditions which ensure a single-valued solution for a single implicit equation. For simultaneous implicit equations we extend the theorem as follows: If F and G are continuous and have continuous first partial derivatives in a range of values of u, v, x, y and the Jacobian J† does not vanish at the point u_0, v_0, x_0, y_0, where

$$J = \begin{vmatrix} \dfrac{\partial F}{\partial x} & \dfrac{\partial F}{\partial x} \\ \dfrac{\partial G}{\partial y} & \dfrac{\partial G}{\partial y} \end{vmatrix} \tag{6.43}$$

and

$$F(u_0, v_0, x_0, y_0) = 0 \qquad G(u_0, v_0, x_0, y_0) = 0 \tag{6.44}$$

there exists a unique pair of values x_0 and y_0 such that

$$x_0 = r(u_0, v_0) \qquad y_0 = s(u_0, v_0) \tag{6.45}$$

and r and s are continuous.

[1] I am not arguing against the employment of these criteria to assist in the programming. If neither alternative works, then something is wrong with the equipment or the equation. At this point one should calculate which of the two alternatives should work and program it. Only then can the programmer begin to trouble-shoot the equipment to locate the malfunction.

† The Jacobian is sometimes written $J \equiv \partial(F, G)/\partial(u, v)$.

Assuming that the conditions of the theorem are met, how are Eqs. (6.42*a*) solved for z and y? The answer is that we may use either equation to solve for x and then use the other equation to solve for y.

In other words, if the correct values are u_0, v_0, x_0, y_0, then we may solve the two equations

$$F(u_0, v_0, x_1, y_1) = \frac{x_1}{A} \qquad G(u_0, v_0, x_1, y_1) = \frac{y_1}{A} \tag{6.46}$$

or

$$F(u_0, v_0, x_2, y_2) = \frac{y_2}{A} \qquad G(u_0, v_0, x_2, y_2) = \frac{x_2}{A} \tag{6.47}$$

Figure 6.10 is the computer program for equation set (6.46).[1] By comparison, Fig. 6.11 is the computer program for equation set (6.47).

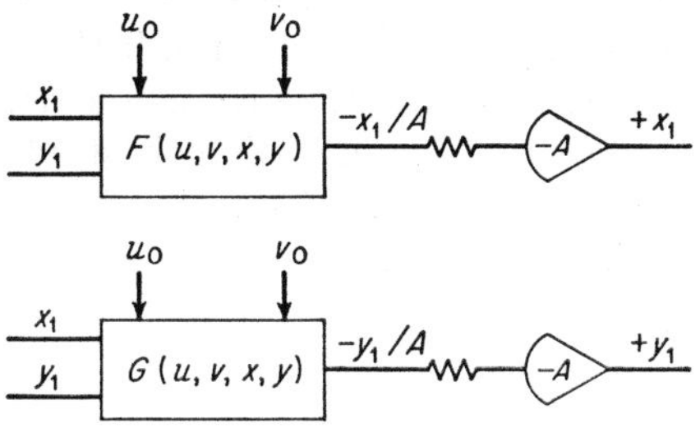

FIG. 6.10 Computer diagram for Eqs. (6.46).

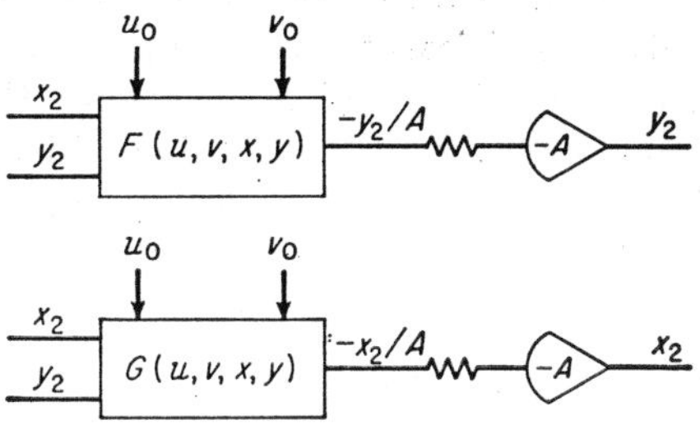

FIG. 6.11 Computer diagram for Eqs. (6.47).

For Fig. 6.10 we can get a measure of the error by a Taylor expansion around the nominal values x_0, y_0. If we neglect higher-order terms,

$$\begin{aligned} F(u_0, v_0, x_0, y_0) &= -\frac{x_1}{\mathrm{A}} \approx F(u_0, v_0, x_0, y_0) + \frac{\partial F}{\partial x}\Delta x + \frac{\partial F}{\partial y}\Delta y \\ G(u_0, v_0, x_0, y_0) &= -\frac{y_1}{A} \approx G(u_0, v_0, x_0, y_0) + \frac{\partial G}{\partial x}\Delta x + \frac{\partial G}{\partial y}\Delta y \end{aligned} \tag{6.48}$$

where $\Delta x_1 = x_1 - x_0 \qquad \Delta y_1 = y_1 - y_0$

and all partial derivatives are evaluated at (u_0, v_0, x_0, y_0). The solution for each error term is

$$\Delta x_1 = \frac{\frac{1}{A}\left(-x_1\frac{\partial G}{\partial y} + y_1\frac{\partial F}{\partial y}\right)}{J} \qquad \Delta y_1 = \frac{\frac{1}{A}\left(x_1\frac{\partial G}{\partial x} - y_1\frac{\partial G}{\partial x}\right)}{J} \tag{6.49}$$

[1] The discussion is in no way restricted because each operational amplifier is assumed to have the same gain, $-A$.

For the computer program of Fig. 6.11 the error terms are

$$\Delta x_2 = \frac{\dfrac{1}{A}\left(y_2 \dfrac{\partial G}{\partial y} - x_2 \dfrac{\partial F}{\partial y}\right)}{J} \tag{6.50}$$

$$\Delta y_2 = \left(\frac{1}{A} x_2 \frac{\partial F}{\partial x} - y_2 \frac{\partial G}{\partial x}\right) \tag{6.51}$$

As we can see, the error terms are somewhat different for the two circuits. One circuit may thus be preferable to the other on this basis. However, when simultaneous equations are programmed, stability is the overriding consideration for choosing one circuit and not another. Because of limitations in the computer elements it is not unlikely that one of these circuits is stable and the other is not.

For the case of two unknowns which has just been discussed we have only two different ways of programming the computer. In general, if there are n variables and n equations, there will be $n!$ computer arrangements. For $n > 3$ the task of finding which, if any, of these configurations is stable is a far from trivial undertaking. We describe in Prob. 8.15 an ingenious method for stabilizing simultaneous algebraic equations.

6.6 The Resolver[1]

Because of the widespread occurrence of trigonometric functions in analog-computer problems a special device called a resolver has been designed to generate sines and cosines of arbitrary angles. In Sec. 5.2.1 we showed that these trigonometric functions may be generated by the technique of generalized integration. The primary advantage of the resolver is that in most cases it employs markedly less equipment.

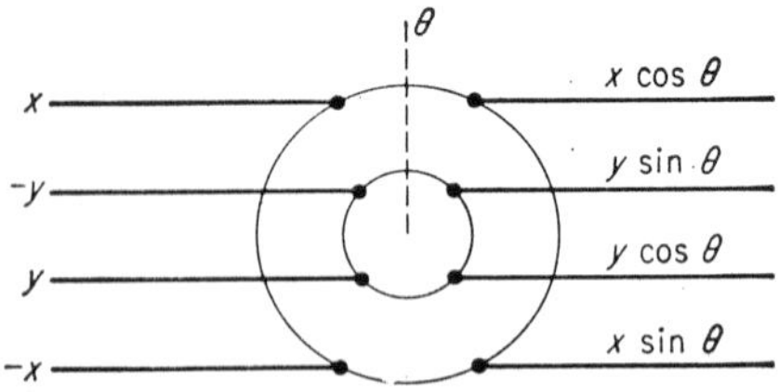

FIG. 6.12 Symbolic representation of a resolver.

Relatively few resolver applications require the solution of implicit-function equations. However, in order to present a coherent picture of the resolver, we shall discuss its principal applications.

Figure 6.12 is a computer diagram of the resolver. Given the variables θ, x, and y, the resolver can generate the four functions

$$x \sin \theta \qquad x \cos \theta \qquad y \sin \theta \qquad y \cos \theta \tag{6.52}$$

The precise reason why the functions of Eq. (6.52) have been chosen to be

[1] For additional information about the resolver see Sec. 7.6.1 and Fig. 1.1.

generated by a special device will be made obvious in our discussion on rotation. Resolvers are commonly constructed so that the range of values on the angle is approximately

$$-\pi \leq \theta \leq \pi \tag{6.53}$$

When the variation of angle is greater than Eq. (6.53) and $d\theta/dt$ is available, the most convenient way to generate sines is by the method of generalized integration.[1] Otherwise relays can extend the operation of the resolver over any angular variation.

6.6.1 *Coordinate Transformations*

Equation set (6.52) shows that the resolver may be used without modification to transform from polar to rectangular coordinates. If, as shown in Fig. 6.13, a vector is given in terms of its polar coordinates r and θ, its projections onto x and y axes represent the rectangular[2] coordinates

$$x_1 = r \cos \theta \qquad y_1 = r \sin \theta \tag{6.54}$$

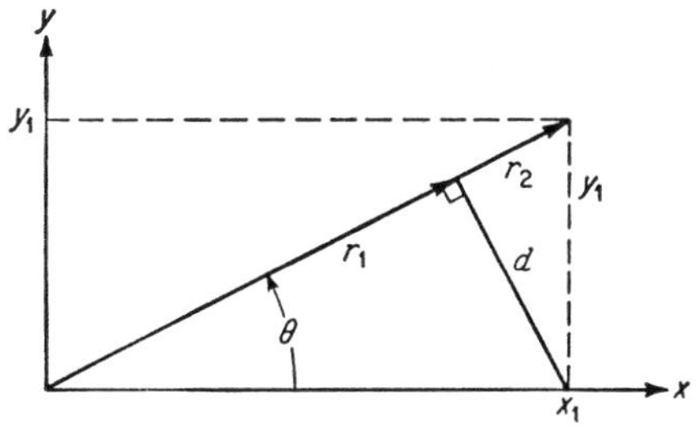

FIG. 6.13 A diagram used to obtain the relationship for rectangular to polar transformation.

Equations (6.54) reveal that, given r and θ, it is a simple matter to obtain x_1 and y_1. However, the converse is not true. If x_1 and y_1 are known, the polar coordinates cannot be obtained directly on the resolver. Instead an implicit-function equation must be solved. To obtain the essential equation we draw a line d which is perpendicular to r, as shown in Fig. 6.13. Noting that d is common to two triangles, we get

$$d = x_1 \sin \theta_0 = y_1 \cos \theta_0 \tag{6.55}$$

Also, the line d divides the line r into two component parts, r_1 and r_2. By

[1] Resolvers have been built to operate with either a θ or $d\theta/dt$ input. In such a case the $d\theta/dt$ input can be used if θ exceeds the range of Eq. (6.53).

[2] The terms cartesian coordinates (after the inventor, Descartes) and euclidean n-space are synonymous with rectangular coordinates. The coefficient n in the latter is the number of coordinates in the space. For Fig. 6.13, $n = 2$.

simple trigonometric relationships

$$r = r_1 + r_2 \qquad (6.56)$$
$$r = x_1 \cos \theta_0 + y_1 \sin \theta_0$$

From Eq. (6.55) we obtain the equation

$$-x_1 \sin \theta_0 + y_1 \cos \theta_0 = 0 \qquad (6.57)^1$$

which is solved implicitly for θ. When this is done, r is obtained explicitly from Eq. (6.56). Utilizing the procedure outlined in Sec. 6.3 we obtain the computer program shown in Fig. 6.14.

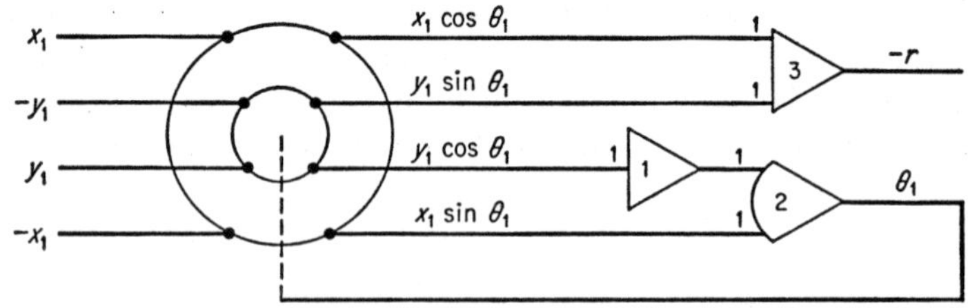

FIG. 6.14 A computer diagram for rectangular to polar coordinates.

We apply Eq. (6.26) to secure a measure of the error in θ as well as a qualitative indication of the region where the error may be excessive. Thus

$$\begin{aligned} F(x_1, y_1, \theta) &= y_1 \cos \theta - x_1 \sin \theta \\ \frac{\partial F(x_1, y_1, \theta)}{\partial \theta} &= -y_1 \sin \theta_0 - x_1 \cos \theta_0 \end{aligned} \qquad (6.58)$$

By Eq. (6.56)

$$\frac{\partial F}{\partial \theta} = -r = -(x^2 + y^2)^{\frac{1}{2}} \qquad (6.59)$$

Consequently

$$\frac{\partial^n F}{\partial \theta^n} = \begin{cases} 0 & n \text{ even} \\ (-1)^{(n+1)/2} r & n \text{ odd} \end{cases} \qquad (6.60)$$

By Eq. (6.26)

$$|\Delta\theta| \approx \frac{|\theta|}{Ar} = \frac{|\theta|}{A(x^2 + y^2)^{\frac{1}{2}}} \qquad (6.61)$$

Equation (6.61) shows that, although the error increases with θ, the significant term is r. The region of primary concern is in the neighborhood of $r = 0$. Commercial resolvers which can operate in the polar mode[2] are designed to offset this effect by an automatic-gain-control circuit which

[1] See Sec. 6.4 for some of the problems associated with programming this equation.

[2] The polar mode is the transformation from rectangular to polar coordinates. A switch in one position adds the circuitry to let the resolver operate in this mode. In the other position the switch enables the resolver to solve equation set (6.52).

increases the operational amplifier gain A when r becomes small. The gain is controlled so that the product Ar remains large and effectively constant. However, when $r = 0$, Eq. (6.57) will not yield a unique value of θ because at this point any value of θ will satisfy the equation.

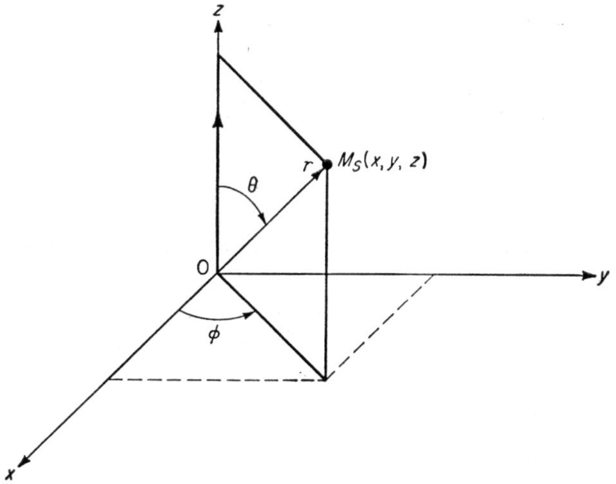

FIG. 6.15 Spherical coordinates.

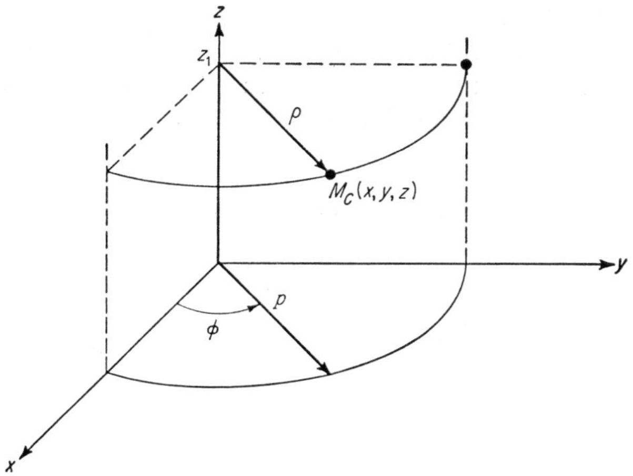

FIG. 6.16 Cylindrical coordinates.

In three dimensions the most common coordinate systems are rectangular (x, y, z), cylindrical (ρ, ϕ, z), and spherical (r, θ, ϕ). Figures 6.15 and 6.16 show the relationship between the rectangular and the other coordinate systems.

The equations for the transformation from spherical to rectangular coordinates are

$$\begin{aligned} x &= r \cos \theta \sin \theta \\ y &= r \sin \theta \sin \theta \\ z &= r \cos \theta \end{aligned} \tag{6.62}$$

from which come the inverse relationships

$$\begin{aligned} r &= (x^2 + y^2 + z^2)^{\frac{1}{2}} \\ \phi &= \cos^{-1} \frac{x}{(x^2 + y^2)^{\frac{1}{2}}} = \sin^{-1} \frac{y}{(x^2 + y^2)^{\frac{1}{2}}} \\ \theta &= \cos^{-1} \frac{z}{(x^2 + y^2 + z^2)^{\frac{1}{2}}} \end{aligned} \tag{6.63}$$

The equations for the rectangular coordinates in terms of the cylindrical coordinates are

$$\begin{aligned} x &= \rho \cos \phi \\ y &= \rho \sin \phi \\ z &= z \end{aligned} \tag{6.64}$$

In this case the inverse is

$$\begin{aligned} \rho &= (x^2 + y^2)^{\frac{1}{2}} \\ \phi &= \cos^{-1} \frac{x}{(x^2 + y^2)^{\frac{1}{2}}} = \sin^{-1} \frac{y}{(x^2 + y^2)^{\frac{1}{2}}} \\ z &= z \end{aligned} \tag{6.65}$$

A brief examination of the equations and diagrams will reveal that by the addition of the same coordinate, z, to both systems we have gone from the two-coordinate rectangular-polar transformation $[(x, y) \leftrightarrow (\rho, \phi)]$ to the three-coordinate rectangular cylindrical transformation $[(x, y, z) \leftrightarrow (\rho, \phi, z)]$.

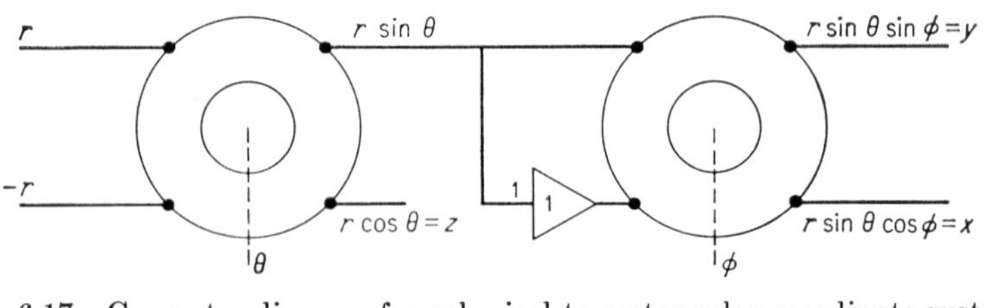

FIG. 6.17 Computer diagram for spherical to rectangular coordinate system.

Hence the discussion on the rectangular $\leftrightarrow$ polar transformation is equally applicable to the rectangular $\leftrightarrow$ cylindrical transformation.

Turning our attention to the rectangular $\leftrightarrow$ spherical coordinate-system transformations, we see that with two resolvers in Fig. 6.17 one may readily make the transformation from spherical to rectangular coordinates.

On the other hand, in going from rectangular to spherical coordinates the angles ϕ and θ are determined implicitly and the length r is obtained explicitly. The equations to be solved are

$$\begin{aligned} -x \sin \phi + y \cos \phi &= 0 \\ x \cos \phi \cos \theta + y \sin \phi \cos \theta - z \sin \theta &= 0 \\ r = x \cos \phi \sin \theta + y \sin \phi \sin \theta &+ z \cos \theta \end{aligned} \tag{6.66}$$

Figure 6.18 is the computer program which solves Eqs. (6.66) for (ϕ, θ, r).

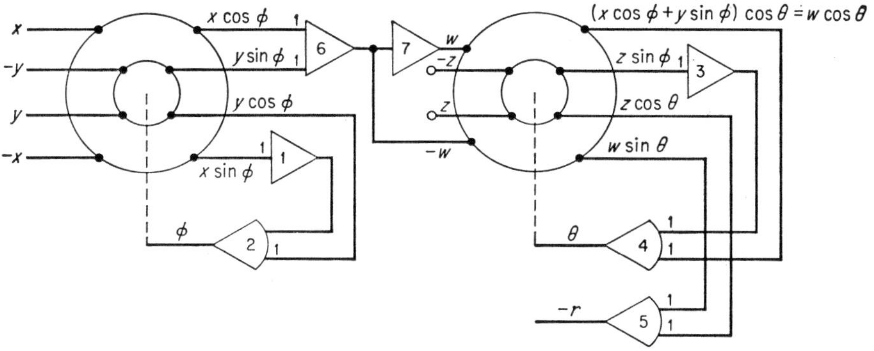

FIG. 6.18 Computer diagram for rectangular to spherical transformation.

The reader might wonder why equation set (6.63) is not used to get these variables. If we rewrite the equations as the computer would program them, we obtain

$$\begin{aligned} r &= (x^2 + y^2 + z^2)^{\frac{1}{2}} \\ (x^2 + y^2)^{\frac{1}{2}} \cos \phi - x &= 0 \\ r \cos \theta - z &= 0 \end{aligned} \tag{6.67}$$

Equation set (6.67) requires not only the two resolvers but also several multipliers.

6.6.2 Rotations[1]

In our discussion of coordinate transformations we showed how a point in space may be represented in terms of different sets of variables. We shall now examine the relationship between the coordinates in two rectangular systems where one system is rotating relative to the other. This situation can arise when we analyze the motion of a rigid body for which it is convenient to express the position in inertial coordinates and angular rates in body coordinates. Problems relating to the motion of aircraft, missile, and space vehicles all use two coordinate systems.

We shall deal with right-handed rectangular coordinate systems which are defined as follows: If we imagine the right hand grasping an axis (Fig. 6.19) with the thumb pointing in the direction of increasing positive values,

[1] For additional reading on this subject consult an advanced text on dynamics. Two excellent references are [74], pp. 93–184, and [77], pp. 288–290, 308–333. Reference 78 covers the subject thoroughly and considers computer applications as well.

the fingers will point in the direction of a positive rotation, i.e., a positive increase of angle. Also, by convention a positive rotation about the z axis moves the x axis toward the former position of the y axis. An x-axis rotation moves the y axis toward the former z axis, and a y-axis rotation moves the z axis toward the former x axis.

Consider now a rotation about a single axis. In the three-dimensional system of Fig. 6.20, the z axis, which is not shown, is perpendicular to the page and coming out toward the reader. A positive rotation θ about the z axis moves x_1 to x_2 and y_1 to y_2. The z axis remains in the same position before and after rotation. The vector r has components along the x_1, y_1,

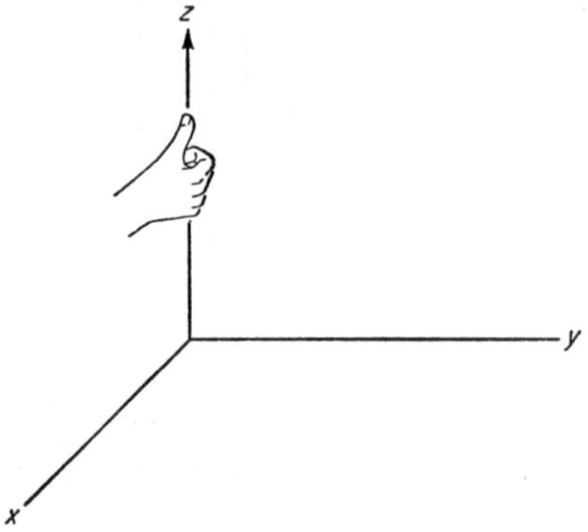

FIG. 6.19 An illustration of a right-handed rotation system.

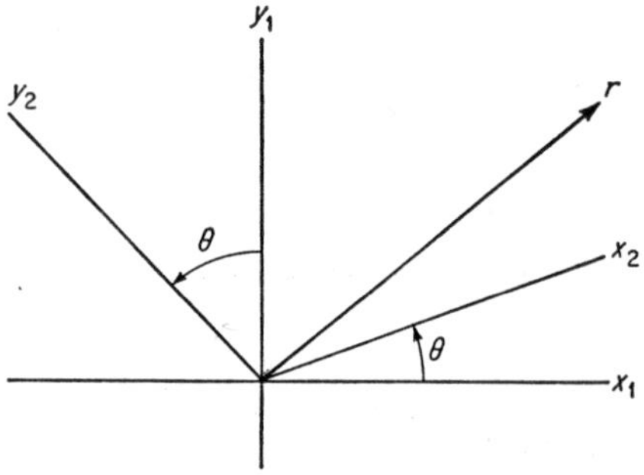

FIG. 6.20 Rotation about the z axis (not shown).

and z_1 (not shown) axes which are x_{1r}, y_{1r}, z_{1r}. In terms of these components the projection of the vector on the axes of the other coordinate system is

$$\begin{aligned} x_{2r} &= x_{1r} \cos \theta + y_{1r} \sin \theta \\ y_{2r} &= -x_{1r} \sin \theta + y_{1r} \cos \theta \\ z_{2r} &= z_{1r} \end{aligned} \tag{6.68}$$

The transformation from the subscript-2 to the subscript-1 coordinate system is

$$\begin{aligned} x_{1r} &= x_{2r} \cos \theta - y_{2r} \sin \theta \\ y_{1r} &= x_{2r} \sin \theta + y_{2r} \cos \theta \\ z_{1r} &= z_{2r} \end{aligned} \tag{6.69}$$

A set of simultaneous equations like equation set (6.68) may be expressed in matrix[1] notation. Because they will be useful to us in what follows, we

[1] Matrices have several useful properties, to which we scarcely allude, in the solution of simultaneous algebraic equations. The interested reader should consult Ref. 79 for a thorough discussion of the subject. Less comprehensive but very well written are Refs. 80, pp. 30–175, and 81, pp. 69–90.

pause momentarily in our discussion of rotations to attempt to render matrices more understandable to the reader who is first encountering them. In matrix form, equation set (6.68) is written

$$\begin{bmatrix} x_{2r} \\ y_{2r} \\ z_{2r} \end{bmatrix} = \begin{bmatrix} \cos\theta & \sin\theta & 0 \\ -\sin\theta & \cos\theta & 0 \\ 0 & 0 & 1 \end{bmatrix} \cdot \begin{bmatrix} x_{1r} \\ y_{1r} \\ z_{1r} \end{bmatrix} \tag{6.70}$$

or

$$C = A_z(\theta) \cdot B$$

where A, B, and C are matrices.[1]

We can best describe matrix multiplication by identifying each element in a matrix by two subscripts which indicate the position of the element in the matrix. For example, in matrix A the term a_{ij} is in the ith row and the jth column. If we want to take the product of two matrices as shown in Eq. (6.70),

$$\begin{bmatrix} c_{11} \\ c_{21} \\ c_{31} \end{bmatrix} = \begin{bmatrix} a_{11} & a_{12} & a_{13} \\ a_{21} & a_{22} & a_{23} \\ a_{31} & a_{32} & a_{33} \end{bmatrix} \begin{bmatrix} b_{11} \\ b_{21} \\ b_{31} \end{bmatrix} \tag{6.71}$$

then from Eq. (6.71)

$$c_{ij} = \sum_{k=1}^{n} a_{ik} b_{kj} \tag{6.72}$$

Equation (6.72) shows that we multiply the elements in the ith row of the first matrix A by the elements in the jth column of the second matrix B to obtain an element in the ith row and jth column. In particular we have

$$c_{21} = a_{21}b_{11} + a_{22}b_{21} + a_{23}b_{31} \tag{6.73}$$

We note in passing that $A \cdot B \neq B \cdot A$.

Matrix A may be regarded as a linear operator operating on B to produce C. Our special notation in Eq. (6.70) records that the A matrix results from a rotation θ around the z axis. The operation represented by the A matrix is called an orthogonal transformation. In essence, if a vector before and after transformation has the same magnitude and direction, the transformation is said to be orthogonal. In this case we may write

$$x_{1r}^2 + y_{1r}^2 + z_{1r}^2 = x_{2r}^2 + y_{2r}^2 + z_{2r}^2 \tag{6.74}$$

A transformation, like matrix A, is orthogonal if

$$\begin{aligned} &\sum_{j=1}^{3} a_{ij}^2 = 1 \qquad i = 1, 2, 3 \\ &\sum_{j=1}^{3} a_{ij}a_{kj} = 0 \qquad i = 1, 2, 3;\ k = 1, 2, 3;\ i \neq k \end{aligned} \tag{6.75}$$

A rotation around any axis is an orthogonal transformation. Equation

[1] Single-column or -row matrices like B and C are also called vectors.

set (6.75) is, therefore, a necessary set of conditions which may be used to verify if the rotation equations have been formulated correctly.

Equation set (6.70) is a rotation about a single axis and requires the use of a single resolver, as we show in Fig. 6.21. Only in special instances will one axial rotation be sufficient to rotate one coordinate system into another.

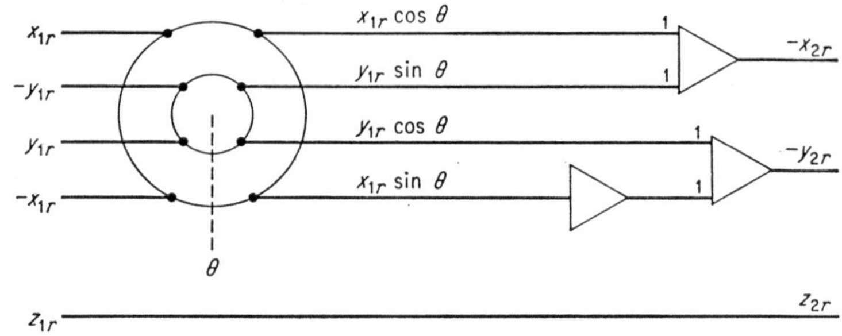

FIG. 6.21 Computer diagram for transformation from one set of axes to another for a single rotation.

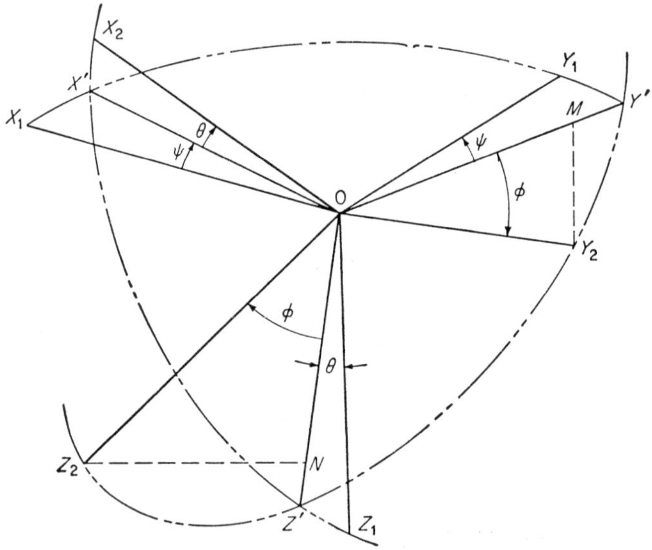

FIG. 6.22 Euler angles.

However, Leonhard Euler (1707–1783) showed that the transformation can always be made in three angular rotations. These three angles are called the *eulerian angles* and arise in the following manner.

Consider Fig. 6.22, with two coordinate systems, (x_1, y_1, z_1) and (x_2, y_2, z_2). We begin with an angular rotation ψ about the z_1 axis, which gives the intermediate coordinate system (x^1, y^1, z_1). Next we rotate an angle θ about the y^1 axis to give the coordinates (x_2, y^1, z^1). Finally, we rotate an angle ϕ about the x_2 axis. This final rotation completes the transformation because the coordinate system is now (x_2, y_2, z_2).

$$\begin{bmatrix} x_2 \\ y_2 \\ z_2 \end{bmatrix} = C_{x_2}(\phi) \cdot B_{y_1}(\theta) \cdot A_{z_1}(\psi) \cdot \begin{bmatrix} x_1 \\ y_1 \\ z_1 \end{bmatrix} \tag{6.76}$$

where

$$A_{z_1}(\psi) = \begin{bmatrix} \cos\psi & \sin\psi & 0 \\ -\sin\psi & \cos\psi & 0 \\ 0 & 0 & 1 \end{bmatrix}$$

$$B_{y_1}(\theta) = \begin{bmatrix} -\sin\theta & 0 & \cos\theta \\ 0 & 1 & 0 \\ \cos\theta & 0 & \sin\theta \end{bmatrix}$$

$$C_{x_2}(\phi) = \begin{bmatrix} 1 & 0 & 0 \\ 0 & \cos\phi & \sin\phi \\ 0 & -\sin\phi & \cos\phi \end{bmatrix}$$

$$C_{x_2}(\phi) \cdot B_{y_1}(\theta) \cdot A_{z_1}(\psi) = \begin{bmatrix} C\psi C\theta & C\theta S\psi & -S\theta \\ C\psi S\theta S\phi - S\psi C\phi & S\psi S\theta S\phi - C\psi S\phi & C\theta S\phi \\ C\psi S\theta C\phi + S\psi S\phi & S\psi S\theta C\phi - C\psi S\phi & C\theta C\phi \end{bmatrix}$$

For example, $C\psi = \cos\psi$; $S\psi = \sin\psi$.

The sequence of rotations we have indicated is by no means the only one that can be taken. The only restriction on the eulerian angles is that no two consecutive rotations can be made about the same axis. For example, the sequence might be in order around the z_1, y', and z' axes, or the sequence

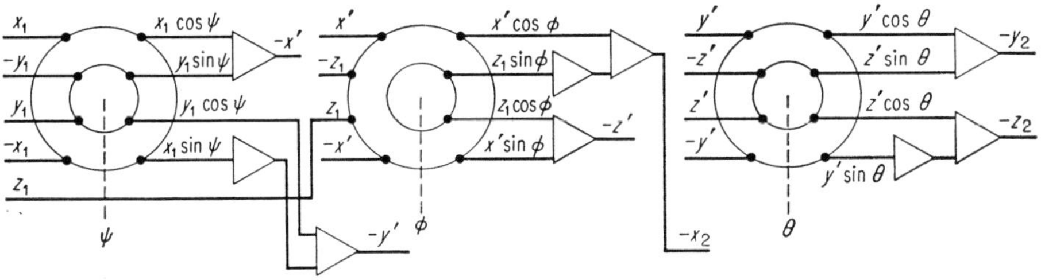

FIG. 6.23 Computer diagram for Euler-angle transformation.

might begin with a rotation about the x_1 axis. The three eulerian angles will depend upon the order of rotations and will in general be different.

Returning to the transformation represented by Eq. (6.76), we show in Fig. 6.23 how three resolvers perform the essential rotations. We have omitted the inverting amplifiers at the resolver inputs to keep the diagram from being cluttered with unnecessary detail.

If the coordinate system x_1, y_1, z_1 is not rotating and if the angular rates about x_2, y_2, and z_2 are, respectively, w_{x_2}, w_{y_2}, w_{z_2}, the dynamic equations relating these rates and the eulerian angles are

$$
\begin{aligned}
w_{x_2} &= \frac{d\phi}{dt} - \frac{d\psi}{dt}\sin\theta \\
w_{y_2} &= \frac{d\psi}{dt}\cos\theta\sin\phi - \frac{d\theta}{dt}\cos\phi \\
w_{z_2} &= \frac{d\psi}{dt}\cos\theta\cos\phi - \frac{d\theta}{dt}\sin\phi
\end{aligned}
\tag{6.77}
$$

or

$$
\begin{aligned}
\frac{d\psi}{dt} &= (w_{y_2}\sin\phi + w_{z_2}\cos\phi)\sec\theta \\
\frac{d\theta}{dt} &= w_{y_2}\cos\phi - w_{z_2}\sin\phi \\
\frac{d\phi}{dt} &= w_{x_2} + \frac{d\psi}{dt}\sin\theta
\end{aligned}
\tag{6.78}
$$

7 Error-reduction Techniques

DEFINITION: An *error* in this chapter, unless otherwise qualified, is the difference between the correct value of a variable and the computer solution.

7.1 Introduction

In the first six chapters we have examined the primary components of the analog computer. We have also discussed the techniques by which these components may be utilized to solve problems. The computer is ostensibly well suited to solve ordinary differential equations. Yet these equations, though presenting no mathematical difficulties, may be programmed and have intolerable errors in the solution.

The source of the trouble is simple enough to identify. The elements of the computer only inexactly approximate the mathematical operations they represent. Thus, the computer is a device which has certain inherent restrictions imposed on its ability to solve even the problems for which it was designed.

A first step in the improvement of computer accuracy is time and amplitude scaling, which we treated in some detail in Chap. 4. These two types of scaling are the most universally applicable methods we have, and in a large number of problems they are sufficient to give the accuracy we seek. However, there are situations where time scaling and amplitude scaling are not enough. This chapter will describe some of these cases and the techniques whereby the inaccuracies are reduced.

Unfortunately, there is not one single technique which is a panacea for the ills which beset the computer. Rather, each particular type of problem has its own remedy. Luckily, all analog computers, no matter what their precision, have the same type of limitations which impair their accuracy. Thus any method that reduces the effects of these limitations on solution error will benefit all computers, although to varying degree. The potpourri of techniques we shall consider should by no means be regarded as a definitive discussion of the subject. Much useful work remains to be done. Hopefully, the contents of this chapter can provide the needed stimulus to some readers.

There are two primary sources of error:

1. Voltages which become too large or remain too small for a sufficiently long portion of the problem
2. Equipment inadequacies

We shall begin the chapter with a discussion of techniques that can be used to cope with voltages that fall outside the usable range. The first of these methods is variable scaling. In Chap. 4 we examined a scaling which remained constant during the problem. Now we extend this concept so that we can change the scale factor during the run to fit our problem needs.

Variable scaling may be programmed directly on the computer diagram. Next we discuss two methods which require a change in the computer equations. These are the perturbation and parameter-influence techniques. As we shall see, both of these methods have applications which range far beyond error reduction.

Quite often we want to read out a voltage at a particular time. The easiest way to accomplish this is with a sample-and-hold circuit (Sec. 5.2.2). However, the inability of the relay to open instantaneously introduces readout errors. Our next subject treats readout errors and the techniques which make them smaller.

Finally, we close the chapter by discussing the shortcomings of several computer elements, including multipliers, resolvers, and dividers. The methods we describe all improve the performance by means of computer programming and not by alteration of the internal structure of the element.

7.2 Voltage-range Difficulties

We justified our need for amplitude scaling in Chap. 4 by showing that the voltage representing any variable has upper and lower bounds. The upper limit $V_{\max}$ occurs when the performance of an element distorts, that is, when an element ceases to perform the operation for which it was designed. The lower limit $V_{\min}$ is less precisely defined and is in part determined by the nature of the particular problem. If the problem voltage remains below $V_{\min}$ a sufficient length of time, the noise in the computer becomes large enough relative to the signal to produce significant errors.

Suppose we have a variable x, with a range $x_{\min} < x < x_{\max}$, represented by a computer variable V, where $V_{\min} < V < V_{\max}$. Harking back to Chap. 4 we recall that x can be scaled to

$$V = kx \tag{7.1a}$$

where

$$k \leq \frac{V_{\max}}{x_{\max}} \tag{7.1b}$$

In this scaling we have made a tacit assumption that there is no difficulty with V_{min}. A more complete statement of the conditions for constant scaling is that Eq. (7.1*a*) is valid subject to the condition of Eq. (7.1*b*), provided that

$$\frac{X_{max}}{X_{min}} \leq \frac{V_{max}}{V_{min}} \tag{7.2}$$

or

$$V_{min} \leq kx \leq V_{max}$$

The ratio V_{max}/V_{min} *may be regarded as a figure of merit for any given computer because the ratio alone fixes the range of variable which the computer can simulate accurately.* At the present time the better computers (REAC, PACE, and EASE) have a $V_{max} \approx 100$ volts and a $V_{min} \approx 0.1$ volt. Hence by Eq. (7.2) their figure of merit is

$$\frac{V_{max}}{V_{min}} = 10^3$$

To reinforce the previous comments about the importance of the figure of merit, let us consider an outlandish example. Suppose one computer has $V_{1,max} = 10^{-1}$ volt, $V_{1,min} = 10^{-5}$ volt, and a second computer has $V_{2,max} = 10^7$ volts, $V_{2,min} = 10^5$ volts. The first computer has the higher figure of merit and is the better machine despite the fact that $V_{2,max} - V_{2,min} > V_{1,max} - V_{1,min}$.

The maximum and minimum voltages may be regarded as the head and foot of a Procrustean bed within which our variables must lie. When we use constant scaling, our scale factor is determined by Eq. (7.1*b*). The overall accuracy of the problem dictates that we use as much of the available voltage range as possible. The reason for this is that many computer errors are approximately constant and independent of the signal voltage. Therefore, we should make the scale factor k as large as possible. However, if Eq. (7.2) is not satisfied, the problem *cannot* be solved when constant scaling is used. We now study variable scaling, a technique that can work when constant scaling fails.

7.2.1 Variable Scaling[1]

The simplest type of variable scaling can be implemented by solving the problem in segments. As we saw in Chap. 2, a problem can be solved on an analog computer in one of two ways. Either the initial conditions may be inserted and the problem run in its entirety or the problem may be stopped and then started again with the outputs of the integrators becoming the new initial conditions. The problem may be stopped and started any number of times by the second method.

[1] Much of the work on variable scaling and readout techniques represents the joint efforts of Clarence Giese, Shozo Matsuno, Robert McGhee, Hans Meissinger, and the author. Wherever possible, I have given additional credit to specific individuals.

From the standpoint of scaling, each segment of a problem run by the second method may be regarded as a different problem. The requirement becomes that Eq. (7.2) must be satisfied separately for each segment.

Take Fig. 7.1 as the plot of a problem variable and assume that the computer figure of merit is 10^3 ($V_{max} = 100$ volts). We note that

$$\frac{x_{max}}{x_{min}} = 10^5$$

Thus it is impossible to run this problem with a constant scale factor. Instead we must break the problem into segments where $x_{max}/x_{min} \leq 10^3$. We begin at t_0 with the scale factor $k_1 = 10^{-1}$ and run until t_1. In the next segment $x_{max} = 10^5$. Therefore the range for which we scale is $10^2 < x < 10^5$ and $k_2 = 10^{-3}$. This segment terminates at t_2. For the remainder of the problem $x_{max} = 10^2$ and the third scale factor is $k_3 = 1$.

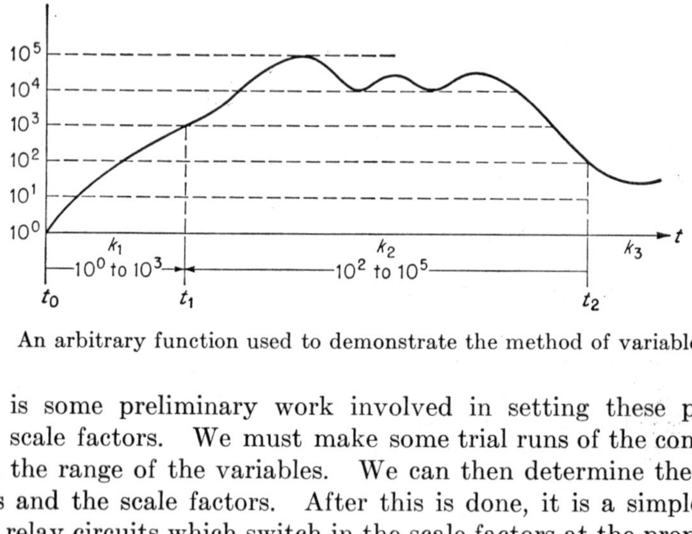

FIG. 7.1 An arbitrary function used to demonstrate the method of variable scaling.

There is some preliminary work involved in setting these piecewise-constant scale factors. We must make some trial runs of the computer to establish the range of the variables. We can then determine the problem segments and the scale factors. After this is done, it is a simple task to program relay circuits which switch in the scale factors at the proper times.

The piecewise-constant scale factors have been discussed in terms of amplitude scaling. However, they may also be used to *time-scale* the problem. Following the procedure described in Chap. 4, we change every integrator input gain at the start of each segment.

If the scale-factor changes affect the input to an electromechanical device, e.g., a servomultiplier, the unit will not be able to respond immediately to a discontinuous change in its voltage. Under these conditions it is desirable to change scales while the computer is in HOLD and wait until all elements settle down to their correct operating positions before running the next segment. When the problem is a deterministic one run exclusively on the computer, we can remain in the hold position a long time without changing the result.[1] However, if the problem involves random processes[2] or certain types of physical simulation, it cannot be put in HOLD at all.

We need not arrest the problem if we use continuously varying instead of piecewise-constant scaling. Our time segments then become infinitesimal. However, when we use time-varying scale factors, we must exercise some care, recognizing that certain processes are not commutative. Thus we have potential difficulty with scale factors which precede integrators.

Let $k = f(t)$. Then we observe after integration that

$$y = \int_0^t kx\,dt \tag{7.3}$$

If our scaling is to be useful, we must be able to divorce the effect of the scale factor from the integration and write

$$y = g(t) \int_0^t x\,dt \tag{7.4}$$

Where k is not a constant, it follows that

$$g(t) \neq k$$

In this circumstance we can obtain $g(t)$ only if we know x and can perform the integration in Eq. (7.3). We can overcome this difficulty. If we wish the scale at the output of an integrator to be ky, then

$$\frac{d}{dt}(ky) = y\frac{dk}{dt} + k\frac{dy}{dt} \tag{7.5}$$

Figure 7.2 is the simple circuit which will solve Eq. (7.5).

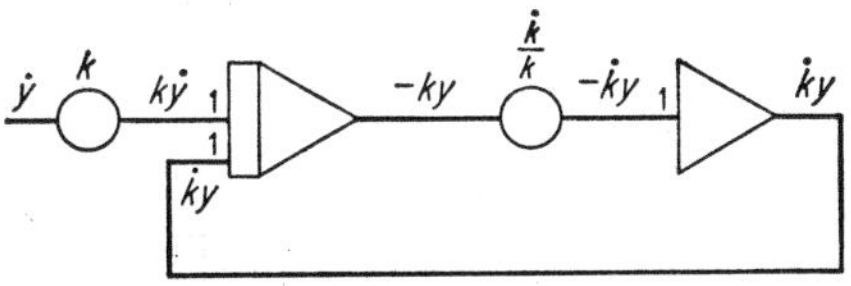

FIG. 7.2 A circuit for continuous variable scaling at the input to an integrator.

CAUTION: If the setting $\dot{k}/k > 0$, the circuit in Fig. 7.2 is the computer program for a divergent solution. Hence the circuit tends to be very sensitive to computer errors. The operator should therefore be very careful in its use.

The choice of the variation of k must be determined by what is happening to the variables. For a monotonically decreasing variable a monotonically increasing k would appear to be the best means of keeping the voltage larger than $V_{\min}$. Either this can be done from the start of the program or the constant scale factor may be used until some point in the problem, when the variable scale factor can be switched in. Which is the better alternative? The answer lies in the problem. We are deliberately vague because one cannot make general statements about the relation between errors in an

[1] The length of time is limited by integrator drift.

[2] Random processes, in general, are correlated in some fashion. If we go into hold, we shall not be able to simulate the noise characteristics properly in a given computer run.

element and the error in the problem without being specific about the problem and the error.

Three possible circuits to be used when y decreases monotonically are shown in Fig. 7.3.[1]

The circuits can be made applicable for monotonically increasing functions by setting $k < 0$. When y is not monotonic, these circuits will not suffice.

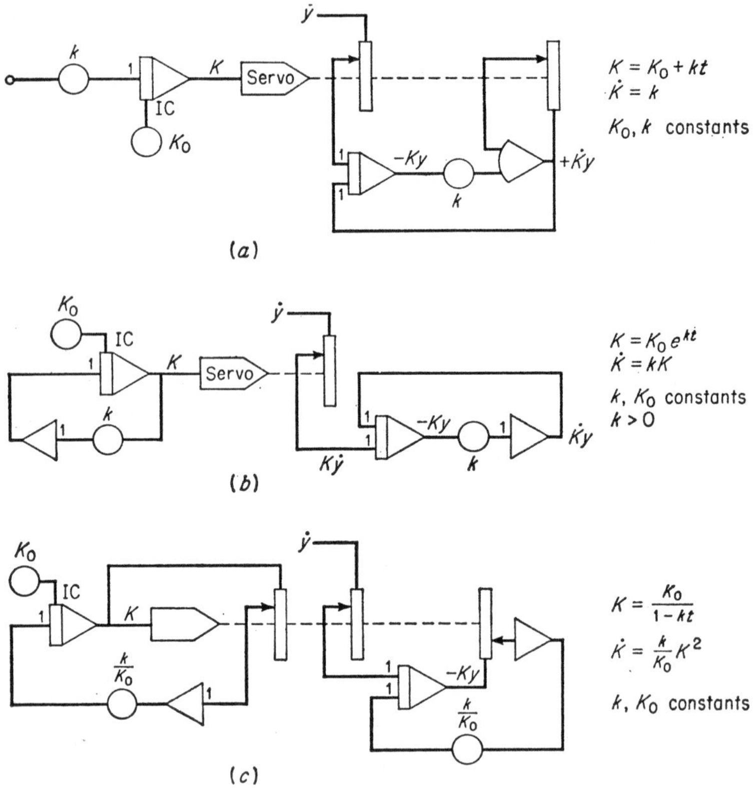

FIG. 7.3 Circuits for a monotonically increasing (decreasing) scale factor when $k > 0$ ($k < 0$).

It may then be necessary to generate dk/dt on a function generator and obtain k by integration.

Where the operation does not involve the dimension of time, there is no particular problem in variable scaling. One must be careful, however, that the scaling is invariant, i.e., that the process is the same after the scaling as it was before. The point may be illustrated in the resolution from rectangular to polar coordinates (see Sec. 6.6). The equations describing this trans-

[1] These circuits were suggested by Shozo Matsuno.

formation are

$$R = (x^2 + y^2)^{\frac{1}{2}} \qquad \theta = \tan^{-1}\frac{y}{x} \tag{7.6}$$

In this mode of operation the resolver becomes inaccurate when the voltage representing R is small. This effect can be minimized by increasing the scaling on R. If R is scaled to $kR(R \to kR)$, then $x \to kx$ and $y \to ky$. The result is

$$\theta_1 = \tan^{-1}\frac{ky}{kx} \qquad R_1 = k(x^2 + y^2)^{\frac{1}{2}} \tag{7.7}$$

θ_1 is the same as θ provided k is positive. However,

$$R_1 = kR$$

Should we need R elsewhere in the problem, we must scale R_1 by $1/k$.

Variable scaling has the feature of working with the computer program without modifying the original equations. Indeed, if one is unaware of the places where the scaling may be a problem, one can first program the computer with fixed scale factors and then add the variable scale factors where they are needed.

However, there is one potential source of difficulty: It may be necessary to apply the scale factor $k(t)$ to more than one variable. In addition, other problem variables usually must be scaled by the reciprocal, $1/k(t)$. One exception to this is the readout circuit, which requires only one scaled variable. When two or more variables are involved, their scaling must be synchronized or errors can be the outcome. Sometimes small deviations from one another in the scale factors will precipitate large solution errors. Should this be so, then our reward for ingenuity is that the cure of variable scaling is worse than the sickness of voltage limitations. To avoid this pitfall we next consider two different approaches in which the computer equations themselves are altered. These are the perturbation technique and the closely related parameter-influence method.

7.3 Perturbation Technique

The perturbation technique, used extensively to solve problems in physics and applied mathematics, revises the equations so that each important variable is measured relative to some nominal value. When this is done, a nonlinear set of equations may be reduced to a form solved more easily by analysis and more accurately on the analog computer. Sometimes the equations to be solved may become linear with all the associated advantages.[1]

To show in a simple way how expansion around a nominal value can assist computer accuracy, consider a situation in which y unscaled varies as shown

[1] See Sec. 2.3.3 for a discussion of these advantages.

in Fig. 7.4. To keep this variable from overload ($V_{max} = 100$ volts), y must be scaled to $\frac{5}{6}y$. On the other hand, if

$$y = y_0 + \Delta y \tag{7.8}$$

where $y_0 = 100$ volts

then $0 \leq |\Delta y| \leq 20$ volts

and the scaling on the new variable can be $5\,\Delta y$.

This type of scaling is of advantage in computer elements where the output error is somewhat independent of the input voltage. Linear elements, like amplifiers and integrators, have output noise and drift, which exhibit this behavior. For example, suppose that an amplifier has as a programmed output $\frac{5}{6}y$ and also has a noise output $n(t)$. Then the combined output is

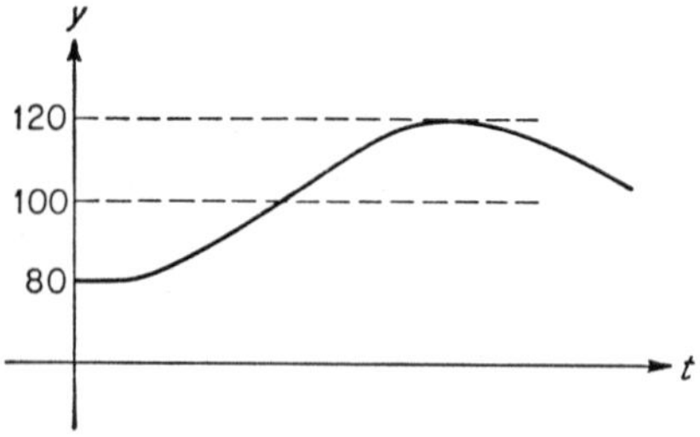

FIG. 7.4 A function perturbed about a fixed nominal value.

$$w = \tfrac{5}{6}y + n(t) \tag{7.9}$$

The computer measurement of y is $\frac{6}{5}w$. Therefore, the error δ_1 is

$$\delta_1 = \tfrac{6}{5}w - y = \tfrac{6}{5}n(t) \tag{7.10}$$

If, instead, we use Eq. (7.8) and program Δy, we can scale the output of the amplifier to be $5\,\Delta y$. Then its actual output becomes

$$u = 5\,\Delta y + n(t) \tag{7.11}$$

If we do not use computer elements to combine y_0 and y to form y,† then our error is only in the computation of y. Thus, our error δ_2 in the computer measurement of y is

$$\delta_2 = y_0 + \frac{u}{5} - y = \frac{n(t)}{5} \tag{7.12}$$

Comparing Eqs. (7.10) and (7.12), we see that perturbation has reduced the error by a factor of 6.

Similarly perturbation will reduce errors of this type in nonlinear devices. Even when the error is a function of the input voltage, this method can improve the solution accuracy, Take, for example, the inputs to the diode function generator, the motor of the servomultiplier, and the motor of the resolver. Here the error is dependent upon the input voltage. However, the error varies in a way unique to each element. If we make the not unreasonable assumption that the error variation over the entire possible range of input voltage is approximately the same as over a smaller region of

† See Example 3 for a further explanation of this point.

input-voltage variation, then our objective is to reduce the output-error level by assuming that the effect of any input voltage is the same. Of course, if the error is markedly different in one region of input voltage as compared with any other region, we might be better off to confine the input voltage to this range of variation. However, experience has shown that perturbation does reduce solution errors in the three nonlinear devices already mentioned.

To demonstrate the advantage of perturbation when applied to servomultipliers, assume that we are programming Fig. 7.5, where ky is the input to the servomotor. The error quantity ϵ is treated as a constant. More

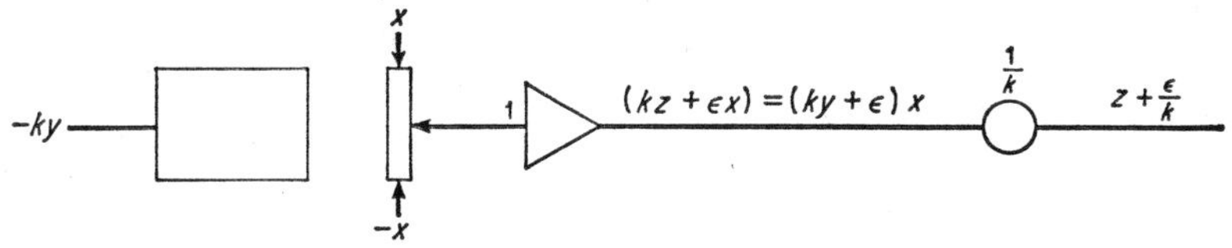

FIG. 7.5 Static error of a servomultiplier.

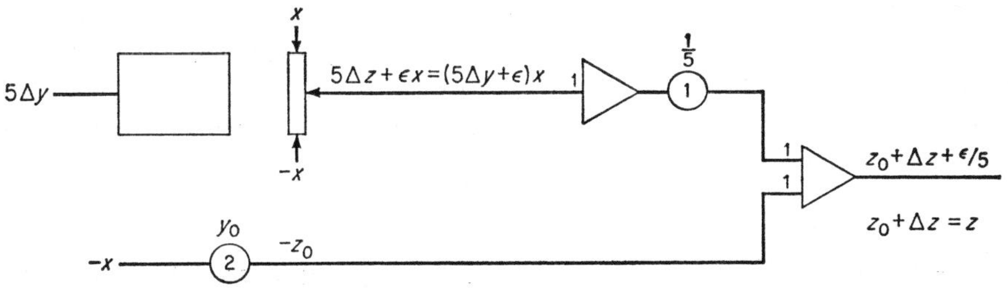

FIG. 7.6 A circuit which reduces the static error of a servomultiplier.

precisely, the error term is $\epsilon f(y)$, where $f(y)$ has about the same characteristics over any reasonable range of y. If y has the form of Fig. 7.4, then $k = \frac{5}{6}$ and the output of the multiplier is $(\frac{5}{6}y + \epsilon)x$. The output of the potentiometer is then

$$z + \tfrac{6}{5}\epsilon x = z + \epsilon_1 \tag{7.13}$$

and the error ϵ_1 is $\frac{6}{5}\epsilon x$.

Equation (7.8) is simulated somewhat differently, as we show in Fig. 7.6. Because y_0 is constant, it may be set on a potentiometer with negligible error. Consequently, z_0 is presumed to be exact. The error ϵ_2 is then associated only with the increment Δy. Comparing Figs. 7.5 and 7.6, we see in the latter that the error is $\epsilon x/5$ compared with $6\epsilon x/5$ in the former. Thus by a judicious selection of the nominal value the error has been reduced by a factor of 6. In general, the error improvement can be shown to satisfy

Eq. (7.14):

$$\frac{\epsilon_1}{\epsilon_2} = \frac{|y_{\max}|}{|\Delta y_{\max}|} \tag{7.14}$$

These points will be expanded in Sec. 7.6, on servomechanisms.[1]

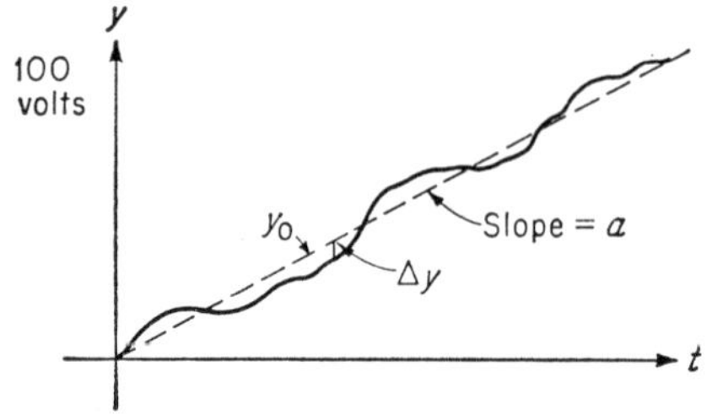

FIG. 7.7 Perturbation of a function about a variable nominal value y_0.

The nominal value need not be fixed. If y varies as shown in Fig. 7.7, a fixed y_0 of 50 would help. However, even better would be a y_0 which varies linearly as shown by the dashed line. In this case the Δy would be smaller than in the case of a fixed y_0 with a commensurately smaller error in the problem. We shall amplify later on the concept of the variable nominal value, but let us now examine some mathematical aspects of perturbation.

7.3.1 Mathematics of Perturbation[2]

In the perturbation technique one considers the effect of an incremental change in each pertinent parameter and variable about some nominal value, which may itself be fixed or varying. A tacit assumption is made that all terms which are higher than the lowest order can be ignored because their effect on the solution is negligible. This is equivalent to saying that the incremental quantity remains small.

7.3.2 Order Explained[3]

To avoid confusion we define order in the following sense. Let a function $f(x)$ be represented by a power-series expansion,

$$f(x) = a_0 + a_1x + \cdots + a_nx^n \tag{7.15}$$

Then, as x is allowed to approach some value asymptotically, one power of x becomes much more significant than all the others. This is said to be the order of x.

In Eq. (7.15), as x becomes very large, the largest-order term becomes much larger than all the rest. As x is varied in this region, the highest-order term x^n is sufficient to describe the behavior. For $x \to \infty$, the inequality

$$x^n \gg x^{n-1} \gg \cdots \gg x$$

[1] Since writing this section on error reduction in multipliers, I found in Ref. 82 that a multiplier has been built incorporating this technique.

[2] See the section by W. M. Gaines in Ref. 49, pp. 25-07 to 25-12, for additional discussion. Some of the material of this section is taken from that reference. Specifically, this includes most of Table 7.1 and the concept expressed by Fig. 7.11.

[3] See Ref. 83, pp. 5–25, for a more complete discussion of order.

holds. Thus Eq. (7.15) may be written

$$f(x) = 0(x^n) \qquad x \to \infty \tag{7.16}$$

Equation (7.16) says that $f(x)$ is of order x^n as x approaches infinity ($x \to \infty$).

Going in the opposite direction, when x becomes small, then smaller powers predominate, or

$$x^0 \gg x^1 \gg x^2 \cdots x^{n-1} \gg x^n \qquad \text{for } x \to 0 \tag{7.17}$$

Then

$$f(x) = 0(x^0) \qquad x \to 0 \tag{7.18}$$

7.3.3 Perturbation Mathematics Revisited

Returning again to our method for constructing perturbation equations, we shall see that, with the assumption of small incremental values reducing the power-series representation of each variable to lowest order, the modified solution will consist of two terms. The first term is the solution of the equations containing nominal values of the variables. The second term is the solution of a new equation containing only incremental differences between the new and nominal values of the perturbed quantities and possibly some of the nominal values as well. This equation, called the *perturbation equation*, will be linear with rare exceptions.

To summarize: Given the equation

$$S = f(w, x, y, z) \tag{7.19}$$

and letting

$$w = w_0 + \Delta w \qquad x = x_0 + \Delta x \qquad y = y_0 + \Delta y \qquad z = z_0 + \Delta z$$

then

$$\begin{aligned} S &= f(w_0, x_0, y_0, z_0) + f_1(\Delta w, \Delta x, \Delta y, \Delta z; \omega_0, x_0, y_0, z_0) \\ &= S_0 + S_1 \end{aligned} \tag{7.20}$$

where

$$S_0 = f(w_0, x_0, y_0, z_0)$$

is the nominal equation.

If Eq. (7.19) is a differential equation, we can perturb coefficients, forcing function, and initial conditions separately or together. As a consequence the dependent variable and its derivatives will also be perturbed.

There is always more computer equipment by about a factor of 2 involved in solving Eq. (7.20) than in solving Eq. (7.19). Because the incremental quantities are small compared with the nominal ones, we can scale to reduce problem inaccuracies. Perturbation equations lend themselves to combined analog-digital computer operation. The digital computer can be assigned the task of solving the nominal equation (whose variables should be changing slowly) while the analog computer simultaneously solves the perturbed equation.[1]

[1] This method has already been used by the MIT Instrumentation Laboratory, Cambridge, Mass., to calculate trajectories for the Polaris missile. The digital computer calculated a nominal path, and the analog computer solved the incremental equations.

The benefit of this combined operation may be demonstrated by assuming that in Eq. (7.19) the analog computer solves for S with an error of 1 per cent. In order to perform a typical calculation we assume

$$S_1 = 0.01S_0 \tag{7.21}$$

If the computer were to solve for S directly, it would have an error of 1 per cent. Instead, suppose that S_0 is known exactly and that the computer need solve only for S_1 in order to determine S. In this circumstance the computer determines S_1 on an expanded amplitude scale, which results in an increased overall computer accuracy. Let us make a pessimistic assumption that the error of the perturbed solution is also 1 per cent. Because the perturbation is linearized and minimizes the need for nonlinear components, which are the most inaccurate parts of a computer, this error will in general be smaller than the error in the solution of the original equation. Let S' be the erroneous solution.

Hence

$$\begin{aligned} S' &= S_0 + S_1(1 \pm 0.01) = S_0 + 0.01S_0(1 \pm 0.01) \\ S' &= 1.01S_0(1 \pm 0.0001) \\ S' &= S(1 \pm 0.0001) \end{aligned} \tag{7.22}$$

By knowing S_0 we have been able to reduce the error by a factor of 100.

We have stressed the accuracy of the perturbation technique, but an important point which should not be overlooked is that the perturbation equation is usually linear. When this is the case, we have converted the problem to the solution of one nonlinear equation (nominal) and one linear equation (perturbation). We can then use the principle of superposition and the weighting function to generalize the results around the nominal value.

For example, if the nonlinear equation is solved directly, we must make one computer run for each set of initial conditions and forcing function. In the perturbation technique we solve the nominal equation in one computer run. Then for the linear perturbation equation we can obtain a normalized set of solutions† for application with initial conditions which deviate from nominal values. From this equation will also come a weighting function for use with the forcing-function deviation from nominal.

Example 1. Let

$$Ly = g(t) \tag{7.23}$$

be a second-order differential equation with initial conditions $\alpha_1 y^{(1)}$, $\alpha_0 y^{(0)}$,† and with L representing a differential operator. Using nominal initial conditions $\beta_1 y^{(1)}$ and $\beta_0 y^{(0)}$

† See Sec. 3.2 for a description of what constitutes a *normalized set* of equations and also a definition of this notation.

and a nominal forcing function $f(t)$, we obtain the nominal equation

$$Ly_1 = f(t) \tag{7.24}$$

and the perturbation equation

$$L_2y_2 = g(t) - f(t) \tag{7.25}$$

where

$$y = y_1 + y_2$$

We assume that Eq. (7.25) is linear and has the normalized solutions $f_0^{(1)}(t)$ and $f_0^{(2)}(t)$, as well as the weighting function $h_2(t_2, t_1)$. The solution to Eq. (7.23) may then be written

$$y(t) = y_1(t) + (\alpha_0 - \beta_0)f_0^{(0)}(t) + (\alpha_1 - \beta_1)f_0^{(1)}(t) + \int_0^t [g(\tau) - f(\tau)]h_2(t, \tau)\, d\tau \tag{7.26}$$

Two approaches, (1) incremental substitution and (2) Taylor series expansion, will produce the same set of perturbation equations. To illustrate the incremental method with an example, suppose that

$$z = f(x, y) = xy \tag{7.27}$$

When the nominal values of x and y are

$$x = x_0(t) \qquad y = y_0(t)$$

the nominal equation is

$$z_0(t) = x_0y_0 \tag{7.28}$$

If we now wish to consider the effect on z of an incremental change in x and y around x_0 and y_0, where

$$x_1 = x_0 + \Delta x \qquad y_1 = y_0 + \Delta y \tag{7.29}$$

then by incremental substitutions

$$z_1 = x_1y_1 = (x_0 + \Delta x)(y_0 + \Delta y) \tag{7.30}$$

$$z_1 \approx x_0y_0 + x_0\,\Delta y + y_0\,\Delta x \qquad \Delta x \to 0,\ \Delta y \to 0 \tag{7.31}$$

The term $\Delta x\,\Delta y$ is omitted from Eq. (7.31) because it is assumed to be small of higher order than the linear terms in Δx and Δy. The presumption is also that $\Delta x \ll x_0$ and $\Delta y \ll y_0$.

A comparison of Eqs. (7.28) and (7.31) shows that z_1 can be expressed as a sum of two terms where

$$z_1 = z_0 + \Delta z \tag{7.32}$$

z_0 is the nominal quantity and

$$\Delta z = x_0\,\Delta y + y_0\,\Delta x \tag{7.33}$$

is the perturbation equation, which is a linear equation in the incremental variables Δx and Δy.

One point worth noting is that the nominal quantities x_0 and y_0 appear in the perturbation equation. Invariably one or more nominal quantities will so appear. In a differential equation the nominal solution may become a forcing function in the perturbation equation, as we show in Example 2.

Example 2. Given

$$\frac{d^2y_0}{dt^2} + a_1\frac{dy_0}{dt} + a_0y_0 = f(t) \tag{7.34}$$

we increase a_0 an incremental amount to $a_0 + \Delta a$. Equation (7.34) now becomes

$$\frac{d^2y_1}{dt^2} + a_1\frac{dy_1}{dt} + (a_0 + \Delta a_0)y_1 = f(t) \tag{7.35}$$

Assuming that the effect of the incremental change on the solution is small, we can write

$$\frac{d^2(y_0 + \Delta y)}{dt^2} + a_1\frac{d}{dt}(y_0 + \Delta y) + (a_0 + \Delta a_0)(y_0 + \Delta y) = f(t) \tag{7.36}$$

Since

$$(a_0 + \Delta a_0)(y_0 + \Delta y) \approx a_0y_0 + a_0\,\Delta y + y_0\,\Delta a_0 \qquad \Delta a_0 \to 0,\ \Delta y \to 0 \tag{7.37}$$

then

$$\begin{aligned}\frac{d^2y_0}{dt^2} + a_1\frac{d}{dt}y_0 + a_0y_0 &= f(t)\\ \frac{d^2\Delta y}{dt^2} + a_1\frac{d(\Delta y)}{dt} + a_0\,\Delta y &= -y_0\,\Delta a_0\end{aligned} \tag{7.38}$$

The computer diagram for Eq. (7.38) is shown in Fig. 7.8.

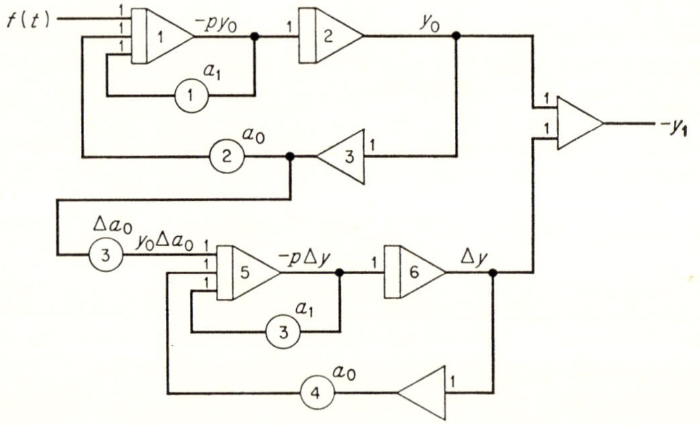

FIG. 7.8 Computer diagram for perturbation of a differential-equation coefficient.

In Table 7.1 are some first-order approximations which can be used in conjunction with the incremental-substitution method. The third column gives a measure of the error.

The question is often posed by the inexperienced about the proper way to make an approximation. "How do I know what the highest-order term should be in the perturbation equation?" is the common plaint. There is no ready answer for this except: "Make the order of the incremental term as low as possible but high enough so that the solution makes sense in the context of the problem."

TABLE 7.1 SOME APPROXIMATE FIRST-ORDER RELATIONSHIPS

	Expression	Approximation $x \ll 1$	Next term in series
1	e^x	$1 + x$	$\frac{x^2}{2}$
2	$\sin x$	x	$-\frac{x^3}{3!}$
3	$\cos x$	1	$-\frac{x^2}{2}$
4	$\tan x$	x	$\frac{x^3}{3}$
5	$\log (1 + x)$	x	$-\frac{x^2}{2}$
6	$(1 + x)^n$	$1 + nx$	$\frac{n(n-1)x^2}{2}$
7	$\frac{1}{1+x}$	$1 - x$	x^2
8	$(1 + x_1)(1 + x_2)$ (x_1 and x_2 are both small)	$1 + x_1 + x_2$	x_1x_2
9	$\sin (x + a)$ (a is a constant)	$x \cos a + \sin a$	$-\frac{x^2}{2} \sin a$
10	$\cos (x + a)$	$\cos a - x \sin a$	$-\frac{x^2}{2} \cos a$
11	$\tan (x + a)$	$\tan a + x \sec^2 a$	$- x^2 \sec^2 a$

Suppose we wish to find the perturbation equation for

$$I = x - \sin x \qquad \text{when } x \to 0 \tag{7.39}$$

From Table 7.1

$$\sin x = x - \frac{x^3}{3!} + \cdots \tag{7.40}$$

Two approximations to Eq. (7.39) are

$$I \approx x - x = 0 \tag{7.41}$$

or

$$I \approx x - \left(x - \frac{x^3}{3!}\right) = \frac{x^3}{3} \tag{7.42}$$

Equation (7.41) states that, when x is very small, I behaves like a constant, viz., zero: $I = 0(0)$. When x is somewhat larger, as Eq. (7.42) shows,

$$I = 0(x^3)$$

One would like to choose Eq. (7.41), but that may not always give the correct answer. For example, suppose I is used in the following manner. The first time I exceeds a constant M, a relay switch closes and changes the value of one of the problem parameters. If Eq. (7.41) is used, the switch will never close and the computer solution will be correct only for the initial interval of time when $I < M$.

In setting up the perturbation equations we should use approximations which yield valid results. For many problems we can rely upon our knowledge of the physical system to provide the answer. If we do not know what these approximations should be, we begin by including only first-order increments. We solve the perturbation equation so obtained on the computer. Next we add second-order terms to each of the perturbed quantities and solve anew on the computer. If the addition of the second-order terms does not affect the solution, we have rather conclusive evidence that the first-order perturbation is sufficiently accurate. It is, of course, quite possible that some of the perturbed quantities need to include higher-order terms and others do not.

As an alternative method for obtaining the perturbation equation, we take a Taylor series expansion of Eq. (7.27) around the nominal values x_0 and y_0:

$$z_0 + \Delta z = f(x_0 + \Delta x, y_0 + \Delta y) \tag{7.43}$$

Neglecting all terms of order Δ^2 or higher,

$$f(x_0 + \Delta x, y_0 + \Delta y) \approx f(x_0, y_0) + \left(\frac{\partial f}{\partial x}\right)_{x_0, y_0} \Delta x + \left(\frac{\partial f}{\partial y}\right)_{x_0, y_0} \Delta y$$

$$\Delta x \to 0, \Delta y \to 0 \tag{7.44}$$

and

$$z_0 = f(x_0, y_0)$$

$$\Delta z = \left(\frac{\partial f}{\partial x}\right)_{x_0, y_0} \Delta x + \left(\frac{\partial f}{\partial y}\right)_{x_0, y_0} \Delta y \tag{7.45}$$

where for $z = xy$

$$\begin{aligned} z_0 &= x_0 y_0 \\ \Delta z &= y_0 \, \Delta x + x_0 \, \Delta y \end{aligned} \tag{7.46}$$

Equations (7.31) and (7.44) are identical. The incremental method and Taylor series expansion will always give the same result for any arbitrary number of perturbed variables.

Example 3.[1] The motion of a mass particle in a viscous medium in the absence of gravity is

[1] This example, including the figures, is taken in its entirety from Ref. 84. In this paper the authors have also included a more detailed example in which the perturbation method is used to analyze a rocket in flight.

$$\frac{dv}{dt} = -\frac{k}{m}v \tag{7.47}$$

where v is the velocity of the particle, m is the mass, and k is the retarding force per unit velocity.

For an initial velocity v_i, the solution is

$$v = v_i e^{-(k/m)t} \tag{7.48}$$

If we were to determine the change in velocity due to a perturbation of mass, say 0.1 per cent, the difference in computer solutions for the two masses would be imperceptible. Instead we set up the perturbation equation, applying the Taylor-series-expansion method.

We rewrite Eq. (7.47):

$$\frac{dv}{dt} = F(v, m) \tag{7.49}$$

Then by Eq. (7.44)

$$\frac{d(v_0 + \Delta v)}{dt} = \left(\frac{\partial F}{\partial v}\right)_{v_0, m_0} \Delta v + \left(\frac{\partial F}{\partial m}\right)_{v_0, m_0} \Delta m + F(v_0, m_0) \tag{7.50}$$

where the nominal values are v_0 and m_0 and

$$\Delta m = 0.001 m_0$$

Evaluating the partial derivatives, we obtain

$$\frac{dv_0}{dt} = -\frac{k}{m_0} v_0 \tag{7.51a}$$

$$\frac{d\Delta v}{dt} = -\frac{k}{m_0}\Delta v + \frac{kv_0}{m_0^2}\Delta m \tag{7.51b}$$

The computer simulation of Eqs. (7.51) is shown in Fig. 7.9. Note that the solution of Eq. (7.51a) is a forcing function on the perturbation equation (7.51b). Although in the figure we combine the nominal and perturbation solutions at the input to amplifier 4 to

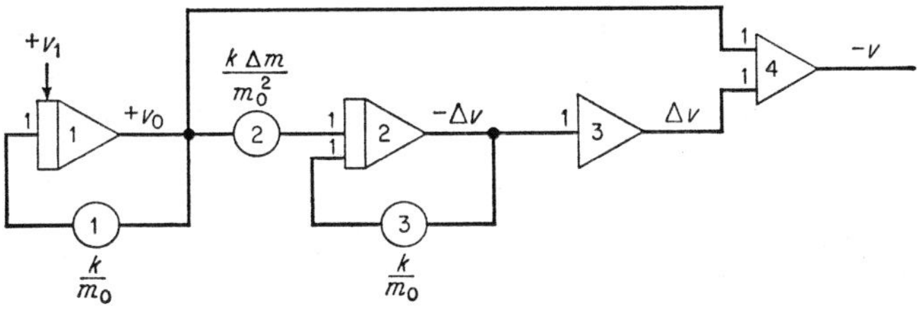

FIG. 7.9 Computer diagram of the perturbation equation set (7.51).

obtain the overall solution, it is only to show the reader the relation between these variables. In practice, this way of getting the solution would net us nothing. The nominal solution v_0 is so much larger than Δv that the latter will have a voltage at the input to amplifier 4 depressed to a magnitude too small for observation. After all, this is why we turned to the perturbation technique in the first place. What we must do is record v_0 and Δv separately so that we can scale Δv to a respectable level. If we want to combine the two to get the overall solution, we cannot do the tabulation on the analog computer.

Incidentally, the solution to Eqs. (7.51) is, by Eq. (7.48),

$$v_0 = v_1 e^{-(k/m_0)t}$$
$$\Delta v = \frac{v_0 k \, \Delta m}{m_0^2} t \tag{7.52}$$

and the overall solution v is

$$v = v_0 + \Delta v = \left(1 + \frac{k \, \Delta m}{m_0} t\right) v_i e^{-(k/m_0)t} \tag{7.53}$$

Figure 7.10 shows the solutions of Eqs. (7.52) in dimensionless plots. Observe that the maximum perturbation is less than 0.04 per cent of the initial velocity. Such an effect could never be noticed on an analog computer solving Eq. (7.47) only.

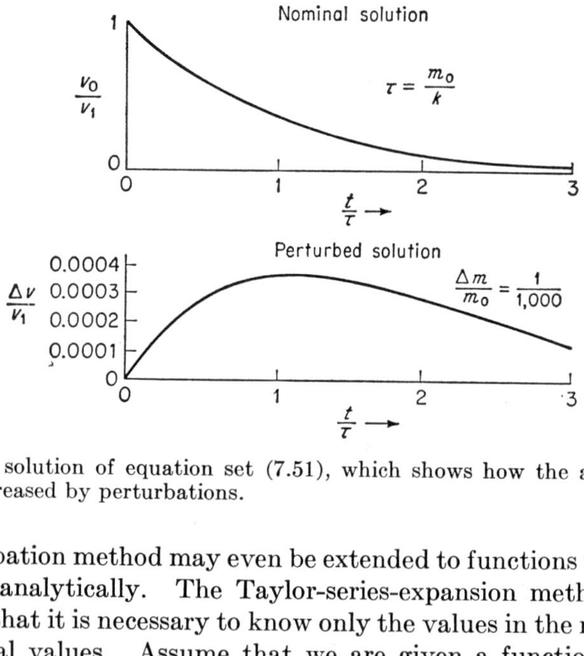

FIG. 7.10 The solution of equation set (7.51), which shows how the accuracy of the computer is increased by perturbations.

The perturbation method may even be extended to functions which cannot be expressed analytically. The Taylor-series-expansion method shows us in Eq. (7.45) that it is necessary to know only the values in the neighborhood of the nominal values. Assume that we are given a function $z = f(x, y)$ in the graphical form of Fig. 7.11*a*. Taking x_2 and y_2 as the nominal values, we can evaluate $(\partial z/\partial x)_{x_2,y_2}$ as the slope of the line tangent to the y_2 curve at x_2. The other partial derivative can be calculated from Fig. 7.11*b*, which is a cross plot of Fig. 7.11*a*.

Example 4. The perturbation technique is useful in problems involving projectiles which are being guided to fly on a prescribed course. The nominal equations then involve this prescribed course, and the perturbation equations arise from errors which cause deviations from this course. For this example we shall consider a missile which intercepts a target by flying a proportional navigation path.

We shall determine a simplified set of equations which will reduce computer errors and thereby enable us to study more accurately the missile flight dynamics. In Example 5 we shall show what can be done to increase the accuracy of terminal miss measurement, i.e., minimum distance between target and missile.

In such a guidance scheme[1] the missile flies a path to reduce the angular rate of the line of sight (the line between the missile and target) to zero. Using Fig. 7.12 to define our nomenclature, we obtain in two dimensions the important equations relating missile and target motion:

$$\begin{aligned} R\dot{\sigma} &= V_m \sin\gamma - V_T \sin\beta \\ \dot{R} &= -V_m \cos\gamma - V_T \cos\beta \end{aligned} \tag{7.54}$$

where $\dot{\sigma}$ is the line-of-sight rate, measured from an inertial reference.

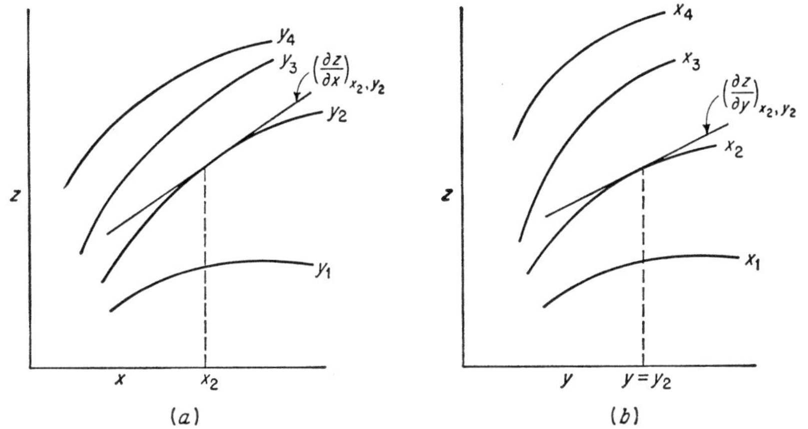

FIG. 7.11 A method for perturbing functions which are known only graphically.

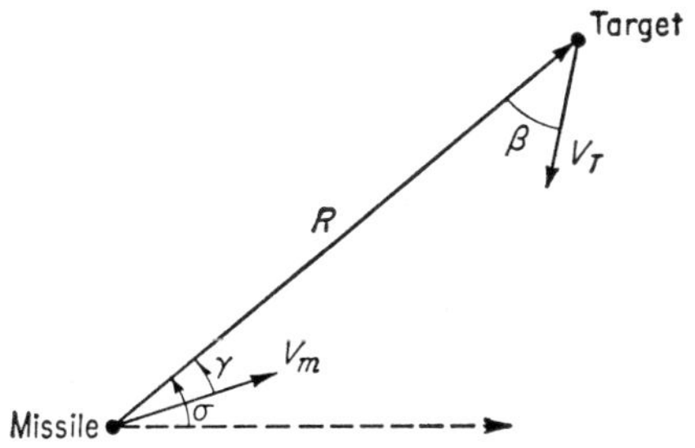

FIG. 7.12 Coordinate system for a homing missile.

Equations (7.54) show that $\dot{\sigma} = 0$ only when the target and missile velocities perpendicular to the line of sight R are equal. If we assume that our missile behaves as it should, the line of sight will not rotate much except during the terminal portion of flight. We can therefore take perturbations around the original line-of-sight angle as well as around the nominal values of the target and missile velocity magnitudes and angles.

We take as our nominal values the initial velocities V_{m_0} and V_{T_0} and the initial angle β_0. We shall assume that the target flies on a straight-line path and that β is perturbed about R, whose nominal orientation angle σ_0 is assumed to be fixed; thus

$$\beta = \beta_0$$

[1] This and other types of homing missiles are discussed in Ref. 85, pp. 445–475, in some detail.

Next, we take

$$\gamma = \gamma_0 + \Delta\gamma \qquad \sigma = \sigma_0 + \Delta\sigma$$

where

$$R\dot{\sigma}_0 = V_{m_0} \sin \gamma_0 - V_{T_0} \sin \beta = 0 \tag{7.55}$$

and we take in addition

$$\Delta\gamma \ll 1$$

This is based on the assumption that V_m is much larger than $\Delta V_T \sin \beta$, the incremental target velocity perpendicular to the line of sight.

Now when we perturb Eqs. (7.54),

$$R\,\Delta\dot{\sigma} = (V_{m_0} + \Delta V_m) \sin (\gamma_0 + \Delta\gamma) - (V_{T_0} + \Delta V_T) \sin \beta \tag{7.56a}$$

$$\dot{R} = -(V_{m_0} + \Delta V_m) \cos (\gamma_0 + \Delta\gamma) - (V_{T_0} + \Delta V_T) \cos \beta \tag{7.56b}$$

Expanding Eqs. (7.54) and making the small-angle approximations indicated in Table 7.1,

$$R\,\Delta\dot{\sigma} \approx V_{m_0} \sin \gamma_0 + \Delta\gamma\, V_{m_0} \cos \gamma_0 + \Delta V_m \sin \gamma_0 - V_{T_0} \sin \beta - \Delta V_T \sin \beta \tag{7.57}$$

Using Eq. (7.55),

$$R\,\Delta\dot{\sigma} = \Delta\gamma\, V_m \cos \gamma_0 + \Delta V_m \sin \gamma_0 - \Delta V_T \sin \beta \tag{7.58}$$

In a similar fashion we can simplify Eq. (7.56*b*), where

$$\dot{R} = -V_m \cos \gamma_0 - V_m\, \Delta\gamma \sin \gamma_0 - V_T \cos \beta \tag{7.59}$$

If the angle γ_0 is small, that is, $V_m \gg V_T$, we may carry the simplification even further and obtain as our final result

$$R\,\Delta\dot{\sigma} = \Delta\gamma\, V_m \cos \gamma_0 - \Delta V_T \sin \beta \tag{7.60a}$$

$$\dot{R} = -V_m \cos \gamma_0 - V_T \cos \beta \tag{7.60b}$$

If we can imagine a line passing through the target and moving in space parallel to the original line of sight, Eq. (7.60*a*) represents the relative velocity of the missile perpendicular to this line. Thus, we define the term V_p as

$$V_p = R\,\Delta\dot{\sigma} \tag{7.61}$$

and a perpendicular distance y_p between the line and the missile position as

$$y_p = \int_0^t V_p\, dt \tag{7.62}$$

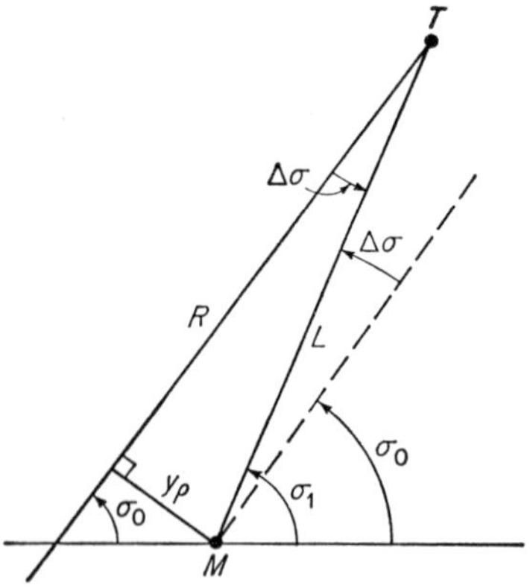

FIG. 7.13 Coordinate system for a homing missile when the angle σ is perturbed.

This now allows us to make the perturbation on the angle $\Delta\sigma$ which we show in Fig. 7.13. Note that $\Delta\dot{\sigma}$ and $\Delta\sigma$ are used solely to obtain interceptor acceleration commands. They do not change the direction of R. We have drawn the relative positions of the missile and the target at some time t after the start of flight. The line L is the present line of sight, and its magnitude from target to missile is

$$L = (R^2 + y_p^2)^{\frac{1}{2}} \tag{7.63}$$

From the diagram we see that the incremental change in the line of sight due to $\Delta\sigma$ is

$$y_p = R \tan \Delta\sigma \tag{7.64}$$

However, because $\Delta\sigma \ll 1$

$$y_p \approx R\,\Delta\sigma \tag{7.65}$$

We have now completed the perturbation of the geometric relationships.

Before constructing the computer block diagram we need an equation specifying how the missile responds to line-of-sight changes. This equation is

$$\Delta\dot{\gamma} = KG(p)\dot{\sigma}_m \tag{7.66}$$

where:

1. $\Delta\dot{\gamma}$ is the turning rate of the missile. It is obtained by lifting surfaces. The missile can also maneuver perpendicular to the line of sight by the use of propulsion. In this case the command would be on V_m and $\Delta\gamma$ would remain fixed at its initial value.
2. $\dot{\sigma}_m$ is the measured line-of-sight rate. It is the output of the missile tracking loop and is somewhat different from the actual line-of-sight rate $\dot{\sigma}$.
3. $KG(p)$ is an operator describing missile behavior, excluding the tracking loop.

The resulting computer diagram is Fig. 7.14. We observe that, if the tracking loop and $KG(p)$ are linear operators, the entire computer simulation is linear.

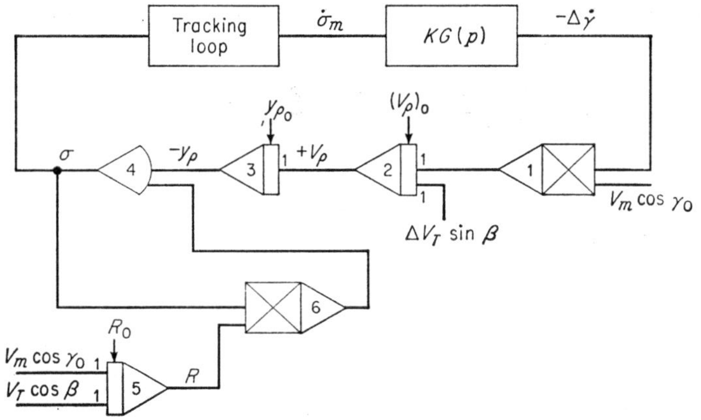

FIG. 7.14 Computer diagram for the perturbed homing-missile equations.

Additional modifications can be made to improve the accuracy of the computer and will be discussed in the sections covering readout and division. These modifications are illustrated in Example 5.

7.4 Parameter-influence Coefficients[1]

Equation (7.44) shows that in the perturbation equation there are partial derivatives of the function with respect to the variables being perturbed. Furthermore, these partial derivatives are independent of the incremental changes of the variables. As we examine this equation, we see that with these partial derivatives we can construct the perturbation solution for any incremental changes which result in linear perturbation equations.

These partial derivatives are called either parameter-influence coefficients or sensitivity coefficients. Whereas a perturbation equation gives the solution for a specific incremental change of each variable, a set of parameter influences can be used to calculate the solution in the entire linear perturba-

[1] We have more to say about this subject in Sec. 8.5.

tion region of these variables. These parameter influences can be generated by the computer, as we show by the following example.[1]

Consider the inhomogeneous linear differential equation

$$\frac{d^2x}{dt^2} + \eta \frac{dx}{dt} + \lambda x = f(t) \tag{7.67}$$

where (1) η and λ can be either constant or time-varying and (2) the initial conditions are

$$x(0) = a \qquad \left(\frac{dx}{dt}\right)_0 = b$$

We first obtain the parameter-influence coefficient which results from an incremental change in λ. To this end we differentiate each term in Eq. (7.67) with respect to λ.† Hence

$$\frac{\partial}{\partial\lambda}\left(\frac{d^2x}{dt^2}\right) + \frac{\partial}{\partial\lambda}\left(\eta\frac{dx}{dt}\right) + \frac{\partial}{\partial\lambda}(\lambda x) = \frac{\partial}{\partial\lambda}[f(t)]$$

or

$$\frac{\partial^3 x}{\partial\lambda\,\partial^2 t} + \eta\frac{\partial^2 x}{\partial\lambda\,\partial t} + x + \lambda\frac{\partial x}{\partial\lambda} = 0 \tag{7.68}$$

Interchanging the order of differentiation and adopting the notation $u = \partial x/\partial\lambda$,

$$\frac{\partial^2 u}{\partial t^2} + \frac{\partial u}{\partial t} + \lambda u = -x \tag{7.69}$$

Note that the initial conditions of Eqs. (7.68) are

$$u(0) = 0 \qquad \left(\frac{du}{dt}\right)_0 = 0$$

because the initial conditions of x do not depend upon λ.

Similarly, the parameter-influence coefficient of η can be obtained from the equations

$$\frac{\partial^2 v}{\partial t^2} + \eta\frac{\partial v}{\partial t} + \lambda v = -\frac{\partial x}{\partial t} \tag{7.70}$$

where

$$v = \frac{\partial x}{\partial \eta}$$

with initial conditions

$$v(0) = 0 \qquad \left(\frac{dv}{dt}\right)_0 = 0$$

[1] This example and the related figures were taken from Ref. 86.

† In Sec. 2.2.5 we indicated that

$$\frac{\partial x}{\partial\lambda}, \ldots, \frac{\partial}{\partial\lambda}\left(\frac{d^{n-1}x}{dt^{n-1}}\right)$$

are continuous.

The solutions to Eqs. (7.69) and (7.70) will yield the parameter-influence coefficients. The complete solution $x(t, \lambda, \eta)$ is

$$x(t, \lambda, \eta) = x(t, \lambda_0, \eta_0) + v\,\Delta\eta + u\,\Delta\lambda \tag{7.71}$$

where (1) $x(t, \lambda_0, \eta_0)$ is the solution to the nominal equations (2) $\lambda = \lambda_0 + \Delta\lambda$ and $\eta = \eta_0 + \Delta\eta$.

If the nominal equation is called the master equation and the equations

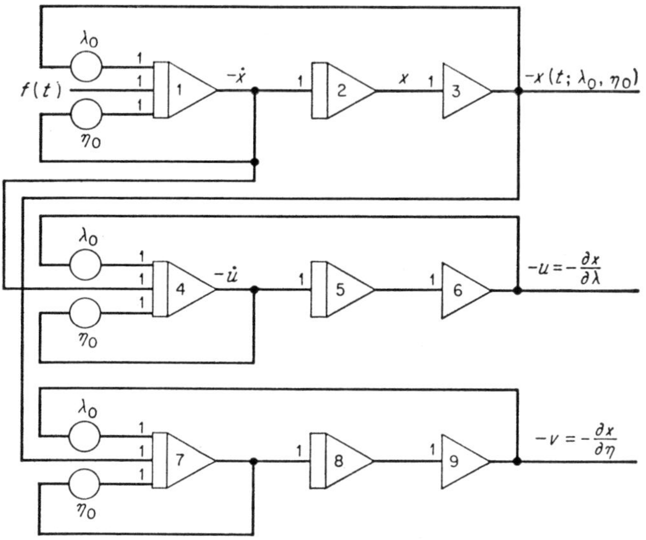

FIG. 7.15 A circuit for generating parameter-influence coefficients.

producing the parameter-influence coefficients are called the slave equations, then we can make the following statements about their relationship:

1. Each slave equation has a forcing function which is a term in the nominal equation.

2. The homogeneous form of the master and slave equations is the same only when linear equations are perturbed.

Figure 7.15 is the computer diagram for the simultaneous generation of the parameter-influence coefficients. Each coefficient requires an amount of equipment roughly equal to that used in the nominal solution. Where several coefficients must be determined simultaneously, the amount of equipment required can exceed the computer capacity. In such a case we can generate the parameter-influence coefficients in several computer runs.

The linear example we have considered is particularly easy to run sequentially. All we need is the nominal solution and one slave circuit. Then each computer run will produce one coefficient. The slave circuit remains the same with only the forcing function changing with every run.

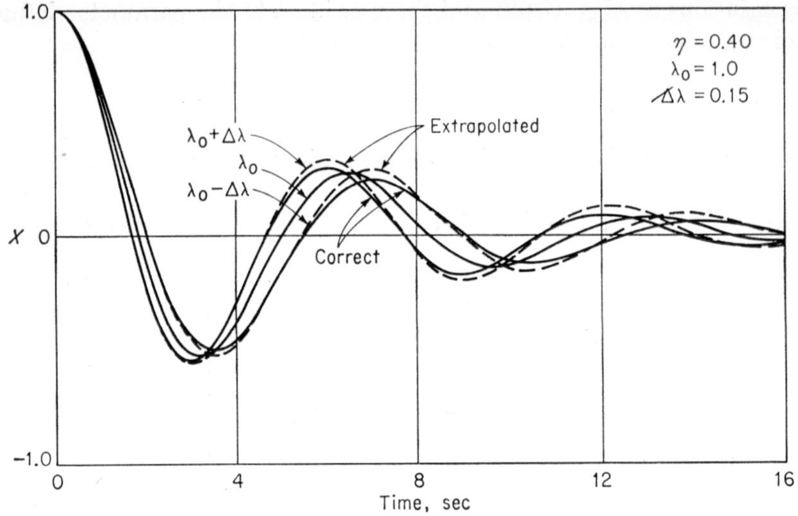

FIG. 7.16 Comparison of correct and extrapolated solutions using parameter-influence coefficients in Eq. (7.67).

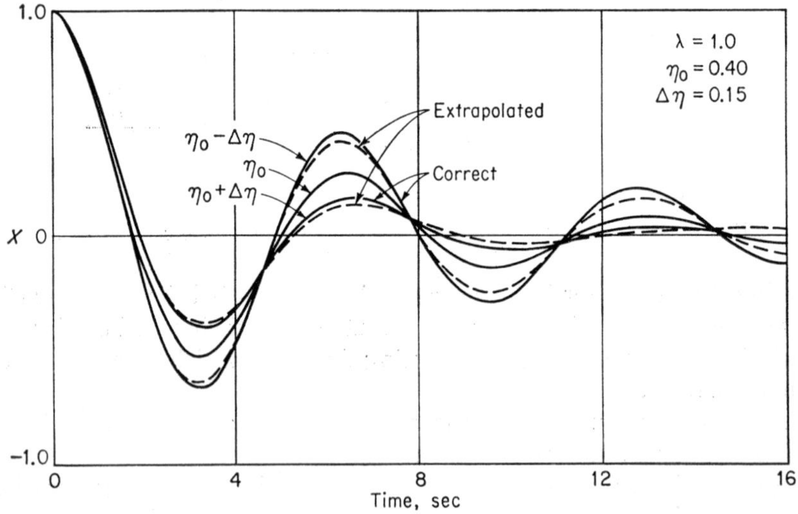

FIG. 7.17 Comparison of correct and extrapolated solutions using parameter-influence coefficients in Eq. (7.67).

The results of the parameter-influence method are shown in Figs. 7.16 and 7.17. Here extrapolated and exact solutions are compared for the perturbation of one parameter. In these examples the error increases with the independent variable. This point should forewarn the reader that the magnitude of the parameter increment resulting in a linear perturbation equation is in part determined by the range of the independent variable.

However, within the framework of this limitation the parameter-influence method has several important applications. We have already seen that in conjunction with the perturbation technique it can be used to improve computer accuracy. In Sec. 8.5 we shall show how parameter-influence coefficients can assist materially in determining the maximum and minimum of functions.

Not to be overlooked is the fact that these coefficients yield important information about the behavior of the solution close to the nominal values of the parameters. Good control systems are designed to make these coefficients as small as possible. Since physical components always deviate to some degree from their nominal value, we want to minimize this effect.

Finally, these coefficients are important from the standpoint of computer accuracy. If any parameter has a large sensitivity coefficient, then small changes in the value of this parameter due to computer inaccuracies will result in large solution errors. As a matter of course, the operator at the start of a problem should vary each parameter slightly, say 5 per cent. Should any large sensitivity coefficients appear, the operator should then attempt to reprogram the problem to minimize their effect, e.g., by use of more accurate components.

7.4.1 Effect of Running the Problem "Forward or Backward"[1]

If a problem run in one direction of the independent variable results in a stable solution, the problem running in the other direction will lead to an unstable solution. Sensitivity coefficients for initial conditions associated with unstable solutions tend to be larger than their counterparts associated with a stable solution. Errors in the setting and generation of these parameters may be regarded as perturbations from the correct (nominal) values. If we assume that these errors are more or less independent of the magnitude of the parameters, then they will be essentially the same no matter how the problem is run. However, because it has the larger sensitivity coefficients, the unstable solution will have larger *solution* errors. Therefore, when one has the option, one should program the stable solution.

7.5 Readout Techniques

Often in a problem the value of a variable at a particular value of the independent variable is of prime interest. Trajectory problems associated with guided missiles fall in this category. More than anything else we want to know how close the vehicle comes to its chosen destination, i.e., where it is at a particular time. Another example is a weighting function generated directly from the differential equation. We get one discrete value of the weighting function by starting the computer run at one time and measuring a voltage at another.

[1] In Sec. 4.6 this topic is discussed in some detail.

There are two principal sources of error in the measurement of an instantaneous voltage. Either it may distort or have too much noise added by becoming too large or too small, or the voltage may be changing so rapidly that a small error in the measurement of the independent variable can result in a large error in the measurement of the voltage.

Both of these errors can be handled by variable scaling. However, there are alternative methods which may be more accurate and which can be applied at the readout point, leaving the rest of the computer program unaffected.

7.5.1 Extrapolation Technique

Consider first the situation when the voltage becomes either excessively large or too small in the region of measurement. We handle this situation by an extrapolation technique.

When a function is analytic,[1] it may be uniquely described in the region of analyticity by a Taylor series expansion. For such an expansion we need only know the function and its derivatives at a single point. In other words, for analytic functions we can record the function and its derivatives at x_1 and by the Taylor series expansion extrapolate the value to x_2. We do not evaluate functions in this manner because infinite series are laborious to evaluate unless they converge rapidly. It is always easier to solve the problem directly on the computer; besides the function may not always be analytic.

However, if the interval between x_1 and x_2 is small and the function is well-behaved, the extrapolation may be approximated by a few terms. The simplest approximation is the linear extrapolation

$$f(x_2) \approx f(x_1) + \left(\frac{df}{dx}\right)_{x=x_1} (x_2 - x_1) \tag{7.72}$$

For more accuracy we may add another term:

$$f(x_2) \approx f(x_1) + \left(\frac{df}{dx}\right)_{x=x_1} (x_2 - x_1) + \frac{1}{2}\left(\frac{d^2f}{dx^2}\right)_{x=x_1} (x_2 - x_1)^2 \tag{7.73}$$

An extrapolation which requires more terms than Eq. (7.73) is too unwieldy to be practical. We should then find a new value of the independent variable x_3 (where $x_1 < x_3 < x_2$) for which Eq. (7.73) holds. Our extrapolation would now be from x_3 to x_2.

We may either record the function and its derivatives and calculate the value at the extrapolated point or program the computer to calculate its own extrapolation to the solution.

[1] A function is analytic in a region of the independent variable if the function and all its derivatives are everywhere continuous.

By dropping the higher-order terms we introduce some error in the extrapolation. However, in Eq. (7.73) if d^2f/dx^2 is bounded, we may use this characteristic to bound the error.

For instance, suppose we have the following values for $f(x_1)$ and its derivatives:

$$f(x_1) = a \qquad \left(\frac{df}{dx}\right)_{x=x_1} = b \qquad \left(\frac{d^2f}{dx^2}\right)_{x=x_1} = c$$

Substituting these values in Eq. (7.73), we get $f_1(x_2)$. The bounds on d^2f/dx^2 are

$$\alpha \leq \frac{d^2f}{dx^2} \leq \beta \tag{7.74}$$

Using the lower bound α for d^2f/dx^2 while keeping $f(x_1)$ and $(df/dx)_{x=x_1}$ the same as before, we get $f_\alpha(x_2)$ as the solution of Eq. (7.73). For the upper bound β, the solution is $f_\beta(x_2)$. If the correct solution at x_2 is $f(x_2)$, both it and $f_1(x_2)$ must fall between the two bounded solutions.

Therefore,

$$f_\alpha(x_2) \leq [f(x_2), f_1(x_2)] \leq f_\beta(x_2) \tag{7.75}$$

From Eq. (7.75) we may bound the error ϵ in the approximation as the maximum difference:

$$\epsilon = |f_1(x_2) - f(x_2)| \leq \max\,[|f_1(x_2) - f_\alpha(x_2)|, |f_1(x_2) - f_\beta(x_2)|] \tag{7.76}$$

7.5.2 Slope-reduction Technique

One way of reading and recording a voltage at a specific time is by using the sample-and-hold circuit of Fig. 5.20. We know that there is some error because the output and input voltages of the integrator are not the same. If they differ markedly, we can reduce the error by time scaling to slow down the problem. Here we shall be concerned with the error arising because the relay opens at the wrong time.

For the sake of argument, suppose our relay should open when its input x is zero. Because of electrical inductance and inertia in the mechanical parts the relay opening will be delayed for some small time. Also the input voltage x may be corrupted by noise which can cause x to become zero before or after its proper time. The combined effect of the relay lag and the input noise can make the relay open either too soon or too late. Unless the noise is large, the likelihood is strong that the opening will be too late.

We may assume that this time error will be small, in which event a Taylor series expansion at the correct opening time t_1 becomes

$$f_1(t_2) = V_1(t_1) + k_1(t_2 - t_1) \tag{7.77}$$

where V_1 = desired value at t_1
k = slope of $f_1(t)$ at t_1
t_2 = time at which relay opens

The error in measurement is

$$k_1(t_2 - t_1)$$

and is dependent upon two quantities: k_1, the slope of $f_1(t)$ at t_1, and $t_2 - t_1$, the error in relay opening time. There is not much we can do to improve the relay characteristics, but we can change the slope of the function passing through the relay contacts.

Suppose we generate a function which in the vicinity of t_1 is

$$f_2(t_2) = V_2(t_1) + k_2(t_2 - t_1) \tag{7.78}$$

where $V_2(t_1)$ is known exactly. If the difference voltage $f_1(t_1) - f_2(t_2)$ is inserted at the input to the sample-and-hold circuit, then

$$f_3(t_2) = f_1(t_2) - f_2(t_2) = V_1(t_1) - V_2(t_1) + (k_1 - k_2)(t_2 - t_1) \tag{7.79}$$

After this operation we add the quantity $V_2(t_1)$, so that

$$f_4(t_2) = f_3(t_2) + V_2(t_1) = V_1(t) + (k_1 - k_2)(t_2 - t_1) \tag{7.80}$$

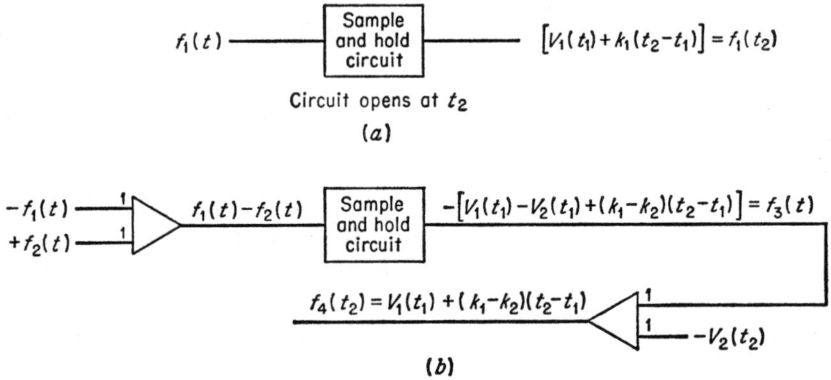

FIG. 7.18 (*a*) A sample-and-hold circuit to record a voltage at time t_2; (*b*) a modified readout circuit which is more accurate than (*a*).

In Fig. 7.18*a* and *b* we show the different computer circuits for programming Eqs. (7.77) and (7.80). The readout accuracy of Fig. 7.18*b* is better than that of Fig. 7.18*a* when

$$|k_2 - k_1| < k_1 \tag{7.81}$$

The condition of Eq. (7.81) may be restated as

$$\begin{aligned} 0 < k_2 < 2k_1 \qquad & \text{for } k_1 \geq 0 \\ 2k_1 < k_2 < 0 \qquad & \text{for } k_1 \leq 0 \end{aligned} \tag{7.82}$$

In the usual circumstance k_1 is not known before the problem is solved by the computer. One way to program the circuit of Fig. 7.18*b* is to run the problem several times and adjust k_2 so that the slope $k_2 - k_1$ is a minimum in the region of the readout time. Another procedure is to let the computer generate its own $f_2(t)$ function. This has special appeal in noise problems where k_1 will change each run. Example 5 illustrates how the computer may be programmed to choose its own $f_2(t)$.

7.6 Improvement of Equipment Accuracy

As we have stressed several times, each computer element has inherent errors apart from those which can be corrected by scaling. These errors can be reduced either by a change of the equations or by compensation of the computer equipment. In the latter case we shall consider only those compensations which can be made on the program board and not those which require modification of the equipment itself.

In any computer problem we may safely assume that the errors contributed by the linear elements (potentiometers, amplifiers, integrators) are an order of

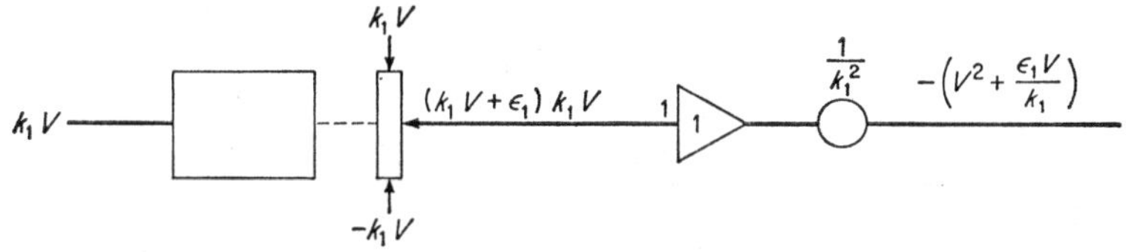

FIG. 7.19 A servomultiplier circuit for squaring a variable.

magnitude better (a factor of 10 *or more) than the nonlinear elements.* We have taken advantage of this characteristic already in the perturbation technique, where the accuracy was improved both by linearization and by scaling. Now we shall show that the accuracy of a nonlinear operation may be improved by a conversion to an equivalent combination of linear and nonlinear operations. We term this process *decomposition* and shall illustrate it by the example of squaring.

Suppose we wish to generate V^2, having V available. Then, as we indicated in Sec. 7.3, the error on the variable V positioning the motor is more or less independent of the value of V. In the subsequent derivation we shall show that, because V goes both to the input of the servomotor and to the multiplying potentiometer, amplitude scaling will improve the accuracy. However, this improvement will not be so great as when the scaling is used in conjunction with decomposition.

Figure 7.19 shows the circuit for generating V^2 directly on a servomultiplier. We note

$$V^2 = (V_0 + \Delta V)^2 = V_0 + 2\,\Delta V\,V_0 + (\Delta V)^2 \tag{7.83}$$

where V_0 is a constant. Unlike perturbation techniques we retain all the terms of Eq. (7.83). We observe that the terms V_0^2 and $2V_0\,\Delta V$ can be produced using a potentiometer and are essentially without error. The only error comes from the term $(\Delta V)^2$. The computer program for Eq. (7.83) is shown in Fig. 7.20.

In Fig. 7.19 the error ϵ_2 at the output of amplifier 1 is

$$\epsilon_2 = \frac{\epsilon_1 V}{k_1} \tag{7.84}$$

and in Fig. 7.20 the error ϵ_3 at the output of amplifier 2 is

$$\epsilon_3 = \frac{\epsilon_1\,\Delta V}{k_2} \tag{7.85}$$

Thus

$$\frac{\epsilon_3}{\epsilon_2} = \frac{k_1\,\Delta V}{k_2 V} \tag{7.86}$$

Even when the scaling in both circuits is the same ($k_1 = k_2$), we see from Eq. (7.86) that there is a reduction of error when we program Eq. (7.83)

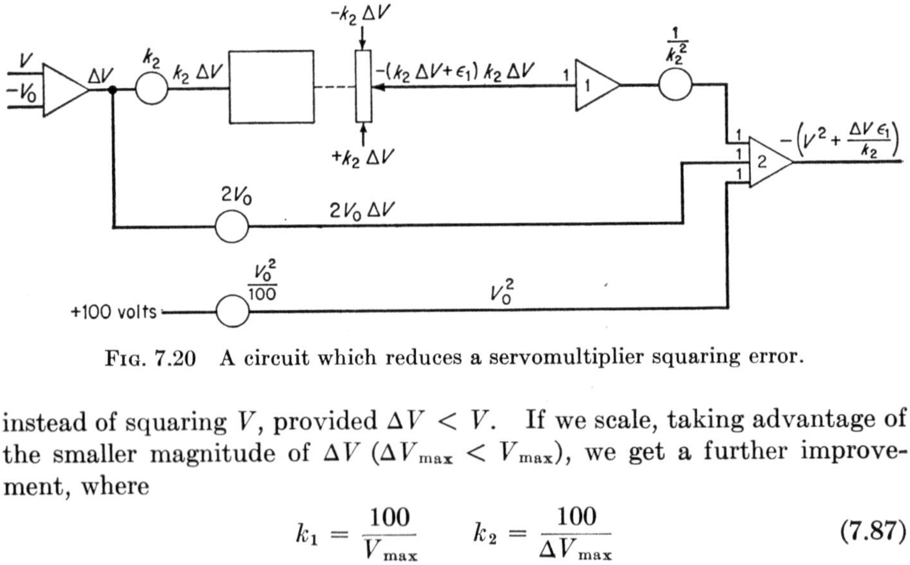

FIG. 7.20 A circuit which reduces a servomultiplier squaring error.

instead of squaring V, provided $\Delta V < V$. If we scale, taking advantage of the smaller magnitude of ΔV ($\Delta V_{\max} < V_{\max}$), we get a further improvement, where

$$k_1 = \frac{100}{V_{\max}} \qquad k_2 = \frac{100}{\Delta V_{\max}} \tag{7.87}$$

and, consequently,

$$\frac{\epsilon_3}{\epsilon_2} = \frac{\Delta V_{\max}}{V_{\max}}\,\frac{\Delta V}{V} \tag{7.88}$$

7.6.1 *Servomultiplier*

Because the servomultiplier is an electromechanical device it has dynamic lags. One way to overcome this is by time-scaling the problem so that it runs slower. However, we would like to run the problem as fast as possible.

We can do this by a simple compensation scheme shown in Fig. 7.21. To take the product xy we normally use only servo 1. However, because of dynamic lags the position of the arm of the follow-up potentiometer (and all other ganged potentiometers) will have a dynamic error ϵ_1. This error is small but may be significant in a sensitive problem. There will also be a static error between the follow-up potentiometer and the other ganged potentiometers, which we shall neglect in this discussion.

The difference $x - x_f = \epsilon_1$ (where x_f is the voltage on the arm of the follow-up potentiometer) is fed to one of the potentiometers of servo 2 while y is fed to the input. The error ϵ_2 is different from ϵ_1 but is of the same order.

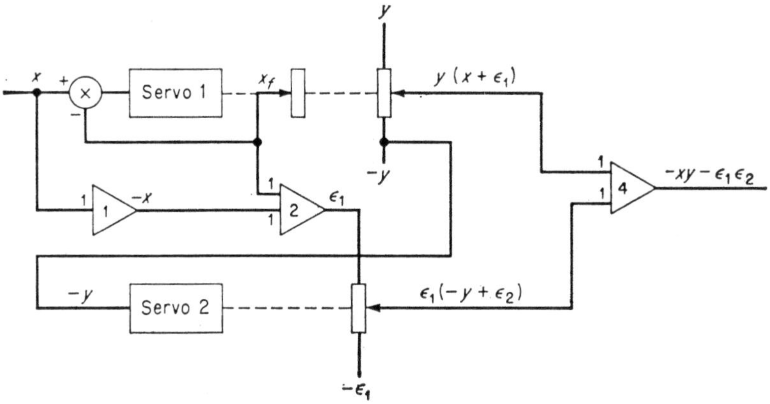

FIG. 7.21 A circuit which improves the dynamic accuracy of a servomultiplier.

When the terms are combined, the output of amplifier 4 is $-xy - \epsilon_1\epsilon_2$, and in general

$$\epsilon_1\epsilon_2 \ll y\epsilon_1 \tag{7.89}$$

Although this technique uses two servos instead of one, it may produce a dramatic improvement in response. In Fig. 7.22 we have shown this by taking x to be a step function and y to be a constant. This is the most severe case for x. In having a constant value of y as an input, servo 2 will have only a static positioning error. A variable y would cause dynamic lags in the servo, thereby making ϵ_2 somewhat larger than it is in this experiment. Our principal justification for using a constant y is that it makes easier a comparison of errors in the compensated and uncompensated multipliers.

Resolver. There are two types of error arising in the resolver whose partial cure we shall describe here. These are (1) static and dynamic errors in the generation of sines and cosines and (2) noise in the circuit resulting from small R in the transformation from rectangular to polar coordinates.

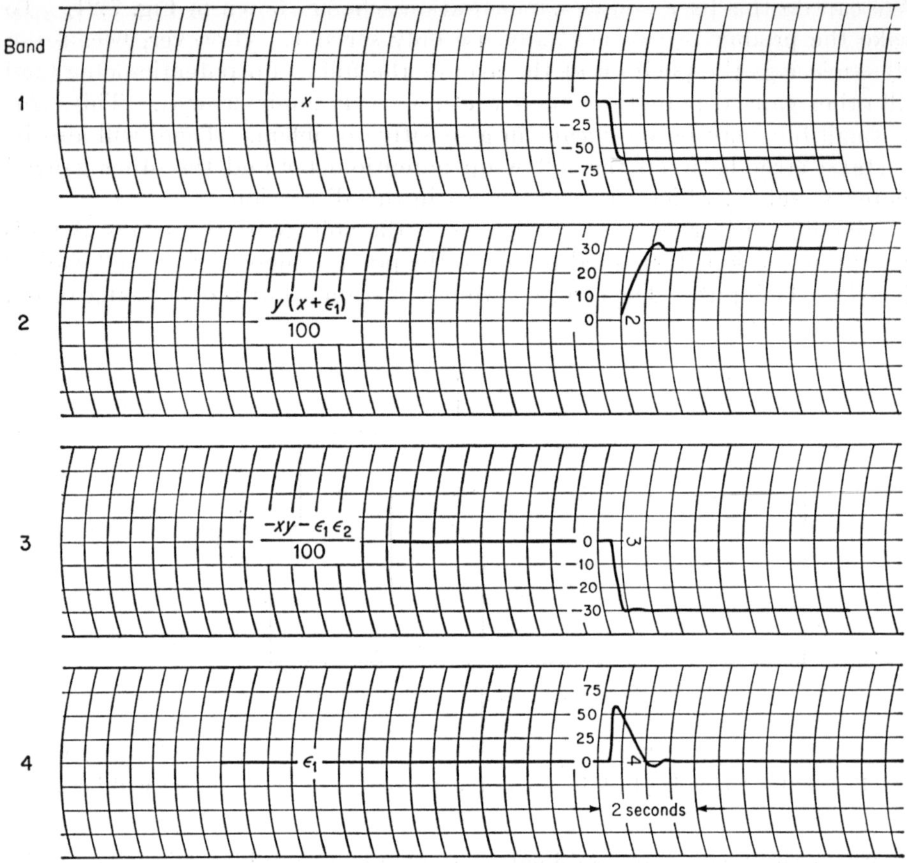

FIG. 7.22 A typical result of using the circuit in Fig. 7.21. The x input is a step input shown in band 1, and $y = -50$ volts. The output should be $xy/100 = 30$ volts. The uncompensated output is shown in band 2. The compensated output closely approximates a step and is shown in band 3.

One compensation scheme follows the principle described for the servomultiplier and is shown in Fig. 7.23. In this figure we have represented the resolver in an unconventional manner to clarify how the term ϵ_1 arises. Without compensation, resolver 1 alone would generate the terms

$$y \sin (x + \epsilon_1) \qquad \text{and} \qquad y \cos (x + \epsilon_1)$$

where ϵ_1 is the error in resolver angle. Assuming ϵ_1 is small, we make the following trigonometric expansions:

$$\begin{aligned} y \sin (x + \epsilon_1) &\approx y \sin x + y\epsilon_1 \cos x \\ y \cos (x + \epsilon_1) &\approx y \cos x - y\epsilon_1 \sin x \end{aligned} \tag{7.90}$$

We add a servomultiplier, whose error is ϵ_2, and by using the circuit shown in Fig. 7.23 we get $f_1(t)$ and $f_2(t)$ at the outputs of amplifiers 4 and 6 respectively, where

$$\begin{aligned} f_1(t) &= (y + \epsilon_2)\epsilon_1 \cos (x + \epsilon_1) - y \sin (x + \epsilon_1) \\ f_2(t) &= -(y + \epsilon_2)\epsilon_1 \sin (x + \epsilon_1) - y \cos (x + \epsilon_1) \end{aligned} \tag{7.91}$$

Making the same small-angle approximations as in Eqs. (7.90), we get

$$\begin{aligned} f_1(t) &= -y \sin x - \epsilon_1^2(y + \epsilon_2) \sin x + \epsilon_1\epsilon_2 \cos x \\ f_2(t) &= -y \cos x - \epsilon_1\epsilon_2 \sin x - (y + \epsilon_2)\epsilon_1^2 \cos x \end{aligned} \tag{7.92}$$

If we compare the compensated and uncompensated resolver, the scales

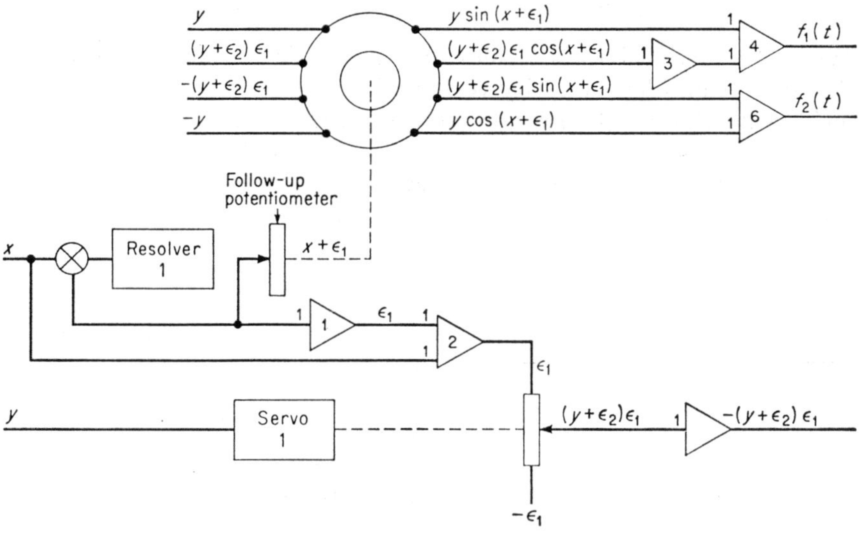

FIG. 7.23 A circuit which improves the dynamic accuracy of a resolver.

are tipped in favor of the former if for the generation of the sine function

1. $y \gg \epsilon_2$
2. $y \cos x \gg \epsilon_1 \sin x$

and for that of the cosine function

1. $y \gg \epsilon_2$
2. $y \sin x \gg \epsilon_1 \cos x$

7.6.2 *Noise in Rectangular-to-Polar Coordinate Transformations*

In transformations from rectangular to polar coordinates we encounter difficulty when the rectangular components are small (see Sec. 6.6). Under these circumstances there can be an excessive amount of noise in the loop generating the angle. This noise occurs because of the wiping action of the arm contacts on the resolver potentiometer. There are two ways in which this noise may be reduced. In the first case one adds an integrator to the implicit-function loop. Integration of the noise sharply reduces its effect. This circuit is shown in Fig. 7.24.

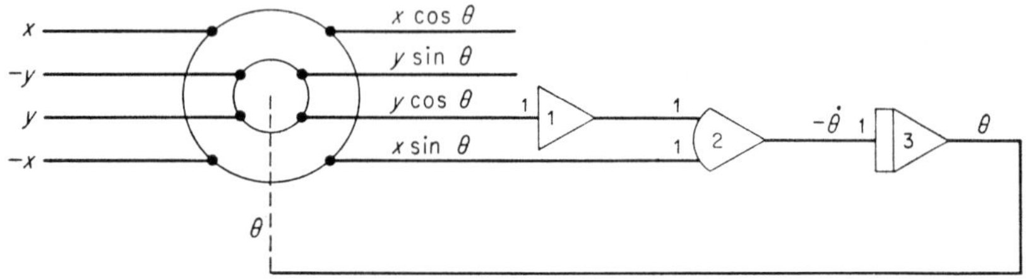

Fig. 7.24 A circuit which reduces the effect of noise in polar resolution.

Where the derivatives of x and y are available, we may use a different technique. Consider a velocity vector $\dot{R}$ in a plane. Then its rates $\dot{x}$ and $\dot{y}$ coincide with the x and y axes, as shown in Fig. 7.25.

The quantities we wish to compute are

$$\begin{aligned} \sigma &= \tan^{-1}\frac{y}{x} \\ R &= (x^2 + y^2)^{\frac{1}{2}} \end{aligned} \tag{7.93}$$

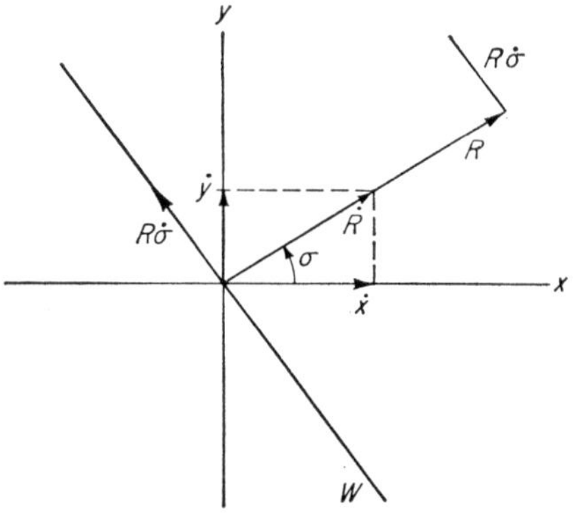

Fig. 7.25 A two-dimensional coordinate system.

We formulate the equations in terms of a coordinate system with one axis along R and the other perpendicular to R, which we designate W. Then the velocities along the R and W axes respectively will be $\dot{R}$ and $R\dot{\sigma}$, where

$$\begin{aligned} \dot{R} &= \dot{x}\cos\sigma + \dot{y}\sin\sigma \\ R\dot{\sigma} &= -\dot{x}\sin\sigma + \dot{y}\cos\sigma \end{aligned} \tag{7.94}$$

The use of these equations to solve for R and σ is shown in Fig. 7.26. Although this circuit requires a division by R, it shows a marked improvement in stability and noise behavior over the standard circuit for rectangular-to-polar transformation. The improvement results because

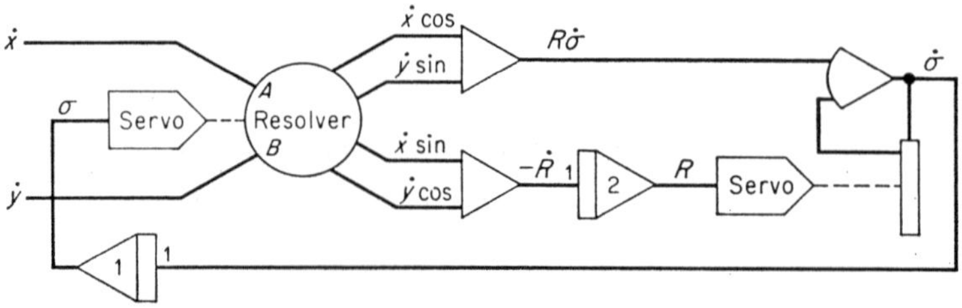

Fig. 7.26 A polar resolution based on the coordinate system of Fig. 7.25, which reduces noise.

resolver-potentiometer wiper noise is integrated by integrator 1 and servo-potentiometer wiper noise is integrated by integrator 2. It therefore permits operation for smaller values of R.

7.6.3 *Division*

We know that, whenever we program the computer to divide by zero, we are assigning to it an impossible task. Even a mathematically well-behaved function, like $(\sin x)/x$, is beyond the computer's scope in the neighborhood of $x = 0$. Should the singularity in the division occur at the beginning of the problem, we can get the computer past this hurdle.

If we must perform the operation

$$z = \frac{y(t)}{x(t)} \tag{7.95}$$

where $x \to 0$ for $t \to 0$, we may add an error term to x so that

$$z \approx \frac{y}{x + e^{-at}} \tag{7.96}$$

If a is large, the decay of the error term e^{-at} will be rapid and the error in z will be small.

Bessel's equation,

$$\frac{d^2y}{dt^2} + \frac{1}{t}\frac{dy}{dt} + \left(1 - \frac{m^2}{t^2}\right)y = 0 \tag{7.97}$$

has a singularity at $t = 0$.

We can avoid the singularity by starting the computer solution at some other value of t. If the computer begins the solution at $t = 0$, it must be programmed to solve the approximation to Bessel's equation,

$$\frac{d^2y}{dt^2} + \frac{1}{t + e^{-at}}\frac{dy}{dt} + \left[1 - \frac{m^2}{(t + e^{-at})^2}\right]y = 0 \tag{7.98}$$

Even when the singularity occurs at other than the start of the problem we may add an error term to eliminate the singularity. We break the problem in segments so that it always begins at the division singularity.

Suppose we are running a problem between t_1 and t_2 where the singularity occurs at t_s, where $t_1 < t_s < t_2$. We break the problem into two parts, running forward from t_s to t_2 and backward from t_s to t_1.

Variable scaling is the other method by which the division operation will be improved. It serves to keep the denominator as large as possible in the vicinity of zero, but it is of no use at zero. Variable scaling finds its greatest utility when the singularity occurs at the end of the problem run. Very often if the division error can be kept small until a short time before the end, it will have no effect on the solution. Such is the case in Example 5, which is a continuation of Example 4. In this latter example we show how variable scaling in conjunction with a coordinate transformation can improve the solution accuracy.

Example 5. We continue the problem begun in Example 4 to show how we can improve the solution accuracy by reducing the variable slope at readout time and by changing the division circuit scaling.

If we plot the parameter y_p versus the range R, Fig. 7.27 is a typical result. The distance d between the target and missile is

$$d = (R^2 + y_p^2)^{\frac{1}{2}}$$

Where $dR/dt \gg dy_p/dt$, as we assume in this case, the minimum of d occurs approximately when $R = 0$. Actually the minimum occurs when d is perpendicular to the relative velocity vector $\mathbf{V}_r - \mathbf{V}_m$. In this region of R the slope of y_p is roughly constant.

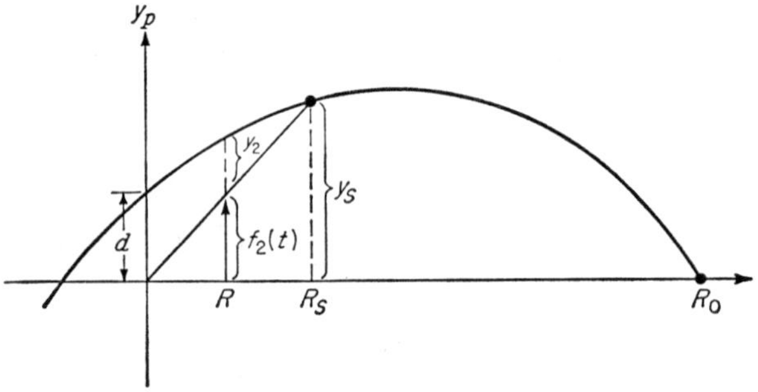

Fig. 7.27 The solution to the homing-missile problem (see Example 5) and a nominal solution $f_2(t)$.

To implement readout accuracy we generate the function $f_2(t)$, also shown in Fig. 7.27 and described by the equation

$$f_2(t) = \frac{y_s}{R_s} R \qquad \text{for } R \leq R_s \tag{7.99}$$

where R_s is a fixed and known value of R and $y_s = y_p$ at $R = R_s$. Then

$$y_p = y_2 + f_2 = y_2 \qquad \text{at } R = 0 \tag{7.100}$$

Our improved readout will come from using y_2 because $dy_2/dt < dy_p/dt$ in the neighborhood of $R = 0$. However, to avoid switching transients, it is convenient to generate dy_2/dt and integrate. Differentiating Eq. (7.100), we get

$$\frac{dy_2}{dt} = \frac{dy_p}{dt} - \frac{df_2}{dt} = V_p - \frac{y_s}{R_s}\frac{dR}{dt} \tag{7.101}$$

We may also modify the scaling on the division circuit by using y_2. Thus we get

$$\sigma_1 = \frac{y_p}{R} = \frac{y_2 + f_2(t)}{R} = \frac{y_2}{R} + \frac{y_s}{R_s} \qquad \text{for } 0 \leq R \leq R_s \tag{7.102}$$

Because y_s/R_s is a constant we write

$$\sigma_1 = \frac{y_2}{R} + \sigma_s \tag{7.103}$$

Now we scale both the denominator and the numerator by a factor k and obtain the equation to be programmed:

$$\sigma_1 = \frac{ky_2}{kR} + \sigma_s \tag{7.104}$$

When we compare this with scaling on y, where

$$\sigma_1 = \frac{ky_p}{kR} \tag{7.105}$$

we observe the following advantage in using Eq. (7.103). Because $y_2 < y_p$, the high-gain amplifier representing the divider output will reach its upper voltage limit for a smaller R when y_2 is the numerator. Hence this method of obtaining σ will allow us with accuracy to go closer to the terminal point, $R = 0$.

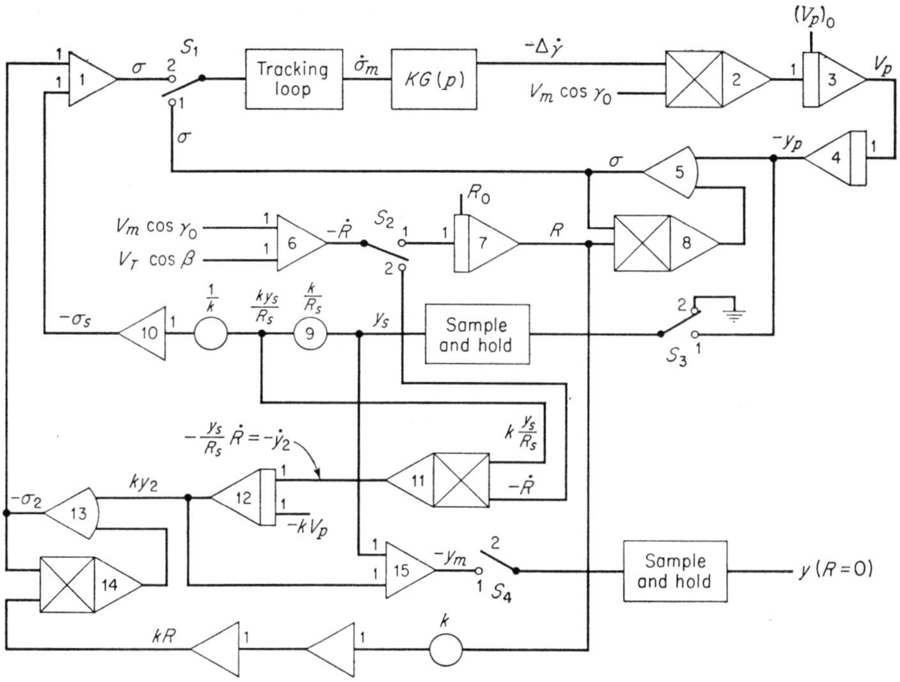

FIG. 7.28 A circuit which improves the accuracy of homing-missile miss measurements by perturbation around a nominal value shown in Fig. 7.27.

Figure 7.28 is the end product of our scheme to increase computer accuracy. At the start of the problem all the switches are in position 1. In this position the circuit is the same as that of Fig. 7.14. Note that switches S_1, S_2, and S_3 are involved in the problem program and S_4 is in the readout circuit. At a predetermined value R_s the relays move switches S_1, S_2, and S_3 to position 2. In moving to its new position, switch 3 (at the output of integrator 4) causes the sample-and-hold circuit to hold its output fixed at y_s. Because R_s is known, we may use a preset potentiometer which has an output y_s/R_s. Switch S_2 is involved in the production of y_2. This switching will introduce a transient at the output of multiplier 11. However, the smoothing effect of integrator 12 will eliminate the transient from y_2. At the time of the switching of S_1, both positions have the same voltage. Hence, no transient arises to mar the signal.

At the time $R = 0$ switch S_4 moves to position 2 and the output of the sample-and-hold circuit stores the miss distance of the missile, y. It is to the accurate measurement of this one quantity that all our effort in Examples 4 and 5 has been directed.

7.6.4 Limiter

Sometimes we can deliberately introduce a small error in the problem to forestall a much larger one. If the small error keeps the solution within acceptable accuracy, we come out of this venture with a considerable gain. The vehicle for keeping the error within bounds is the limiter. What the limiter does is to prevent the voltage from becoming too large.

One application is in keeping the output of an amplifier from overloading. An amplifier has the nasty habit of remaining in overload some time after the input falls below the level that put it in overload in the first place.

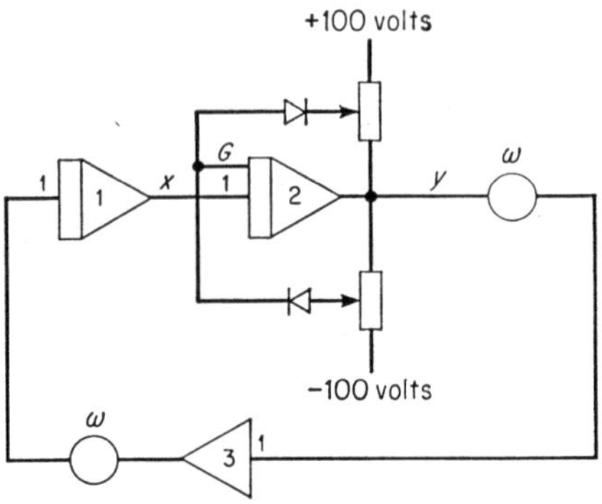

FIG. 7.29 A circuit which uses a limiter to improve the performance of an unstable oscillator solution.

As a result an amplifier can hang up at a high voltage and badly distort the solution. A limiter shunting the amplifier will keep it out of overload. It is also true that, while it is limiting, the amplifier is not producing the correct output voltage. However, if the output is limited for only a short time, the solution error may be small. In any event, limiting is preferable to overloading. Incidentally, a likely candidate for protection by the limiter is the amplifier used in a division circuit, as, for instance, amplifier 4 in Fig. 7.14.

Another application of the limiter is shown in Fig. 7.29. Without the limiter this is a circuit for generating sin ωt and cos ωt. In reality if we program the computer so that the output of integrator 2 should be $A \sin \omega t$, stray effects in the program board and imperfections in the elements are apt to make the solution be like

$$y = Ae^{\alpha t} \sin \omega t \tag{7.106}$$

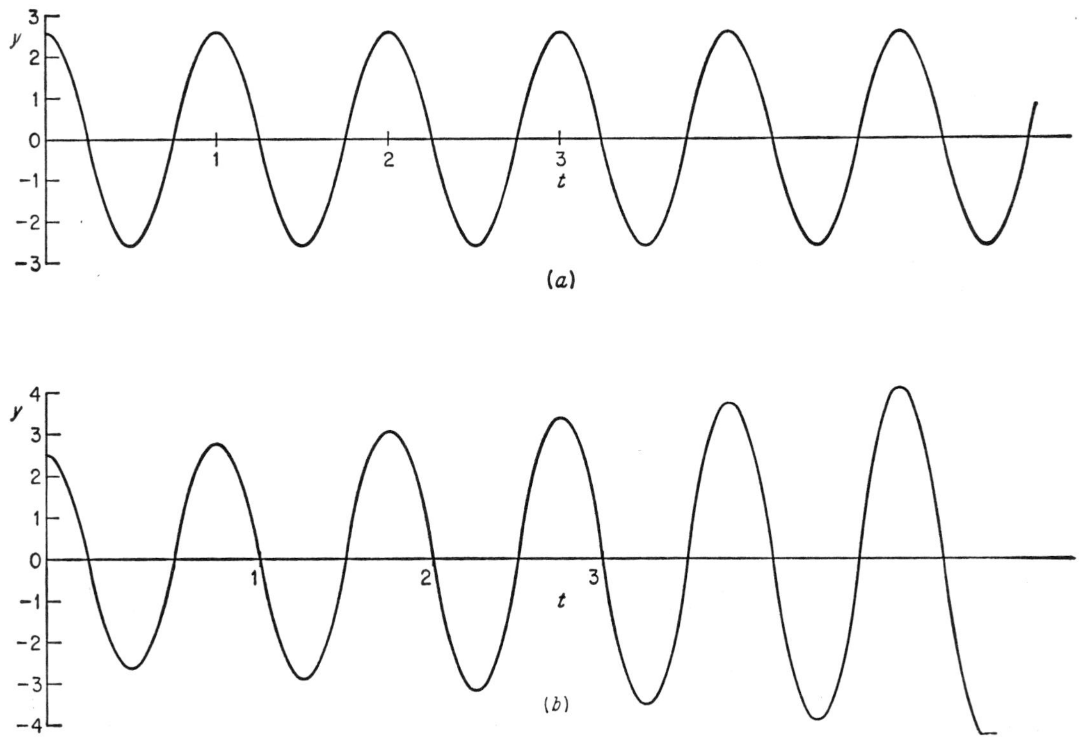

FIG. 7.30 (*a*) The response of Fig. 7.29. Without the limiter the response would be $x = ke^{at} \cos \omega t$, where $a = 0.1$. (See part *b*.) (*b*) A plot of the function $x = ke^{at} \cos \omega t$, where $a = 0.1$.

Because α is small, y departs from the correct solution only when t is large. In this case for positive α the magnitude of y will become large. The limiters prevent this from happening. Indeed the limiters can be set so that the magnitude of y does not exceed A. We introduce some distortion in y, that is, higher harmonics of $\sin \omega t$, but if α is small this error will be small. The difference in the limited and unlimited y is shown in Fig. 7.30*a* and *b*, respectively. We have deliberately chosen a large α to exaggerate the effects of limiting so that they may be readily observed.

When α is negative, the limiter is of no help. However, by connecting the output of amplifier 3 to the input of integrator 2 we can with the proper gain make α positive.

7.6.5 Diode Function Generator

The diodes in a diode function generator are subject to drift because of a variation in ambient temperature. This drift is principally manifested in a shift of the diode break point. We can overcome this difficulty by placing the diodes in a temperature-controlled chamber. However, if we must subject the diodes to the vagaries of temperature fluctuations, we can minimize the error by scaling the function-generator input as high as possible. In this manner the percentage of drift relative to the input signal is at its smallest amount.

The diode break points are set by voltage-divider networks which bias the diodes with some fraction of the reference voltage. Thus if this reference voltage changes, so also will the diode break points shift, with a corresponding change in the function-generator input and output *absolute* voltages. If the drift is large in a given computer run, its effect will be an error in the solution. However, reference-voltage supplies are well regulated and the drift is more likely to be a slow, long-term phenomenon. Therefore we can assume it to be quasi-static, where the reference voltage remains constant during any run. In this case reference drift has no effect on the solution.

The reason for this is that all voltages derived from the reference voltages, like diode break points and initial conditions, are set and measured relative to the reference voltage. Any percentage change in the reference voltage is reflected in the same percentage change in these other voltages. Likewise the solution shows the same percentage change in voltage. Thus if it is measured relative to the reference voltage, the solution will be the same in problem-variable dimensions no matter what the reference voltage is. *If the same reference voltage is used everywhere in the problem, the solution will be invariant with reference voltage.*

8 *Optimization Techniques: Gradient Methods for Finding Maxima and Minima*

8.1 Prologue

In previous chapters we have discussed how several classes of problems may be programmed on the computer. Once the problem has been set up, we still have the important task of varying the parameters until we find the set of values that is the solution. It is true that the solution can in principle be obtained if the problem is set up correctly. However, the most elegant programming may be effectively wasted by poor scheduling of computer runs. The cost of a haphazard attempt to find a solution is measured in extra computer running time, excessive data output, and even in the possibility that the solution may not be found at all.

In many problems we may vary the parameters to control the behavior of two or more variables. As we go from one set of parameter values to another set, the performance of one variable may be improved while that of the other variable may be degraded. Which set of parameter values results in a "better" solution? Clearly, this question can be answered only if we have some quantitative basis on which to compare the two results. To meet this requirement we must represent the solution as a single-valued function of the pertinent parameters. For reasons stated in Sec. 8.2 this function is called a distance. A process of variation of problem parameters by which the distance is made to approach an extremum (maximum or minimum) as closely as possible will be defined as optimization.

The selection of a distance function from the infinite number of possibilities that will fulfill the purpose of the problem can involve much skill and judgment. However, once the choice is made, the computer has a task well within its decision-making capacities. For the maximum- (minimum-) seeking case the computer must vary the parameters so that after the change the distance function is larger (smaller) than before. *The computer can be programmed to carry out the optimization systematically without any human assistance.*

We begin the chapter with a discussion of distance. The term as we use it is a mathematical one having properties different from distance which refers to spatial separation in the physical world. We shall delineate these differences. In so doing we shall show how a mathematical distance function may be formulated for any computer problem.

Next we have a brief interlude spent in discussing the analytic mathematical techniques for finding extrema. The two most useful methods are the calculus of variations and Lagrangian multipliers. We shall discuss their general areas of applicability and limitations.

Finally, we examine the computer techniques which may be used to find extrema. These include (1) the method of steepest descent, (2) relaxation method, and (3) random search. All of these are in the category of gradient methods. We shall consider their use both when the parameters are free to assume any values and when they have constraints. These gradient methods have a wide range of applicability, encompassing the areas of (1) algebraic equations, (2) boundary-value problems, (3) eigenvalue problems, (4) process control, and (5) the young but burgeoning field of adaptive control systems.

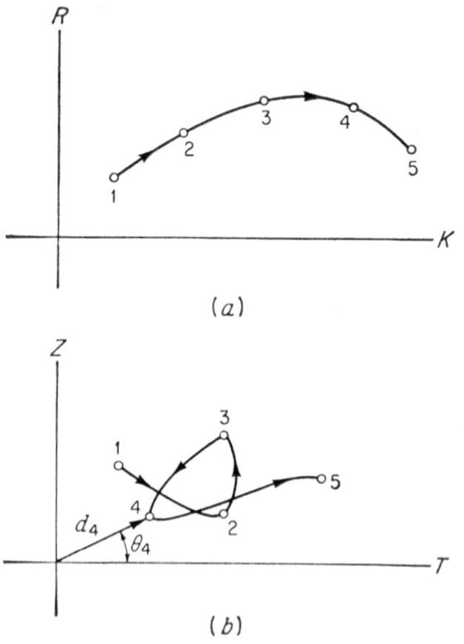

FIG. 8.1 (*a*) A path in parameter space. (*b*) A path in solution space caused by the variation in parameter space in (*a*). Like numbers in the two figures correspond.

8.2 Distance Function[1]

Suppose in a chemical-processing problem the parameters are R, the ratio of two chemicals, and K, the temperature of the process, whereas the variables representing the solution are T, the time to produce a unit batch, and Z, the percentage of corrosive by-product. These quantities can be the rectangular coordinates of the parameter and solution spaces. An ordered pair of parameter values (R_1, K_1) plotted as a point in the parameter space will have its outcome mapped as a single point (T_1, Z_1) in the solution space. In Fig. 8.1 we show several points in the parameter and solution spaces. The points in each space with the same number correspond

[1] In the literature this is also called an objective function, criterion function, or risk function.

to one another. The arrows indicate the direction of variation. A point in the solution space may be represented by its polar coordinates, d and σ. The quantity d is the distance from the origin and at a particular point, for instance point 1, is

$$d_1 = (Z_1^2 + T_1^2)^{\frac{1}{2}} \tag{8.1}$$

where $$Z_1 = Z(R_1, K_1) \qquad T_1 = T(R_1, K_1)$$

We see that this distance function contains two quantities that we would like to make as small as possible. In minimizing d we shall achieve a desirable objective. Observe that we have compressed all the available information in the solution space to a single scalar quantity, called a distance function.

We are in no way constrained to choose Eq. (8.1) as our distance function. Should Z and T not go to zero simultaneously, other distance functions can have minima resulting in different values of parameters and problem solutions. In choosing a specific function and asserting that the minimum of the function is the optimum design, we consign to Z and T their ultimate values even before the problem is solved. Of course, with the final values of Z and T also goes a particular set of parameter values.

An alternative distance function is one which weights one of the solutions relative to the other. Thus,

$$d = [Z^2 + (aT)^2]^{\frac{1}{2}} \tag{8.2}$$

where a is a constant.

Another distance function which has certain advantages discussed in Example 1 is

$$d_1 = |Z| + |T| \tag{8.3}$$

These distance functions and others which are suitable for our use satisfy the mathematical definition of distance, stated as follows: If u, v, and w are points in a space, then $d(u, v)$ is a real number representing the distance from u to v. The distance has the properties:

1. $d(u, u) = 0$
2. $d(u, v) \leq d(u, w) + d(v, w)$

Property 2 is called the triangle inequality. From these two properties we may derive two others:

3. $d(u, v) \geq 0$
4. $d(u, v) = d(v, u)$

A space having these properties is called a *metric space.*

There are a few points worth considering about distance functions before we leave them. First, there can be any number of dimensions in the parameter and solution spaces. Furthermore, each space can have more or less dimensions than the other. Although when we go to four or more dimensions distance loses its physical meaning, there is no conceptual difficulty in constructing distance functions for higher dimensional systems. For example, for the solution variables $x_1, x_2, \ldots, x_n$, a distance function like Eq. (8.1) would be

$$d = (x_1^2 + x_2^2 + \cdots + x_n^2)^{\frac{1}{2}} \tag{8.4}$$

Second, we can measure our distance from any point other than the origin. Let us choose as the optimum values Z_0 and T_0. If we minimize a distance function like Eq. (8.3), its form becomes

$$d_1 = |Z - Z_0| + |T - T_0| \tag{8.5}$$

Finally, distance functions can be maximized as well as minimized. If our two solution functions were (1) yield and (2) profit per unit batch, we would want the distance measured from the origin to be as large as possible.

One important question remains unanswered. How do we vary the parameters to reach an extremum of a distance function? We shall devote the rest of the chapter to this topic after a brief discussion of the analytic methods which may be employed.

8.3 A Mathematical Interlude[1]

If at any point other than an end point of the variable a function of one variable, $f(x)$, has an extremum, it must at that point satisfy the condition

$$\frac{\partial f}{\partial x} = 0 \tag{8.6}$$

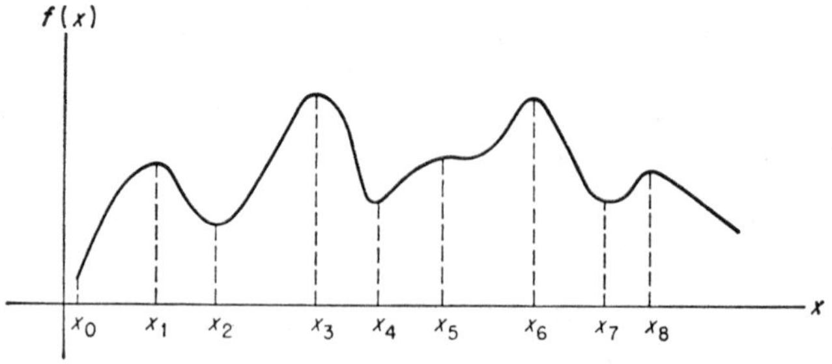

FIG. 8.2 An arbitrary function.

Figure 8.2 shows why this is only a necessary condition. Maxima appear at x_1, x_3, x_6, and x_8, and minima occur at x_2, x_4, and x_7. At x_5 the condition

[1] For a brief but readable and entertaining discussion of the shortcomings of these mathematical techniques, see Ref. 87, pp. 13–36. In the rest of the book Bellman discusses dynamic programming, a technique developed by the author for finding extrema on the digital computer. A more detailed discussion of maxima and minima is found in Ref. 88, and the calculus of variations is treated in Ref. 101.

of Eq. (8.6) is met. However, in a small neighborhood around x_5, $f(x_5)$ is neither the largest nor the smallest value. The point is called an inflection point.

This is but one of the spurious solutions that can result from our attempt to find extrema by Eq. (8.6). When there are several maxima and minima, the largest maximum and the smallest minimum are called *absolute* or *global*. All others are *relative* or *local* maxima and minima.[1] We may seek an absolute maximum but solve Eq. (8.6) for a relative maximum. Even more distressing, the global extremum may not be a bound on the value of the function. $f(x_0)$ is the smallest value of $f(x)$. Yet because it occurs at an end point in the range of x we have no mathematical formulation for finding it. This is one of the greatest shortcomings in mathematical methods for finding the largest or smallest value of a function, a limitation not shared by the computer.

For a function of n variables, $f(x_1, x_2, \ldots, x_n)$, a necessary condition for a relative extremum at a point $x_1 = \alpha_1$, $x_2 = \alpha_2$, . . . , $x_n = \alpha_n$ is that all first partial derivatives vanish at this point; that is,

$$\frac{\partial F}{\partial x_1} = 0 \qquad \frac{\partial F}{\partial x_2} = 0 \qquad \cdots \qquad \frac{\partial F}{\partial x_n} = 0 \tag{8.7}$$

If the quantity

$$G = \sum_{i=1}^{n} \sum_{j=1}^{n} \frac{\partial^2 F}{\partial x_i \, \partial x_j} \Delta x_i \, \Delta x_j \tag{8.8}$$

is positive for all real (nonzero) values of Δx_i and Δx_j, G is said to be a positive-definite quadratic form and $f(\alpha_1, \alpha_2, \ldots, \alpha_n)$ is a maximum. When G is negative for all real values of Δx_i, Δx_j, it is called negative definite and $f(\alpha_1, \alpha_2, \ldots, \alpha_n)$ is a minimum. Otherwise G is indefinite and $f(\alpha_1, \alpha_2, \ldots, \alpha_n)$ is neither a maximum nor a minimum.

8.3.1 Constraints and Lagrange Multipliers

Where the parameters are not independent, we find a constrained maximum. If we seek the extrema of $f(x_1, x_2, \ldots, x_n)$ subject to the two conditions

$$g(x_1, x_2, \ldots, x_n) = 0 \qquad h(x_1, x_2, \ldots, x_n) = 0 \tag{8.9}$$

we can, in principle, eliminate two of the parameters and then find the value of each of the others that will satisfy Eq. (8.7). More often than not this reduction of parameters cannot be carried out. In this case we resort to the method of Lagrange multipliers.

[1] A function of one or more variables, $f(x_1, \ldots, x_n)$, which has only one extremum is called a concave surface if it has a maximum and a convex surface if it has a minimum.

We begin by constructing a new function to be minimized:

$$Q(x_1, x_2, \ldots, x_n; \lambda_1, \lambda_2) = f(x_1, x_2, \ldots, x_n) + \lambda_1 g(x_1, x_2, \ldots, x_n) + \lambda_2 h(x_1, x_2, \ldots, x_n) \quad (8.10)$$

where λ_1 and λ_2 are called Lagrange multipliers. Then we solve the set of equations

$$\frac{\partial Q}{\partial x_1} = 0 \qquad \frac{\partial Q}{\partial x_2} = 0 \qquad \cdots \qquad \frac{\partial Q}{\partial x_n} = 0 \quad (8.11)$$

in conjunction with equation set (8.9). As we can see, the number of equations equals the number of parameters plus the number of Lagrange multipliers.

This method may be used for any number of constraint equations provided they are less than the number of parameters. One distinct drawback of the method of Lagrange multipliers is that the constraint equations must be equalities. A constraint expressed as

$$g(x_1, x_2, \ldots, x_n) \geq 0 \quad (8.12)$$

cannot be treated in this form by the method of Lagrange multipliers.

8.3.2 Calculus of Variations [101]

A definite integral of the form

$$I = \int_{x_1}^{x_2} F\left(y, \frac{dy}{dt}, \ldots, \frac{d^n y}{dt^n}, \alpha\right) d\alpha \quad (8.13)$$

as a necessary condition for an extremum must satisfy the Euler differential equation

$$\frac{\partial F}{\partial y} - \frac{d}{dx}\left(\frac{\partial F}{\partial y^{(1)}}\right) + \frac{d^2}{dx^2}\left(\frac{\partial F}{\partial y^{(2)}}\right) + \cdots + (-1)^n \frac{d^n}{dx^n}\left(\frac{\partial F}{\partial y^{(n)}}\right) = 0 \quad (8.14)$$

where

$$y^{(r)} = \frac{d^r y}{dx^r}$$

When, for the integral in Eq. (8.13), we must find an extremum subject to the conditions

$$J_r = \int_{x_1}^{x_2} G\left(y, \frac{dy}{dt}, \ldots, \frac{d^n y}{dt^n}, \alpha\right) d\alpha \qquad r = 1, 2, \ldots, m; m < n \quad (8.15)$$

we again use the method of Lagrange multipliers. To this end we construct a new function:

$$Q\left(y, \frac{dy}{dt}, \ldots, \frac{d^n y}{dt^n}, \lambda_1, \lambda_2, \ldots, \lambda_m\right) = F\left(y, \frac{dy}{dt}, \ldots, \frac{d^n y}{dt^n}\right) + \sum_{i=1}^{m} \lambda_i G_i\left(y, \frac{dy}{dt}, \ldots, \frac{d^n y}{dt^n}\right) \quad (8.16)$$

The resulting equation to be satisfied is

$$\frac{\partial Q}{\partial y} - \frac{d}{dx}\left(\frac{\partial Q}{\partial y^{(1)}}\right) + \frac{d^2}{dx^2}\left(\frac{\partial Q}{\partial y^{(2)}}\right) + \cdots + (-1)^n \frac{d^n}{dx^n}\left(\frac{\partial Q}{\partial y^{(n)}}\right) = 0 \quad (8.17)$$

Equation (8.17) can be solved with the prerequisite boundary conditions and equation set (8.15).

8.3.3 The Summing Up

We have glossed over an important area of applied mathematics. Particularly in the field of mathematical physics these concepts have done yeoman service. Nonetheless they are almost useless for finding extrema of any but the simplest distance functions. The equations from which the extrema are to be found usually cannot be solved analytically. For this reason we turn with gratitude and relief to the computer to help us find the solutions.

8.4 A Physical Analogy

Now we ask the question which is the crux of this chapter: "How is the computer to vary the parameters so that the distance function is minimized?"[1] The answer is best couched in terms of geometric concepts. A distance function involving n parameters can be represented in an $(n + 1)$-dimensional euclidean space with the parameters as n of the coordinates and the distance function as the remaining coordinate. So that the reader can better visualize our arguments, we shall consider the minimization of a distance function d of two parameters, x and y; thereby we operate in a three-dimensional rectilinear coordinate space (x, y, d). The ideas we develop here will then be extended to optimizations where several parameters are involved.

If the vertical axis (altitude) is the distance function, Fig. 8.3 can represent a plot in the parameter–distance-function space. It looks very much like some typically hilly terrain. We complete the picture by placing a person on this hill. As the scene unfolds, he wants to get to the valley (minimum). Unfortunately the area is shrouded in fog and he cannot see where he must go. To reach the valley our peripatetic wanderer need only take a path such that the elevation always decreases.[2] *Under this constraint the path will always end at a minimum.*[3] Furthermore, a minimum will be reached for any number of parameters if the variation is such that it always

[1] In the rest of the chapter we shall show by example how each technique can find minima. In each case the procedure for finding the maxima will be indicated.

[2] He can get to the maximum altitude by selecting a path which always increases the elevation.

[3] See Ref. 1, p. 134. In this book Polya discusses maxima and minima with a style and clarity that will stimulate the mathematician and applied scientist alike.

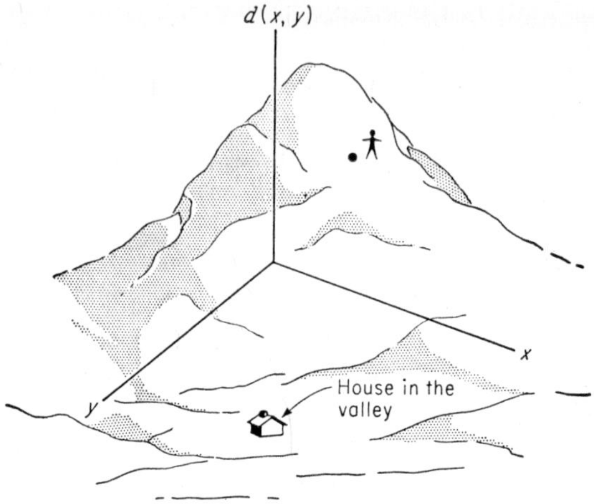

FIG. 8.3 A man on a mountain trying to find his way home.

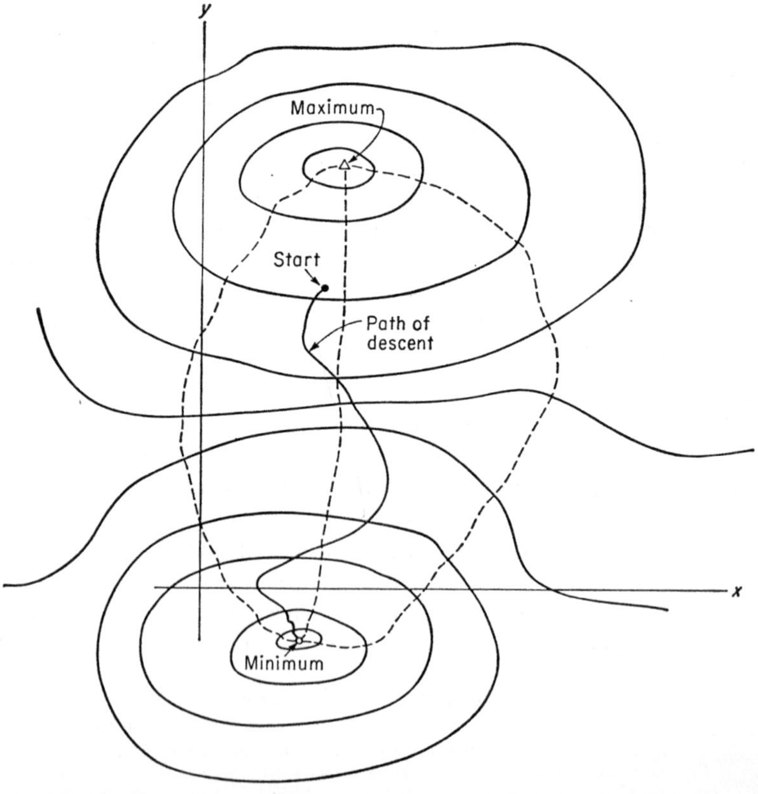

FIG. 8.4 A contour map. Solid lines are contours, and dashed lines are gradients. The path of descent always crosses contour lines of decreasing magnitude.

reduces the magnitude of the distance function. Such a procedure of minimization is called a gradient method.[1]

The projection of the path onto the xy plane in Fig. 8.4 shows how the parameters are varied to achieve the minimization. The solid lines are contour lines (lines of constant elevation), and the dashed lines are gradient lines (lines indicating the direction of maximum slope at every point). Gradient lines are always perpendicular to contour lines. The path drawn is but one of an infinite number which could be taken to reach the minimum designated by the star. However, the path taken along a gradient is unique and is called a path of steepest descent.[2] The name is well chosen. For an incremental change of the parameters the decrease of the distance function is greatest if the path is along the gradient.

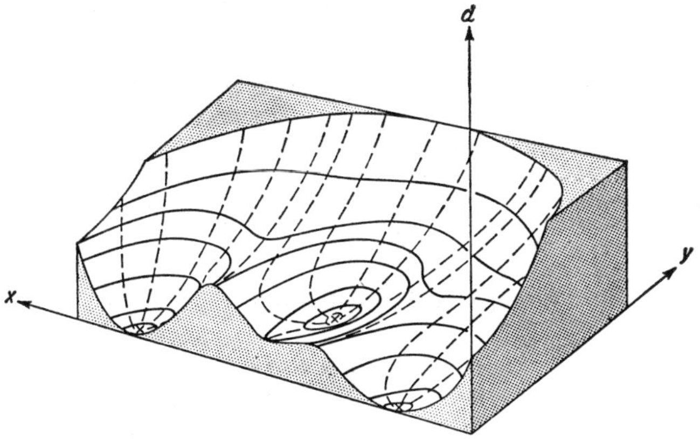

FIG. 8.5 A terrain map with three minima indicated by the crosses.

The primary advantage of the gradient method is that it gives to the computer a well-defined course of action. "Change the parameters how you will," it tells the computer, "but make certain that each change makes the distance function smaller than before." This is a simple-minded criterion and, as we shall show, can be programmed so that the computer varies the parameters automatically along a minimization path.

The gradient method has one disadvantage which can be demonstrated by assuming a parameter–distance-function dimensional plot, as in Fig. 8.5. There are three minima in this figure marked by $\times$'s. If we want a par-

[1] There is extensive literature on this subject. For a discussion of the various aspects of gradient method of optimization see Refs. 55 and 89 to 92.

[2] The path of steepest descent is used in a different context in association with the saddle-point method. Here we use the steepest-descent path away from the saddle point of a function as the path along which to obtain the value of the integral asymptotically. See Refs. 83, pp. 39–41, and 111, pp. 77–101. The method has nothing to do with finding maxima or minima.

ticular one of the minima, we cannot guarantee that the computer solution will converge to that minimum from an arbitrary starting set of parameter values.

The three major gradient methods are (1) steepest descent, (2) relaxation, and (3) random search. We shall consider each of these techniques in turn, with special emphasis on how they may be implemented on the computer.

8.5 Method of Steepest Descent[1]

We have already discussed how we may change the parameters to minimize a function of two variables. By a simple extension of this principle the function $F(x_1, \ldots, x_n)$ may be minimized in a region[2] R of the n-dimensional rectilinear parameter space if one starts from an arbitrarily selected point in the region and follows a path which leads in a direction of decreasing values of F.

The rate of change dF/dt is greatest if the path chosen in the parameter space is tangential to the gradient vector in this same space.

We prove this statement by starting with

$$\frac{dF}{dt} = \frac{\partial F}{\partial x_1}\frac{dx_1}{dt} + \frac{\partial F}{\partial x_2}\frac{dx_2}{dt} + \cdots + \frac{\partial F}{\partial x_n}\frac{dx_n}{dt} \tag{8.18}$$

Equation (8.18) may be written in vector notation as an inner product,

$$\frac{dF}{dt} = (\text{grad } \mathbf{F}) \cdot \left(\frac{d\mathbf{s}}{dt}\right) \tag{8.19}$$

where

$$\text{grad } \mathbf{F} = \frac{\partial F}{\partial x_1}\mathbf{i}_1 + \frac{\partial F}{\partial x_2}\mathbf{i}_2 + \cdots + \frac{\partial F}{\partial x_n}\mathbf{i}_n$$

$$\frac{d\mathbf{s}}{dt} = \frac{dx_1}{dt}\mathbf{i}_1 + \frac{dx_2}{dt}\mathbf{i}_2 + \cdots + \frac{dx_n}{dt}\mathbf{i}_n$$

An inner product attains its maximum when the two vectors are parallel (collinear) and pointing in the same direction. The minimum, always negative, occurs when the collinear vectors point in opposite directions.

To ensure collinearity of the velocity vector $d\mathbf{s}/dt$ with grad $\mathbf{F}$, components of $d\mathbf{s}/dt$ and grad $\mathbf{F}$ must satisfy the relationship

$$\frac{dx_1}{dt} : \frac{dx_2}{dt} : \cdots : \frac{dx_n}{dt} = \frac{\partial F}{\partial x_1} : \frac{\partial F}{\partial x_2} : \cdots : \frac{\partial F}{\partial x_n} \tag{8.20a}$$

or

$$\frac{dx_i/dt}{dx_j/dt} = \frac{\partial F/\partial x_i}{\partial F/\partial x_j} \tag{8.20b}$$

[1] The approach we use in presenting the concept of steepest descent was suggested by Ref. 93. For additional material on this subject see Refs. 94 and 95 (C. B. Tompkins, chap. 18, Methods of Steep Descent, pp. 448–479).

[2] The region is bounded by the range of each parameter x_i.

Equation (8.20*b*) may be alternatively written as

$$\frac{dx_i}{dt} = c\,\frac{\partial F}{\partial x_i} \qquad i = 1, 2, \ldots, n \tag{8.21}$$

When c is a negative constant, $dF/dt < 0$, and the path is of steepest descent. Otherwise it is a path of steepest ascent.

Having progressed this far in our discussion, we shall illustrate how the method of steepest descent is applied in determination of the roots of a polynomial of nth degree and in the solution of the general quadratic equation. These two examples will illuminate some characteristics of the method not as yet covered. They are, therefore, a continuation of our discussion of steepest descent and should not be omitted by the reader who wants a complete understanding of the subject.

Example 1. Determination of the Roots of a Polynomial of *n*th Degree [96]. A polynomial of nth degree may be written as

$$W(z) = a_n z^n + \cdots + a_1 z + a_0 = u + iv \tag{8.22}$$

where z is the complex variable, written as

$$z = x + iy = re^{i\theta}$$

The roots z_i of this polynomial satisfy the equation

$$W(z_i) = 0 \qquad i = 1, 2, \ldots, n \tag{8.23}$$

From the fundamental theorem of algebra we know that there are exactly n values of z which will satisfy Eq. (8.23).

The two dimensions of our solution space are

$$f_1 = u \qquad f_2 = v \tag{8.24}$$

with u and v functions of either x and y or r and θ. The distance functions we select to be minimized are

$$d_1 = u^2 + v^2 \tag{8.25a}$$
$$d_2 = |u| + |v| \tag{8.25b}$$

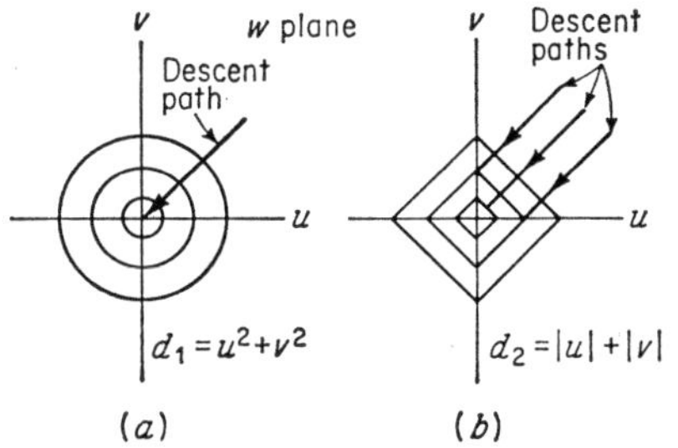

FIG. 8.6 Descent paths for two different distance functions in Example 1.

It is quite apparent that the minimum value for d_1 as well as d_2 is zero. This can occur only when u and v simultaneously become zero, which, as Eq. (8.23) reveals, happens only for z's which are roots of the polynomial. It can be shown that there are no other minima of either d_1 or d_2 for finite values of z.[1] Hence the minimization of either d_1 or d_2 by steepest descent will determine one of the roots of the polynomial.

The nature of the steepest-descent paths for d_1 and d_2 may be fully surmised by an examination of the mapping of contours and gradient lines in the solution plane, as shown in Fig. 8.6*a* and *b*. We choose this instead of the parameter plane because the contours and gradients can be drawn without any foreknowledge of the polynomial coefficients. In the case of d_1 the contours and gradients are a set of concentric circles

[1] This is proved for d_1 in Ref. 96 and for d_2 in Ref. 97.

and of radii, respectively. For d_2 the contours are a set of nested squares intersected by parallel sets of gradient lines in each quadrant of the w plane. The steepest-descent paths in this case converge to the origin by following, at first, the gradient to an intercept with either the u or v coordinate axis and then by continuing along this axis to the origin. The only paths which reach the origin without detour are along the lines $v = \pm u$.

These two contour-gradient mappings illustrate an important point which is characteristic of gradient methods. *The solution may be obtained in different ways depending upon the form of the distance function.*

Equations of Steepest Descent. At first glance one might be tempted to use the rectangular coordinates x and y as the parameters to be optimized along the steepest-descent path. However, the equations are considerably simpler when the polar coordinates r and θ are used instead. If we use $z = re^{i\theta}$ in Eq. (8.22) and separate the real and imaginary parts, we obtain[1]

$$u = \sum_{k=0}^{n} a_k r^k \cos k\theta \tag{8.26a}$$

$$v = \sum_{k=0}^{n} a_k r^k \sin k\theta \tag{8.26b}$$

For a steepest-descent path the condition of Eq. (8.20*b*) becomes

$$\frac{dr/dt}{r\,d\theta/dt} = \frac{\partial F/\partial r}{(1/r)\partial F/\partial \theta} \tag{8.27}$$

This is equivalent to a useful pair of simultaneous equations:

$$\frac{dr}{dt} = -a\frac{\partial F}{\partial r} \qquad r\frac{d\theta}{dt} = -\frac{a}{r}\frac{\partial F}{\partial \theta} \tag{8.28}$$

where a is an arbitrary positive constant which determines the rate of descent along the path. As we have already emphasized, the negative sign is essential to establish a descending path.

By applying Eq. (8.27) to the functions d_1 and d_2 we obtain the computer equations for a steepest-descent path. In the case of d_1

$$\begin{aligned} \frac{dr}{dt} &= -a\frac{\partial d_1}{\partial r} = -2a\left(u\frac{\partial u}{\partial r} + v\frac{d\theta}{\partial r}\right) \\ r\frac{d\theta}{dt} &= -\frac{a}{r}\frac{\partial d_1}{\partial \theta} = -\frac{2a}{r}\left(u\frac{\partial u}{\partial \theta} + v\frac{\partial v}{\partial \theta}\right) \end{aligned} \tag{8.29}$$

In the case of d_2

$$\begin{aligned} \frac{dr}{dt} &= -\frac{\partial d_2}{\partial r} = -a\left(\frac{\partial |u|}{\partial r} + \frac{\partial |v|}{\partial r}\right) \\ r\frac{d\theta}{dt} &= -\frac{a}{r}\frac{\partial d_2}{\partial \theta} = -\frac{a}{r}\left(\frac{\partial |u|}{\partial \theta} + \frac{\partial |v|}{\partial \theta}\right) \end{aligned} \tag{8.30}$$

By the chain rule

$$\frac{\partial |u|}{\partial r} = \frac{\partial |u|}{\partial u}\frac{\partial u}{\partial r} = (\text{sign } u)\frac{\partial u}{\partial r} \tag{8.31}$$

Therefore

$$\begin{aligned} \frac{dr}{dt} &= -a\left[(\text{sign } u)\frac{\partial u}{\partial r} + (\text{sign } v)\frac{\partial v}{\partial r}\right] \\ r\frac{d\theta}{dt} &= -\frac{a}{r}\left[(\text{sign } u)\frac{\partial u}{\partial \theta} + (\text{sign } v)\frac{\partial v}{\partial \theta}\right] \end{aligned} \tag{8.32}$$

[1] Equations (8.26*a*) and (8.26*b*) are correct only when all coefficients a_k are real. If any a_k is complex, the equations change slightly (see Ref. 96).

By the Cauchy-Riemann equations for analytic functions of a complex variable,

$$r\frac{\partial u}{\partial r} = \frac{\partial v}{\partial \theta} = \sum_{k=0}^{n} ka_k r^k \cos k\theta \tag{8.33}$$

$$\frac{\partial u}{\partial \theta} = -r\frac{\partial v}{\partial r} = -\sum_{k=1}^{n} ka_k r^k \sin k\theta \tag{8.34}$$

Equations (8.26*a*), (8.26*b*), (8.33), and (8.34) substituted into either Eqs. (8.29) or (8.32) provide the necessary equations for the computer programming. Note that in the case of Eqs. (8.32), Eqs. (8.26*a*) and (8.26*b*) are used only to determine the proper sign of each term. The similarity of Eqs. (8.26*a*) and (8.26*b*) to Eqs. (8.33) and (8.34), respectively, simplifies considerably the number of required multipliers and resolvers.

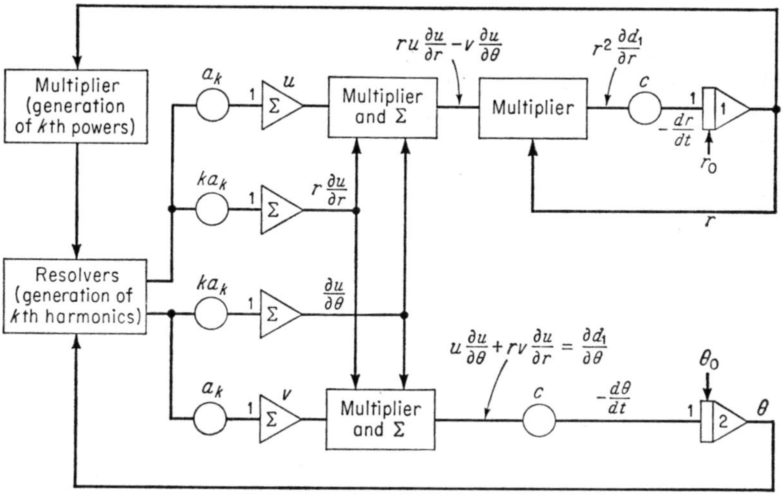

FIG. 8.7 Computer program to solve for roots of polynomials using d_1 [Eq. (8.25*a*)].

Figures 8.7 and 8.8 show in block diagram the computer setup for minimization of d_1 and d_2. To simplify computer programming we let

$$c = r^2 c_1$$

The starting values of r and θ appear as initial conditions on the proper integrators.

We show in Fig. 8.9*a* and *b* a number of steepest-descent trajectories in the parameter plane (x, y) for d_1 and d_2 applied to the polynomial equation

$$W = z^3 - 5z^2 + 8z - 6 = 0 \tag{8.35}$$

having the roots

$$z_1 = 3.0 \qquad z_{2,3} = 1.0 \pm j1.0$$

Figure 8.9*a* shows the paths for the minimization of d_1 converging to the polynomial roots z_1 and z_2. Plotted against a backdrop of contours of d_2, these trajectories can be seen as gradient lines.

Figure 8.9*b* exhibits descent paths for the d_2 minimization. Two sets of contours, u = const and v = const, are shown in the diagram to demonstrate the nature of descent which proceeds in a generally diagonal direction through the network of u and v curves until acquisition of either the $u = 0$ or $v = 0$ curve, which is then followed into the root.

Figure 8.9*c* shows a number of trajectories of the d_1 and d_2 minimizations for contrast. In one instance the trajectories originating from the same point P_2 descend to different minima owing to the different geometric structure of the d_1 and d_2 surfaces.

Of the two distance functions, d_2 is the better one for finding minima. The fact that in Fig. 8.9*c* the descent paths seem longer than those for d_1 is of little consequence. The d_1 minimization has the inherent shortcoming that, as the minima are approached, the rates of change dr/dt and $d\theta/dt$ become increasingly small since the first-order derivatives of d_1 tend to zero. The d_2 minimization does not have this distressing characteristic. Its gradient is always steeper than that of d_1 in the neighborhood of a minimum. Hence, the computer program minimizing d_2 will find the parameter values at the minima with less error than the program minimizing d_1.

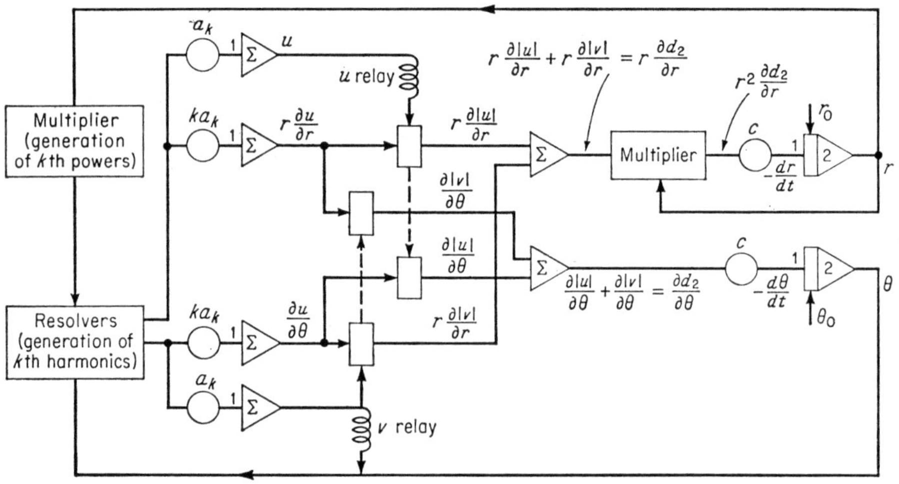

FIG. 8.8 Computer program to solve for roots of polynomials using d_2 [Eq. (8.25*b*)].

Tracking Variable Minima (Plotting Root Loci). The distance functions d_1 and d_2, as we have considered them, are functions only of the parameters x and y or r and θ. Thus a three-dimensional plot of these distance functions and the parameters has an unchanging terrain because the polynomial coefficients are held constant. Now, if we were to change one or more of the coefficients, we would also change the entire appearance of the terrain, especially the position of the minima.

Let us presume that from some initial value of r and θ the computer has minimized one of the distance functions, while the polynomial coefficients are held constant. If nothing is done to the distance function, the computer will "lock on" to and retain indefinitely the parameter values associated with the minimum. The reason is that these minima are stable solutions of the steepest-descent differential equations.

Suppose after "lock-on" the minima are displaced because of a change in the polynomial coefficients. The computer will seek a new minimum by varying the parameters along a path of steepest descent. If the coefficients are varied continuously and slowly enough, the computer will follow the path of the "locked-on" minimum in the parameter space. Similarly the path of an extremum can be tracked in any dimension parameter space, with the one restriction that the extremum must occur at a discrete point.

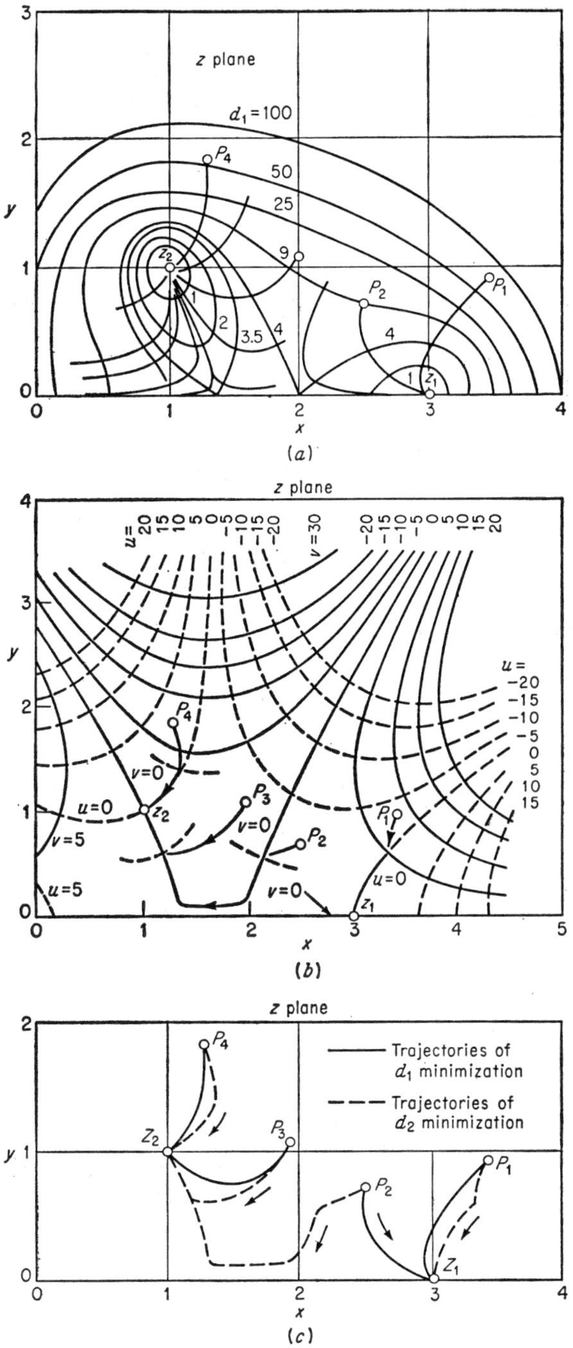

FIG. 8.9 (a) Parameter space trajectories for minimization of d_1 in Example 1; (b) parameter space trajectories for minimization of d_2 in Example 1; (c) a comparison of trajectories in parameter space when d_1 and d_2 are minimized in Example 1.

This type of extremum tracking applied to polynomials is called a root-locus plot. Control-system engineers have used a special root locus to obtain stability of servo systems.[1] The root loci to which engineers refer arise from the variation of k in a polynomial of the form

$$a_n s^n + a_{n-1} s^{n-1} + \cdots + a_1 s + a_0 + k(b_m s^m + b_{m-1} s^{m-1} + \cdots + b_1 s + b_0) = 0 \quad (8.36)$$

The root loci plotted by steepest-descent tracking are in no way restricted to any particular variation of coefficients.

Figure 8.10 illustrates a root locus obtained by minimization of d_2 for the equation

$$z^3 - 5z^2 + 8z - K = 0 \quad (8.37)$$

and compares it with the numerically calculated root locus. K is the coefficient which was varied.

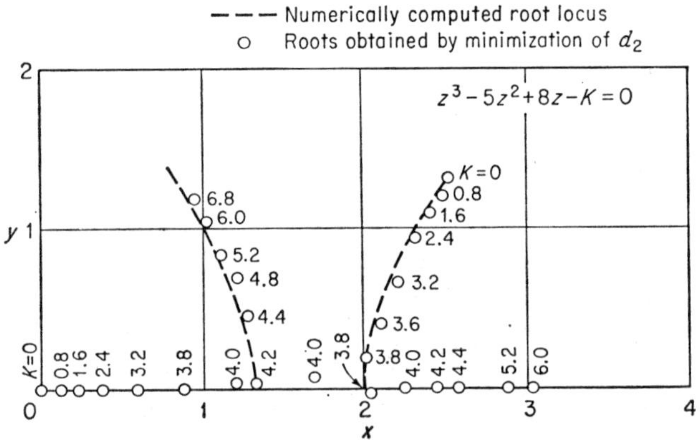

FIG. 8.10 A root-locus plot using d_2 minimization.

Example 2.[2] There are a variety of occasions in applied mathematics and engineering in which a function of the form

$$f(x_1, x_2, \ldots, x_n) = a_0 + \sum_{i=1}^{n} a_i x_i + \sum_{i=1}^{n} \sum_{j=1}^{n} a_{ij} x_i x_j \quad (8.38)$$

arises. It is called a nonhomogeneous quadratic form[3] or a general quadratic equation in n variables when

$$f(x_1^1, x_2^1, \ldots, x_n^1) = 0 \quad (8.39)$$

The equations of conic sections, quadratic regression analysis, and potential and kinetic energy in a conservative system with two degrees of freedom are all illustrations of the general quadratic equations.

Our problem, then, is to apply steepest-descent techniques to solve for the value of $x_1, x_2, \ldots, x_n$ in Eq. (8.38) that will satisfy Eq. (8.39). We shall designate the solutions

[1] Almost any recent book in the field of feedback control systems covers the application of root-locus techniques to servo systems. See, for example Ref. 52, pp. 237–263.

[2] This example was taken in large part from Ref. 98. We follow the example in using a d_1 to be minimized although a d_2 function would have been better (see Example 1)

[3] When $a_i = 0$, it is called a homogeneous quadratic form.

as $x_1^{(1)}, x_2^{(1)}, \ldots, x_n^{(1)}$. There is one obvious and important difference between this problem and Example 1. Whereas in the case of the roots of the polynomial there were n discrete minima, in this problem there are an infinite number of minima, whose location is a line in any parameter plane (x_i, x_j). In both examples we use the equations of steepest descent to reach a minimum. However, any minimum is a steady-state solution of the steepest-descent equations. We must, therefore, add some auxiliary terms to get the computer to plot the minimum line in the (x_i, x_j) plane after having reached it by steepest descent.

The distance function has only one dimension in the solution space and hence has the form

$$d_1 = [f(x_1, x_2, \ldots, x_n)]^2 \tag{8.40}$$

Then by Eq. (8.21)

$$\frac{dx_1}{dt} = -cf\frac{\partial f}{\partial x_1} \qquad \frac{dx_2}{dt} = -cf\frac{\partial f}{\partial x_2} \qquad \frac{dx_n}{dt} = -cf\frac{\partial f}{\partial x_n} \tag{8.41}$$

As we have already pointed out, Eqs. (8.41) will take us from an arbitrary initial set of parameter values to a particular set of values, $x_1^{(1)}, x_2^{(1)}, \ldots, x_n^{(1)}$, which are a solution of Eq. (8.39). If all but two parameters (x_i, x_j) are held fixed at this particular solution, the minimum, $d_1 = 0$, is a line in the (x_i, x_j) plane. The computer will plot the entire solution of Eqs. (8.41) in the (x_i, x_j) plane when we add proper terms to the velocities dx_i/dt and dx_j/dt so that they are not zero along the curve $d_1 = 0$ [or the identical curve $f(x_1, x_2, \ldots, x_n) = 0$].

We shall acquire these additional terms by utilizing an important property of curves. We begin by defining s as the arc length along the curve $d_2 = 0$ in the (x_i, x_j) plane. Since $f(x_i, x_j) = 0$ everywhere on the curve $d_1 = 0$,

$$\frac{df}{dt} = \frac{\partial f}{\partial x_i}\frac{dx_i}{dt} + \frac{\partial f}{\partial x_j}\frac{dx_j}{dt} = 0 \tag{8.42}$$

From Eq. (8.42) the two component derivatives dx_i/dt and dx_j/dt along the curve s are related by equations

$$\frac{dx_i/dt}{dx_j/dt} = -\frac{\partial f/\partial x_j}{\partial f/\partial x_i} \tag{8.43}$$

Equation (8.43) may also be written as two equations:

$$\frac{dx_i}{dt} = \mp\frac{\partial f}{\partial x_j} \tag{8.44}$$

$$\frac{dx_j}{dt} = \pm\frac{\partial f}{\partial x_i} \tag{8.45}$$

In Eqs. (8.44) and (8.45) there is an ambiguity of sign. Since these equations are used to move the path along the contour $f = 0$, we need only remember that, if the positive sign is chosen for one equation, the negative sign must be chosen for the other. The particular combination of signs will determine which of the two directions along $f = 0$ is taken. Observe that Eqs. (8.44) and (8.45) can also be used to trace the path along any contour $f = c$, where $c \neq 0$.

In summary we use Eqs. (8.41) to minimize $f(x_1, x_2, \ldots, x_n)$ from an arbitrary starting point in the parameter space. Then, holding all other parameters fixed at the values they attained when $f = 0$, we vary x_i and x_j by Eqs. (8.44) and (8.45) to plot the minimum in the (x_i, x_j) plane.

8.5.1 Parameter Constraints[1]

The two examples we have just considered have one important property in common. The parameters are allowed to vary unrestrictedly along a steepest-descent path. However, particularly in physical systems, our parameters are not so unfettered. Constraints are placed on parameters individually or in some functional relationship with other parameters and may be expressed as equalities or inequalities. Because the concepts associated with inequality constraints are easier to grasp, we shall present them first. We shall then show how these computer techniques can be extended to find extrema with equality constraints.

Suppose we want to minimize the distance function $f(x_1, x_2, \ldots, x_n)$ subject to the constraints

$$\begin{aligned} g_1(x_1, x_2, \ldots, x_n) &\geq 0 \\ g_2(x_1, x_2, \ldots, x_n) &\geq 0 \\ \cdots\cdots\cdots &\cdots \\ g_m(x_1, x_2, \ldots, x_n) &\geq 0 \end{aligned} \tag{8.46}$$

Observe that we do not restrict our discussion by making the right side of the constraint equations zero. If a constraint is given as

$$h_i(x_1, x_2, \ldots, x_n) \geq A_i \tag{8.47}$$

we may rewrite it as

$$g_i(x_1, x_2, \ldots, x_n) = h_i(x_1, x_2, \ldots, x_n) - A_i \geq 0 \tag{8.48}$$

If the distance function and the constraints are linear combinations of the parameters, it is called a linear programming problem. Otherwise it is referred to as a nonlinear programming problem.[2] The linear programming problem has only one optimum to which the computer will converge from any starting point, a property not usually shared by the nonlinear programming problem. Any constraint may involve some or all of the parameters. On the other hand, some of the parameters may not be constrained at all.

8.5.2 Geometric Explanation

The parameters $x_1, x_2, \ldots, x_n$ to be optimized are the coordinates of an n-dimensional euclidean space. Each constraint is a surface called a boundary in some subspace of this space. The boundary is defined by the equality constraint, for example, $g_i = 0$, and the parameters in the constraint equation are the coordinates of the subspace in which the boundary lies. On one side of this boundary, where $g_i > 0$, is the permissible region

[1] Several of the ideas in this section were suggested by Ref. 99. For one of the early papers on the subject see also Ref. 100.

[2] In linear and nonlinear programming problems the distance function is usually called the objective function.

of the parameters of the subspace. On the other side of the boundary ($g_i < 0$) is the forbidden region where points are excluded from the solution by the constraint. Any boundary which lies wholly within a forbidden region defined by two or more other constraints is superfluous.

When the starting point in the n-dimensional space is within the permissible region, the trajectory is along a steepest-descent path until this path intersects a boundary. If this path were extended across the boundary, it would move into the forbidden region. Instead, from the point of intersection onward, the path is constrained to be along the boundary. Thus in the subspace in which the boundary lies the path is tangential to the boundary $g_i = 0$, whereas in the subspace of the remaining parameters the path is tangential to the gradient of the distance function $f(x_1, x_2, \ldots, x_n)$.

The path continues along the boundary until the steepest-descent trajectory is directed into the permissible region. At this point the path moves off the boundary along the gradient and the inequality $g_i > 0$ holds again. Of course it is possible that while tangential to g_i the path may encounter boundaries imposed by other constraints. In this circumstance these other constraints will modify the path just as g_i did. Constrained paths have the following characteristics: (1) When the solution exists on a boundary, the optimum is not an extremum in the mathematical sense [see Eq. (8.6)]. (2) When the solution is in the interior of the permissible region for a descent path which has been along one or more boundaries, the optimum is not necessarily the same as the one that would be obtained for an unrestrained distance-function minimization.

The word picture of motion in the hyperspace which we have painted serves to delineate the manner in which the computer should be programmed. We recapitulate these requirements: (1) When the inequalities are satisfied, the computer should solve the conventional steepest-descent equations. (2) At the intersection of a steepest-descent path and a boundary the computer should keep the parameters from assuming forbidden values. (3) If initially the parameters have values which make one or more of the constraint equations negative, the computer must move the parameters to the permissible region as quickly as possible.

We can achieve all of these programming objectives in the following manner. First, we construct a distance function d_c related to the constraints and of the form

$$d_c = \sum_{i=1}^{m} [p_i(x_1, x_2, \ldots, x_n)]^2 \tag{8.49}$$

where

$$\begin{aligned} p_i &= 0 \qquad \text{for } g_i \geq 0 \\ p_i &= g_i \qquad \text{for } g_i < 0 \\ i &= 1, 2, \ldots, m \end{aligned}$$

We program a new distance function d_t to be minimized, where

$$d_t = af(x_1, x_2, \ldots, x_n) + bd_c \tag{8.50}$$

Both a and b are constants and for reasons soon made apparent we require $b \gg a$.

Thus the set of differential equations programmed for minimization is

$$\frac{dx_j}{dt} = -\frac{\partial d_t}{\partial x_j} \qquad j = 1, 2, \ldots, m \tag{8.51}$$

or

$$\frac{dx_j}{dt} = -a\frac{\partial f}{\partial x_j} - 2b \sum_{i=1}^{m} p_i \frac{\partial p_i}{\partial x_j} \qquad j = 1, 2, \ldots, m \tag{8.52}$$

If the optimization were along a path of steepest ascent, we would change from $-a$ to a in Eq. (8.52). We may write the set of equations embodied in Eq. (8.52) as a single vector equation:

$$\frac{d\mathbf{x}}{dt} = -a\,\nabla\mathbf{f} - 2b \sum_{i=1}^{m} p_i\,\nabla p_i \tag{8.53}$$

where

$$\frac{d\mathbf{x}}{dt} = \frac{dx_1}{dt}\mathbf{i}_1 + \frac{dx_2}{dt}\mathbf{i}_2 + \cdots + \frac{dx_n}{dt}\mathbf{i}_n \tag{8.54}$$

$$\nabla\mathbf{f} = \text{grad}\,\mathbf{f} = \frac{\partial f}{\partial x_1}\mathbf{i}_1 + \frac{\partial f}{\partial x_2}\mathbf{i}_2 + \cdots + \frac{\partial f}{\partial x_n}\mathbf{i}_n \tag{8.55}$$

$$\nabla p_i = \frac{\partial p_i}{\partial x_1}\mathbf{i}_1 + \frac{\partial p_i}{\partial x_2}\mathbf{i}_2 + \cdots + \frac{\partial p_i}{\partial x_n}\mathbf{i}_n \tag{8.56}$$

$\mathbf{i}_1, \mathbf{i}_2, \ldots, \mathbf{i}_n$ are unit vectors along the coordinates $x_1, x_2, \ldots, x_n$, respectively.

Equation (8.53) meets all the requirements prescribed for the computer to find a constrained minimum. A problem beginning with all inequality constraints met would start on a steepest-descent trajectory because $d_c = 0$. The moment a constraint, say g_1, becomes negative, a typical parameter x_3 affected by these constraints has the steepest-descent equation changed to

$$\frac{dx_3}{dt} = -a\frac{\partial f}{\partial x_3} - 2bp_1\frac{\partial p_1}{\partial x_3} \tag{8.57}$$

Now two terms are controlling dx_3/dt. The one, $-a\,\partial f/\partial x_3$, will move x_3 along a steepest-descent path of f to decrease g_1 (or p_1). The other, $-2bp_1\,\partial p_1/\partial x_3$, will move x_3 along a steepest-descent path of g_1^2 to increase g_1 to zero. Together these two terms have an opposite effect which will cause dx_3/dt to take x_3 to some point outside the boundary, i.e., where $g_1 < 0$. If $b \gg a$, the second term will predominate and dx_3/dt will vary so x_3 is essentially at the boundary $g_1 = 0$. The behavior of all other parameters in the constraint g_1 follows a pattern similar to the one described for x_3.

The trajectory will move along the boundary until the steepest-descent path of f is directed inward toward the permissible space. Now dx_3/dt and the other derivatives are such that $g_1 > 0$ and $p_1 = 0$. This restores the computer program to the solution along the gradient of f alone. Eventually by this process, with the possibility that other constraints may also be involved, the computer will find the parameter values for the constrained extremum.

The computer will solve the problem equally well even when the initial parameter values make one or more of the constraints negative. However, because the constant b is made very large (usually the gain of an operational amplifier), we can get excessive computer voltages during the transient stage when the distance function d_c is being minimized. In order to protect the computer equipment we use limiters so that

$$0 \leq \left| 2b \sum_{i=1}^{m} p_i \frac{\partial p_i}{\partial x_j} \right| \leq K \tag{8.58}$$

where K is as large a constant as is permitted by the computer.

This method breaks down when $\nabla g_i = 0$† at the boundary even if d_c is like Eq. (8.25*b*). In this event we no longer have the condition where in Eq. (8.52) the second term is large compared with the first. As a result x_3 moves beyond the boundary, driving g_i negative until ∇g_i is no longer zero.

8.5.3 *Equality Constraints*

If we wish to minimize the distance function $f(x_1, x_2, \ldots, x_n)$ subject to the constraints

$$g_i(x_1, x_2, \ldots, x_n) = 0 \qquad i = 1, 2, \ldots, m; m < n \tag{8.59}$$

we program the computer to minimize the distance function d_t, whose equation is

$$d_t = af(x_1, x_2, \ldots, x_n) + b \sum_{i=1}^{m} g_i^2(x_1, x_2, \ldots, x_n) \tag{8.60}$$

where $b \gg a$.

Now the path to a constrained extremum will always be along the boundary defined by Eq. (8.59).

8.5.4 *Tracking Constrained Minima*

Once the computer solution has come to rest at a constrained minimum, the distance function f and the constraints g_i may be changed and the computer will seek the constrained minimum as it changes position in the parameter space. In Example 1 we described how the computer tracks a moving minimum without parameters constraints. Except that the path may not be one of steepest descent because of the constraints, the phenomenon in

† $\nabla g_i = 0$ only when

$$\left(\frac{\partial g_i}{\partial x_1}\right)^2 + \left(\frac{\partial g_2}{\partial x_2}\right)^2 + \cdots + \left(\frac{\partial g_i}{\partial x_n}\right)^2 = 0$$

hat is, when all partial derivatives are zero simultaneously.

principle is the same when constraints are applied. We shall therefore not describe anew the mechanism of computer tracking. Of course, as in Example 1, the distance function and constraints must be changed slowly enough to be tracked by the computer.

Example 3. We show the behavior of constraints by considering the polynomial $w(z)$ of Example 1, where

$$w(z) = z^3 - 5z^2 + 8z - 6 = u + iv \tag{8.61}$$

and

$$z = x + iy = re^{i\theta} \qquad x = r\cos\theta \qquad y = r\sin\theta \tag{8.62}$$

We now minimize the distance function

$$d = u^2 + u^2 \tag{8.63}$$

subject to the restraints

$$g_1 \geq 0 \qquad g_2 \geq 0$$

where

$$g_1 = (x - 1.5)^2 + 4(y - 1)^2 - 1 \qquad g_2 = 0.4x + 0.5y - 1 \tag{8.64}$$

Utilizing the constraints we construct a new distance function d_1 to be minimized:

$$d_1 = d + k(h_1{}^2 + h_2{}^2) \qquad k \gg 1 \tag{8.65}$$

where

$$h_1 = \begin{cases} g_1 & \text{when } g_1 < 0 \\ 0 & \text{when } g_1 \geq 0 \end{cases}$$

$$h_2 = \begin{cases} g_2 & \text{when } g_2 < 0 \\ 0 & \text{when } g_2 \geq 0 \end{cases}$$

The mathematical operations are more easily performed if both the distance function and restraints are expressed in terms of the polar coordinates r and θ. As a result we obtain from Eqs. (8.26a) and (8.26b)

$$\begin{aligned} u &= r^3\cos 3\theta - 5r^2\cos 2\theta + 8r\cos\theta - 6 \\ v &= r^3\sin 3\theta - 5r^2\sin 2\theta + 8r\sin\theta \end{aligned} \tag{8.66}$$

and

$$\begin{aligned} g_1 &= (r\cos\theta - 1.5)^2 + 4(r\sin\theta - 1)^2 - 1 \\ g_2 &= 0.4r\cos\theta + 0.5r\sin\theta - 1 \end{aligned} \tag{8.67}$$

The descent equations are

$$\begin{aligned} \frac{dr}{dt} &= -\left[\frac{\partial d}{\partial x} + 2k\left(h_1\frac{\partial h_1}{\partial r} + h_2\frac{\partial h_2}{\partial r}\right)\right] \\ \frac{r\,d\theta}{dt} &= -r\left[\frac{\partial d}{\partial \theta} + 2k\left(h_1\frac{\partial h_1}{r\,\partial\theta} + h_2\frac{\partial h_2}{r\,\partial\theta}\right)\right] \end{aligned} \tag{8.68}$$

Evaluating the component parts of Eqs. (8.68), we have

$$\frac{\partial d}{\partial r} = 2u\frac{\partial u}{\partial r} + 2v\frac{\partial v}{\partial r} \tag{8.69}$$

$$\frac{\partial u}{\partial r} = 3r^2\cos 3\theta - 10r\cos 2\theta + 8\cos\theta \tag{8.70}$$

$$\frac{\partial v}{\partial r} = 3r^2\sin 3\theta - 10r\sin 2\theta + 8\sin\theta \tag{8.71}$$

$$\begin{aligned} \frac{\partial h_1}{\partial r} &= \frac{\partial g_1}{\partial r} = 2\cos\theta(r\cos\theta - 1.5) + 8\sin\theta(r\sin\theta - 1) && \text{for } g_1 < 0 \\ \frac{\partial h_1}{\partial r} &= 0 && \text{for } g_1 > 0 \end{aligned} \tag{8.72}$$

$$\frac{\partial h_2}{\partial r} = \frac{\partial g_2}{\partial r} = 0.4 \cos \theta - 0.5 \sin \theta \qquad \text{for } g_2 < 0$$
$$\frac{\partial h_2}{\partial r} = 0 \qquad \text{for } g_2 \geq 0 \tag{8.73}$$

$$\frac{\partial d}{r\,\partial \theta} = \frac{2}{r}\left(u \frac{\partial u}{\partial \theta} + v \frac{\partial v}{\partial \theta}\right) \tag{8.74}$$

$$\frac{\partial u}{\partial \theta} = -3r^3 \sin 3\theta + 10r^2 \sin 2\theta - 8r \sin \theta$$
$$\frac{\partial v}{\partial \theta} = 3r^3 \cos 3\theta - 10r^2 \cos 2\theta + 8r \cos \theta \tag{8.75}$$

$$\frac{\partial h_1}{\partial \theta} = \frac{\partial g_1}{\partial \theta} = 2r \sin \theta(r \cos \theta - 1.5) + 8r \cos \theta(r \sin \theta - 1) \qquad \text{for } g_1 < 0$$
$$\frac{\partial h_1}{\partial \theta} = 0 \qquad \text{for } g_1 \geq 0 \tag{8.76}$$

$$\frac{\partial h_2}{\partial \theta} = -0.4r \sin \theta + 0.5 \cos \theta \qquad \text{for } g_2 \leq 0$$
$$\frac{\partial h_2}{\partial \theta} = 0 \qquad \text{for } g_2 > 0 \tag{8.77}$$

In Fig. 8.11 we show a partial block diagram of these equations, designed to show how e computer programs the constraints. Operational amplifier 1 has as its input $-g_1$.

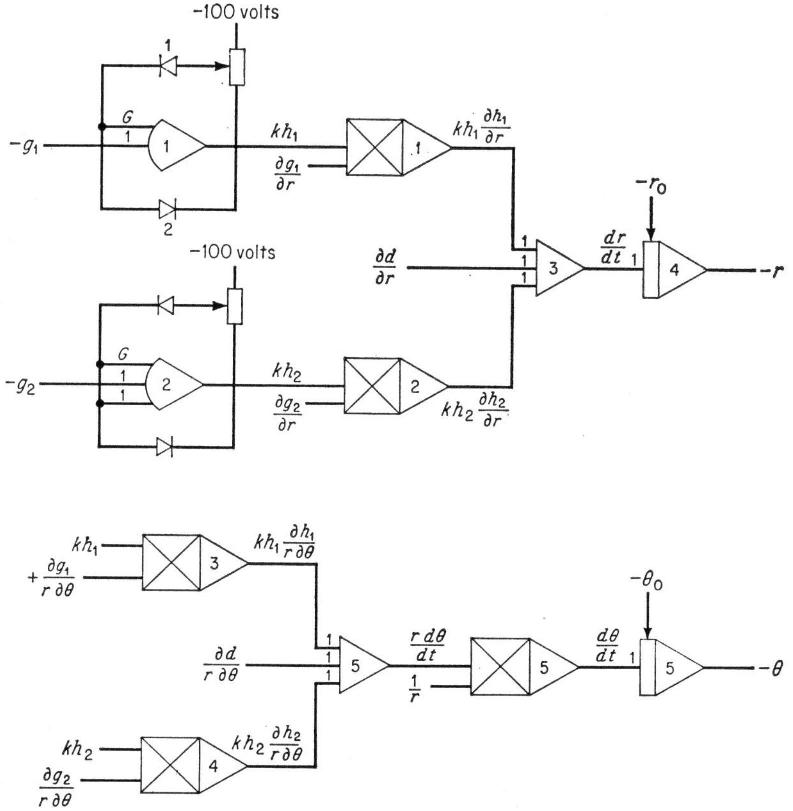

FIG. 8.11 A program for optimization with constraints. (See Example 3.)

Without limiters its output would be kg_1. However, diode 1 limits the output at som large positive value and diode 2 keeps the output voltage from falling below zero— precisely the condition for kh_1. One further point of interest is that the input to multiplier 1 is $\partial g_1/\partial r$ instead of $\partial h_1/\partial r$. We note that without exception

$$h_1 \frac{\partial g_1}{\partial r} = h_1 \frac{\partial h_1}{\partial r} \tag{8.78}$$

The rest of the circuit consists of terms which combine to give r and θ at the outputs o integrators 4 and 5. In the complete program the quantities in the diagram which ar inputs to elements but not shown as outputs of other elements, for example, $\partial d/(r\ \partial\theta)$, $\partial g_2/\partial r$, are generated as functions of r and θ.

Figure 8.12 shows contour lines, constraint boundaries, and trajectories in a portion c the parameter space. The shaded region is the forbidden one. See also Fig. 8.9*a* fo paths without parameter constraints.

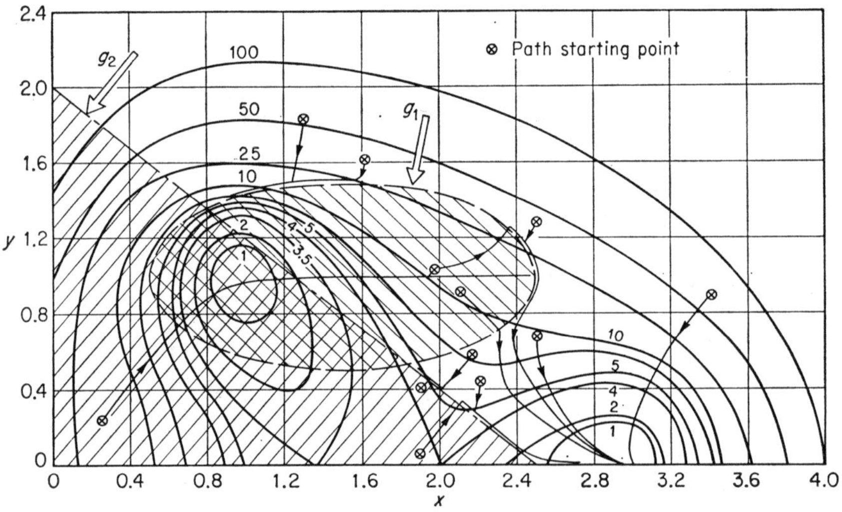

FIG. 8.12 Steepest-descent paths with constraints. (See Example 3.)

8.5.5 *Discrete Approximation of Steepest Descent*

We have discussed methods of formulating a set of first-order differentia equations, whose solution results in a continuous variation of the parameter until an equilibrium (optimum) condition is reached. These differentia equations cannot be formulated in at least two important instances, whic are:

1. The distance function is not known analytically.
2. The value of the distance function cannot be determined immediatel after a parameter change.

If the distance function is not known analytically, we cannot find th expression for the partial derivatives which constitute the gradient. Unles we know these partial derivatives, we obviously cannot program the steepest descent equations. Now consider the second case. Suppose we have a dis tance function $F(x_1, x_2, \ldots, x_n, t)$ defined as

$$F(x_1, x_2, \ldots, x_n) = \int_0^{t_1} |g(x_1, x_2, \ldots, x_n, t)|\, dt \tag{8.79}$$

where t_1 is a constant. After the parameter change has been made, $|g(x_1, x_2, \ldots, x_n)|$ must be integrated from $t = 0$ to $t = t_1$ before the new value of F can be evaluated. The partial derivatives of F which comprise the gradient also require the same length of time to evaluate. Thus for the parameter x_i

$$\frac{\partial F}{\partial x_i} = \int_0^{t_1} (\text{sign } g) \frac{\partial g}{\partial x_i}\, dt \tag{8.80}$$

It makes no sense to write

$$\frac{dx_i}{dt} = -c \frac{\partial F}{\partial x_i} \tag{8.81}$$

because we cannot simultaneously wait a time t_1 to calculate dx_i/dt and vary this same quantity continuously. We may, however, approximate a continuous steepest-descent path if we time-scale the computer so that the integration of Eq. (8.79) is on a much faster time scale than the differentiation of Eq. (8.80).

Lacking the wherewithal to change the parameters continuously, we must vary them instead in discrete increments. We adopt the following procedure in calculating the incremental changes in parameter values. At a point A in the parameter space the partial derivatives $(\partial F/\partial x_1)_A$, $(\partial F/\partial x_2)_A$, $\ldots$, $(\partial F/\partial x_n)_A$ are evaluated. Next, we assume that the gradient does not change. As a result we move in a straight-line path, i.e., a path in which the vector dx/dt does not change direction in parameter space from point A to point B, so that the parameter values at B become

$$\begin{aligned} x_{1B} &= x_{1A} - k\left(\frac{\partial f}{\partial x_1}\right)_A \\ &\cdots\cdots\cdots\cdots \\ x_{nB} &= x_{nA} - k\left(\frac{\partial f}{\partial x_n}\right)_A \end{aligned} \tag{8.82}$$

where the subscripts A and B indicate the value at A and B, respectively. At B the process is continued until by a series of straight-line paths the minimum is approached. The value k determines the length of the straight-line segment and dictates the ultimate closeness to the minimum. In the final steps k is made increasingly smaller to reduce the distance function below a predesignated level.

We can justify Eqs. (8.82) by the following reasoning. For simplicity and without any loss of generality we shall assume that the distance function F is dependent upon only two parameters, x and y. The equation for an incremental change in the function ΔF because of discrete changes in the parameters Δx and Δy is[1]

$$\Delta F = \frac{\partial F}{\partial x}\Delta x + \frac{\partial F}{\partial y}\Delta y + \epsilon(|\Delta x| + |\Delta y|) \qquad (8.83a$$

where $\epsilon \to 0$ when $(\Delta x, \Delta y) \to 0$. If Δx and Δy are sufficiently small,

$$\Delta F \approx \frac{\partial F}{\partial x}\Delta x + \frac{\partial F}{\partial y}\Delta y \qquad (8.83b$$

By arguments similar to those used to derive Eq. (8.21), the discret steepest-descent equations become

$$\Delta x = -k\frac{\partial F}{\partial x} \qquad \Delta y = -k\frac{\partial F}{\partial y} \qquad (8.84a$$

Thus, if x_{1A} and y_{1A} are the values at point A, then at point B

$$x_{1B} = x_{1A} + (\Delta x)_A \qquad y_{1B} = y_{1A} + (\Delta y)_A \qquad (8.84b$$

where $(\Delta x)_A$ and $(\Delta y)_A$ are evaluated at point A.

A distance function which requires a finite time to evaluate after a pa rameter change is the solution of a differential equation. The ter $g(x_1, x_2, \ldots, x_n, t)$ in Eq. (8.79) definitely falls in the category of being th solution to some differential equation. As a result we can use the paramete influence technique (see Sec. 7.4) and program the computer to give th partial derivative of $g(x_1, \ldots, x_n, t)$ with respect to any parameter. Thu a relationship like $\partial F/\partial x_i$ in Eq. (8.80) can be calculated by the compute

If we are disinclined to use the extra equipment required for the param ter-influence method or if the equations cannot be expressed analyticall we must approximate the partial derivatives comprising the gradien Again for brevity of notation we shall describe the procedure for obtainin approximate partial derivatives where only two parameters are involve

Suppose we already have $F(x_0, y_0)$. Next we obtain two other values o F for the pairs of parameter values, (x_a, y_0) and (x_0, y_b), which are close t the point (x_0, y_0). Then the difference approximations to the parti derivatives become

$$\frac{F(x_a, y_0) - F(x_0, y_0)}{x_a - x_0} \approx \left(\frac{\partial F}{\partial x}\right)_{x_0,y_0}$$
$$\frac{F(x_0, y_b) - F(x_0, y_0)}{y_b - y_0} \approx \left(\frac{\partial F}{\partial y}\right)_{x_0,y_0} \qquad (8.85a$$

From the starting parameter values (x_0, y_0) we move along a straight-lin path in the (x, y) parameter plane to the new values (x_1, y_1) by the relatio

$$x_1 = x_0 - c\left(\frac{\partial F}{\partial x}\right)_{x_0,y_0} \qquad y_1 = y_0 - c\left(\frac{\partial F}{\partial y}\right)_{x_0,y_0} \qquad (8.85b$$

[1] F is assumed to be a differentiable function. For a discussion of what constitutes differentiable function, see Ref. 102, p. 163.

The positive constant c is arbitrary. Since the number of steps required to reach a minimum is directly related to c, one should make c as large as possible. However, if c is too large, the computer may oscillate around the minimum without ever reaching it. Probably the best compromise is to reduce c as the minimum is approached.

The difference between continuous and discrete steepest descent is illustrated for a two-parameter space in Fig. 8.13 in the minimization of a dis-

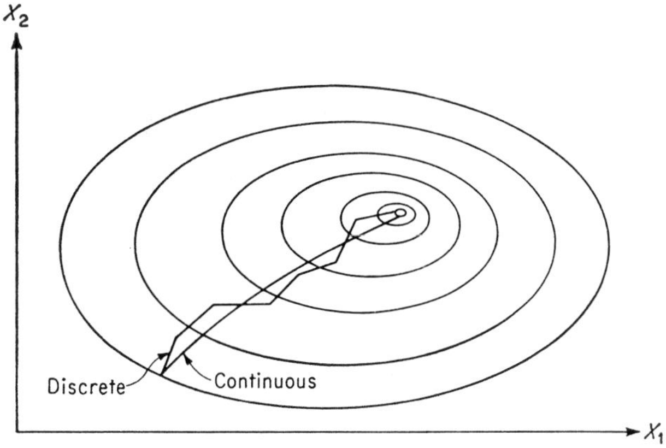

FIG. 8.13 A comparison of discrete and continuous steepest descent.

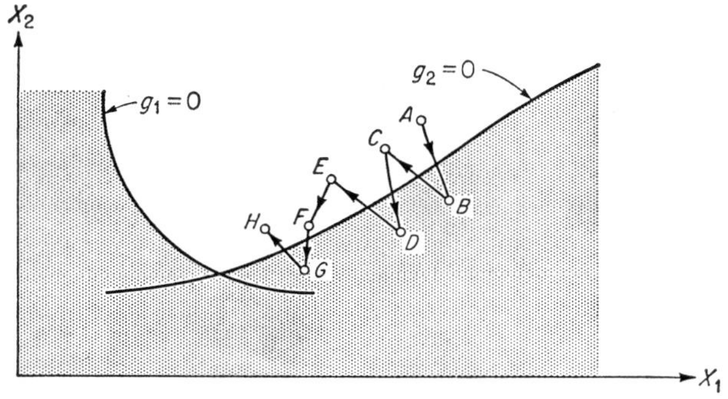

FIG. 8.14 A discrete steepest-descent path with constraints. The shaded region is excluded by the constraints.

tance function. The closed curves represent contour lines of the distance function.

We have special but not insurmountable computer problems when the discrete-steepest-descent method is used to minimize a distance function F with inequality constraints. As long as the parameter increments keep the parameters in the permissible region, the minimization path is the path expressed by Eqs. (8.82). However, the parameters may take a discrete step which moves them across a boundary and into a forbidden region. As shown in Fig. 8.14, the movement could be from point A to point B, where the shaded region is the unallowable one. At this point we cease using the distance function F and take one of two courses of action which will move the parameter values back into the permissible part of the space. One alternative is to find the gradient at point B for a constraint distance function whose minimum is at the boundary. Equation (8.49) is such a distance function. Then with the particular function determining the path a large enough incremental step is taken to return to the permissible region.

If we observe that, once in the forbidden region we wish only to recross the boundary, we may operate with the alternative distance function

$$G = \sum_{i=1}^{m} h_i \tag{8.86}$$

where
$$h_i = \begin{cases} g_i & \text{when } g_i \leq 0 \\ 0 & \text{when } g_i > 0 \end{cases}$$

When the inequality constraints are of the form $g_i \geq 0$, they are violated when they are less than zero. Thus, we move out of a forbidden region by increasing G. This means that we take a *discrete-steepest-ascent* path on G from point B. For $g_i \leq 0$ a *discrete-steepest-descent* path will move the parameter back to the permissible region. Once in this region the point again moves along a discrete-steepest-descent path specified by F.† Starting at point A, the parameters might take a zigzag path across a boundary, as shown in Fig. 8.14.

This technique requires a great deal of calculation. A much simpler procedure which achieves the same results is the relaxation method, described below.

However, the discrete steepest descent yields one useful piece of information. In its application one observes the influence of the parameters, i.e., the sensitivity of the minimizing function to the parameter changes, at the discrete points of evaluation of the partial derivatives.

8.6 Relaxation[1]

Relaxation methods were known to mathematicians in the early nineteenth century, having been initially developed by Gauss around 1823. After a promising beginning the method lapsed into obscurity. When Southwell redeveloped the technique in England in the 1930s, he was totally unaware of Gauss' original work. Since that time the method has been exploited by applied mathematicians particularly in the solution of partial differential equations.

† A device which minimizes a function in this fashion with constraints is described in Refs. 103 and 104.

[1] See Refs. 105 and 95 (chap. 17, pp. 428–442) by George E. Forsythe for additional information on relaxation methods.

When the minimization must be performed by discrete straight-line segments in the parameter space, relaxation requires little of the calculation inherent in steepest descent and, hence, is easier to implement. In the relaxation method, as in discrete steepest descent, the initial path is along a straight line tangent to the gradient at the starting point. However, unlike discrete descent, whose parameters are changed only a relatively small amount, relaxation continues the straight path until a local minimum is reached.[1] At the local minimum the path is moved orthogonally to the original path and continued until another local minimum is reached.[2] This process is repeated until the actual minimum is reached. A modified relaxation method which achieves the same result but is easier to implement on the computer begins by varying one of the parameters and holding the others fixed. The parameter is varied until a local minimum is reached.

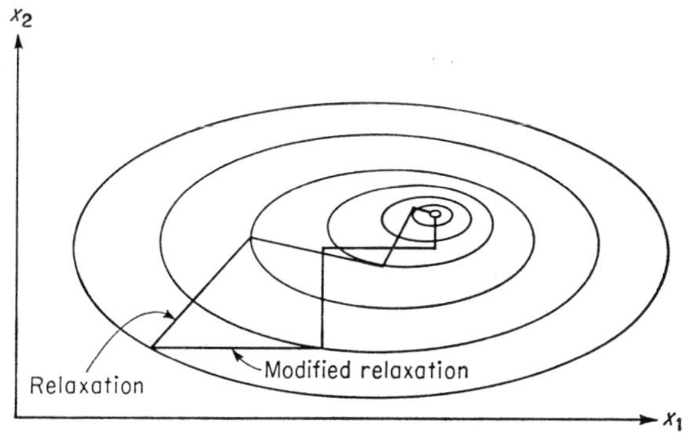

FIG. 8.15 A comparison of relaxation and modified relaxation.

The iterative process continues with each parameter varied in turn until the minimum is reached. The process of minimization can often be performed in fewer steps by making the switch to the next parameter if the local minimum is not reached in a specified number of steps.[3]

By way of contrast, Fig. 8.15 shows the path for the relaxation method of the same distance function whose minimization in Fig. 8.11 is by steepest descent.

Example 4. Determination of Roots of Polynomials by Relaxation [106]. One way of determining the roots of a polynomial is by using relaxation to minimize d_1 or d_2 as specified by Eqs. (8.25*a*) and (8.25*b*), where u and v are defined by Eqs. (8.26*a*) and (8.26*b*). It is a straightforward operation to change first r and then θ until a minimum

[1] The local minimum is the point on the straight-line path at which the distance function assumes its lowest value.

[2] Two paths are orthogonal if their vector inner (dot) product is zero.

[3] Hans Wittsenhausen of Electronic Associates, Inc., disclosed this fact to me in a private conversation, after he had used the technique successfully in minimizing several distance functions.

(root) is reached. We shall consider in this example a somewhat different formulation of the problem where the amplitude of a carrier sine wave is the distance function to be minimized.

The polynomial in Eq. (8.22) may be represented as a sum of n vectors by introducing polar coordinates. This yields

$$w(z) = a_n r^n e^{in\theta} + a_{n-1} r^{n-1} e^{i(n-1)\theta} + \cdots + a_1 r e^{i\theta} + a_0 = Ne^{i\phi} \tag{8.87}$$

where $\quad z = x + iy = re^{i\theta} \qquad N = (u^2 + v^2)^{\frac{1}{2}} \qquad \phi = \tan^{-1} \dfrac{v}{u}$

In Fig. 8.16 the individual vectors of a fourth-degree polynomial are added to give the resultant vector $Ne^{i\phi}$, whose modulus N must be reduced to zero. Note that N fulfills the requirements of a distance function and is similar to d_1.

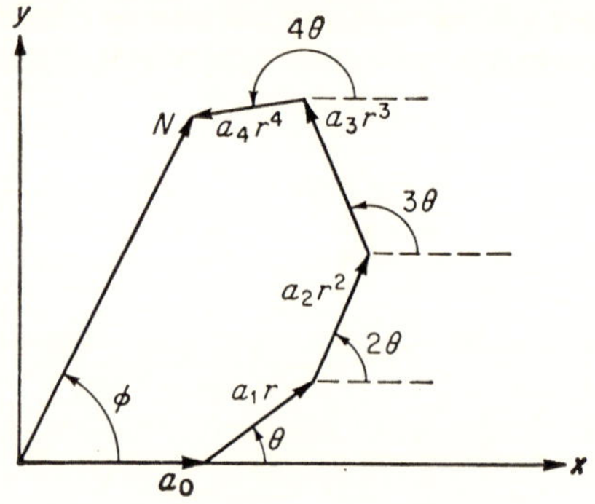

FIG. 8.16 A vector polygon.

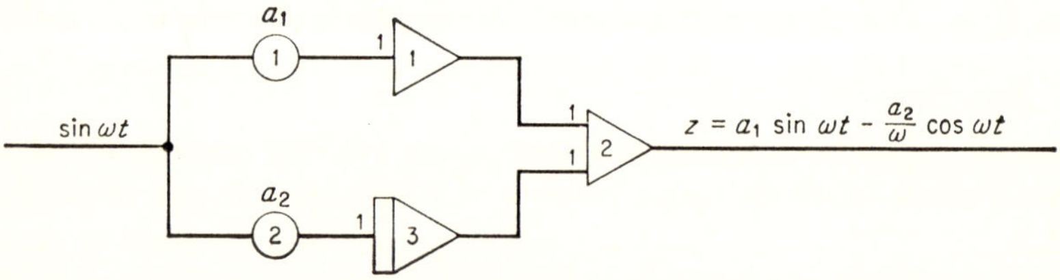

FIG. 8.17 A linear operator.

On the computer the vectors are obtained in the following manner. A sine wave of fixed frequency is passed through n identical cascaded networks, each of which shifts the phase of the input signal by an angle θ and multiplies its magnitude by r. These networks may, therefore, be interpreted as complex operators[1] multiplying their respective input signals by $z = re^{i\phi} = x + iy$. Each network requires basically two amplifiers in parallel with an integrator, as we show in Fig. 8.17. Here the input to the two potentiometers is $\sin \omega t$, and the output of amplifier 2 is

$$\begin{aligned} z &= a_1 \sin \omega t - \frac{a_2}{w} \cos \omega t = \sin(\omega t + \theta) \\ &= re^{i\theta} \sin \omega t = (x + iy) \sin \omega t \end{aligned} \tag{8.88}$$

where $\quad r = \left[a_1^2 + \left(\dfrac{a_2}{w}\right)^2\right]^{\frac{1}{2}} \qquad \theta = \tan^{-1} \dfrac{-a_2/w}{a_1}$

[1] Operators are discussed in detail in Chap. 3.

Thus, the operator shown in Fig. 8.17 is

$$x = a_1 \qquad y = -\frac{a^2}{\omega}$$

If the n intermediate voltages are added properly (see Fig. 8.18), the resultant output of the computer is $Ne^{i\phi} \sin \omega t$. The quantity equals zero only if $N = 0$ and hence, by Eqs. (8.87), only if $x + iy$ is a root of the polynomial. As Fig. 8.16 shows, N is measured as the amplitude of the sinusoidal output of the polynomial circuit.

The value of N is reduced to zero by adjusting x and y, the gain factors of the amplifiers and integrators in the network, alternately in a minimizing sequence. Thus, if the process is started at $z_0 = x_0 + iy_0$, the coordinate x is varied, with y held constant until a relative minimum is reached. The value of N may be observed on a recorder, meter, or any other sine-wave peak-reading device. Next x is held constant, while y is varied, until a new relative minimum is reached, and this process is continued until N decreases to zero. Note that the phase angle ϕ is not required and, hence, is not measured in the nulling process.

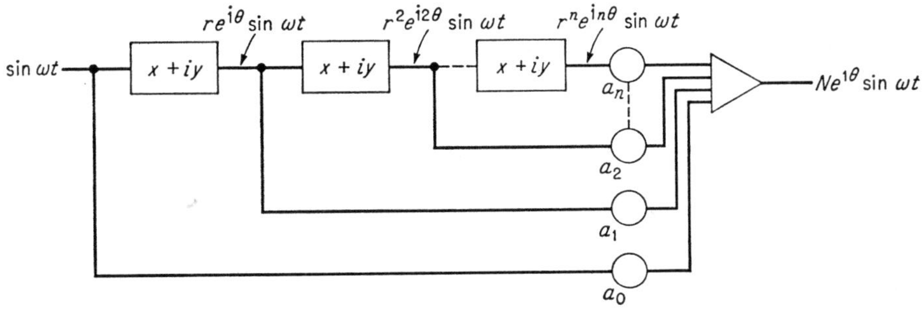

FIG. 8.18 A circuit for finding roots of polynomials by the relaxation method.

8.6.1 *Step Size*[1]

In discrete-gradient methods there is always the possibility that the distance function will be larger after an incremental change of parameters than before. This occurs because the step was large enough to carry the parameters beyond the minimum value of the distance function along the path. What we shall now consider is the establishment of a criterion for step size in the modified relaxation method which ensures that the step is always sufficiently small to reduce the distance function.

We begin with the distance function d, where

$$d = \sum_{i=1}^{n} f_i^2 \tag{8.89}$$

The coordinates f_i of the solution space can be functions of several variables. However, to focus attention on the technique we shall consider them only a function of the parameter α, which we are changing to reduce d. We may

[1] The ideas expressed in this section are taken mainly from Ref. 107.

denote $\mathbf{F}(\alpha)$ as a vector in the solution space, where

$$\mathbf{F}(\alpha) = f_1(\alpha)\mathbf{i}_1 + f_2(\alpha)\mathbf{i}_2 + \cdots + f_n(\alpha)\mathbf{i}_n \tag{8.90}$$

To obtain a path which reduces the distance function we make vector $\mathbf{F}(\alpha + \Delta\alpha)$ (the vector after an incremental change of parameter $\Delta\alpha$) orthogonal to the gradient of $\mathbf{F}$ evaluated at α. The condition can be expressed in terms of the vector inner product as

$$\sum_{i=1}^{n} \frac{\partial f_i}{\partial \alpha} \cdot f_i(\alpha + \Delta\alpha) = 0 \tag{8.91}$$

As a first step in solving Eq. (8.91) for $\Delta\alpha$, we expand $f_i(\alpha + \Delta\alpha)$ in a Taylor

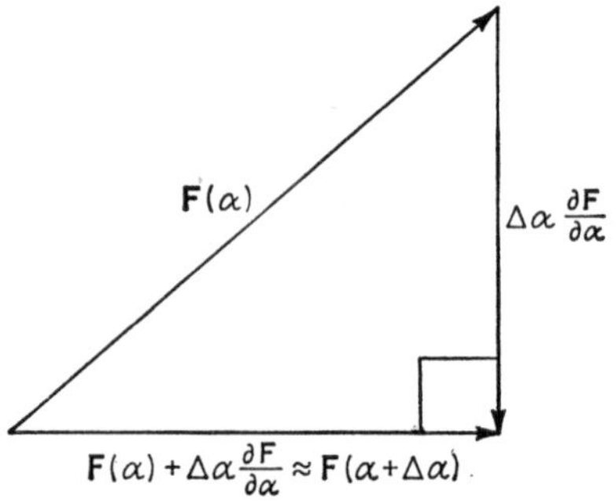

FIG. 8.19 A vector polygon which shows that the step-size criterion of Eq. (8.93) reduces the distance function.

series and keep only the first-order terms. Thus

$$\sum_{i=1}^{n} \frac{\partial f_i}{\partial \alpha} \cdot \left[f_i(\alpha) + \Delta\alpha \frac{\partial f_i}{\partial \alpha} \right] = 0 \tag{8.92}$$

Solving Eq. (8.92) for α, we get as the desired relationship for the incremental parameter change

$$\Delta\alpha = - \frac{\sum_{i=1}^{n} f_i \frac{\partial f_i}{\partial \alpha}}{\sum_{i=1}^{n} (f_i)^2} \tag{8.93}$$

By Eq. (8.92) the vector $\mathbf{F}(\alpha) + \Delta\alpha\ \partial\mathbf{F}/\partial\alpha$ is orthogonal to the vector $\partial\mathbf{F}/\partial\alpha$. The vector diagram is shown in Fig. 8.19. Since a leg of a right triangle is never larger than its hypotenuse, we obtain

$$\mathbf{F}(\alpha + \Delta\alpha) = \mathbf{F}(\alpha) + \Delta\alpha \frac{\partial \mathbf{F}}{\partial \alpha} \leq \mathbf{F}(\alpha) \tag{8.94}$$

The equality in Eq. (8.94) holds only for $\partial \mathbf{F}/\partial \alpha = 0$, which is a minimum of the distance function along the path.

We can also show that the distance function is reduced because at $\alpha + \Delta\alpha$ we have

$$d(\alpha + \Delta\alpha) = \sum_{i=1}^{n} \left(f_i + \Delta\alpha \frac{\partial f_i}{\partial \alpha} \right)^2$$
$$= \sum_{i=1}^{n} f_i^2 + 2 \sum_{i=1}^{n} \frac{\partial f_i}{\partial \alpha} \left(f_i + \Delta\alpha \frac{\partial f_i}{\partial \alpha} \right) \Delta\alpha - \sum_{i=1}^{n} \left(\frac{\partial f_i}{\partial \alpha} \right)^2 (\Delta\alpha)^2 \tag{8.95}$$

By Eq. (8.92)

$$d(\alpha + \Delta\alpha) = d(\alpha) + 0 - \sum_{i=1}^{n} \left(\frac{\partial f_i}{\partial \alpha} \right)^2 (\Delta\alpha)^2 \tag{8.96}$$

From Eq. (8.96) it follows that

$$d(\alpha + \Delta\alpha) \leq d(\alpha) \tag{8.97}$$

We note in conclusion that Eq. (8.93) arises from a Taylor series expansion of $f_i(\alpha + \Delta\alpha)$ where all but the first-order term was discarded. Therefore, this equation holds in general only when the increment $\Delta\alpha$ is sufficiently small. Observe in Eq. (8.93) that $\Delta\alpha$ becomes large in the vicinity of $d = 0$.

8.6.2 Constraints

Parameter constraints have no effect on the modified relaxation method as long as the parameter is in the permissible region. Should a particular incremental change move the parameter across a boundary, the minimization along the path will cease. For the next step an incremental change is made in the parameter opposite in sign (but not necessarily of the same magnitude) to the previous incremental value. This procedure is continued until the parameter is again in the permissible zone. Then the reduction of the distance function is carried out along the next parameter path.[1]

8.7 Random Search [109]

Of all the methods of optimization the random-search method is the least sophisticated mathematically. This blunderbuss technique of exploring a large part of the region instead of moving along a prescribed path may seem to rely too heavily on a brute-force approach. However, a computer technique should be evaluated not on the basis of the mathematical niceties but rather the ease with which we get the answer. Any method which removes the human operator from direct involvement in the running of the problem is worth considering, for the machine is always faster and more accurate.

[1] Reference 108 describes a problem which was solved by this method.

When the distance function F can be evaluated without delay after a parameter change, the random-search method is a rapid way of approaching a minimum. The minimization process begins with the parameters set at some nominal value, and the magnitude of the distance function F_0 is stored in the computer. Next each parameter is changed randomly around its nominal value. Whenever the value F_1 is less than the stored value F_0, the computer changes the nominal parameter settings to the values they have for F_1 and stores F_1. The parameters are then changed randomly around their new nominal values. This leapfrogging random search is continued until the minimum is reached.

This method works equally well when there are inequality parameter constraints. Now, however, the computer will change parameter values

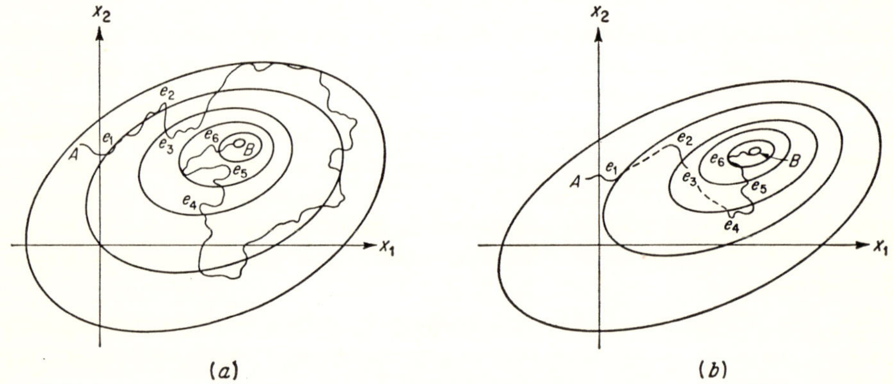

FIG. 8.20 (*a*) A typical path in parameter space due to a random variation of parameters; (*b*) a plot of (*a*), which indicates by the solid lines the portion of the parameter variation that reduces the distance function.

only when the distance function is less then before and simultaneously all the inequality constraints are satisfied.

The actual path taken in two dimensions might typically be the one shown in Fig. 8.20*a*. From starting point A the computer moves to B on the random path shown. The solid-line segments of Fig. 8.20*b* show the parameter values stored in the computer, i.e., the segments along which F is being reduced. The dashed lines indicate the discrete jumps from one set of parameter values to another and correspond to portions of the path in Fig. 8.20*b* where F is increasing.

Random search is useful in minimizing functions which can be expressed as algebraic equations. However, where the effect of a parameter change requires a finite time to evaluate, the method appears to be at a severe disadvantage compared with steepest descent or relaxation. For whereas the latter two methods have a logical procedure for finding the direction in which to change parameters, the random-search method blindly guesses what to use as parameter values. Hence it might have several useless runs for each one that reduces the distance function. Where each run may take several seconds or longer, this would be too great a waste of computer equipment and time.

There are two important features that differentiate the random search from the other gradient methods we have considered. First, from a given set of initial parameter values the path taken each computer run will be different. In the other methods the parameters move along a unique path from any starting point. Second, because each path is different, the computer solution may terminate at any one of the minima from an initial set of parameter values.

8.8 Optimization of Statistical Problems

The cardinal principle of optimization by gradient methods is that the distance functions must be reduced with each parameter change. This rule applies equally well for statistical and deterministic problems. However, in statistical problems the distance function is a random variable which poses certain difficulties because its value for a fixed set of problem parameters changes with each measurement. The characteristics and measurement of random variables are described in considerable detail in Chaps. 9 to 11. In this section we shall indicate briefly how such distance functions are optimized.

We can always formulate the distance function so that the random-variable quantity to be optimized is the expectation value (also called the mean or average). This is a number which is a constant for any given set of parameters. However, because the random variable changes from one measurement to the next, the mean requires an infinite number of measurements to be calculated precisely. More practically, the mean is evaluated with an error that decreases as $1/\sqrt{n}$, where n is the number of measurements.

Assume that we begin our optimization at a particular point in the parameter space. As a first step we compute the sample mean of the measurements of the random variable. Since there are only a finite number of samples, this mean differs from the true mean. Next we proceed to another point in the parameter space and take a new set of measurements. Suppose the second sample mean is larger than the first. What can we say about the relative magnitude of the true means? The strongest assertion that we can make is that, if we assume that the second true mean is larger than the first, we have a certain probability of being correct. This probability is, not surprisingly, related to the errors in the estimation of the means. *Therefore, in a statistical problem we shall sometimes direct the computer to change the parameters incorrectly.* How often a false move is made depends upon the errors in our estimation of the mean.

Because each measurement of a random variable represents computer running time, our primary concern in a statistical problem is to find the optimum value of a distance function in as few computer runs as possible. It does not necessarily follow that this is accomplished by reducing the number of false moves, for what counts is the total number of measurements of the distance function which are needed to find the minimum.

Qualitatively, one way to proceed could be by decreasing the distance between steps in the parameter space and increasing the number of samples. We note that, if at step 1 the parameter values are $a_{11}, a_{12}, \ldots, a_{1n}$ and at step 2 the parameter values are $a_{21}, a_{22}, \ldots, a_{2n}$, then the distance between these two sets of parameter values is $[(a_{11} - a_{21})^2 + (a_{12} - a_{22})^2 + \cdots + (a_{1n} - a_{2n})^2]^{\frac{1}{2}}$. This method has the advantage of increasing measurement accuracy and simultaneously decreasing the interval between measurements as we approach the optimum.

An even simpler procedure has been proposed by Keifer and Wolfowitz [112]. We describe the technique for a one-dimensional parameter space although it can be extended to multidimensional ones as well. Let the parameter be x for which the distance function is $f(x)$. Then on a given measurement the quantity obtained is

$$y(x_i) = f(x_i) + \epsilon_i \tag{8.98}$$

where ϵ_i is the measurement error. Assume that we are at x_n; then a pair of measurements are taken at $x_n \pm C_n$. The next value, x_{n+1}, in the iteration is

$$x_{n+1} = x_n + \frac{a_n}{2C_n}[y(x_n + C_n) - y(x_n - C_n)] \tag{8.99}$$

where a_n is positive for maximizing $f(x)$ and negative otherwise and C_n is positive. Then, as $n \to \infty$, if we satisfy the conditions

$$a_n, C_n \to 0 \qquad \sum_{i=1}^{n} a_n \to \infty \qquad \sum_{i=1}^{n} \frac{a_i^2}{C_i^2} < \infty$$

Kiefer and Wolfowitz show that $f(x)$ will converge to its optimum value with probability 1.

In another optimization method we take measurements at several points in the parameter space. Then with these data we employ standard regression analysis to fit a surface to the distance function, which is a function of the parameters. We can then obtain steepest-descent (-ascent) equations from this analytic expression and program the computer to find the optimum. The technique for finding the surface is discussed in Chap. 10.

8.9 Optimization of Dynamic Systems

Thus far we have described gradient methods which can be applied to distance functions, whose extrema occur at a set of constant parameter values. In a distance function derived from a static system the optimum parameter values are constants. For instance, in Example 1, the polynomial coefficients are fixed and, as a result, so are the optimum values of r and θ.

On the other hand, consider a distance function of the form of Eq. (8.79). The requirement that the parameters remain constant during the entire range of integration imposes a set of parameter constraints which in turn determines the nature of the distance-function extrema. If, instead, we broaden our class of parameter functions from constants to continuous variables, we can in general expect to find smaller (larger) minima (maxima).

Recently, Bryson [146, 147] and Kelley [55, 148], working independently in conjunction with colleagues, successfully formulated the equations that determine the values of continuously variable parameters which optimize dynamic distance functions. Employing a calculus-of-variations technique, they derive the equations whose solution is the optimum set of parameter values. Because invariably these equations cannot be solved analytically, they resort to gradient methods where the distance function is reduced in discrete steps.

Unfortunately these methods require storage beyond the practical capacity of the current analog computers. As a result, this type of optimization is now solved exclusively on the digital computer. However, on the assumption that the storage of the hybrid computer will ultimately place this type of problem in the domain of the analog-computer user, we shall briefly sketch the method here.[1]

8.9.1 Optimization Equations

Consider the system equations

$$\begin{bmatrix} \dfrac{dx_1}{dt} \\ \cdots \\ \dfrac{dx_n}{dt} \end{bmatrix} = \begin{bmatrix} f_1[x_1, \ldots, x_n; u_1(t), \ldots, u_m(t); t] \\ \cdots\cdots\cdots\cdots\cdots \\ f_n[x_1, \ldots, x_n; u_1(t), \ldots, u_m(t); t] \end{bmatrix} \tag{8.100}$$

in matrix notation, where for simplicity we write

$$\frac{dx}{dt} = [f[x(t); u(t); t]] \qquad x(t_0),\ t_0 < t < T \text{ are initial conditions} \tag{8.101}$$

The quantities dx/dt and $x(t)$ are n-dimensional vectors, and $u(t)$ is an m-dimensional set of optimization parameters. The system operates over the regions $t_0 < t < T$ with a vector set of given initial conditions $x(t_0)$.

[1] I follow closely the notes of a seminar, Optimization of Dynamic Systems, which was taught at Harvard University July 22–Aug. 2, 1963. I wish to thank Dr. Bryson, the organizer of this seminar, for permission to use these notes prior to their inclusion in a book that he and some associates are now writing.

We take as our distance function the scalar

$$d = \phi[x(T), T] + \int_{t_0}^{T} \theta[x(t), u(t), t]\, dt \tag{8.102}$$

where $\phi[x(T), T]$ is a terminal condition to be optimized.

Now, treating Eq. (8.101) as a constraint, we employ Lagrangian multipliers and obtain as our distance function to be optimized

$$d = \phi[x(t), T] + \int_{t_0}^{T} \left(\theta[x(t), u(t), t] + \lambda^T(t)\left\{f[x(t), u(t), t] - \frac{dx}{dt}\right\}\right) dt \tag{8.103}$$

where $\lambda^T(t) = [\lambda_1, \lambda_2, \ldots, \lambda_n]$ and is the transpose of the vector $\lambda(t)$.

For convenience we define a scalar:

$$H[x(t), u(t), T] = \theta[x(t), u(t), t] + \lambda^T(t) f[x(t), u(t), t] \tag{8.104}$$

Now, integrating the last term of Eq. (8.102) by parts,

$$d = \phi[x(T), T] - \lambda^T(T)x(T) + \lambda^T(t_0)x(t_0) + \int_{t_0}^{T} \left\{H[x(t), u(t), t] + \frac{d\lambda^T}{dt} x(t)\right\} dt \tag{8.105}$$

Next we find the perturbation in d caused by small variations in the parameters, $u(t)$, around their nominal value:

$$\delta d = \left[\left(\frac{\partial \phi}{\partial x} - \lambda^T\right)\delta x\right]_{t=T} + [\lambda^T\, \delta x]_{t=t_0} + \int_{t_0}^{T} \left[\left(\frac{\partial H}{\partial x} + \frac{d\lambda^T}{dt}\right)\delta x + \frac{\partial H}{\partial u}\, \delta u\right] dt \tag{8.106}$$

where $\partial\phi/\partial x$ and $\partial H/\partial u$ are n-dimensional vectors and $\partial H/\partial u$ is an m-dimensional vector. Note that all partial derivatives are evaluated for the nominal values of $x(t)$ and $u(t)$.

We do not know at this point what the variations $\delta x(t)$ are. Hence, we select the Lagrangian multipliers to cause the coefficients of $\delta x(t)$ to go to zero in Eq. (8.106). Thus we have the set of equations

$$\frac{d\lambda^T}{dt} = \frac{\partial H}{\partial x} = -\frac{\partial \theta}{\partial x} - \lambda^T \frac{\partial f}{\partial x} \tag{8.107}$$

with the boundary condition, at $t = T$,

$$\lambda^T(T) = \left(\frac{\partial \phi}{\partial x}\right)_{t=T}$$

Equation (8.106) now reduces to

$$\delta d = \lambda^T(t_0)\, \delta x(t_0) + \int_{t_0}^{T} \frac{\partial H}{\partial u}\, \delta u\, dt \tag{8.108}$$

Examining this equation, we see that $\lambda^T(t_0)$, a constant, is the gradient of d with respect to changes in initial conditions. The functions $\partial H/\partial u$ are weighting functions for the differential equations which have δd as a solution, with δu as forcing functions.

An extremum occurs when $\delta d = 0$ for any arbitrary δu. If we assume $\delta x(t_0) = 0$, this can occur only when

$$\frac{\partial H}{\partial u} = 0 \qquad t_0 < t < T \tag{8.109}$$

or equivalently

$$\left(\frac{\partial H}{\partial u_1}\right)^2 + \left(\frac{\partial H}{\partial u_2}\right)^2 + \cdots + \left(\frac{\partial H}{\partial u_m}\right)^2 = 0$$

8.9.2 Computer Optimization Procedure

We program the computer to proceed by the following steps to achieve Eq. (8.109):

1. Choose a vector set of time-varying functions $u(t)$.

2. With $x(t_0)$ given, determine $x(t)$ by integrating $dx/dt = f(x, u, t)$ from t_0 to T.

3. Determine the Lagrangian multipliers $\lambda(T)$ by integrating equation set (8.107) backward from T to t with the proper boundary conditions at T. We get the partial derivatives $\partial\theta/\partial x$, $\partial f/\partial x$ by running $dx/dt = f(x, u, t)$ backward from T to give us the values of x.

4. Determine the values of $\partial H/\partial u = \partial\theta/\partial u + \lambda^T(t)\, \partial f/\partial x$.

5. Taking $\partial H/\partial u$ as a gradient, change the estimate of the parameter $u(t)$ by the amount

$$\Delta u(t) = -K(t) \left[\frac{\partial H}{\partial u}(t)\right]^T$$

The quantity $K(t)$ is a positive scalar chosen at the discretion of the programmer. If we substitute Δu into Eq. (8.108), we get

$$\delta d = -\int_{t_0}^{T} K(t) \frac{\partial H}{\partial u} \left(\frac{\partial H}{\partial u}\right)^T dt \tag{8.110}$$

Because δd is always a nonpositive quantity, this operation will reduce the value of d.

6. Store $u(t) + \Delta u(t)$ and use it as the next estimate of the parameters.

7. Repeat steps 1 to 6 until $\partial H/\partial u$ is small for all values of t or until

$$\int_{t_0}^{T} \frac{\partial H}{\partial u} \left(\frac{\partial H}{\partial u}\right)^T dt$$

is small.

9 Estimation and Test of Hypotheses[1]

9.1 Introduction[2]

During the last decade, successful applications of statistical theory to diverse engineering problems have made this theory an essential part of engineering mathematics. Indicative of its importance, statistics has now become an auxiliary field of study in the engineering curriculum of the major universities and is regarded as a prerequisite to useful and creative work in many fields.

Only part of the recent impetus in the training of engineers in this area is attributable to its importance in solving problems: partly it arises from the realization that this knowledge is also useful in the *formulation* of a large variety of problems. Whereas some thirty years ago the engineer confined himself primarily to mathematical models of deterministic systems, presently such terms as *correlation, variance, spectral density* mark his conversation and literature, resulting in the verbalization of new concepts and the development of highly imaginative models which often culminate in the design of new systems, deterministic and nondeterministic.

Correspondingly, the simulation engineer, who is expert in one of the most useful mathematical tools of systems analysis, has had to learn more and more about these new ideas and problems and has been called upon, more and more, to simulate *random* systems, systems having random inputs and/or outputs. Examples of such cases are numerous; they arise in missile-performance problems, in oceanography, in the simulation of aircraft flight characteristics, in problems of equipment reliability, and recently in nonlinear system specification [122]. We shall touch upon the role of simulation in some of these areas. Before doing so, however, we must understand and apply the rudimentary notions of statistics—the mathematics of data analysis and interpretation.

[1] This chapter was written by A. Levine.

[2] Some basic concepts of probability and statistics are discussed in Appendix 4.

9.2 Delineation of Problem Areas

Statistical theory has evolved into a large body of knowledge, too large to treat fully in a few chapters of an elementary text. Hence we must demarcate and present those areas which are most useful in general engineering practice and in simulation. To illustrate those parts of statistical theory treated herein, consider the common problem of analyzing the output of a system when the input is supplied by a noise generator. The output of such a system may appear as in Fig. 9.1.

Suppose we are interested in the output $\overset{>}{x}$ at the n times $T_1, T_2, \ldots, T_n$. We then obtain the sample $\overset{>}{x} = (x_1, x_2, \ldots, x_n)$, where x_i is the output at time T_i. If we repeat the simulation, the new sample of the system output, $\overset{>}{x}' = (x_1', x_2', \ldots, x_n')$ for times $T_1, T_2, \ldots, T_n$, would not necessarily be

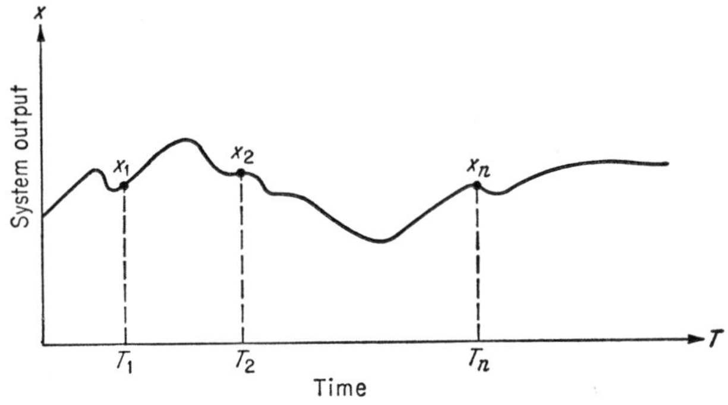

FIG. 9.1 A system response.

the same as the first sample, $\overset{>}{x} = (x_1, x_2, \ldots, x_n)$. If we perform the experiment several times, we may very well obtain a different sample each time, although no change has taken place in the *equipment*.

This type of experiment poses two major problems to the engineer. The first is connected with testing the model of his system. The engineer develops a mathematical model (be it in equation or circuit diagram form) of a system. His theory tells him that the system should behave in a certain way under certain conditions. He goes to the simulator to seek experimental corroboration of this theory. In the deterministic case, the decision as to whether or not to accept the theory is fairly easy. The engineer runs the experiment and compares the system output with the predicted output. If the two are "reasonably" close, that is, within experimental error, the theory is accepted. In the case of random outputs, the problem of corroboration is not so easy to resolve. The sample output $(x_1, x_2, \ldots, x_n)$ is different each time the experiment is run. How, then, are the data to be interpreted in order to corroborate the theory? This problem is extremely important, for it is clear that the theory stands or falls on the nature of the data. This simple fact is often lost sight of in simulation studies; hence the crucial importance of data interpretation to the success or failure of a simulation is ill-considered and much is lost through slipshod methods of interpretation.

The second problem arises from the reliability (or unreliability) of the computer. If some random effect, like an intermittent connection, should occur during a deterministic experiment, the fault might easily be discovered by rerunning the experiment several times. A random effect would appear in the output where no random effect should be; consequently the operator would realize that the computer was malfunctioning. The problem is a far more difficult one when dealing with random outputs, for the random nature of the output often masks errors created by unreliable components. These errors may, nevertheless, be important enough to force a wrong decision in rejecting or accepting a theory.

To summarize, there are, at the outset, two major problem areas:

1. Interpreting the data in order to test the validity of a mathematical model or to discover some pertinent characteristics of the model
2. Detecting random computer malfunctions when running experiments having random outputs

The latter problem will be investigated in Chap. 10. The present chapter is devoted to a discussion of the interpretation of data.

Prob. 9.1.[1] We define a statistic t_n to be any function of $\vec{x}$, where n is equal to the number of sample points in $\vec{x}$. For example, let $\vec{x} = (x_1, x_2, \ldots, x_{10})$. Let

$$x_i = \begin{cases} 0 & \text{with probability } \frac{1}{2} \\ 1 & \text{with probability } \frac{1}{2} \end{cases}$$

The x_i might represent the ith outcome of a toss of a coin where $x_i = 0$ if heads occurs and $x_i = 1$ if tails occurs. An example of a statistic would be

$$t_n = \bar{x} = \sum_{i=1}^{10} \frac{x_i}{10} \qquad \text{sample mean}$$

Another would be

$$t'_n = \frac{x_l - x_s}{2} \qquad \text{mid-range}$$

where x_l is the largest sample and x_s is the smallest.

(*a*) Is $\bar{x}$ a random variable? For $n = 3$, list the possible outcomes for $\bar{x}$ together with their probabilities of occurrence.

[1] The problems have been incorporated as an integral part of the text in the hope that practice with the multitude of notions introduced in this chapter as they arise will facilitate the absorption of the material by the reader. The problems should not be skipped since subsequent discussions assume that they have been completed and the points of importance assimilated by the reader.

(*b*) Is t'_n a random variable? Repeat part *a* for the mid-range.

(*c*) Is a statistic always a random variable? If so, why? If not, why not?

Answer. (*a*) The quantity $\bar{x}$ is indeed a random variable, for each of its possible values occurs only with a given probability. For example, for $n = 3$, all the possible outcomes for $\bar{x}$ are

$$\bar{x} = \begin{cases} \frac{1}{3}(0+0+0) = 0 \\ \frac{1}{3}(0+0+1) = \frac{1}{3} \\ \frac{1}{3}(0+1+0) = \frac{1}{3} \\ \frac{1}{3}(1+0+0) = \frac{1}{3} \\ \frac{1}{3}(0+1+1) = \frac{2}{3} \\ \frac{1}{3}(1+0+1) = \frac{2}{3} \\ \frac{1}{3}(1+1+0) = \frac{2}{3} \\ \frac{1}{3}(1+1+1) = 1 \end{cases}$$

We note that there are three possible ways to get $\bar{x} = \frac{1}{3}$; hence,

$$P(\bar{x} = \tfrac{1}{3}) = (\text{probability that } \bar{x} = \tfrac{1}{3}) = \tfrac{3}{8}$$

Similarly, $P(\bar{x} = 0) = \frac{1}{8}$, $P(\bar{x} = \frac{2}{3}) = \frac{3}{8}$, $P(\bar{x} = 1) = \frac{1}{8}$. It is to be emphasized here that *before the experiment* the symbol $\bar{x}$ represents a random variable. After the experiment, when the computation has been completed, $\bar{x}$ is a *number*. Similarly, before the experiment a sample value x_i is a random variable; after the experiment it is a number.

9.3 Data and the Model

The most common problems associated with the interpretation of the output data from random processes are those of *point estimation, interval estimation,* and *test of hypotheses.* The following are paragraphs intended to introduce these subjects and to compare them.

9.3.1 Point Estimation

The point-estimation problem arises as follows: The *form* of the probability-density function is known, but the parameters of this function are unknown and are to be determined from the data. For example, in many cases it is well known that the system output is normally distributed; i.e., the probability-density function of the output has the form

$$p(x; \mu, \sigma^2) = \frac{\exp[-(x-\mu)^2/2\sigma^2]}{\sqrt{2\pi}\,\sigma} \tag{9.1}$$

However, the parameters of this function, namely, the theoretical mean μ and the theoretical variance σ^2, are in many instances unknown and are to be determined. If one has no prior knowledge as to the values of these parameters, then the only source of knowledge about them is the data. One seeks to *estimate* μ and σ^2 from the *data.* The aim of the analysis of the data is to supply estimates which are *close* to the true values rather than exactly *equal* to them. These estimates are called *point estimates.* For example, the well-known point estimate of the theoretical mean μ is the *sample mean* $\bar{x}$. In general let us denote the point estimate of some parameter θ by $\hat{\theta}$. Then the point estimate of μ is denoted by $\hat{\mu}$, and one possible such estimate is

$$\hat{\mu}_1 = \bar{x} = \frac{\sum_{i=1}^{n} x_i}{n} \tag{9.2}$$

where $\vec{x} = (x_1, x_2, \ldots, x_i, \ldots, x_n)$
x_i = ith system output or ith sample point
n = number of observations

Another point estimate of μ is the mid-range $\hat{\mu}_2$:

$$\hat{\mu}_2 = \frac{x_l - x_s}{2} \tag{9.3}$$

where x_l = largest observed value of output
x_s = smallest observed value of output

Still another is

$$\hat{\mu}_3 = \frac{\sum_{i=1}^{n} x_i}{n + 1{,}000} \tag{9.4}$$

where x_i and n are defined as above. It is the object of the theory of point estimation to aid us in choosing the "best" point estimate among all possible point estimates.

The problem of point estimation can now be stated in general terms: Given the form of the probability density $f(x; \theta, \theta_1, \theta_2, \ldots, \theta_k)$, with k unknown parameters $\theta_1, \theta_2, \ldots, \theta_k$, the problem is to determine the "best" estimates $\hat{\theta}_1, \hat{\theta}_2, \ldots, \hat{\theta}_k$ of $\theta_1, \theta_2, \ldots, \theta_k$, respectively, on the basis of the sample $\vec{x}$. It remains for us, in the following paragraphs, to define what we mean by "best."

Prob. 9.2. Suppose it is known that the output x of a system has the density function

$$p(x) = \frac{1}{\pi[1 + (x - \theta)^2]} \qquad -\infty \leq x \leq \infty$$

(*a*) Which quantity in $p(x)$ would you estimate with the data in order to specify $p(x)$ completely?

(*b*) Is the parameter θ a moment of $p(x)$? NOTE: $p(x)$ is known as the Cauchy distribution.

Answers. (*a*) θ; (*b*) no.

Prob. 9.3. Given $\vec{x} = (1, 2, -3, 5, -1)$, conjecture on how you would use $\vec{x}$ to estimate θ in Prob. 9.2.

Prob. 9.4. Is a point estimate a statistic? (See Prob. 9.1.)

9.3.2 Interval Estimation

It is important to emphasize that the point estimate $\hat{\theta}$ is not necessarily equal to the parameter θ. Instead

$$\theta = \hat{\theta} + \epsilon$$

where ϵ is undetermined. As it stands, $\hat{\theta}$ is an estimate, but one can well wonder how *good* an estimate! The object of the theory of interval estimation is to put bounds, $\pm\delta$, on the value of ϵ, for if

$$-\delta < \epsilon < \delta$$

where δ is some known number, then it can be asserted that

$$\hat{\theta} - \delta < \theta < \hat{\theta} + \delta$$

Hence we have a measure "δ" of how "close" $\hat{\theta}$ is to θ. Only in this way can the point estimate $\hat{\theta}$ be made meaningful to the engineer.

Prob. 9.5. Let

$$x_i = i\text{th sample point in } \vec{x} = (x_1, x_2, \ldots, x_{10})$$

Let the x_i be mutually independent and

$$p(x_i) = \frac{1}{\sqrt{2\pi}\,\sigma} e^{-(x_i-\mu)^2/2\sigma^2} \qquad \text{for each } i$$

Let

$$\bar{x} = \frac{\sum_{i=1}^{10} x_i}{10}$$

be an estimate for the theoretical mean μ. Show that:

(*a*)
$$V(\bar{x}) = \text{variance of } \bar{x} = \frac{\sigma^2}{10}$$

Answer

$$V(\bar{x}) = V\left(\frac{\sum_{i=1}^{n} x_i}{n}\right) = \frac{1}{n^2} V\left(\sum_{i=1}^{n} x_i\right) = \frac{1}{n^2}\sum V(x_i) = \frac{n\sigma^2}{n^2} = \frac{\sigma^2}{n}$$

(*b*)
$$E(\bar{x}) = \mu$$

Answer

$$E(\bar{x}) = E\left(\frac{\sum_{i=1}^{n} x_i}{n}\right) = \frac{1}{n} E\left(\sum_{i=1}^{n} x_i\right) = \frac{1}{n}\sum_{i=1}^{n} E(x_i) = \mu$$

Let $\sigma = 1$; then show that

(*c*)
$$\Pr\left(-1.95 < \frac{\bar{x} - \mu}{1/\sqrt{10}} < 1.95\right) = 95 \text{ per cent}$$

(*d*)
$$\Pr\,(\bar{x} - 0.617 < \mu < \bar{x} + 0.617) = 95 \text{ per cent}$$

(*e*) Find δ such that $\bar{x} - \delta < \mu < \bar{x} + \delta$ with 95 per cent probability.
(*f*) Find δ such that $\bar{x} - \delta < \mu < \bar{x} + \delta$ with 50 per cent probability.

Note that we cannot ascertain that μ lies between $\bar{x} - \delta$ and $\bar{x} + \delta$ with certainty but only to a certain probability.

(*g*) Given $\overset{>}{x} = (1, 2, 0, -3, 6, -2, 0, -1, 4, -3)$, find limits L_1 and L_2 such that $L_1 < \mu < L_2$ with 95 per cent probability.

Answers. (*c*) Let

$$y = \frac{\bar{x} - \mu}{\sigma/\sqrt{n}} = \frac{\bar{x}}{\sigma/\sqrt{n}} - \frac{\mu}{\sigma/\sqrt{n}}$$

The average value of y, $E(y)$, is then

$$E(y) = E\left(\frac{\bar{x}}{\sigma/\sqrt{n}} - \frac{\mu}{\sigma/\sqrt{n}}\right) = E\left(\frac{\bar{x}}{\sigma/\sqrt{n}}\right) - E\left(\frac{\mu}{\sigma/\sqrt{n}}\right)$$

$$= \frac{E(\bar{x})}{\sigma/\sqrt{n}} - \frac{\mu}{\sigma/\sqrt{n}} = \frac{\mu - \mu}{\sigma/\sqrt{n}} = 0$$

The variance of y, $V(y)$, is

$$V(y) = V\left(\frac{\bar{x}}{\sigma/\sqrt{n}} - \frac{\mu}{\sigma/\sqrt{n}}\right) = V\left(\frac{\bar{x}}{\sigma/\sqrt{n}}\right) = \frac{V(\bar{x})}{\sigma^2/n}$$

But from part *a*, $V(\bar{x}) = \sigma^2/n$. Hence

$$V(y) = 1$$

Since each x_i is normally distributed, $\bar{x}$ is normally distributed. Hence y is normally distributed. Thus, the distribution of y is $N(0, 1)$; that is, y has a normal distribution with zero mean and unit variance. The density function of y is, therefore,

$$p(y) = \frac{1}{\sqrt{2\pi}} \exp\left(-\frac{y^2}{2}\right)$$

Hence,

$$\text{Probability that } (-1.95 < y < 1.95) = \Pr(-1.95 < y < 1.95) = 0.95$$

is obtained from the table of the error function (Appendix 6). In this case, $\sigma = 1$, $n = 10$. Hence

$$\Pr(-1.95 < y < 1.95) = \Pr\left(-1.95 < \frac{\bar{x} - \mu}{1/\sqrt{10}} < 1.95\right) = 0.95$$

(*d*)

$$0.95 = \Pr\left(-1.95 < \frac{\bar{x} - \mu}{1/\sqrt{10}} < 1.95\right) = \Pr\left(-1.95\frac{1}{\sqrt{10}} < \bar{x} - \mu < 1.95\frac{1}{\sqrt{10}}\right)$$

$$= \Pr(-\bar{x} - 0.617 < \mu < -\bar{x} + 0.617) = \Pr(\bar{x} + 0.617 > \mu > \bar{x} - 0.617)$$

(*e*)

$$\delta = 0.617$$

9.3.3 Test of Hypotheses

The techniques of hypothesis testing are essential to testing the theory or model of a system. For example, on the basis of the theory we may predict that the output of a system has a continuous frequency function with mean 5 volts; or we might predict that one system is better than another, etc. These predictions constitute *hypotheses* to be tested. It is the object of certain types of statistical analysis to test such hypotheses.[1]

[1] The above paragraphs introduce the three topics of this chapter. The following sections will deal with each in detail.

Prob. 9.6. Let the assumptions of Prob. 9.5 hold. Suppose we hypothesize that $\mu = 0$. Conjecture as to how you would test this hypothesis with the data of Prob. 9.5*g*. How does this problem differ from the point-estimation and interval-estimation problem?

9.4 Point Estimation

Since "interval estimation" and "test of hypotheses" cannot be separated from point estimation, we shall treat point estimation first.

9.4.1 Maximum-likelihood Estimate

The above paragraphs indicate how the point-estimation problem arises. Let us supplement these remarks with an example by taking 10 samples from a table of random numbers.[1] This table is formed by sampling from a population having a normal distribution with theoretical mean μ equal to 0 and the theoretical variance σ^2 equal to 1, that is, $N(0, 1)$.

Hence, sampling from the table is equivalent to sampling from the normal population. To obtain a random sample from the table, choose a page at random, then choose a row and column at random, and begin reading down the column.

Let us take a typical sample of size 10:

$$\vec{x}_{10} = (0.241, 0.022, -0.853, -0.501, +0.439, -1.85, -0.255, -0.423, +0.857, -0.260) \qquad (9.5)$$

Suppose we are faced with the problem, commonly encountered in practice, wherein we know the form of the distribution; for example, in this case we know that the population is normal, but one of the population parameters, for example, μ, is unknown to us. The problem is to estimate the theoretical mean μ from the data $\vec{x}_{10}$. There are many possible estimates. We listed two possible estimates of μ in Sec. 9.3.1, viz., the sample mean $\bar{x}$ and the mid-range. There are many others. We wish to find the *best* estimate of μ from all the possible estimates of μ. In order to decide which estimate is "best," we must establish some criterion or criteria by which estimates can be compared. To begin, we note that our estimate of μ, denoted by $\hat{\mu}$, must be dependent on our sample $\vec{x}_n = \vec{x}_{10}$. Note too that $\hat{\mu}$ is a random variable (Prob. 9.1), whereas μ is a particular number, albeit unknown. Certainly we shall assign a different value to μ for different samples. For example, our present sample is $\vec{x}_{10}$, as indicated above, but for a different position in the table of random numbers we might have obtained $\vec{x}_{10} = (1.6, 0.9, 2.4, 0.8, 3, 9.2, 0.2, 1.9, 0.5, 1.3)$. For the first sample our estimate of μ would be somewhere near 0; for the latter sample we would be likely to estimate μ near 1 or 2. Thus, given a sample $\vec{x}_n$, it is clear that some of the uncertainty about μ is dispelled, that some values of μ will seem

[1] See Ref. 136, appendix II, for a typical table of random numbers.

more reasonable (or likely) than others. Hence, we can assign a conditional probability density to μ *given* the sample $\vec{x}_n$. Let $p(\mu|\vec{x}_n)$ represent this conditional density. Our object is to choose a value $\hat{\mu}$, the estimate of μ, which will maximize this conditional density. The reader should immediately ask why we should want to do such a thing, for it is here that the criterion is established for choosing an estimate of μ. The reason for so doing is that we wish to choose the most probable value of μ on the basis of the sample $\vec{x}_n$. This is an eminently sensible approach, for it is only on the basis of the data that we can estimate μ. Hence, it is clear that we wish to choose the most probable value of μ *given* $\vec{x}_n$, which is equivalent to saying that we wish to find the value of μ for which $p(\mu|\vec{x}_n)$ is a maximum. Using Bayes' rule, we note that

$$p(\mu|\vec{x}_n) = \frac{p(\mu)p(\vec{x}_n|\mu)}{p(\vec{x}_n)} \tag{9.6}$$

where $p(\mu)$ = a priori density of μ
$p(\vec{x}_n)$ = a priori density of $\vec{x}_n$
$p(\vec{x}_n|\mu)$ = conditional probability of $\vec{x}_n$ given μ

From the total-probability rule

$$p(\vec{x}_n) = \sum_\alpha p(\mu_\alpha)p(\vec{x}_n|\mu_\alpha) \tag{9.7}$$

Let us assume that prior to the experiment we have not the slightest idea what μ might be; that is, a priori all values of μ are equally likely. Then $p(\mu_\alpha) = p(\mu) = \text{const}$ for all α. Hence

$$p(\mu|\vec{x}_n) = \frac{p(\mu)p(\vec{x}_n|\mu)}{\sum_\alpha p(\mu_\alpha)p(\vec{x}|\mu_\alpha)} = \frac{p(\mu)p(\vec{x}_n|\mu)}{\sum_\alpha p(\mu)p(\vec{x}_n|\mu_\alpha)} = \frac{p(\vec{x}_n|\mu)}{C} \tag{9.8}$$

Note that C is a constant independent of μ. Hence the *a priori* conditional distribution of $\vec{x}$ is related to the *a posteriori* conditional distribution of μ by a constant.

Recall that our object was to choose a value of μ, designated by $\hat{\mu}$, which maximizes $p(\mu|\vec{x}_n)$. By Eq. (9.8), if we find a value $\hat{\mu}$ which maximizes $p(\vec{x}_n|\mu)$, then we have achieved our object. Let us see how we go about solving this problem in the example proposed at the start of this section. Recall that we were sampling from a normal population with unknown mean μ and unknown variance σ^2. To make matters simple, assume that we know σ^2 and $\sigma^2 = 1$. We had a sample of size 10, the random variables $\{x_i\}$ being mutually independent. Therefore,

$$p(\vec{x}_n|\mu) = \left(\frac{1}{2\pi}\right)^{n/2} \exp\left[-\frac{\sum_{i=1}^{10}(x_i-\mu)^2}{2}\right] \tag{9.9}$$

We wish to find $\hat{\mu}$ which will maximize $p(\vec{x}_n|\mu)$. We note that

$$\log p(\vec{x}_n|\mu) = \frac{n}{2}\log 2\pi - \frac{\sum_{i=1}^{10}(x_i-\mu)^2}{2} \tag{9.10}$$

If we find $\hat{\mu}$ which maximizes $\log p(\vec{x}_n|\mu)$, this is equivalent to finding the value which maximizes $p(\vec{x}_n|\mu)$. Hence, we solve for μ in

$$\frac{\partial}{\partial\mu}\log p(\vec{x}_n|\mu) = 0 \tag{9.11}$$

Hence,

$$\hat{\mu} = \frac{\sum_{i=1}^{10} x_i}{n} = \bar{x} \tag{9.12}$$

We note that for $\vec{x}_{10}$ in Eq. (9.5)

$$\hat{\mu} = -0.259$$

The estimate obtained by the above procedure is called the Maximum Likelihood Estimate (MLE). Such estimates are preferred by statisticians not only because they satisfy the criterion for a best estimate stated herein, but also because they have other desirable properties. One of these properties is described in Sec. 9.4.2.

Prob. 9.7. Let $\vec{x}_4 = (1, 7, 1, 0)$. Find and evaluate the MLE of θ if

(*a*) $p(x) = \dfrac{1}{\sqrt{2\pi}\,\theta} e^{-x^2/2\theta^2} \qquad -\infty < x < \infty$

(*b*) $p(x) = \dfrac{e^{-\theta x}}{\theta} \qquad x \geq 0$

(*c*) $p(x)$ is the Cauchy density function (see Probs. 9.2 and 9.3).

Answer. (*a*)

$$p(x_1, x_2, x_3, x_4) = \prod_{i=1}^{4} p(x_i) = \frac{1}{(2\pi)^2\theta^4}\exp\left(-\sum_{i=1}^{4}\frac{x_i^2}{2\theta^2}\right)$$

Hence

$$\log p(x_1, x_2, x_3, x_4) = \log (2\pi)^{-2} - 4\log\theta - \frac{\sum_{i=1}^{4} x_i^2}{2\theta^2}$$

and

$$\frac{\partial}{\partial\theta}\log p(x_1, x_2, x_3, x_4) = -\frac{4}{\theta} + \frac{\sum_{i=1}^{4} x_i^2}{\theta^3}$$

Setting the partial derivative equal to zero and solving for θ, we have

$$\hat{\theta} = \sqrt{\tfrac{1}{4} \sum_{i=1}^{4} x_i^2} = \tfrac{1}{2}\sqrt{51}$$

Prob. 9.8. If $p(x; \mu, \sigma^2) = e^{-(x-\mu)^2/2\sigma^2}/\sqrt{2\pi}\,\sigma$ and if μ is unknown, then the MLE estimate of σ^2 is

$$s^2 = \sum_{i=1}^{n} \frac{(x_i - \bar{x})^2}{n}$$

Find the value of s^2 for the data given in Prob. 9.7.

Prob. 9.9. Imagine that we are firing a missile at a target. Let there be n such firings. For each firing i, let

$$x_i = \begin{cases} 0 & \text{if there is a miss} \\ 1 & \text{if there is a hit} \end{cases}$$

Let the results of 10 firings be

$$\overset{>}{x}_{10} = (1, 0, 0, 1, 1, 1, 0, 0, 1, 0)$$

Estimate p_H, the probability of hitting the target on any given trial, with an MLE estimate.

Answer. Let n = number of firings or samples. Then the probability of getting y hits out of a particular sequence of n firings is

$$f(y) = p_H{}^y(1 - p_H)^{n-y}$$

where

$$y = \sum_{i=1}^{n} x_i$$

Note that

$$\log f(y) = y \log p_H + (n - y) \log (1 - p_H)$$

To find the MLE of p_H, solve

$$\frac{\partial}{\partial p_H} \log f(y) = \frac{y}{p_H} - \frac{n - y}{1 - p_H} = 0$$

or

$$y(1 - p_H) - (n - y)p_H = 0$$

or

$$\hat{p}_H = \frac{y}{n} = \frac{\sum_{i=1}^{n} x_i}{n} = \bar{x}$$

Hence $\bar{x}$ is an MLE of p_H. For the given $\overset{>}{x}_{10}$ above,

$$\bar{x} = \tfrac{1}{2}$$

9.4.2 Efficiency

There are other criteria than that of maximum likelihood by which an estimate $\hat{\theta}$ can be selected for a parameter θ. One may compare estimates on the basis of which one comes closest to the true value θ most often. To state the matter precisely, let $\hat{\mu}_1$ and $\hat{\mu}_2$ be estimates of μ, the theoretical mean of a population distributed $N(\mu, \sigma^2)$. Let us assume that[1]

$$E(\hat{\mu}_1) = \mu \qquad E(\hat{\mu}_2) = \mu$$

for all sample sizes n; that is, both estimates, *on the average*, are equal to μ, so that we do not make any distinction between them on this basis. Suppose now that the variance of $\hat{\mu}_1$, $V(\hat{\mu}_1)$, is less than $V(\hat{\mu}_2)$, the variance of $\hat{\mu}_2$. If $\hat{\mu}_1$ and $\hat{\mu}_2$ are normally distributed, the fact that

$$V(\hat{\mu}_1) < V(\hat{\mu}_2)$$

implies that $\hat{\mu}_1$ will come closer to μ *more often* than will $\hat{\mu}_2$ (why?).[2] In this sense, $\hat{\mu}_1$ is said to be *more efficient* than $\hat{\mu}_2$. For a moment, let Q be the smallest value of the variance which can be obtained by any unbiased estimate of μ. Then we can define the efficiency of an estimate $\hat{\mu}$ of μ by

$$\text{Efficiency} = \frac{Q}{V(\hat{\mu})} \tag{9.13}$$
$$V(\hat{\mu}) = \text{variance of } \hat{\mu}$$

To be specific, we note from Prob. 9.5 that

$$V(\bar{x}) = \frac{\sigma^2}{N} \tag{9.14}$$
$$E(\bar{x}) = \mu$$

where N = number of samples
$\bar{x}$ = *sample* mean (estimate of μ)
μ = theoretical mean
σ^2 = variance of population

Another possible estimate of μ which proves convenient in data reduction is the sample median which is the middle value in any group of observations. For example, if

$$\overset{>}{x}_{11} = (0, 5, 5, 7, 10, 10, 10, 8, 5, 5, 0)$$

then on reordering the samples according to size,

0, 0, 5, 5, 5, 5, 7, 8, 10, 10, 10

the median is 5, the middle or sixth value. If

$$\overset{>}{x}_{6} = (1, 3, 4, 2, 3, 2)$$

[1] If an estimate $\hat{\mu}$ of μ is such that, for all n (n = sample size), $E(\hat{\mu}) = \mu$, then $\hat{\mu}$ is said to be an unbiased estimate of μ.

[2] To answer this question, note that the normal distribution is completely specified if the mean and variance are specified, and that the variance is a measure of the *dispersion* of the population about the mean.

then on reordering according to size we have

$$1,\ 2,\ \underbrace{2,\ 3},\ 3,\ 4$$

and the median is

$$\frac{2 + 3}{2} = 2.5$$

The variance of the median is [136, p. 72]

$$\frac{\sigma^2}{NE}$$

where σ^2 = variance of population
N = number of samples
E = a number ≥ 1 which is a function of number of samples (see Appendix 5)

It is well known that the variance of any unbiased estimate of μ must be larger than or equal to the variance of $\bar{x}$ [124, pp. 481–487]. Hence, for Q in Eq. (9.13), we can use [see Eq. (9.14) also]

$$Q = \frac{\sigma^2}{N} \tag{9.15}$$

Hence, the efficiency of the median is

$$\text{Efficiency} = \frac{\sigma^2/N}{\sigma^2/NE} = E \tag{9.16}$$

A table of E versus N is given in Appendix 5.

Note that in the above example the sample sizes chosen in calculating the sample median and sample mean were assumed to be the same, namely, N. Let us assume now that these sample sizes are not the same but that

N_1 = sample size when finding the sample median
N_2 = sample size when finding the sample mean

Then the ratio of the variances of the two estimates is

$$\frac{\sigma^2/N_2}{\sigma^2/N_1E} = \frac{N_1E}{N_2} \tag{9.17}$$

Suppose we wish to choose N_1 and N_2 so that the two estimates (sample mean and median) were equally *precise,* i.e., so that the variances of these two estimates were equal, so that

$$\frac{N_1E'}{E} = 1$$

That is, E represents the fraction of *additional* samples needed when the median is computed over that needed for the sample mean in order that the two estimates be equally precise.

The importance of the quantity E becomes apparent when it is realized that the criterion for selecting an estimate may not be choosing the most precise one for the smallest number of samples but rather may be the cost of computing the estimate. For example, even though $\bar{x}$ is the most precise estimate of the theoretical mean, it is much more difficult (for large samples) to compute than the median. It may pay to take more samples for the median in order to obtain the same stated precision as the mean. It may pay in the sense that the cost of the additional samples is more than made up by the decrease in cost of data reduction. As an example, consider the following problem.[1]

If data collected for determining the median cost 8 cents per observation and data collected for computing the mean cost 15 cents per observation, would it be more economical to take a sample of 100 cases and measure $\bar{x}$ or a sample sufficiently large to provide a median as precise as $\bar{x}$?

To provide an answer to this question, we must determine the number of samples needed to provide a median as precise as $\bar{x}$. By Eq. (9.17),

$$1 = \frac{N_1 E}{N_2}$$

where N_2 = number of samples used for $\bar{x}$
N_1 = number of samples used for median

In this example

$$N_2 = 100$$

and N_1 is unknown, but we require by the above equation that

$$N_1 E = N_2 = 100$$

Using Appendix 5, we note that E for samples greater than 20 is 0.637. Hence

$$N_1(0.637) = 100$$

or

$$N_1 = 157 \text{ samples}$$

The cost of $\bar{x}$ is

$$0.15(100) = \$15$$

The cost of the median is

$$0.08(157) = \$12.56$$

Hence, on the basis of *cost* in this example the median is superior to the mean.

[1] This example is prob. 5 in the problems section of Ref. 136, chap. 6.

9.4.3 Interval Estimation

Let us continue with the example of Sec. 9.4.1, where

$$p(x) = \frac{e^{-\frac{1}{2}(x-\mu)^2}}{\sqrt{2\pi}}$$

We found that $\bar{x}$ was an MLE of μ, and for the sample obtained from the table of random numbers, $\bar{x} = -0.259$. This number is our estimate of μ, the theoretical mean. A natural question to ask is: "How good is this estimate?" or "How much *confidence* do we have that this estimate is *close* to the true value μ?" The question as to how much "confidence" we can have in $\bar{x}$ as an estimate of μ or indeed, in general, of $\hat{\theta}$ as an estimate of θ is extremely important, for an estimate can be a "best" possible one and yet may lie far from the true value. It is in the determination of "confidence" that the interval-estimation problem enters.

The notion of "confidence" relates to the faith we have that the estimate lies close to the true value. We translate this notion into mathematical terms.

Let

$$\begin{aligned} P_\delta &= \Pr\,(\hat{\theta} - \delta \le \theta \le \hat{\theta} + \delta) \\ &= \text{probability that } \theta, \text{ the true value of the parameter estimated,} \\ &\quad\ \text{lies in interval } (\hat{\theta} - \delta,\ \hat{\theta} + \delta) \end{aligned} \tag{9.18}$$

Now let us ask: "What is the δ such that $P_\delta = 95$ per cent?" That is, we wish to find the interval $(\hat{\theta} - \delta, \hat{\theta} + \delta)$ such that we are sure that the number θ lies in this interval with probability 95 per cent. To state the matter still another way, if we take 100 different samples $\{\overset{>}{x}\}$ and for each one find the δ and each time maintain that the θ lies in the interval $(\hat{\theta} - \delta, \hat{\theta} + \delta)$, then we shall be wrong "on the average" 5 times out of 100. In other words, if we follow the above procedure for finding δ, then out of a *large* number of times in which we find this δ, the statement that "θ lies in the interval $(\hat{\theta} - \delta, \hat{\theta} + \delta)$" will be wrong 5 per cent of the time. In the example at hand let us see how to find δ such that $P_\delta = 95$ per cent when $\bar{x}$ is an estimate of μ.

In this case

$$P_\delta = \Pr\,(\bar{x} - \delta < \mu < \bar{x} + \delta)$$

To find P_δ note that

$$\Pr\,(-\delta < \bar{x} - \mu < \delta) = \Pr\,(\bar{x} - \delta < \mu < \bar{x} + \delta) \tag{9.19}$$

Hence all we need do is to find the probability on the left side of Eq. (9.19). We know from the central limit theorem that $\bar{x}$ is $N(\mu, \sigma^2/n)$, where σ^2 is the variance of the underlying normal population and n is the number of sample points in $\overset{>}{x}$. In this particular example, $\bar{x}$ is $N(\mu, 1/n)$. It follows that $\gamma = \bar{x} - \mu$ is $N(0, 1/n)$.† Since

$$P_\delta = 95 \text{ per cent} = \Pr(-\delta < \gamma < \delta) \tag{9.20}$$

the problem reduces to finding δ such that

$$0.95 = \int_{-\delta}^{\delta} \frac{\sqrt{n}\, e^{-n\gamma^2/2}}{\sqrt{2\pi}}\, d\gamma = \int_{-\sqrt{n}\delta}^{\sqrt{n}\delta} \frac{e^{-t^2/2}}{\sqrt{2\pi}}\, dt \tag{9.21}$$

The quantity δ is easily found from the table in Appendix 6 to be 0.62 for the case $n = 10$. Hence the two 95 per cent confidence limits on μ are

$$L_1 = \bar{x} - 0.62 \qquad L_2 = \bar{x} + 0.62$$

But from the data $\overset{>}{x}_{10}$ (see Sec. 9.4.1) we found that $\bar{x} = -0.259$. Hence $L_1 = -0.879$, $L_2 = 0.361$. We can now state with 95 per cent certainty, or confidence, that μ lies between L_1 and L_2.

Note that we have chosen an interval such that we are sure μ lies in this interval with 95 per cent probability. This probability P_δ is a measure of our "faith" and is called the *confidence level*, or significance level. The interval $(\hat{\theta} - \delta, \hat{\theta} + \delta)$, or more often the number δ, is called the 95 per cent confidence interval. If we wish greater faith, for example, a 99 per cent confidence interval, then, of course, δ would be a different (larger) size. Our choice of a confidence level, such as 95 per cent, reflects how certain we wish to be that the theoretical mean lies in the confidence interval.

Prob. 9.10. (*a*) **Normal Distribution.** In the example of the text, with

$$p(x) = \frac{e^{-(x-\mu)^2/2}}{\sqrt{2\pi}}$$

the given sample $\overset{>}{x}_{10}$, and $\bar{x} = -0.259$, find 99 per cent confidence limits for μ.

(*b*) **Normal Distribution.** The quantity

$$\frac{s^2}{\sigma^2} = \sum_{i=1}^{n} \frac{(x_i - \bar{x})^2}{(n-1)\sigma^2} \qquad p(x_i) = N(\mu, \sigma^2) \qquad i = 1, 2, \ldots, n$$

is χ_n^2-distributed.[1] Hence, two values $\chi^2_{n\alpha/2}$ and $\chi^2_{n(1-\alpha/2)}$ can be found such that

$$\Pr\left(\chi^2_{n\alpha/2} < \frac{\sigma^2}{s^2} < \chi^2_{n(1-\alpha/2)}\right) = 1 - \alpha$$

or, equivalently,

$$\Pr\left(\frac{s^2}{\chi^2_{n(1-\alpha/2)}} < \sigma^2 < \frac{s^2}{\chi^2_{n\alpha/2}}\right) = 1 - \alpha$$

where

$$\chi_1^2 = \frac{s^2}{\chi^2_{n(1-\alpha/2)}} \qquad \chi_2^2 = \frac{s^2}{\chi^2_{n\alpha/2}}$$

† See Prob. 9.5*c*.

[1] See Ref. 125, p. 219, for a derivation of this distribution.

represent the *end points* of a confidence limit about σ^2 of significance level $1 - \alpha$. Find 95 and 99 per cent confidence intervals on σ^2 using the above data. (The χ_n^2 table is located in Appendix 7.)

(*c*) If μ is known to be 2, find 95 per cent confidence intervals on σ^2 using the above data. [Recall that

$$\frac{s^2}{\sigma^2} = \frac{\sum_{i=1}^{n} (x_1 - \mu)^2}{n\sigma^2}$$

is χ_n^2-distributed if μ is known.]

(*d*) Recall that

$$t = \frac{(\bar{x} - \mu)\sqrt{n-1}}{s}$$

has the t distribution. Using the t distribution with the data of Sec. 9.4.1, find 95 per cent confidence limits on μ assuming σ^2 unknown. HINT: Use the t table (Appendix 11) to find two numbers $t_{0.05/2}$ and $t_{0.975}$ $(0.975 = 1 - 0.05/2)$ such that

$$\Pr\,(t_{0.025} < t < t_{0.975}) = 1 - 0.05$$

where

$$\Pr\,(t_\beta < t) = \Pr\,(t > t_{1-\beta}) = 1 - \beta$$

Use this equality to find two numbers a, b such that

$$P(a < \mu < b) = 1 - 0.05$$

Prob. 9.11. Dichotomous Population. Let

$$x_i = \begin{cases} 1 & \text{with probability } p \\ 0 & \text{with probability } 1 - p \end{cases}$$

$$\overset{>}{x} = (1, 0, 0, 1, 0)$$

Find confidence intervals on p using three different confidence levels of your own choice.

Answer. See Appendix 8.

9.4.4 Special Example: Dichotomous Population. Confidence Limits on p in the Binomial Distribution for Large Sample Sizes

A common example encountered in simulation problems is establishing a confidence interval for p, the probability of "success," when there are only two possible outcomes to an experiment. The quantity p may represent the probability of hit of a missile, the probability of a system overloading, etc. Before proceeding to general analytical methods, it is important to emphasize that there *exist* tables and curves which give confidence intervals for a given sample and for a specified confidence level. These curves are obtained in much the same way that the values of δ (Prob. 9.11) were obtained. Two curves (Figs. A8.1 and A8.2) are examples. Their use is illustrated in the appendix. We add the caution that the methods described in the following discussion are applicable only when the total number of samples is greater than 30. The reason for this restriction lies in the fact that the normal density function is used to approximate the binomial density [115, pp. 682–683]. When the sample size becomes greater than 30, the error becomes insignificant for ordinary applications. For a smaller number, straightforward use of the binomial distribution will suffice [125]. Hence, the methods described in this section are advantageous when the sample size is large (≥ 30) and when curves *do not* exist for the confidence level in which one is interested.

Maximum Confidence Intervals and Estimation of Sample Size. To begin, we shall describe in a formal way how to obtain a δ such that p lies in the interval $(\bar{x} - \delta, \bar{x} + \delta)$ *at least* 95 per cent of the time. The interval so obtained is more conservative than the usual confidence interval.

To find this "conservative" confidence interval, use the following procedure:[1]

1. Let $\bar{x}$ be the estimate of p. Let $V_{\max}(\bar{x}) = 0.25/N$ be the maximum value of the variance, where N represents the total number of samples. Then, for a confidence level of size Y per cent, find the limits of

$$\frac{1}{\sqrt{2\pi}} \int_{-\delta\sqrt{V_{\max}(\bar{x})}}^{\delta/\sqrt{V_{\max}(\bar{x})}} e^{-t^2/2}\, dt = \frac{Y \text{ per cent}}{100} \tag{9.22}$$

from the standard tables of the error function (see Appendix 6).

2. As an example, assume $Y = 95$ per cent, $N = 100$. Hence, to find δ we evaluate the limits of

$$\frac{1}{\sqrt{2\pi}} \int_{-\delta/0.05}^{\delta/0.05} e^{-t^2/2}\, dt = 0.95 \tag{9.23}$$

Using Appendix 6, we find $\delta = 10^{-1}$. If this "error" or "confidence" is acceptable, we need proceed no further. However, suppose this error is too large. We would then wish to find the exact 95 per cent confidence interval which will certainly be equal to or smaller than the δ found above. The method for finding more exact confidence intervals is given in cookbook fashion below, but before proceeding, the natural question arises as to why we even bother with the above procedure when we have a better one all along. The answer is simple. Note that the δ found above is independent of the actual sample taken; i.e., it does not depend on the p but only on N, the total number of samples. Hence, we have struck upon a method of evaluating approximately the amount of data we need for a given size of confidence interval. For example, suppose we are about to begin experimenting and we wish to know about how long the experiment will take if we know that each run will take about one second. We want a 95 per cent confidence interval in which $\delta = 0.01$. Certainly 100 samples will not permit us to meet these conditions. Using expression (9.23) above with $Y = 95$ per cent, $\delta = 0.01$, and N unknown, we find that N is about 10^4. If we know that each run will take about one second, the total running time is conservatively estimated to be about eight minutes. It is in this fashion that the above technique finds its use.

[1] For details, see Appendix 9.

For convenience, curves of δ_{max} versus "amount of data" for specified confidence intervals are given in Fig. A9.1.

A More Exact Method for Determining Confidence Intervals. Now let us proceed to the more sensitive method of determining confidence intervals. To demonstrate the method, we must operate on actual data. (Note that

TABLE 9.1 CASE 1; NEW METHOD; $P_{t_1} = 0.82$

Tactical situation	True probability of hit P_t	Run									
		1	2	3	4	5	6	7	8	9	10
1	0.9	1	1	1	1	1	1	1	1	1	1
2	0.8	1	1	1	1	1	1	1	1	1	1
3	0.95	1	1	1	1	1	1	1	1	1	1
4	0.75	0	1	1	1	1	0	0	1	0	1
5	0.9	1	1	0	1	1	1	1	1	1	1
6	0.85	1	1	1	1	1	1	0	1	0	1
7	0.8	1	1	1	1	1	1	1	1	0	1
8	0.7	1	1	1	1	0	1	0	0	1	1
9	0.65	1	0	1	1	1	0	1	1	1	0
10	0.9	1	1	1	1	1	0	1	1	1	1

TABLE 9.2 CASE 2; OLD METHOD; $P_{t_2} = 0.79$

Tactical situation	True probability of hit P_t	Run									
		1	2	3	4	5	6	7	8	9	10
1	0.9	1	1	1	1	1	0	1	1	1	1
2	0.8	1	0	0	1	0	0	1	1	1	1
3	0.9	1	1	1	1	1	1	1	1	1	1
4	0.65	0	1	1	0	0	1	0	1	1	1
5	0.9	1	1	1	1	1	1	1	0	1	1
6	0.8	1	0	1	1	1	1	0	1	1	0
7	0.8	1	1	1	1	0	1	1	1	1	1
8	0.65	1	0	1	0	0	1	1	1	1	0
9	0.65	1	0	1	0	0	1	1	1	1	0
10	0.85	1	1	1	1	1	1	1	1	1	1

the fact that the δ found above was independent of the data suggests that we should certainly do better when the actual data are taken into account.) We have generated a synthetic body of data with the aid of a table of random numbers [126, pp. 219–225]. The results are given in Tables 9.1 to 9.4. To make the presentation more palatable, we envision the following situation: A manufacturer of a given missile is undecided as to whether or not to make a change in his missile system. Should he retain his "old system" or should he change to the "new system"? He decides to choose that system which will give him a higher probability of hitting the target. However, the missile's probability of hit varies with the tactical situation, e.g., whether it is fired from a moving vehicle or a stationary one, whether or not the enemy target is moving toward or away from the missile carrier, etc.

TABLE 9.3 CASE 1; NEW METHOD; $P_{t_1} = 0.82$

Tactical situation	True probability of hit	Run									
		1	2	3	4	5	6	7	8	9	10
1	0.9	1	1	1	1	1	0	1	1	1	1
2	0.8	1	1	1	1	0	0	1	1	0	1
3	0.95	1	1	1	1	0	1	1	1	1	1
4	0.75	1	1	1	0	1	0	1	1	1	1
5	0.9	1	1	1	1	1	1	1	1	1	1
6	0.85	0	0	1	1	1	1	1	1	1	1
7	0.8	1	1	0	1	1	1	0	1	1	0
8	0.7	1	0	1	1	0	1	1	1	1	0
9	0.65	0	1	1	1	1	1	1	1	1	1
10	0.9	1	1	1	1	1	1	0	1	1	1

TABLE 9.4 CASE 2; OLD METHOD; $P_{t_2} = 0.79$

Tactical situation	True probability of hit	Run									
		1	2	3	4	5	6	7	8	9	10
1	0.9	1	0	1	1	1	1	1	1	1	1
2	0.8	0	1	1	1	1	1	1	1	0	1
3	0.9	1	1	1	1	1	1	1	1	1	1
4	0.65	1	1	1	0	0	1	0	0	1	1
5	0.9	0	1	1	1	1	1	1	1	1	1
6	0.8	1	1	1	1	1	1	1	1	0	0
7	0.8	1	1	1	1	1	1	1	1	0	0
8	0.65	0	1	0	1	0	0	0	1	1	1
9	0.65	1	1	0	0	0	1	0	1	0	1
10	0.85	1	1	1	1	0	1	1	1	1	1

Let us assume that there are 10 different tactical situations. Listed in the table are the true probabilities of hit associated with each tactical situation. Of course, the manufacturer does not know these "true" probabilities. If he did, there would be no dilemma. He would choose the missile which has the higher overall true probability P_t:

$$P_t = \sum_{1}^{10} \frac{P_j}{10}$$

where P_j = true probability of hit for tactical situation j. Involved here is a question of test of hypotheses. We shall come to this question in a subsequent section. For the moment let us imagine that all the manufacturer wishes to do is estimate P_{t_1}, the overall true probability of hit of his new missile, and to establish a confidence interval about P_{t_1}. He fires 10 times per tactical situation and obtains the data of Table 9.1, where a 1 is a hit and a 0 is a miss. Note that in Table 9.1, $N = 100$. To find the 95 per cent confidence interval we use the following expression:

$$\frac{\bar{x} - L}{\sqrt{\dfrac{L(1 - L)}{N}}} = y \tag{9.24}$$

where $\bar{x}$ is the sample mean, N is the number of samples, y is the number such that

$$\frac{1}{\sqrt{2\pi}} \int_{-y}^{y} e^{-t^2/2}\, dt = \frac{Y}{100} \tag{9.25}$$

and Y is the confidence level in per cent. To find the confidence interval, solve the above quadratic equation (9.24) for L. The values found, L_1 and L_2, constitute the confidence interval (L_1, L_2).† As an example, let us find the 95 per cent confidence interval for the case of Table 9.1. Calculate

$$\bar{x} = \bar{x}_{\text{Table 9.1}} = \frac{\sum_{i=1}^{10} \sum_{j=1}^{10} x_{ij}}{100} = 0.85 \tag{9.26}$$

where x_{ij} represents the ith run for the jth tactical situation. For $Y = 95$ per cent we find from the tables of the error function (erf) that $y = 1.96$. Using Eq. (9.24),

$$\begin{aligned} L &= \frac{2(0.85)100 + (1.96)^2}{2[100 + (1.96)^2]} \pm \frac{1}{2}\sqrt{\left[\frac{2(0.85)100 + (1.96)^2}{100 + (1.96)^2}\right]^2 - 4\,\frac{100(0.85)^2}{100 + (1.96)^2}} \\ &\approx 0.835 \pm 0.07 \end{aligned} \tag{9.27}$$

That is, $\delta = 0.07$. Comparing this result with that found using the procedure described first, we find that $\delta = 0.1$ is, indeed, a conservative estimate. Now we shall compare the approximate result $\delta = 0.07$ with that computed from *exact* methods. It is here that the figures of Appendix 8 prove useful. The curves in these figures were calculated using exact methods (i.e., a binomial distribution for $\bar{x}$ rather than a normal one) and hence can be used to check our results. In doing so, we shall also learn how to use these curves. Note that Fig. A8.1 has the label "Confidence coefficient, $1 - \alpha = 0.95$." This labeling means that the figure is to be used to find 95 per cent confidence intervals *only*. Figure A8.2 has a similar label-

† The derivation behind this procedure is given in Appendix 10.

ing, except that 0.99 replaces 0.95. This figure is to be used to find 99 per cent confidence intervals *only*. To check our result we use, naturally, the first figure. Since we had a total of 100 runs, we use the curves marked by 100. Our percentage of hits was 0.85. Hence, we find c/n (at the top) equal to 0.85. We read the intersection of $c/n = 0.85$ with the two curves marked 100. In our case they read about 0.07 and 0.24. The confidence interval δ is

$$\delta = \frac{0.92 - 0.76}{2} = 0.08$$

Hence, the approximation we have obtained is very good.[1] In addition, the calculated center of the interval is 0.835, and that indicated in the figure is 0.84.

A good approximation to (9.27) is the expression

$$L_1 = \bar{x} - \frac{y}{n}\sqrt{n\bar{x}(1-\bar{x})} \qquad (9.28)$$

$$L_2 = \bar{x} + \frac{y}{n}\sqrt{n\bar{x}(1-\bar{x})}$$

This approximation can be used when $n \gg y$ (see Appendix 10). In the example above we find that $(y/n)\sqrt{n\bar{x}(1-\bar{x})} = 0.0699$. Hence, the *size* of the confidence interval is approximated very well but the center of the interval is shifted.

It is important to emphasize once again the fact that all the above approximate methods are applicable for large-sample testing, i.e., when $n \geq 30$. The small-sample cases can be treated simply by using the binomial distribution directly.

9.5 Test of Hypothesis

In the problem of the missile manufacturer described in the preceding section, an estimate of P_H, the probability of hit, was desired and obtained. However, in many simulation problems we are interested not in *estimating* parameters per se but in evaluating the effect of two different modes of system operation on the system performance. Thus the missile manufacturer wants to know which system, the "old" or the "new," gives him a missile with a higher probability of hit. He must make a decision between three assumptions:

1. $P_{H1} > P_{H2}$
2. $P_{H1} < P_{H2}$
3. $P_{H1} = P_{H2}$

[1] It is suggested that the reader, for practice, or the skeptic, for conviction, find 95 per cent confidence intervals with the data in the other tables.

Each assumption concerns the nature of the binomial density function representing the frequency of a given number of successful hits. We shall call these assumptions hypotheses and shall define a statistical hypothesis as follows:

DEFINITION: A statistical hypothesis is an assumption about the frequency function of a random variable.

The hypotheses we select as "true" will depend on the outcome of an experiment. For example, if we found that the old missile system produced 30 hits out of 30 trials whereas the new system failed to hit at all in 30 trials, we would certainly prefer hypothesis 1 to the others. However, this example presents an extreme case. Clearly we must have some procedure for selecting a hypothesis when the data do not present us with such a clear-cut choice. The procedure we choose is called the *test* of the hypothesis.

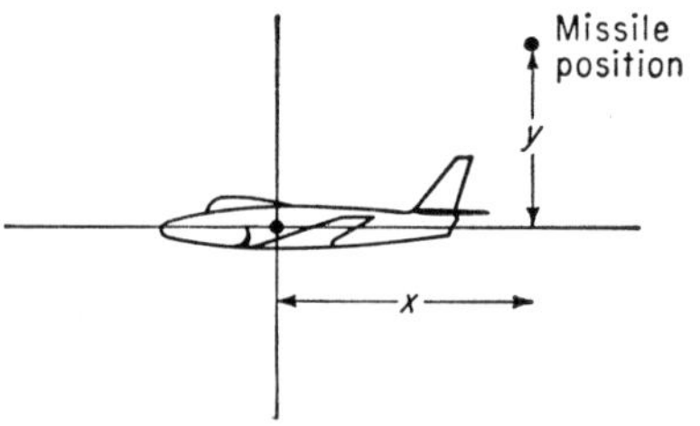

FIG. 9.2 A drawing in the vertical plane of the relative positions of an aircraft and missile.

DEFINITION: A test of a statistical hypothesis is a procedure for deciding whether to accept or reject the hypothesis on the basis of the data.

We shall return to our missile manufacturer's dilemma in a subsequent section. However, to simplify the discussion, consider a normally distributed random variable x with the unknown mean μ and $\sigma^2 = 1$. To present a concrete example, consider the airplane pictured in Fig. 9.2. Let x represent the distance from zero that a missile falls along the abscissa. Assume for the sake of simplicity sake that the mean μ of x is either zero or 10 kilofeet. We shall denote these two hypotheses by H_0 and H_1, respectively. If the mean is zero, then, of course, we keep the missile we have. If $\mu = 10$ kilofeet, then we must scrap the present missiles and begin our search for a better one. Thus, we are not simply faced with a choice between two alternatives which affect us in the same manner, for the consequences attendant upon the selection of H_0 as opposed to H_1 are vastly different. The acceptance of H_0 represents an acceptance of the *status quo;* the acceptance of H_1 means great additional expenditures of time and money. Hence, we must beware of the possibility of rejecting H_0 when it is true. This type of error, of rejecting H_0 *when* it is *true,* is called the Type I error. We wish to make the *probability* of this error small. How we accomplish this task is the topic of subsequent paragraphs.

Another type of error we may make, called the Type II error, is that of *accepting* H_0 when it is *false.* However, since rejecting H_0 when it is true results in catastrophe, whereas accepting H_0 when it is false only means an irritated customer, we do not mind if the probability of the Type II error is high as long as the probability of the Type I error is small. Hence, whatever the test of hypotheses we make, we require that the *probability* of *Type I error* be a small, fixed quantity, henceforth denoted by α.

Our task is to select a "best" test. In order to achieve our end, we must first know what we are talking about when we refer to a "test" of a hypothesis. The definition above is a generalization which leads us to the concept but does not indicate how to employ it. To understand what we mean specifically by the term *test*, suppose we hypothesize that the theoretical mean of a distribution is zero. If one were to collect the sample $\overset{>}{x}_3 = (0, 0, 10^{-6})$ from this population, one would be inclined to accept this hypothesis; on the other hand, if the sample obtained were $\overset{>}{x}{}'_3 = (10^6, 10^{10}, 10^{12})$ then, to say the least, one would be skeptical of the validity of this hypothesis. Note that $\overset{>}{x}_3$ and $\overset{>}{x}{}'_3$ are two points in 3-space. Given one of these two points, the choice between rejecting and accepting the hypothesis is not difficult. The idea to be emphasized here is that there are two regions in the 3-space;[1] one region, called the acceptance region, is a volume such that, if the sample $\overset{>}{x}_3$ should fall into it, then the hypothesis would be accepted; in addition, there is a *rejection region*—a region such that, if the sample should fall into it, the hypothesis would be rejected. This latter region is commonly called the *critical region*. The question which remains is: "Where is the boundary between the regions?" The choice of the boundary specifies the *critical region*. A test of a hypothesis is defined as the *selection* of the critical region. To illustrate the fact that it is possible to choose any one of a number of critical regions, consider Example 1, which also contains some important additional ideas on testing hypotheses.

Example 1. Suppose that missiles are fired (always under the same conditions) at a ground target. The target is a large circular patch of ground. It is important to know what the probability is that a missile will fall (or be fired into) the target area. Assume mutual independence of the firings.

$$
\begin{aligned}
p &= \text{probability of success} \\
q &= \text{probability of failure} \\
p + q &= 1
\end{aligned}
$$

Assume

$$P(S = x) = \binom{n}{x} p^x q^{n-x} = \frac{n!}{x!(n-x)!} p^x q^{n-x} \tag{9.29}$$

where n = number of missile firings
x = number of successes

On the basis of data or otherwise, the statistical hypothesis (H_0) is made that $p = \frac{1}{4}$. This is equivalent to making the assumption (statistical hypothesis) that $q = \frac{3}{4}$ or that $E(S) = n/4$.

Let us assume that $n = 3$. Define H = hit, M = miss; $H = 1$, $M = 0$. The sample space, or all possible outcomes of three firings, is 000, 001, 010, 100, 011, 101, 110, 111. Since $P(111)$ is smaller than the probability of any other sample point, we *might* choose to reject the hypothesis if we obtain this sample point. Or we might choose to reject

[1] In general, there will be n data points; hence the space will be n-dimensional. This n-dimensional space is called the sample space.

the hypothesis if $S = 1$. Whatever criterion we choose, the "rejection values" constitute the critical region of the test.

DEFINITION: The critical region of a test of a statistical hypothesis is that part of the sample space which corresponds to the rejection of the hypothesis being tested.

Let H_1 denote the alternative hypothesis ($p = \frac{1}{2}$). We propose the problem of testing H_0 against H_1.

In order to decide whether selection of a particular critical region (cr) is wise, consider the consequences of the choice. If the hypothesis (H_0) is true and the observed value (x) falls in the critical region, then H_0 will be rejected. As stated above, this kind of error is called the Type I error. If H_1 is true and x does not fall into the cr, then H_0 will be accepted. As above, this error is called a Type II error. Table 9.5 summarizes the consequences of the possible decisions.

TABLE 9.5

	H_0 true	H_1 true
x in cr (reject H_0)............	Type I error	Correct decision
x not in cr (accept H_0)........	Correct decision	Type II error

DEFINITION: The size of the Type I error is the probability that the sample point will fall in the cr when H_0 is true; the size of the Type II error is the probability that the sample point will not fall in the cr when H_1 is true.

Test I. Assume cr, $x \leq 1$. Then

$$\alpha = \text{Type I error} = \sum_{x=0}^{1} \binom{3}{x} \left(\tfrac{1}{4}\right)^x \left(\tfrac{3}{4}\right)^{3-x} = \tfrac{27}{32}$$

$$\beta = \text{Type II error} = \sum_{x=2}^{3} \binom{3}{x} \left(\tfrac{1}{2}\right)^x \left(\tfrac{1}{2}\right)^{3-x} = \tfrac{1}{2}$$

Test II. Assume cr, $x = 0$.

$$\alpha = \tfrac{27}{64} \qquad \beta = \tfrac{7}{8}$$

Test III. Assume cr, $x = 1$.

$$\alpha = \tfrac{27}{64} \qquad \beta = \tfrac{5}{8}$$

As is easily seen, there are many critical regions to choose from. Suppose we restrict this choice by selecting a critical region from only those which have a specified Type I error. For example, if we specify

$$\alpha = \tfrac{27}{64}$$

then in the choice above, test I is immediately eliminated. It is conceivable that many choices of a critical region may remain even after we impose this restriction on the choice. In the example above, two choices remain, tests II and III. Our task is to select the "best" test from these two tests. In general, our task is to select a "best" test from all the available tests having the same specified α. One requirement for a best test immediately comes to mind, namely, to select the test which, for a given α, has the smallest probability of Type II error associated with it.

PRINCIPLE: Among all tests possessing the same size of Type I error, choose one for which the size of Type II error is as small as possible.

Hence, we wish to choose the test i such that $\beta_i \leq \beta_j$ for any other test j. Equivalently we choose the test i such that $1 - \beta_i \geq 1 - \beta_j$ for all tests j.

Continuing the above example, consider the alternative hypothesis (H_1) as being $1 \geq p > \frac{1}{4}$. For this more general hypothesis, it is necessary to consider the calculation of the size of Type II error as a function of p. The size of this error will be denoted as $\beta(p)$. Note that $\phi(p) = 1 - \beta(p)$. Thus, seeking the test that minimizes the Type II error $\beta(p)$ is equivalent to seeking for a test which maximizes the function $\phi(p)$.

Test II: $$\phi_{II}(p) = (1 - p)^3 \tag{9.30}$$
Test III: $$\phi_{III}(p) = 3p(1 - p)^2 \tag{9.31}$$

The function $\phi(p)$ is called the *power function.* This terminology leads to the following definition.

DEFINITION: The power function of $\phi(\theta)$ of a test is the function of the parameter θ which gives the probability that the sample point will fall in the critical region of the test when θ is the true value of the parameter.

In accordance with the principle stated above regarding the selection of a test, we note in Fig. 9.3 that the power curve of test III lies above the power curve for test II for all values of $p > \frac{1}{4}$; consequently, test III is to be preferred over test II in *that* range of p. Since H_1 is $1 \geq p > \frac{1}{4}$, test III is preferred to Test II.

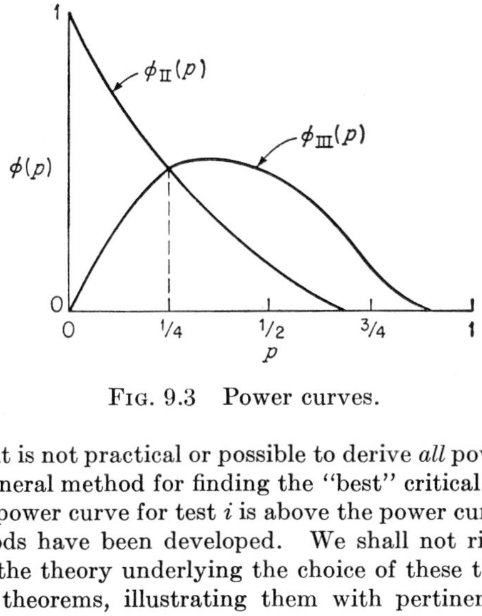

FIG. 9.3 Power curves.

In many problems, it is not practical or possible to derive *all* power functions. Hence, we must have some general method for finding the "best" critical region, i.e., the critical region i such that the power curve for test i is above the power curve for any other test j.

Such general methods have been developed. We shall not risk the patience of the reader by developing the theory underlying the choice of these tests. Instead we shall quote the important theorems, illustrating them with pertinent examples, and then indicate various applications in simulation problems. Before proceeding, the following problems are suggested for study in order to fix ideas.

Prob. 9.12. Let

$$p(x) = \frac{e^{-\lambda}\lambda^x}{x!}$$

The distribution $p(x)$ is called the Poisson distribution. For H_0: $\lambda =$ (*a*) 0.2, (*b*) 2, (*c*) 20 and H_1: $\lambda = 4$, find various critical regions for a fixed α and various β's and for a sample size $n = 3$. [HINT: We note that, if each x_i is Poisson-distributed, then $y = x_1 + x_2 + x_3$ is Poisson-distributed; i.e.,

$$p(y) = \frac{e^{-3\lambda}(3\lambda)^y}{y!}$$

If we let $\bar{x} = y/3$, then

$$p(\bar{x}) = \frac{e^{-3\lambda}(3\lambda)^{3\bar{x}}}{(3\bar{x})!}$$

Hence we can establish our critical region by some numbers c_1 and c_2 such that we reject H_0 if

$$c_2 \leq \bar{x} \leq c_1$$

or

$$\bar{x} \leq c_2 \qquad \bar{x} \geq c_1$$

or

$$\bar{x} \leq c_1$$

and so forth.]

Prob. 9.13. Make a sketch of the best such critical region in three dimensions.

Prob. 9.14. Using the best critical region you have found in Prob. 9.12, test H_0 using the data

$$\overset{>}{x}_3 = (2, 6, 1)$$

Prob. 9.15. Sketch the power function $\phi(\lambda)$ in Prob. 9.12.

9.5.1 Simple Hypotheses

In Example 1, in which we were concerned with the probability of hit of a missile, we stated several hypotheses. In particular, one hypothesis was

$$H_0: p = \tfrac{1}{4}$$

Note that p is a parameter of the *binomial* distribution. This hypothesis specifies the binomial frequency function completely, since it specifies the value of the only parameter of that distribution. A hypothesis in which the frequency function is completely determined is called a simple hypothesis. In contrast, consider the hypothesis

$$H_0: \tfrac{1}{4} \leq p \leq 1$$

This hypothesis is not simple, because it does not specify the value of p, only its range. It is called a composite hypothesis. As another example, consider the density function

$$p(x; \mu, \sigma) = \frac{1}{\sqrt{2\pi}\,\sigma} \exp\left[-\frac{(x-\mu)^2}{2\sigma^2}\right]$$

The hypothesis

$$H_0: \mu = 0$$

when σ is known, is a simple hypothesis. However,

$$H_0: \mu = \sigma$$

is a composite hypothesis. We shall first be concerned with simple hypotheses.

Consider again the simple hypothesis

$$H_0: p = \tfrac{1}{4}$$

Note in Fig. 9.3 that $\phi_{\text{III}}(p)$ does not lie above $\phi_{\text{II}}(p)$ for *all* values of p. For values of $p < \frac{1}{4}$, II is the better test. If $\phi_{\text{III}}(p)$ were greater than $\phi_{\text{II}}(p)$ for *all* values of p, *then* III would be called a *most powerful test.* The following theorem, called the Neyman-Pearson lemma, asserts the basis for choosing a critical region which gives a most powerful test.

NEYMAN-PEARSON LEMMA: If a critical region C having Type I error α and a positive constant k exists such that inside C

$$\frac{p_0}{p_1} \leq k \tag{9.32}$$

and outside C

$$\frac{p_0}{p_1} \geq k$$

where p_0 and p_1 are the density functions assuming H_0 and H_1 are true, respectively, then the choice of C yields a most powerful test.

Example 2. Let the population under consideration have the density

$$p(x) = \frac{1}{\sqrt{2\pi}} e^{-(x-\mu)^2/2}$$

Let $$H_0: \mu = \delta_0 \qquad H_1: \mu = \delta_1$$

where δ_0 and δ_1 are two real numbers.

Let n represent the number of samples, all independent. Then

$$p_0 = \prod_{i=1}^{n} \frac{e^{-\frac{1}{2}(x_i-\delta_0)^2}}{\sqrt{2\pi}} \tag{9.33}$$

$$p_1 = \prod_{i=1}^{n} \frac{e^{-\frac{1}{2}(x_i-\delta_1)^2}}{\sqrt{2\pi}} \tag{9.34}$$

Hence $$\frac{p_0}{p_1} = \exp\left[(\delta_0 - \delta_1)\sum_{i=1}^{n} x_i + \frac{n}{2}(\delta_1^2 - \delta_0^2)\right] \tag{9.35}$$

The inequality

$$\log \frac{p_0}{p_1} \leq \log k \tag{9.36}$$

yields

$$\bar{x} = \frac{1}{n}\sum_{i=1}^{n} x_i \leq \frac{1}{\delta_0 - \delta_1}\left[\frac{\log k}{2} + \frac{\delta_0^2 - \delta_1^2}{n(\delta_1 - \delta_0)}\right] \tag{9.37}$$

The choice of k dictates the size of the right side of the equation. Hence, the right side can be made any value. Thus, there will be a k such that the probability that $\bar{x}$ will be less than the right side can be made equal to α when $\mu = \delta_0$. Hence, the condition in the theorem that a critical region of size α exists is satisfied; moreover the critical region is given by the right tail of the density function of $\bar{x}$ (see Fig. 9.4). Since the inequality will hold for all $\delta_1 > \delta_0$, it follows that for

$$H_0: \mu = \delta_0 \qquad H_1: \mu = \delta_1 \qquad \delta_1 > \delta_0$$

the right tail of the $\bar{x}$ curve should be chosen. If $\delta_1 < \delta_0$, then the inequality is reversed, and the left tail of the $\bar{x}$ curve should be selected.

Example 3. Let

$$f(x) = \binom{n}{x} p^x (1-p)^{n-x}$$

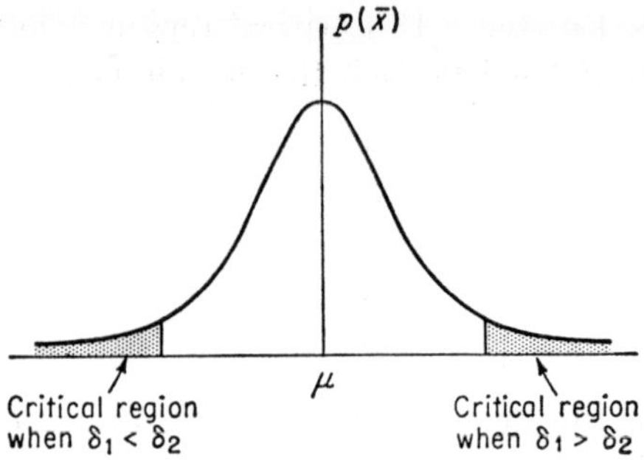

FIG. 9.4 Density function for the sample mean $\bar{X}$, where

$$\bar{X} = \sum_{i=1}^{n} \frac{X_1}{n}$$

showing critical region for the two-tail test. X_i is $N(\mu, \sigma^2)$.

where x represents the number of "successes" in n trials and p is the probability of success.

Let

$$H_0: p = p_0 \qquad H_1: p = p_1$$

Then

$$f_0(x) = \binom{n}{x} p_0^x (1 - p_0)^{n-x}$$
$$f_1(x) = \binom{n}{x} p_1^x (1 - p_1)^{n-x} \tag{9.38}$$

Following the procedure outlined in Example 2, we let

$$\frac{f_0}{f_1} \leq k \tag{9.39}$$

or

$$\frac{p_0^x(1 - p_0)^{n-x}}{p_1^x(1 - p_1)^{n-x}} \leq k$$

Taking the logarithm of both sides and solving the inequality for x, we find

$$x \leq \frac{\log k - n \log \dfrac{1 - p_0}{1 - p_1}}{\log \dfrac{p_0(1 - p_1)}{(1 - p_0)p_1}} \tag{9.40}$$

On noting that x represents the *number* of successes in n trials, it follows that

$$x = \sum_{i=1}^{n} x_i$$

$$x_i = \begin{cases} 0 & \text{if outcome on } i\text{th trial is a failure} \\ 1 & \text{if outcome on } i\text{th trial is a success} \end{cases}$$

Hence, the best critical region, *if there is one*, is a plane in n-dimensional space satisfying the condition

$$\sum_{i=1}^{n} x_i \leq \frac{\log k - n \log \dfrac{1 - p_0}{1 - p_1}}{\log \dfrac{p_0(1 - p_1)}{(1 - p_0)p_1}} \tag{9.41}$$

It is possible in this case that there is no best critical region for a specified α because our sample space is discrete here instead of a continuum as in the preceding example. To illustrate this fact, suppose $\alpha = 0.05$. As can be seen for $n = 3$, $p_0 = \frac{1}{2}$, there is no critical region of this size; hence there can be no best test for $\alpha = 0.05$.

Example 4

$$p(x) = \frac{\lambda^x e^{-\lambda}}{x!}$$

$$H_0\colon \lambda = \lambda_0 \qquad H_1\colon \lambda = \lambda_1$$

To find the best critical region (sample size $= n$) for this test of hypothesis, we note that

$$p_0 = \prod_{i=1}^{n} \frac{\lambda_0^{x_i} e^{-\lambda_0}}{x_i!} \qquad p_1 = \prod_{i=1}^{n} \frac{\lambda_1^{x_i} e^{-\lambda_0}}{x_i!}$$

Then
$$\frac{p_0}{p_1} = \prod_{i=1}^{n} \left(\frac{\lambda_0}{\lambda_1}\right)^{x_i} e^{-\lambda_0+\lambda_1} = \left(\frac{\lambda_0}{\lambda_1}\right)^{\sum_{i=1}^{n} x_i} \exp\left[-n(\lambda_0 - \lambda_1)\right]$$

Using the above procedure, we find the region for which the inequality

$$\log \frac{p_0}{p_1} \leq \log k$$

is satisfied. Since

$$\log \frac{p_0}{p_1} = -n(\lambda_0 - \lambda_1) + \left(\sum_{i=1}^{n} x_i\right) \log \frac{\lambda_0}{\lambda_1}$$

the best choice satisfies the inequality

$$\bar{x} = \frac{1}{n} \sum_{i=1}^{n} x_i \leq \frac{1}{\log \frac{\lambda_0}{\lambda_1}} \left(\frac{\log k}{n} + \lambda_0 - \lambda_1\right)$$

As an exercise describe the best region when $\lambda_0 \geq \lambda_1$. For $n = 2$, sketch the best critical region for a given α.

Prob. 9.16. The distribution of the "time to failure" for a system is given by

$$p(t) = e^{-\lambda t} \qquad t \geq 0$$

We hypothesize

$$H_0\colon \lambda = 4 \text{ hr}^{-1} \qquad H_1\colon \lambda = 10 \text{ hr}^{-1}$$

(*a*) Find a best test for $\alpha = 0.05$.

(*b*) Ten systems were tested. Their times to failure, in hours, were 1, 15, 2, 29, 18, 7, 4, 6, 18, 12. Test H_0 against H_1.

9.5.2 Composite Hypotheses

In many simulation problems we are interested not in estimating performance of the system per se but in evaluating the effects of a modification to the system or in comparing the performances of the system in two different modes of operation. As an example, in Sec. 9.4.4 we posed the problem of the missile manufacturer who was forced to choose between keeping his system as is and accepting a modification to it. He wishes to choose the system which will give him the higher probability of hitting the target. However, the missile's probability of hit varies with the conditions under which the target is intercepted. It would appear, then, that he should select the system which has the larger *overall* probability of hit. However, this is not necessarily the case. If the probability of hit with no modification (P_{t_2}) is 0.5 and the probability of hit with the modification P_{t_1} is 0.505, the missile manufacturer might still prefer P_{t_2} because the difference $P_{t_1} - P_{t_2}$ might not be enough to warrant the *expenditure* entailed in modifying the system. We shall illustrate the handling of this problem below. At the outset, however, it should be noted that the analysis presented is applicable to any dichotomous population, i.e., a population in which only one of two possibilities can occur on a single trial. In the case considered below, the outcomes are hit or miss on a single firing. In the analysis of production lines the outcomes might be *acceptable* or *unacceptable* items off the line; in general, with respect to simulation problems, the outcomes might be *failures* or *successes* in the performance of the system with respect to a given input.

Let us return to the missile manufacturer's dilemma. He wishes to test the hypothesis

$$H_0: P_{t_1} \leq P_{t_2}$$

against the hypothesis

$$H_1: P_{t_1} > P_{t_2}$$

The hypothesis H_0 states that the system with no modification is better than or equivalent to one with a particular modification. One could well ask: Why not test

$$H_0: P_{t_1} > P_{t_2} \qquad \text{against} \qquad H_1: P_{t_1} \leqq P_{t_2}$$

or

$$H_0: P_{t_1} \geq P_{t_2} \qquad \text{against} \qquad H_1: P_{t_1} < P_{t_2}$$

The convention is that we use for H_0 that hypothesis which, if we reject it when it is true, will cost us more than rejecting the alternative hypothesis when *it* is true. Clearly, in the case treated above, if we maintain that there is a difference in the behavior of the system in favor of the modification when in fact there is none, it will cost a great deal more than if we reject $P_{t_1} < P_{t_2}$ when it is true, especially if $|P_{t_1} - P_{t_2}|$ is small. Hence, if we accept $P_{t_1} > P_{t_2}$, this means that we must change the system to include the modification. Such changes are extremely costly. Hence we must accept the hypothesis $P_{t_1} > P_{t_2}$ with great caution and so relegate it to the H_1 hypothesis.[1]

We already know from Tables 9.1 and 9.2 that

$$P_{t_1} = 0.82 \qquad P_{t_2} = 0.79$$

[1] It is to be noted that there are a great many subtle points concerning which hypothesis should be H_0 and which should be H_1. We shall treat another case in subsequent paragraphs, but for details see Ref. 125, chap. 3.

We have the advantage, then, of knowing that H_0 is false and H_1 is true. Hence, any test we use should come to this conclusion. We shall now describe in a formal way a very sensitive test of such hypotheses. We shall also indicate its limitations.[1]

Tables 9.1 and 9.2 will be used first for illustration. We have already found that [see Eq. (9.26)]

$$\bar{x} = \bar{x}_{\text{Table 9.1}} = 0.85$$

We now calculate

$$s_x = \sqrt{\frac{\sum_{i=1}^{n_r} (\bar{x}_i - \bar{x})^2}{n_r}} \tag{9.42}$$

where

$$\bar{x}_i = \frac{\sum_{j=1}^{n_T} x_{ij}}{n_T}$$

and x_{ij} represents the ith run for the jth tactical situation, n_r represents the number of runs per tactical situation, and n_T represents the number of tactical situations. In the case of Table 9.1

$$s_x = 0.108$$

Operating on the data in Table 9.2, we find that

$$\bar{y} = \frac{\sum_{j}^{10} \sum_{i}^{10} y_{ij}}{100} = 0.80 \tag{9.43}$$

$$s_y = \sqrt{\sum_{i=1}^{10} \frac{(\bar{y}_i - \bar{y})^2}{10}} = 0.131 \tag{9.44}$$

where

$$\bar{y}_i = \frac{\sum_{j=1}^{10} y_{ij}}{n_T}$$

and y_{ij} represents the outcome of the ith run for the jth tactical situation in Table 9.2. We use the formula[2]

$$t = \frac{\bar{x} - \bar{y}}{\sqrt{n_x s_x^2 + n_y s_y^2}} \sqrt{\frac{n_x n_y (n_x + n_y - 2)}{n_x + n_y}} \tag{9.45}$$

where n_x = number of runs *per tactical situation* in Table 9.1 = 10
n_y = number of runs *per tactical situation* in Table 9.2 = 10

Hence $$t = 0.885$$

[1] The theory is not described herein, for it is much too involved. Reference 125, chaps. 10 and 11, is recommended.

[2] For a discussion of the t distribution, see Ref. 125, chap. 11.

Before proceeding, we must resolve a question in our minds as to how certain we wish to be of our result. That is, if we reject H_0 in favor of H_1, how certain do we want to be of our choice? Can we chance making one mistake in 10 such problems, one mistake in 100, in 200, etc.? Note that the greatest probability of rejecting H_0 when it is true occurs when $P_{t_1} = P_{t_2}$. As stated, this probability of rejecting H_0 when $P_{t_1} = P_{t_2}$ is called the significance level. For purposes of illustration we shall use the 0.05 level of significance; that is, we shall see what happens when we wish to avoid rejecting H_0 when it is true *at least* 95 per cent of the time.

We now use Student's t table (see Appendix 11). If Y represents the level of significance, then we use column Y. For example, if our level of significance is 0.05, then we use the column marked 0.05, that is, $t_{0.95}$ or $-t_{0.05}$. We use the row denoted by df (degrees of freedom) $= n_x + n_y - 2$, which in the illustrative case is 18. The number located at the intersection of this column and row is 1.734. We accept H_0 if our calculated value of t is less than 1.734. We reject H_0 in favor of H_1 if the calculated t is greater than 1.734. Since our value of t is 0.885, we accept H_0. Note that we have made an error. All this means is that not enough data have been taken to detect the difference between the effects of the system with the modification and without the modification.[1] Certainly we should expect to accept H_0 after taking only 200 pieces of data, for the difference $P_{t_2} - P_{t_1}$ is only 0.03.

Now let us see what happens when we take 400 pieces of data. Consider the additional data of Tables 9.3 and 9.4. Combine these with the data of Tables 9.1 and 9.2, respectively. We find, then, that

$$\bar{x} = 0.84 \qquad s_x = 0.061 \qquad \bar{y} = 0.79 \qquad s_y = 0.888$$

and

$$t = \frac{0.05}{\sqrt{10(0.0037 + 0.0079)}} \sqrt{\frac{100(18)}{20}} = 1.28$$

Using Student's t table, we find that the difference between "no modification" and "modification"[2] is still not detectable at the 0.05 significance level, but it is certainly detected at the 0.1 level. That is, using these data we can say, with a probability of at least 90 per cent, that there is a significant difference between the two modes of system operation. To detect the difference at the 0.05 level would mean taking more data.

There are certain restrictions on the test used above, but they are not serious when the number of samples is of the order of magnitude described here. These restrictions are emphasized in Ref. 125, secs. 11.5.1 and 11.5.2.

[1] From the table we discern that a difference *was* detected at the 0.2 significance level. In other words, with this amount of data we could reject H_0 if we were willing to be wrong one time in every five.

[2] It is emphasized that "no modification" and "modification" are used here for illustrative purposes only and that any other parameter changes can be considered.

Of some slight importance here is the restriction that the test is not applicable when one P_t is near 0 or 1 and the other is near 0.5. In such unusual cases, the "differences" between "no modification" and "modification" should be obvious from observation and can be measured by the crude but effective test of "differences" used in Appendix 12. This method will not be outlined here, for it deals with a special case and is fully explained in Appendix 12.

We have promised to explain how to test

$$H_0: P_{t_1} \geqq P_{t_2} \qquad \text{against} \qquad H_1: P_{t_1} < P_{t_2}$$

To perform this test, just reverse the positions of $\bar{x}$ and $\bar{y}$ in Eq. (9.45). Again, if t is less than the number found in Student's t table, then accept H_0. If it is greater than this value, reject H_0 in favor of H_1.

9.5.3 *Estimation of the Amount of Data*

Another problem of importance is the one of estimating the maximum amount of data necessary to test a hypothesis at a given significance level. Figures A13.1 to A13.3 and A13.5 are plots of P versus $\theta\sqrt{n_T}$ for the parameters n_r and significance level; in these figures

P = probability of accepting H_0 as a function of $\theta\sqrt{n_T}$
θ = true difference, $P_{t_1} - P_{t_2}$
n_T = number of tactical situations
n_r = number of runs per tactical situation (the same for every tactical situation)

Note that there are two abscissas, one for the hypothesis H_1: $P_{t_1} > P_{t_2}$, the other for the hypothesis H_1: $P_{t_1} < P_{t_2}$. To use the graph, you establish *before* the experiment the probability P for a given $\theta\sqrt{n_T}$. For example, suppose you wish to test H_0: $P_{t_1} \leqq P_{t_2}$ against H_1: $P_{t_1} > P_{t_2}$ for a 0.025 significance level. Use Fig. A13.1, the lower abscissa. Suppose you wish to avoid accepting H_0 *at least* 90 per cent of the time when the true difference θ is 0.20 and when $n_T = 9$. Using Fig. A13.1, we find that $n_r = 16$. Hence, we should take about 16 runs per tactical situation if we wish to avoid rejecting a difference of 0.20 with probability about 90 per cent at the 0.025 significance level.[1,2]

9.5.4 *Unequal Runs*

We have so far treated only the case where n_r was the same for each tactical situation. The case where the number of runs is not the same is easily treated. Suppose we had an arrangement of data as in Table 9.6.

[1] The number of tactical situations n_T should be such that $n_r n_T \geq 30$.

[2] Note that, when $\theta > 0$ and H_1: $P_{t_2} > P_{t_2}$, then P is a Type II error. Also, when H_1: $P_{t_1} < P_{t_2}$ and $\theta < 0$, then P is again a Type II error.

TABLE 9.6

Tactical situation	Run								
	1	2	3	4	5	6	7	8	9
1	1	1	0	1	1	0	1	1	1
2	0	1	1	0	1	1			
3	0	0	1						
4	0	1	0						
5	1	1	1						

We can easily make the number of runs per tactical situation the same by chopping off the number of runs greater than three and then treating these runs as components of a new tactical situation. For example, we may take those runs in boxes 4 to 9 of tactical situation 1 and divide them into two new tactical situations, 6 and 7, as shown in Table 9.7.

TABLE 9.7

Tactical situation	Run					
	1	2	3	4	5	6
1	1	1	0			
2	0	1	1	0	1	1
3	0	0	1			
4	0	1	0			
5	1	1	1			
6	1	1	0			
7	1	1	1			

Similarly, we take runs 4, 5, and 6 of tactical situation 2 and treat them as components of a new tactical situation 8, as in Table 9.8.

TABLE 9.8

Tactical situation	Run		
	1	2	3
1	1	1	0
2	0	1	1
3	0	0	1
4	0	1	0
5	1	1	1
6	1	1	0
7	1	1	1
8	0	1	1

Hence, we can use all the theory developed for equal runs for unequal runs as well. Table 9.9 represents another possible example. Using the process of reduction as described below, we shall get one run for 13 tactical situations.[1]

TABLE 9.9

Tactical situation	Run						
	1	2	3	4	5	6	7
1	0	0	1	1	1	0	0
2	1	0	0	1			
3	1	1					

In the case where the tactical situations are randomized, the number of tactical situations can be said to be just *one* since all we are then seeking is the *average* probability of hit. It should be noted that a similar technique is useful in evaluating the amount of data necessary to test hypotheses. In this case, we use Figs. A13.1 to A13.3. Suppose we wish to find n_r in the case $\theta = 0.1$, $n_T = 9$, significance level 0.025, such that $P = 10$ per cent. Using A13.1, we find that we do not have a curve drawn for this case. Let us *pretend* that $n_{T'} = 36$. Then we find $n_r \approx 16$. The total number of runs is $n_{T'}n_r = 576$, so that we may now split up this number among the nine tactical situations, making the n_r per tactical situation equal. If this cannot be accomplished, the number of runs left over should be split up randomly among the tactical situations.

9.5.5 *Another Type of Hypothesis Testing*

In the preceding treatment of determining a priori the number of runs n_r necessary to detect a difference θ_0, at a given significance level, when it exists, we considered the case where H_0 was more "important" than H_1 in the sense that rejecting H_0 when it was true is much more serious than rejecting H_1 when it was true. Suppose we have the hypotheses

$$H_0\colon P_{t_1} < P_{t_2} \qquad H_1\colon P_{t_1} > P_{t_2}$$

where neither hypothesis is more important in the sense given above than the other. In this case we develop a different approach to the problem. All we ask is: "What is the probability of accepting H_0 when $\theta = P_{t_2} - P_{t_1} = \theta_0$ (θ_0 a numerical value) is true?" It should be noted that

$$P(H_1|\theta_0) = 1 - P(H_0|\theta_0)$$

where $P(H_0|\theta_0)$ = probability of accepting H_0 *given* that $\theta = \theta_0$
$P(H_1|\theta_0)$ = probability of accepting H_1 *given* that $\theta = \theta_0$

[1] See Appendix 13 for the definition of s_x in this case.

Note that we do not here establish a significance level; that is, we assume that H_0 will be accepted if $\bar{x}$ (the estimate of P_{t_1}) is less than $\bar{y}$ (the estimate of P_{t_2}). In other words, H_0 is accepted if $\bar{x} - \bar{y} < 0$; otherwise, H_1 is accepted. One evident danger in this approach is that, if θ is small enough, the probability of making a wrong decision will be close to 0.5. Nevertheless, the approach above is useful when one does not wish to weight unduly the importance of H_0.

Figure A13.5 shows a plot of $P(H_0|\theta_0)$ versus $\theta\sqrt{n_T}$ with n_r as a parameter.[1]

9.5.6 *Comments*

A word of caution is important in connection with the example discussed in Sec. 9.5.2. We have given methods for testing particular kinds of hypotheses. These methods were developed for a special *design of experiment*, namely, where n_r is the same for all tactical situations, where the random variables have a binomial distribution, where $n_T n_r \geq 30$, etc. Of course, when the n_r are not the same for each tactical situation, the trick illustrated in the above paragraphs can be used to make them equal. The analysis, conclusions, and tests are valid only in those experiments where these assumptions prevail. In fact, the experiment should be *designed* so that the assumptions do hold.

9.6 Important Special Case for Normal Distributions

In Sec. 9.5.1 we discussed testing the hypotheses

$$H_0\colon \mu = \mu_0 \qquad H_1\colon \mu = \mu_1$$

for

$$p(x) = \frac{e^{-(x-\mu)^2/2\sigma^2}}{\sqrt{2\pi}\,\sigma}$$

when σ was known. In many practical cases of interest, however, σ is unknown. To test the above hypothesis when σ is unknown, the procedure outlined below is to be followed:

1. Establish α.
2. Calculate $t = (\bar{x} - u_0)/(s/\sqrt{n})$.
3. Look up the value of t_α for $n - 1$ degrees of freedom in Student's t table (Appendix 11).
4. If $\mu_1 > \mu_0$, then reject H_0 if $t > t_\alpha$. If $\mu_1 < \mu_0$, then reject H_0 if $t < t_\alpha$.
5. If there is no alternative hypothesis H_1, then reject H_0 if $|t| > t_{\frac{1}{2}\alpha}$.

The reasons underlying these rules may be appreciated by realizing that the distribution of $(\bar{x} - \mu_0)(s/\sqrt{n})$ is a t distribution of the sort illustrated in Fig. 9.5. The theoretical mean of this distribution is zero *if* H_0 is true. If H_1 is true and $\mu_1 > \mu_0$, then the mean is shifted to the right, as indicated

[1] See Sec. A13.2 for an explanation of these curves.

by the dashed curve in Fig. 9.6. The critical region is indicated by the shaded area. Clearly β, the Type II error, is the area indicated in the figure under the dashed curve. If we were to choose the opposite tail as our critical region, β would be larger. Hence the right tail is the correct

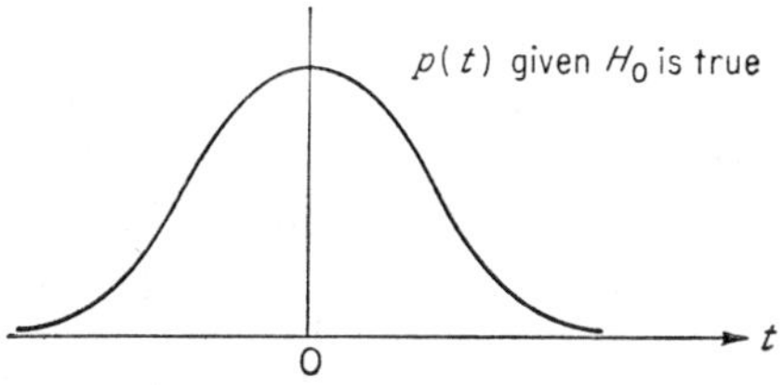

FIG. 9.5 A t-distribution probability density.

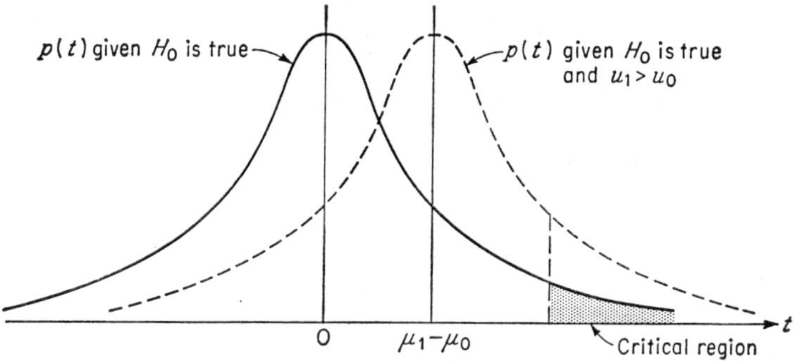

FIG. 9.6 Two t-distribution density functions.

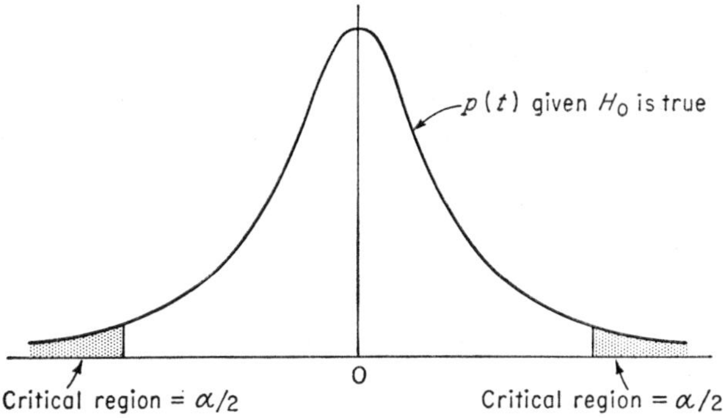

FIG. 9.7 Critical regions distributed equally between the tails.

critical region for this case. If there is no alternative hypothesis, i.e., if we have no prior knowledge as to whether μ_1 is larger or smaller than μ_0, then our ignorance is exhibited by the fact that we distribute α between both tails (Fig. 9.7).

9.7 Another Approach

In the preceding sections we have presented some elements of what is now generally called the *classical* field of hypothesis testing, although the study is only sixty years old. In the elementary classical study, one considers hypotheses H_0 and H_1 and weights H_0 more highly by establishing prior to the experiment the size of the Type I error. The decision is weighted in favor of H_0 by establishing a small Type I error; the smaller the Type I error, the greater is the weighting in favor of finally deciding for H_0. The choice of the weighting (size of Type I error) can be made by considering the total loss W_I incurred by making a Type I error. The *average* loss due to a Type I error would be $(W_I)\alpha$. If $W_I\alpha$ is too large, then, of course, α must be made smaller. Similarly one could establish the total loss W_{II} due to a Type II error β. The average loss due to a Type II error would be $W_{II}\beta$. If this loss were too large one could decrease it by taking more samples.

To be more precise in our statements we should note that

$$\alpha = \text{Pr (choosing } H_1 \textit{ given } H_0 \text{ true)} = P(H_1|H_0)$$
$$\beta = \text{Pr (choosing } H_0 \textit{ given } H_1 \text{ true)} = P(H_0|H_1)$$

Hence $W_I\alpha$ and $W_{II}\beta$ are average losses *given* H_0 and H_1 true, respectively. Suppose we wish to calculate the *total loss* due to our actions without any a priori information, that is, without being *given* that either H_0 or H_1 is true. The possible events that can occur after the data are taken are:

1. H_0 is true *and* we select H_0.
2. H_0 is true *and* we select H_1.
3. H_1 is true *and* we select H_0.
4. H_1 is true *and* we select H_1.

We assume that the loss due to either (1) or (4) occurring is zero since these events are those we strive to attain. Then the total average loss W is

$$W = W_I P(H_0 \text{ is true } \textit{and} \text{ we select } H_1) + W_{II} P(H_1 \text{ is true } \textit{and} \text{ we select } H_0) \quad (9.46)$$

$$W = W_I P(H_0 \text{ is true}) P(H_1|H_0) + W_{II} P(H_0 \text{ is true}) P(H_0|H_1)$$
$$W = W_I \alpha P(H_0 \text{ is true}) + W_{II} \beta P(H_1 \text{ is true})$$

Note that two new quantities are introduced above, the a priori probabilities of H_0 and H_1, respectively. Assuming that such a priori probabilities exist, Eq. (9.46) gives us an expression for the average losses of our test. It is to be noted that, in the classical approach, the α and β are established before the experiment. It is on the basis of the choice of α that the critical region is selected. There is no guarantee that the critical region so chosen is such that the *total average loss is a minimum*. A departure from the "classical approach" is a change in criterion of a "best test"; i.e., a best test is not necessarily one which, for a given α, yields the greatest power but one which provides a critical region that *yields* minimum overall average loss. To see how this is accomplished, let A_0 represent our acceptance region. Let $p_0(x)$ represent the probability density of the random variable when H_0 is true, and let $p_1(x)$ represent the density function when H_1 is true. Then

$$P(H_0|H_1) = \int_{A_0} p_1(x)\,dx \tag{9.47}$$
$$P(H_1|H_0) = 1 - \int_{A_0} p_0(x)\,dx$$

Then Eq. (9.46) becomes

$$W = W_{\mathrm{I}}P(H_0 \text{ true}) - W_{\mathrm{I}}P(H_0 \text{ true})\left[\int_{A_0} p_0(x)\,dx\right] + W_{\mathrm{II}}P(H_1 \text{ true})\int_{A_0} p_1(x)\,dx \tag{9.48}$$

It is easily seen that W is minimized if A_0 is chosen to be the set of points such that

$$\frac{p_0(x)}{p_1(x)} \geq \frac{P(H_1 \text{ true})W_{\mathrm{II}}}{P(H_0 \text{ true})W_{\mathrm{I}}} \tag{9.49}$$

There are several practical difficulties with this approach, among them the assignment of the losses W_{II} and W_{I}, as well as the interpretation of the a priori probabilities $P(H_1$ true) and $P(H_0$ true). Nevertheless, the idea of establishing "minimum average" loss as a principle for selecting a best test as opposed to choosing a test of maximum power is gaining ascendancy. It should be emphasized that there are many principles to choose from; the choice should be made which best suits the needs of the problem at hand.

10 Experimental Design and Detection of Computer Errors[1]

10.1 Introduction

Chapter 9 developed some basic techniques used in solving problems of estimation and hypothesis testing. In so doing, the question posed at the beginning of the chapter concerning the interpretation of data in corroborating a theory was partially answered. In this connection, however, a problem of extreme importance in simulation engineering was only touched upon; that is, how are we to ascertain when enough data have been taken during a series of simulation runs so that a meaningful statistical analysis can be made? By answering this question we aim at eliminating excessive data taking in order to save computer time and to avoid costly data reduction. The first part of this chapter will be a discussion of this problem.

In conjunction with the above discussion, we shall treat the separate problem of determining whether or not the data have been corrupted by a computer malfunction. Computer malfunctions can occur in a statistical problem, but how do we find them? Noise processes are quite unlike any other class of problems solved on the analog computer. The most marked difference is that for a given case no two successive noise runs are or should be alike. Such computer behavior in other types of problems is genuine cause for alarm. For noise problems it is the normal mode of operation. A pattern emerges in the solution only after several repeated computer runs, when the sample average tends to approach the statistical parameter being measured.

In deterministic problems, repeatability of successive solutions is a necessary condition for errorless computer behavior. Such a criterion is useless in a noise problem where no two successive runs are alike. Sometimes the errors, if caused by the breakdown of a key computer element, show up because the statistical pattern of the solution changes radically. Most of the time the errors are so slight that to the untrained observer they become

[1] This chapter was written by A. Levine.

lost in the statistical nature of the solution. Several statistical tests will be described which can ferret out computer inaccuracies.

This latter problem will lead us into the field of nonparametric statistics, an area using techniques we have not heretofore considered. Nevertheless, the underlying notions involved in treating this latter problem are the same as those underlying the former problem discussed above. It is these ideas, ideas introduced in the previous chapter, which can join the solutions of these two problems. However, to make the exposition concrete, we shall discuss the answers to the questions posed above in terms of the design of a hypothetical, small, cheap special-purpose computer which will act as an appendage to the analog computer. This special-purpose computer processes the data automatically and in concurrence with the simulation.

All the data-processing operations of this special computer may also be performed by the logic and memory circuits of the recently marketed hybrid computer (see Sec. 11.4). The ensuing discussions will indicate the logical operations that should be programmed on the hybrid computer to duplicate the functions of the special computer.

The advantage that such a computer seeks is to prevent lost time by stopping the simulation in order to make intermediate statistical analyses either for the purpose of determining if enough data have been taken or for determining whether or not the computer has malfunctioned during a series of runs. The purpose in using this "hypothetical" device to describe the techniques of this chapter is that the reader is introduced to the use of these techniques immediately upon their presentation and in a concrete manner.

At this point, to clarify the discussion, we introduce two different names for this computer; each is indicative of the function under discussion. It is important to emphasize that these names are introduced for ease in exposition and are not meant to suggest that two different computers are being described. Emphasis is placed on this fact in order to establish the idea clearly that the *same notions* lead to the solution of these two different problems. When the computer is used to detect simulator malfunctions, it is called a *probability alarm.* When it is used to ascertain when enough data have been taken for a meaningful statistical analysis, it is called a *probability meter.*

10.2 Probability Meter

As stated above, the computer is used as a probability meter when it is employed to determine when enough data have been taken for the purposes of a *meaningful* statistical analysis. To speak precisely about this matter, it is necessary to know exactly what we mean by a meaningful statistical analysis. The word *meaningful* can be employed precisely only if we state explicitly what analysis of the data we intend to perform. Herein we consider only the problem of determining when enough data have been taken to specify a confidence interval within specified limits and the problem of when enough data have been taken to test a hypothesis to within some significance level. A consideration of these problems yields techniques pertinent not only to their own solution but also to problems of wider interest and applicability. To begin with, we treat the problem of determining when enough data have been taken in order to establish confidence intervals of a specified size and level.

To suggest the general approach to this problem, we first treat in particular some common distributions encountered in simulation work.

10.2.1 Binomial Distribution: Confidence Interval on p

Recall that the binomial distribution arises in problems where the result of each simulation run is either a "success" or "failure." For example, in problems in which air-air missile firings are simulated, the outcome of each firing is a hit or miss. In computer projects involving production-line operation the outcome of each run is an "acceptable" or "unacceptable" item; in problems that test the stability of a new design for an aircraft, missile, or, perhaps, some "black box," the result is "stable" or "unstable."

To demonstrate the techniques used, let us assume that we seek a 95 per cent confidence interval on p, the probability of "success" on a single trial, of size 0.15. We again refer to Fig. A8.1 for pairs of curves bounding

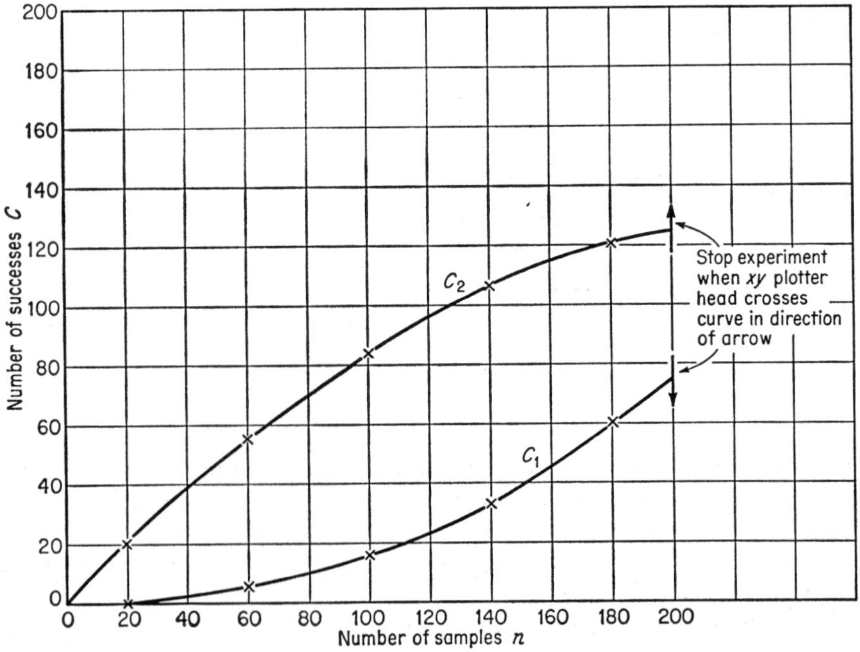

FIG. 10.1 Critical line at the 95 per cent level with a 0.15 confidence interval for a binomial distribution.

the 95 per cent confidence interval as a function of c, the number of successes, and n, the number of trials. For example, for a given n we find the point on the abscissa where the distance between the two curves along the ordinate is 0.15. We use the bottom row of numbers to get a point on curve C_1 and the top row to provide a point on curve C_2 (see Fig. 10.1). Observe that $C_1 + C_2 = n$. The problem will always begin with the condition $C_1 \leq C \leq C_2$. Whenever we reach the condition where $C > C_2$ or $C < C_1$, the data provide less than a 0.15 confidence interval and the simulation is over. To program such a monitor we must plot a function of the two variables C and n which assumes one value in the region $C_1 \leq C \leq C_2$ and another value elsewhere. Techniques for generating functions of two variables are described in Chap. 5.

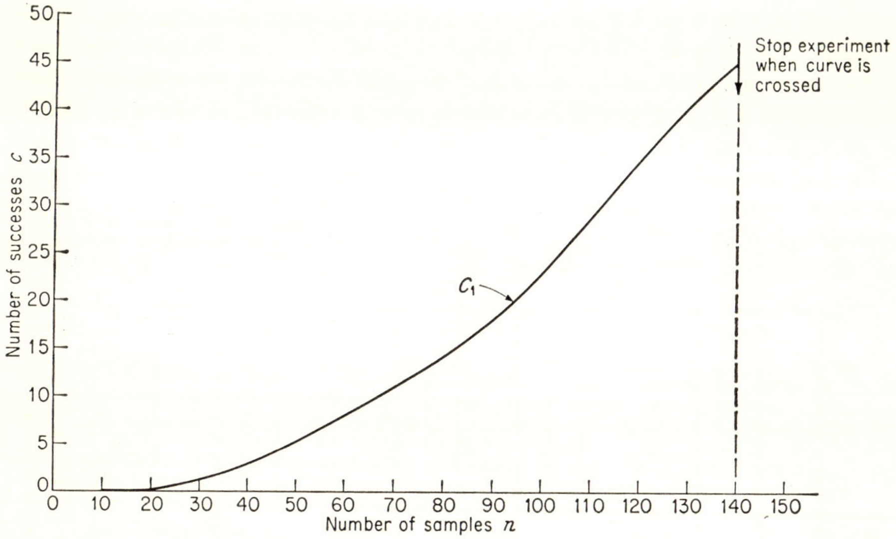

FIG. 10.2 Critical region at the 95 per cent level with a ± 0.1 (20 per cent) confidence interval for the Poisson distribution.

10.2.2 Poisson Distribution Confidence Intervals on λ, the Theoretical Mean[1]

Curves similar to those of Figs. A8.1 and A8.2 can be drawn for the Poisson distribution from Table 10.1. In addition, and of particular interest in this discussion, is that a curve of c, the number of "successes," versus n, the number of samples, can be drawn from this table for a fixed confidence-interval size and level. For example, if we choose the size to be 0.20 and the level to be 0.95, then the resulting curve of C_1 versus n is as shown in Fig. 10.2. We obtain the curve C_1 in the following manner. For a given c there will be a confidence interval, say ΔC. Then we derive n from the relationship $\Delta c/n = 0.2$. For example, let $c = 5$. Then for a 0.95 con-

[1] See Prob. 9.12 and Appendix 4 for a definition of the Poisson distribution.

TABLE 10.1 CONFIDENCE LIMITS FOR THE EXPECTATION OF A POISSON VARIABLE†

$1-2\alpha$	0.998		0.99		0.98		0.95		0.90		$1-2\alpha$
α	0.001		0.005		0.01		0.025		0.05		α
c	Lower	Upper	Lower	Upper	Lower	Upper	Lower	Upper	Lower	Upper	c
0	0.00000	6.91	0.00000	5.30	0.0000	4.61	0.0000	3.69	0.0000	3.00	0
1	0.00100	9.23	0.00501	7.43	0.0101	6.64	0.0253	5.57	0.0513	4.74	1
2	0.0454	11.23	0.103	9.27	0.149	8.41	0.242	7.22	0.355	6.30	2
3	0.191	13.06	0.338	10.98	0.436	10.05	0.619	8.77	0.818	7.75	3
4	0.429	14.79	0.672	12.59	0.823	11.60	1.09	10.24	1.37	9.15	4
5	0.739	16.45	1.08	14.15	1.28	13.11	1.62	11.67	1.97	10.51	5
6	1.11	18.06	1.54	15.66	1.79	14.57	2.20	13.06	2.61	11.84	6
7	1.52	19.63	2.04	17.13	2.33	16.00	2.81	14.42	3.29	13.15	7
8	1.97	21.16	2.57	18.58	2.91	17.40	3.45	15.76	3.98	14.43	8
9	2.45	22.66	3.13	20.00	3.51	18.78	4.12	17.08	4.70	15.71	9
10	2.96	24.13	3.72	21.40	4.13	20.14	4.80	18.39	5.43	16.96	10
11	3.49	25.59	4.32	22.78	4.77	21.49	5.49	19.68	6.17	18.21	11
12	4.04	27.03	4.94	24.14	5.43	22.82	6.20	20.96	6.92	19.44	12
13	4.61	28.45	5.58	25.50	6.10	24.14	6.92	22.23	7.69	20.67	13
14	5.20	29.85	6.23	26.84	6.78	25.45	7.65	23.49	8.46	21.89	14

15	5.79	31.24	6.89	28.16	7.48	26.74	8.40	24.74	9.25	23.10	15
16	6.41	32.62	7.57	29.48	8.18	28.03	9.15	25.98	10.04	24.30	16
17	7.03	33.99	8.25	30.79	8.89	29.31	9.90	27.22	10.83	25.50	17
18	7.66	35.35	8.94	32.09	9.62	30.58	10.67	28.45	11.63	26.69	18
19	8.31	36.70	9.64	33.38	10.35	31.85	11.44	29.67	12.44	27.88	19
20	8.96	38.04	10.35	34.67	11.08	33.10	12.22	30.89	13.25	29.06	20
21	9.62	39.38	11.07	35.95	11.82	34.36	13.00	32.10	14.07	30.24	21
22	10.29	40.70	11.79	37.22	12.57	35.60	13.79	33.31	14.89	31.42	22
23	10.96	42.02	12.52	38.48	13.33	36.84	14.58	34.51	15.72	32.59	23
24	11.65	43.33	13.25	39.74	14.09	38.08	15.38	35.71	16.55	33.75	24
25	12.34	44.64	14.00	41.00	14.85	39.31	16.18	36.90	17.38	34.92	25
26	13.03	45.94	14.74	42.25	15.62	40.53	16.98	38.10	18.22	36.08	26
27	13.73	47.23	15.49	43.50	16.40	41.76	17.79	39.28	19.06	37.23	27
28	14.44	48.52	16.24	44.74	17.17	42.98	18.61	40.47	19.90	38.39	28
29	15.15	49.80	17.00	45.98	17.96	44.19	19.42	41.65	20.75	39.54	29
30	15.87	51.08	17.77	47.21	18.74	45.40	20.24	42.83	21.59	40.69	30
35	19.52	57.42	21.64	53.32	22.72	51.41	24.38	48.68	25.87	46.40	35
40	23.26	63.66	25.59	59.36	26.77	57.35	28.58	54.47	30.20	52.07	40
45	27.08	69.83	29.60	65.34	30.88	63.23	32.82	60.21	34.56	57.69	45
50	30.96	75.94	33.66	71.27	35.03	69.07	37.11	65.92	38.96	63.29	50

If c is the observed frequency or count and m_A, m_B are the lower and upper confidence limits for its expectation m, then

$$\Pr\ (m_A \leq m \leq m_B) \leq 1 - \alpha$$

† Reprinted by permission from A. Hald, "Statistical Tables and Formulas," p. 203, John Wiley & Sons, Inc., New York, 1952.

fidence interval $\Delta c = 11.67 - 1.62 = 10.05$. Thus, for a confidence interval of size 0.2,

$$n = \frac{10.05}{0.2} \cong 50$$

As we can see, Fig. 10.2 applies when the sample mean c/n is small.

For sample means substantially larger than unity, the Poisson distribution approaches the normal. In this event we may use the Student t distribution to provide the confidence intervals on the sample mean. With this information we can then construct a curve like that in Fig. 10.2.

The simulation begins with $c > C_1$, for which the confidence interval is greater than 0.20. By our criterion there are sufficient data when $c < C_1$. The monitor must, therefore, sense when the C_1 line is crossed.

10.2.3 General Techniques[1]

Once curves are obtained such as those illustrated in Figs. A8.1 and A8.2, it is a simple matter to generate curves of c versus n. Hence the problem is to develop a general technique for obtaining curves of confidence interval versus n and the parameter, denoted by θ, to be estimated. For example, with respect to the binomial distribution, $\theta = p$; with respect to the Poisson distribution, $\theta = \lambda$. We shall represent by $\hat{\theta}$ the statistic that will be used to estimate θ. For example, in Sec. 10.2.1 the estimate used was $\hat{\theta} = \bar{x} = c/n$. Let the frequency function (also called the *probability density*) we are dealing with be represented by $p(x;\, \theta)$, let the significance level be $1 - \alpha$, and let $h(\hat{\theta};\, \theta)$ be the density function of $\hat{\theta}$. (Recall that any statistic is a random variable. Why?) Assume for the moment that $\theta = \theta_0$. Once θ_0 and n are fixed, the density $h(\hat{\theta};\, \theta_0)$ is specified. Hence it is possible to find $\hat{\theta}_1$, $\hat{\theta}_2$ such that

$$\begin{aligned} \int_{-\infty}^{\hat{\theta}_1} h(\hat{\theta};\, \theta_0)\, d\hat{\theta} &= 1 - \frac{\alpha}{2} \\ \int_{\hat{\theta}_2}^{\infty} h(\hat{\theta};\, \theta_0)\, d\hat{\theta} &= 1 - \frac{\alpha}{2} \end{aligned} \qquad (10.1)$$

if $h(\hat{\theta};\, \theta_0)$ is a continuous frequency function; otherwise, the integrals are replaced by sum signs (Σ). The results are curves such as those of Figs. A8.1 and A8.2. It is left for the reader to justify the fact that the distance between the lines for a given n and $\hat{\theta}$ is the size of the confidence interval [125].

10.2.4 Normal Distribution. Confidence Intervals on μ, the Theoretical Mean, When σ Is Known

As was shown in the previous chapter, the size of a confidence interval of level $1 - \alpha$ is

$$2 \frac{\sigma}{\sqrt{n}} z_{\alpha/2}$$

[1] For a more complete discussion of the points made in this section, see Ref. 125, sec. 10.2.4.

where $z_{\alpha/2}$ is the number such that

$$\frac{1}{\sqrt{2\pi}} \int_{z_{\alpha/2}}^{\infty} e^{-t^2/2}\, dt = \alpha \tag{10.2}$$

Hence, the size of the confidence interval is *not* dependent on $\hat{\theta} = \bar{x}$ in this case. Consequently, if σ is known, the value of n needed to achieve a confidence interval of a stated size and level can be calculated a priori to the experiment. It is unnecessary in this case to employ the probability meter.

10.2.5 Normal Distribution. Confidence Intervals on σ^2

As was shown in Chap. 9, the confidence *limits* of level $1 - \alpha$ on σ^2 are

$$\frac{s^2}{\chi^2_{(\alpha/2)\mathrm{df}(N-1)}} \quad \text{and} \quad \frac{s^2}{\chi^2_{(1-\alpha/2)\mathrm{df}(N-1)}}$$

where $\mathrm{df}(N - 1)$ is $N - 1$ degrees of freedom.

Hence, the size of the interval is

$$\frac{s^2}{\chi^2_{(\alpha/2)\mathrm{df}(N-1)}} - \frac{s^2}{\chi^2_{(1-\alpha/2)\mathrm{df}(N-1)}}$$

Clearly, then, the size of the interval is a function of s^2 and n. For a constant size interval, it is possible to draw s^2 versus n on xy plotter graph paper and employ the technique for automatically stopping the simulation when the proper size of confidence interval is obtained, as discussed in Sec. 10.2.1. To do so in this case, however, it is necessary to calculate

$$s^2 = \frac{\sum_{i=1}^{n} (x_i - \bar{x})^2}{n - 1} \tag{10.3}$$

during the experiment, which is not a trivial machine computation. To perform this computation would necessitate the incorporation of a great deal of equipment in the probability meter, which is not in the spirit of the design of this device. We must seek elsewhere for a statistic to estimate σ^2—one which, though perhaps not so precise[1] an estimate, involves less computation. The search for such a statistic leads us to a discussion of nonparametric statistics. We now digress to investigate some of the basic results in this area.

10.3 Nonparametric Statistics

10.3.1 Introductory Comments

In the discussions above and in the previous chapter, we have assumed the form of the distribution or density function describing the random vari-

[1] See discussion of efficiency in Sec. 9.4.2.

able under investigation. We sought knowledge about the distribution by trying to obtain information about the parameters of the distribution. For example, we have assumed in some examples that the population was normally distributed. We sought to obtain information concerning the normal distribution by attempting to estimate the mean and/or the variance. However, in many cases, the *form* of the distribution is *not* known. Thus it is not possible to estimate the parameters of the distribution. Still, some information concerning some characteristic of this unknown distribution may be desirable. For example, it may be useful to know whether or not the distribution is symmetrical about some point, say zero.[1] In such a case, we do not necessarily attempt to *estimate* or *evaluate* a *parameter* of the distribution. Other techniques may be employed; because these techniques are not used in the evaluation of a parameter, they are grouped together under the heading of "nonparametric" statistics.

Nonparametric techniques have, in many cases, the important advantage of simplicity. Even when the form of the distribution is known and parametric techniques are applicable, nonparametric tests may be simpler to perform and the subsequent data reduction cheaper and easier. An example of a nonparametric statistic that is simpler and easier to use than its "parametric" counterpart s^2 is the range R. Its use is described below.

10.3.2 Range

As discussed above, the computation of the sample variance s^2 is too complicated for use in the probability meter. In other fields, too, such as industrial quality control, the amount of routine and repeated computation of s^2 becomes burdensome. Another statistic is required which involves only a small amount of computation. Such a statistic is the range. The range R is defined as

$$R_n = |x_{L_n} - x_{s_n}| \qquad (10.4)$$

where x_{L_n} = value of largest of n samples
x_{s_n} = value of smallest of n samples

The range is particularly attractive for use in the probability meter since for small samples it is nearly as efficient for estimating σ as is the sample standard deviation s.

It is important to note that the range is a random variable and hence has a distribution associated with it. However, we shall not delve into the derivation of the distribution of the range, either in general or for specific

[1] Note that the fact that a distribution is symmetrical about some point q does not imply that q is the theoretical mean of the distribution. For example, the Cauchy distribution

$$\frac{1}{\pi(1 + x^2)}$$

is symmetrical about zero, but its theoretical mean is not zero.

cases [125, p. 238]. We shall confine ourselves here to its use in the probability meter. For the moment, we treat one example of great generality, that in which the random output is discrete in time and is normally distributed with an unknown or known mean u and unknown standard deviation σ. In this case the mean $E(R)$ of the range R is given by $\alpha_n\sigma$, and the standard deviation $\sigma(R)$ of R is given by $\beta_n\sigma$, where α_n and β_n are tabulated functions of the number of samples n (Table 10.2).[1]

We may treat the normalized random variable

$$\frac{R - \alpha_n\sigma}{\beta_n\sigma} = \frac{R/\alpha_n - \sigma}{\gamma_n\sigma} \qquad \gamma_n = \frac{\beta_n}{\alpha_n} \tag{10.5}$$

as approximately $N(0, 1)$. Hence two numbers $r_{\alpha/2}$ and $r_{1-\alpha/2}$ can be found from the normal tables such that

$$\Pr\left(r_{\alpha/2} \leq \frac{R - \alpha_n\sigma}{\beta_n\sigma} \leq r_{1-\alpha/2}\right) = 1 - \alpha \tag{10.6}$$

or

$$\Pr\left(\frac{R}{\alpha_n}\frac{1}{1 + r_{1-\alpha/2}\gamma_n} \leq \sigma \leq \frac{R}{\alpha_n}\frac{1}{1 + r_{\alpha/2}\gamma_n}\right) = 1 - \alpha$$

Thus, the size of the $(1 - \alpha/100)$ per cent confidence interval on σ is

$$\frac{R}{\alpha_n}\left(\frac{1}{1 + r_{\alpha/2}\gamma_n} - \frac{1}{1 + r_{1-\alpha/2}\gamma_n}\right)$$

If we wish this size to be less than some specified value S, then we stop the sequence of simulation runs when

$$R \leq \frac{S\alpha_n}{\dfrac{1}{1 + r_{\alpha/2}\gamma_n} - \dfrac{1}{1 + r_{1-\alpha/2}\gamma_n}} = S_n \tag{10.7}$$

In terms of the design of the probability meter, a curve of S_n versus n, derived from Table 10.2, is drawn on xy plotter paper. The voltage on the y coordinate is proportional to the range R, and the voltage on the x coordinate is proportional to the number of runs. When the pickoff head crosses the curve of S_n versus n, then the simulation stops. To obtain a voltage on the y coordinate proportional to the range, the only operations needed in the probability meter are those of "comparison," "retention," and subtraction. These operations are illustrated in the block diagram of Fig. 10.3.

[1] Here we discuss the use of R when the distribution is known; however, we shall also indicate its use when the form of the distribution is not known.

TABLE 10.2 FRACTILES W_P OF THE DISTRIBUTION OF THE RANGE†

n	$m(W)$ α_n	$\sigma(W)$ β_n	σ/m γ_n	Probability, per cent 0.05	0.1	0.5	1.0	2.5	5.0	10.0	20.0	30.0	40.0	50.0	60.0	70.0	80.0	90.0	95.0	97.5	99.0	99.5	99.9	99.95	n
2	1.128	0.893	0.756	0.00	0.00	0.01	0.02	0.04	0.09	0.18	0.36	0.55	0.74	0.95	1.20	1.47	1.81	2.33	2.77	3.17	3.64	3.97	4.65	4.92	2
3	1.693	0.888	0.525	0.04	0.06	0.13	0.19	0.30	0.43	0.62	0.90	1.14	1.36	1.59	1.83	2.09	2.42	2.90	3.31	3.68	4.12	4.42	5.06	5.31	3
4	2.059	0.880	0.427	0.16	0.20	0.34	0.43	0.59	0.76	0.98	1.29	1.53	1.76	1.98	2.21	2.47	2.78	3.24	3.63	3.98	4.40	4.69	5.31	5.56	4
5	2.326	0.864	0.371	0.31	0.37	0.55	0.66	0.85	1.03	1.26	1.57	1.82	2.04	2.26	2.48	2.73	3.04	3.48	3.86	4.20	4.60	4.89	5.48	5.72	5
6	2.534	0.848	0.335	0.47	0.54	0.75	0.87	1.06	1.25	1.49	1.80	2.04	2.26	2.47	2.69	2.94	3.23	3.66	4.03	4.36	4.76	5.03	5.62	5.86	6
7	2.704	0.833	0.308	0.61	0.69	0.92	1.05	1.25	1.44	1.68	1.99	2.22	2.44	2.65	2.86	3.10	3.39	3.81	4.17	4.49	4.88	5.15	5.73	5.96	7
8	2.847	0.820	0.288	0.75	0.83	1.08	1.20	1.41	1.60	1.83	2.14	2.38	2.59	2.79	3.00	3.24	3.52	3.93	4.29	4.61	4.99	5.26	5.82	6.04	8
9	2.970	0.808	0.272	0.88	0.96	1.21	1.34	1.55	1.74	1.97	2.28	2.51	2.71	2.92	3.12	3.35	3.63	4.04	4.39	4.70	5.08	5.34	5.90	6.12	9
10	3.078	0.797	0.259	1.00	1.08	1.33	1.47	1.67	1.86	2.09	2.39	2.62	2.83	3.02	3.23	3.46	3.73	4.13	4.47	4.79	5.16	5.42	5.97	6.19	10
11	3.173	0.787	0.248	1.10	1.20	1.45	1.58	1.78	1.97	2.20	2.50	2.72	2.93	3.12	3.32	3.55	3.82	4.21	4.55	4.86	5.23	5.49	6.04	6.25	11
12	3.258	0.778	0.239	1.21	1.30	1.55	1.68	1.88	2.07	2.30	2.59	2.82	3.01	3.21	3.41	3.63	3.90	4.29	4.62	4.92	5.29	5.54	6.09	6.31	12
13	3.336	0.770	0.231	1.30	1.39	1.64	1.77	1.97	2.16	2.39	2.68	2.90	3.09	3.29	3.48	3.70	3.97	4.35	4.69	4.99	5.35	5.60	6.14	6.36	13
14	3.407	0.762	0.224	1.38	1.48	1.72	1.86	2.06	2.24	2.47	2.75	2.97	3.17	3.36	3.55	3.77	4.03	4.41	4.74	5.04	5.40	5.65	6.19	6.40	14
15	3.472	0.755	0.217	1.46	1.56	1.80	1.93	2.14	2.32	2.54	2.83	3.04	3.23	3.42	3.62	3.83	4.09	4.47	4.80	5.09	5.45	5.70	6.23	6.45	15
16	3.532	0.749	0.212	1.53	1.63	1.88	2.01	2.21	2.39	2.61	2.89	3.11	3.30	3.48	3.67	3.89	4.14	4.52	4.85	5.14	5.49	5.74	6.28	6.49	16
17	3.588	0.743	0.207	1.60	1.69	1.94	2.07	2.27	2.45	2.67	2.95	3.17	3.35	3.54	3.73	3.94	4.19	4.57	4.89	5.18	5.54	5.79	6.32	6.52	17
18	3.640	0.738	0.203	1.66	1.75	2.01	2.14	2.34	2.51	2.73	3.01	3.22	3.41	3.59	3.78	3.99	4.24	4.61	4.93	5.22	5.57	5.82	6.35	6.56	18
19	3.689	0.733	0.199	1.72	1.82	2.07	2.20	2.39	2.57	2.79	3.06	3.27	3.46	3.64	3.83	4.03	4.29	4.65	4.97	5.26	5.61	5.86	6.38	6.59	19
20	3.735	0.729	0.195	1.78	1.88	2.12	2.25	2.45	2.63	2.84	3.11	3.32	3.51	3.69	3.87	4.08	4.33	4.69	5.01	5.30	5.65	5.89	6.41	6.62	20

Example: For $n = 10$ we get $m(W) = 3.078$, $\sigma(W) = 0.797$, and $P(W < 4.79) = 97.5$ per cent.

The range w denotes the difference between the largest and the smallest observation in a sample, $w = w_n = x_{(n)} - x_{(1)}$. The standardized range is $W = w/\sigma = u_{(n)} - u_{(1)}$. The table contains $m(W_n) = \alpha_n$, $\sigma(W_n) = \beta_n$, $\gamma_n = \beta_n/\alpha_n$, and the fractiles W_P for $n = 2(1)20$. It follows that

$$m(w_n) = \alpha_n \sigma$$
$$\sigma(w_n) = \beta_n \sigma$$
$$P(w < \sigma W_P) = P\left(\frac{w}{W_P} < \sigma\right) = P$$

From k samples of n observations each from populations with the same variance we may compute the k ranges $w_1, \ldots, w_k$ and the mean range $\bar{w} = (w_1 + \cdots + w_k)/k$. It follows that $m(\bar{w}) = \alpha_n \sigma$ and $\sigma(\bar{w}) = \beta_n \sigma/\sqrt{k}$. For values of n between 5 and 10, $\bar{w}$ is approximately normally distributed even for small values of k. Hence $\bar{w}/\alpha_n$ will be an unbiased normally distributed estimate of σ with standard error $\gamma_n \sigma/\sqrt{k}$. Confidence limits for σ may be found as

$$P\left(\frac{\dfrac{\bar{w}}{\alpha_n}}{1 + u_{P_2}\dfrac{\gamma_n}{\sqrt{k}}} < \sigma < \frac{\dfrac{\bar{w}}{\alpha_n}}{1 + u_{P_1}\dfrac{\gamma_n}{\sqrt{k}}}\right) = P_2 - P_1$$

† Reprinted by permission from A. Hald, "Statistical Tables and Formulas," pp. 60, 61, John Wiley & Sons, Inc., New York, 1952.

A table of α_n and a diagram for determining β_n may be found in L. H. C. Tippett, On the Extreme Individuals and the Range of Samples Taken from a Normal Population, *Biometrika*, vol. 17, pp. 364–387, 1925. A table of $P_n\{W\}$ for $n = 2(1)20$ and $W = 0.00(0.05)7.25$ has been given by E. S. Pearson in *Biometrika*, vol. 32, pp. 301–308, 1942. (See Tippett for $n > 20$.)

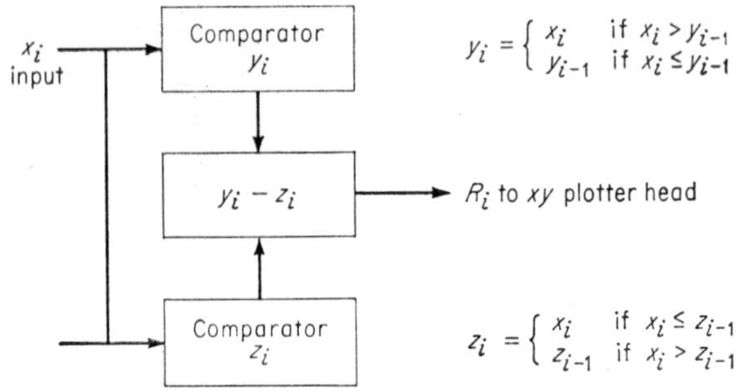

FIG. 10.3 Block diagram of a range tester.

10.3.3 Mean-range Test

In general, the population from which a sample is drawn is not normal. However, if the first and second moments of the population distribution are finite,[1] then the sum of the samples is approximately normally distributed, the approximation being better as the sample size increases. With this fact in mind, we define a new random variable $R^{(k)}$:

$$R^{(k)} = \text{value of range after } k \text{ observations}$$

Then we define

$$\bar{R} = \frac{\sum_{i=1}^{n} R_i^{(k)}}{n} \tag{10.8}$$

as the mean range over n samples of k observations each. The theoretical mean of $\bar{R}$ is $\alpha_n\sigma$, and the standard deviation of $\bar{R}$ is $\beta_n\sigma/\sqrt{k}$, where σ is the standard deviation of the theoretical population. The distribution of $\bar{R}$ for values of n greater than 5 may be replaced by the normal distribution for quite small values of k [115, p. 322]. Assuming that

$$\frac{\bar{R}/\alpha_n - \sigma}{\gamma_n\sigma/\sqrt{k}}$$

is $N(0, 1)$, two numbers $r_{\alpha/2}$ and $r_{1-\alpha/2}$ can be found from the normal tables such that

$$\Pr\left(r_{\alpha/2} \le \frac{\bar{R}/\alpha_n - \sigma}{\gamma_n\sigma/\sqrt{k}} \le r_{1-\alpha/2}\right) = 1 - \alpha \tag{10.9}$$

or

$$\Pr\left[\frac{\bar{R}}{\alpha_n} \frac{1}{1 + r_{1-\alpha/2}\dfrac{\gamma_n}{\sqrt{k}}} \le \sigma \le \frac{\bar{R}}{\alpha_n\left(1 + r_{\alpha/2}\dfrac{\gamma_n}{\sqrt{k}}\right)}\right] = 1 - \alpha$$

[1] This condition is almost always fulfilled in physical experiments. In particular, in dealing with populations of "electrical signals" the second moment is proportional to the power, which in engineering applications is invariably finite.

Thus, the size of the $(1 - \alpha/100)$ per cent confidence interval on σ is

$$\frac{\bar{R}}{\alpha_n}\left(\frac{1}{1 + r_{\alpha/2}\dfrac{\gamma_n}{\sqrt{k}}} - \frac{1}{1 + r_{1-\alpha/2}\dfrac{\gamma_n}{\sqrt{k}}}\right)$$

If we wish this size to be less than some specified value of S, then we stop the sequence of simulation runs when

$$\bar{R} \le \frac{S\alpha_n}{\dfrac{1}{1 + r_{\alpha/2}\dfrac{\gamma_n}{\sqrt{k}}} - \dfrac{1}{1 + r_{1-\alpha/2}\dfrac{\gamma_n}{\sqrt{k}}}} \tag{10.10}$$

and when $n > 5$. The mechanization of Fig. 10.3 must be modified if the

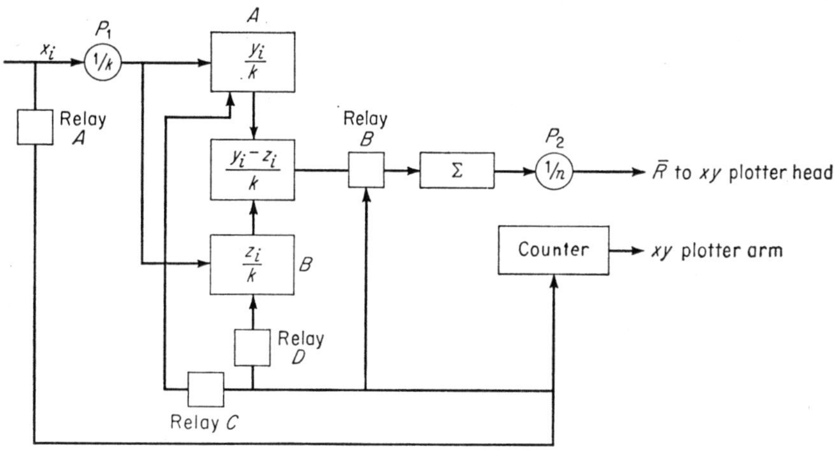

FIG. 10.4 Block diagram of a mean-range tester.

mean-range test is to be incorporated into the probability meter. The modifications are illustrated in Fig. 10.4. The potentiometer P_1 scales x_i, the output sample from the computer, down by $1/k$ or some other value so that the final output $\bar{R}$ does not overload the counters and amplifiers.

The counter releases relay B after every k samples so that $R^{(k)}$ enters the adder every k samples. Also the counter enables relay C and relay D every five samples, clearing the comparators A and B.

Figure 10.5 is a normalized curve of r versus n for $k = 5$ [115, chap. 12] for various significance levels. To draw the proper curve for application, choose the significance level and then multiply r by the desired size of the confidence interval, δ, for example, $r' = \delta r$, where $0 \le \delta \le 1.0$. If the xy plotter head crosses the resulting line r' so that $\bar{R} < r'$, then the simulation ceases.

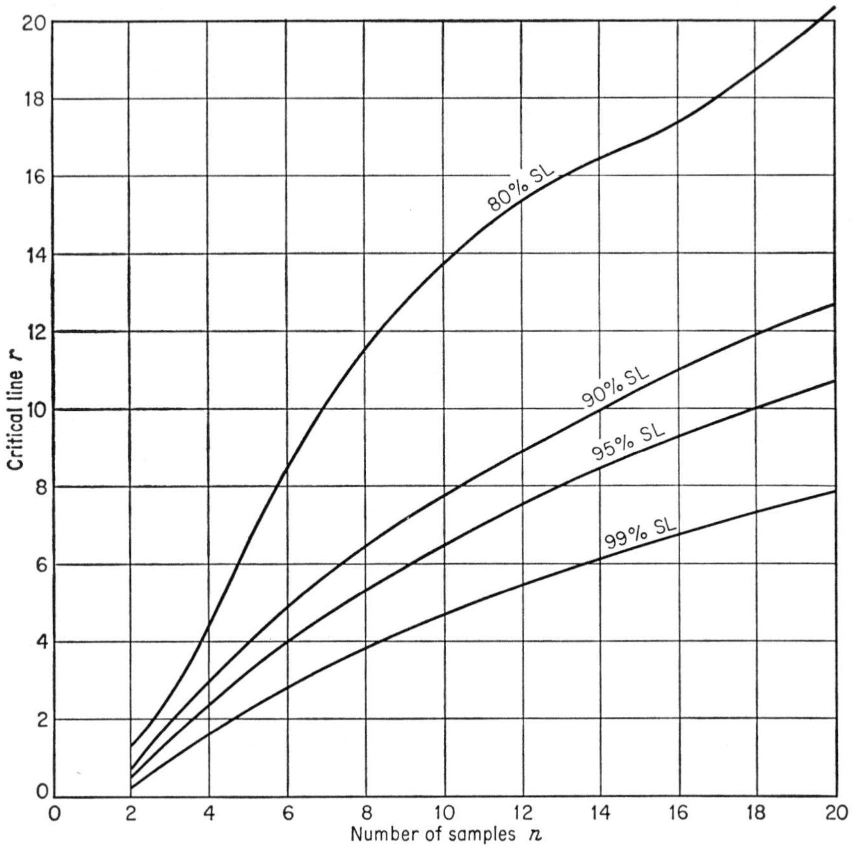

FIG. 10.5 Range-test critical lines with various significance levels for the normal distribution $N(0, 1)$.

10.4 Probability Alarm

10.4.1 Introductory Comments

In the following sections we present for analog computers those statistical tests, and their implementation, which are aimed at discovering computer malfunctions during statistical experiments. The problem of discovering the existence of the malfunctions is treated here as one in hypothesis testing. The "null" hypothesis is that *the computer is performing properly.* The null hypothesis is so stated in order to be in accord with the philosophy established in the previous chapter that it is much less costly to accept the status quo, even though the status quo is not quite up to standard, than to reject the status quo when in fact it is up to standard. Put another way, it is much more costly to stop the simulation and look for a malfunction when in fact there is none than it is to keep running even though there is a malfunction which does not influence the data significantly. All the "tests of hypotheses" described below are nonparametric tests. Each test has its own function, and each has restrictions associated with its use. The user must be careful to employ each test within its capabilities. In general, the more restrictive a test, the "better" it is.

10.4.2 Range Test

We begin by demonstrating the use of the range test in the probability alarm. We do so because the reader is already somewhat familiar with the range as a random variable, because this test serves as a model for subsequent tests, and because we now have an opportunity to round out our discussion of the range R without too much loss of continuity. Recall that the range R is defined as the absolute value of the difference between the smallest sample x_S and the largest sample x_L:

$$R = |x_L - x_S| \tag{10.11}$$

Clearly, if we know the probability distribution of the output, we should be able to predict the "average" range of values we should obtain. If the actual output should fall far from this average, we should suspect either that our knowledge of the distribution function was faulty or that the computer was malfunctioning. Many times we are certain of our knowledge of the distribution. In such an event we can use our knowledge to determine whether or not the computer malfunctions during a run. It should be noted that the range is used instead of some other derived characteristic of the distribution, since it is particularly simple to measure.

To use the test, it is necessary first to find the value of the range—call it r—such that the probability of the range R being less than r is $1 - \alpha$, where α is a specified number. That is, we find r such that

$$\Pr(R < r) = 1 - \alpha$$

The value $1 - \alpha$ is called the significance level of the test. As an illustration let $\alpha = 0.05$. The frequency function of the range is

$$g(R) = n(n-1) \int_a^{b-R} f(u)f(u+R) \left[\int_u^{u+R} f(x)\,dx \right]^{n-2} du \tag{10.12}$$

where n = number of samples
$f(x)$ = known frequency function of output

(It is important to emphasize that the frequency function from which the sample comes must be *known* and must be continuous.) Hence we must find r such that

$$1 - \alpha = \int_0^r g(R)\,dR \tag{10.13}$$

Assume for the moment that we can solve for r as a function of n. We wish to stop the experiment when $R \geq r$. In this case we plot r as a function of n and drive the y coordinate of the xy plotter pen by R until $R > r$. The mechanization of this test is the same as in the probability meter.

Determination of r versus n. The value of r for a given n is, for the two-tail test, determined from the equation

$$1 - \alpha = \text{significance level} = \int_{r_1}^{r_2} g(R)\, dR$$

The first part of the problem is the determination of the significance level. One determines the significance level by how "sure" one wishes to be of one's result; that is, if $\alpha = 0.05$, then the probability of $R > r_2$ or $R < r_1$, when the density is $p(x)$, is 5 per cent. Thus, if R crosses r_1 or r_2 during the experiment, the chances are 95 per cent that something is wrong; if

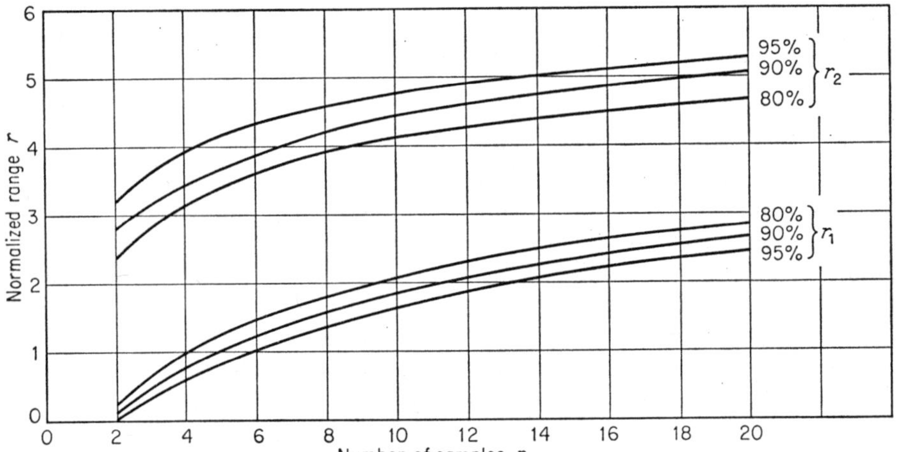

FIG. 10.6 Normalized range for the normal distribution with various significance levels. See Sec. 10.3.3 for a definition of normalized range.

$\alpha = 0.01$, the chances are 99 per cent; if $\alpha = 0.2$, the chances are 80 per cent; etc.

Once the significance level is chosen, the function $g(R)$ must be integrated. Unless $f(x)$ is quite simple, the integration indicated might be difficult to perform. Numerical integration might be the only alternative. Even this may prove too difficult, so that one of the other tests discussed below may have to be used instead. However, in some important cases, the results have been worked out. For example, in the case in which the population is uniformly distributed between 0 and 1, that is,

$$f(x) = \begin{cases} 1 & 0 \le x \le 1 \\ 0 & \text{otherwise} \end{cases}$$

the density of R is

$$g(R) = n(n-1)R^{n-2}(1-R) \tag{10.14}$$

For a given α and n it is a simple matter to find r.

In the most frequent case, that in which the population is $N(0, \sigma^2)$, a normalized plot of r_1 and r_2 versus n has been developed for various significance levels (see Fig. 10.6). If the pickoff head goes above line r_2 or below line r_1 for a given significance level, then this is a warning that the computer is malfunctioning. It is to be noted that n goes only to 20. The plot can be extended for larger n by extrapolating linearly or using the method described below.

10.4.3 Mean-range Test

It is clear that the difficulty with the range test is in the determination of $g(R)$, the frequency function of the range. The reason for this difficulty is the dependence of $g(R)$ on the frequency function $f(x)$ of the sample. The mean-range test provides a test independent of $f(x)$ as long as the first and second moments of $f(x)$ are finite and as long as the variance is known. The mechanization is the same as in the probability meter.

Figure 10.7 is a normalized plot of the curves to be placed on the xy plotter. If the xy plotter head does not remain between the lines (properly

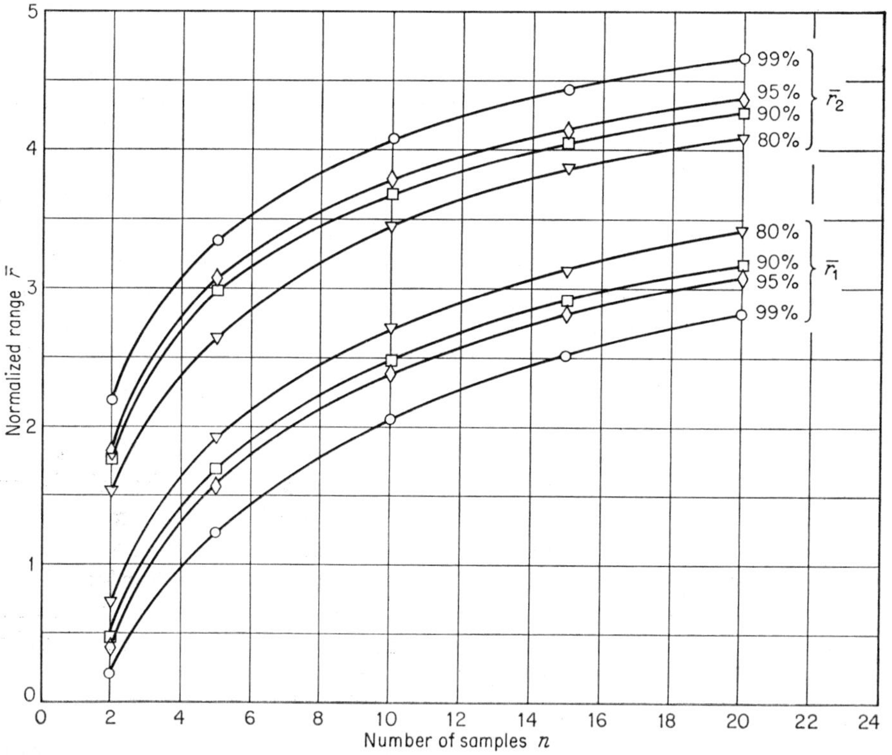

FIG. 10.7 Normalized mean-range test for several significance levels. See Sec. 10.3.3 for a definition of normalized range.

scaled), then one may presume that there is a malfunction in the computer with probability $1 - \alpha$.

It is to be noted that one need not know the frequency function of the sample but only its mean and variance. It is to be noted too that this test is reliable only when the mean range distribution is approximately normal.

10.4.4 The Binomial Test

In Sec. 10.2.1 we discussed the case in which the result of each simulation run is either a "success" or a "failure," i.e., in which the outcomes are conceived as belonging to one of two classes. In some instances, the probability p that an outcome of any single trial will belong to one of the classes is known, or at least an upper bound on p is known. Hence, the probability that a sample will contain a certain proportion p' of members of this class can be calculated. Equivalently we can find p' such that

$$\Pr(\hat{p} > p') < \alpha$$

where α is some predetermined probability level and $\hat{p}$ is a random variable representing the proportion actually obtained during the run. If the proportion *actually* obtained $\hat{p}$ is greater than p', then we must suspect that all is not right with the computer run.

To restate the entire problem in more precise terms, at the outset of the computer runs we assume that the probability of success is p_0; this assumption constitutes our null hypothesis H_0. On the basis of this null hypothesis and the value of the established Type I error α, we calculate our critical region. This region is a function of n, the number of samples. If $\hat{p}$ $(= c/n)$, the proportion of "successes," falls in this region, we must reject H_0. However, we know that H_0 *is* true if the computer is operating properly. Hence rejection of H_0 is tantamount to concluding that the computer is malfunctioning. Therefore the problem of determining whether or not the computer is malfunctioning resolves itself into a test of the hypothesis H_0.

The mechanization of the test for the purpose of the probability alarm is the same as in the probability meter except for the curves drawn on the xy plotter. The curves drawn are the upper and lower bounds (if a two-tailed test is used) of the critical region as a function of n. The calculation of the critical region for fixed α and as a function of n has been demonstrated in Chap. 9. Tables of the binomial distribution have been developed so that the upper and lower bounds can be drawn on the xy plotter for a given p_0 and α such that, if the xy plotter head crosses either curve, then with probability α the computer is malfunctioning. Such tables are usually unnecessary for large samples ($n > 30$), for in these cases

$$z = \frac{x - p_0}{\sqrt{np_0(1 - p_0)}}$$

is $N(0, 1)$. When p_0 is close to $\frac{1}{2}$, the approximation is excellent for $n > 25$.

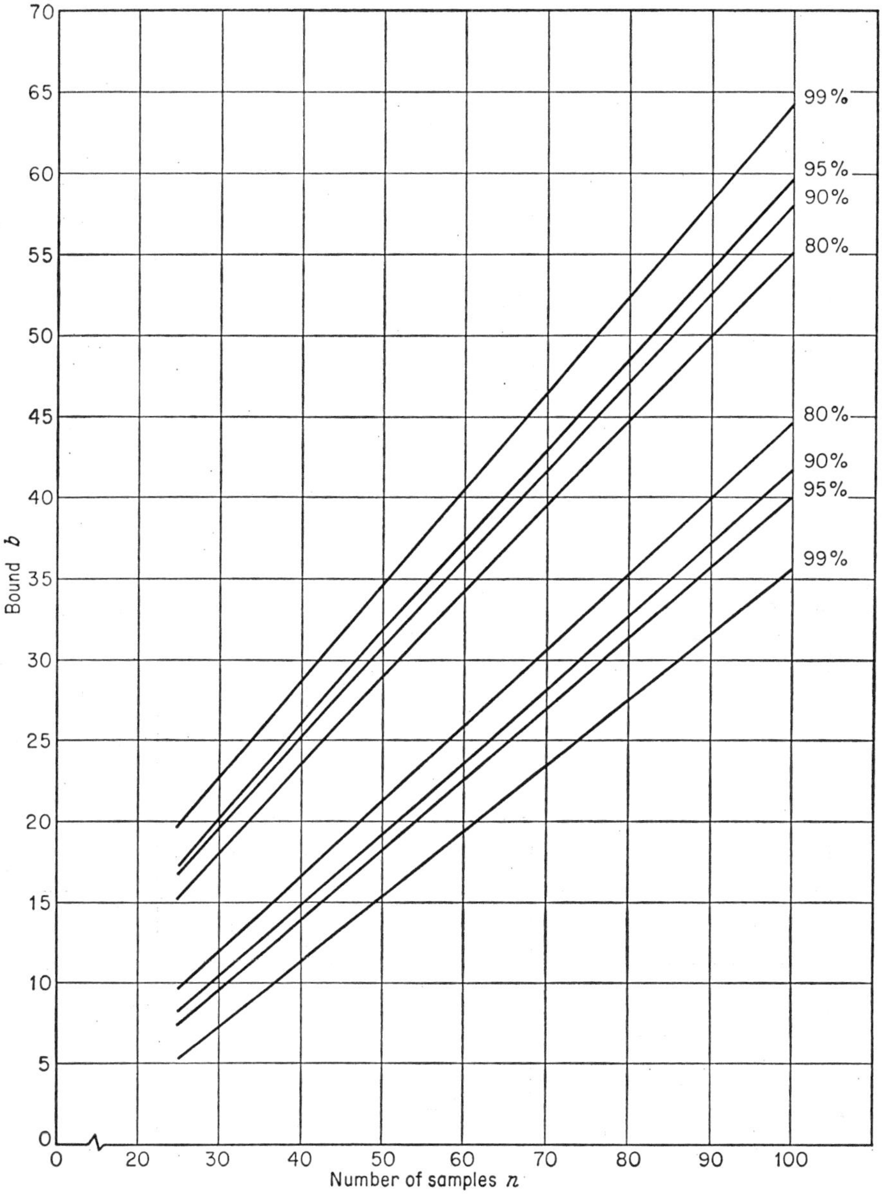

FIG. 10.8 Critical region for the binomial test; $p_0 = \frac{1}{2}$ for several significance levels.

For p_0 near 0 or 1, $np_0(1 - p_0)$ should be greater than 9. Additional corrections are available which make this approximation highly useful.[1] Using this approximation, the curves of Fig. 10.8 show the upper and lower bounds for the common situation in which $p_0 = \frac{1}{2}$. Figure 10.9 represents another

[1] Reference 115, pp. 682–685. See also Ref. 127, pp. 40–41.

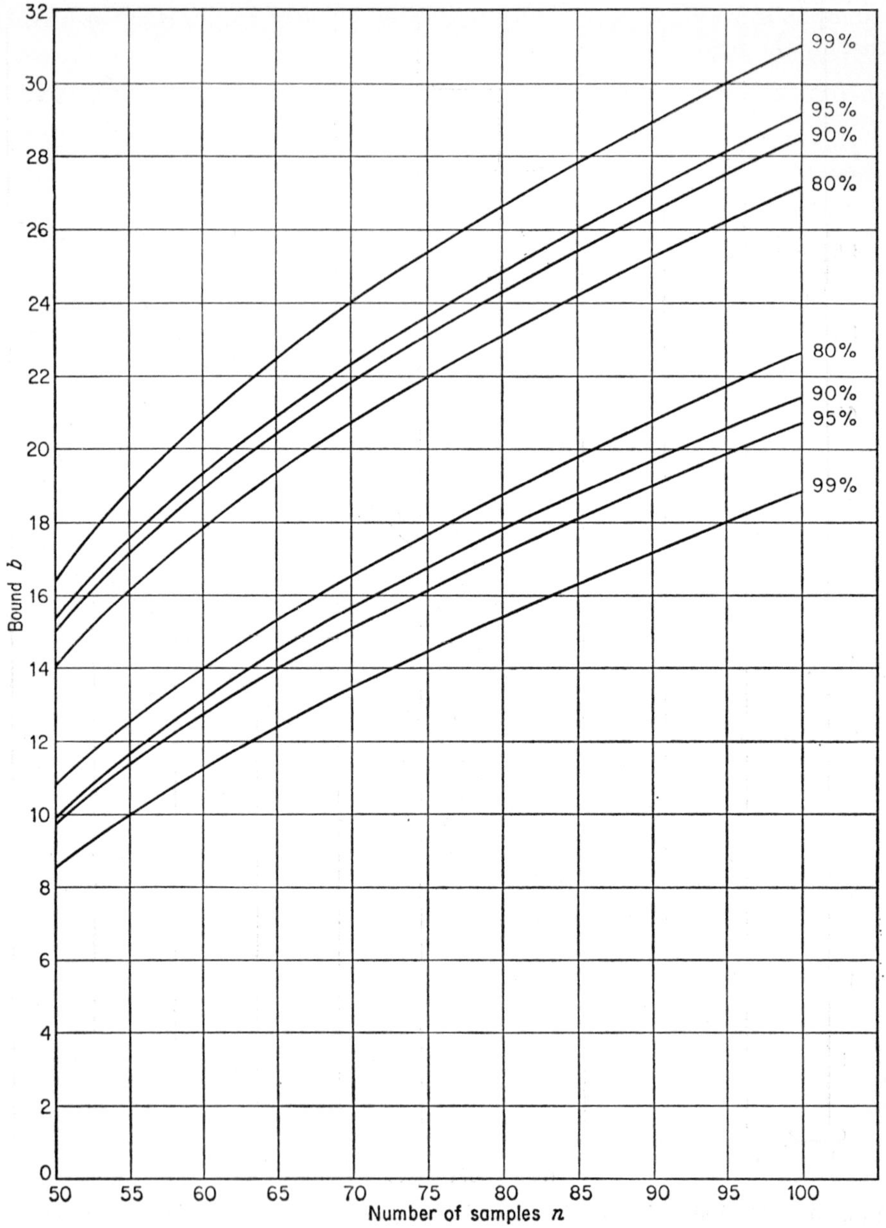

FIG. 10.9 Critical region for the binomial test; $p_0 = \frac{1}{4}$ for several significance levels.

example, in which $p_0 = \frac{1}{4}$. The derivation of these curves is left as an exercise for the reader.

10.4.5 Sign Test

The sign test is an application of the binomial test to more general situations—situations in which the population is not necessarily dichotomous. As an example, suppose we know that the population we are sampling from has a distribution symmetric about some point c. If the value of sample χ_i is greater than c, then we assign the symbol $+$ to this sample; if $\chi_i \leq c$, then we assign the symbol $-$ to it. Since the distribution is symmetric about c, the probability that x_i is "plus" is $\frac{1}{2}$. Hence our null hypothesis is

$$H_0\colon p_0 = \tfrac{1}{2}$$
$$p_0 = \text{probability that a sample outcome is } +$$

As is clear, the sign test gets its name from the fact that it uses plus and minus signs rather than quantitative measures as its data. The appearance of too many or too few $+$ signs indicates that H_0 is not true, i.e., that the data have been corrupted by a computer malfunction. It is clear that the calculations of the critical regions, the mechanization of the test, and the curves of critical region versus n are the same for the sign test as for the binomial test.

It is to be emphasized that knowledge of the population distribution $f(x)$ is not necessary when employing the sign test. All that is needed is knowledge of some point c such that the probability of any sample being greater than c is known.[1]

10.4.6 Runs Test

In Sec. 10.4.5 we compared the total number of positive differences observed with the total number of negative differences observed. Of importance, in many cases, is not the proportion of positive and negative differences but the *order* in which they occur. For example, a proportion of 7 out of 10 $+$ signs may be acceptable at the given significance level, but if the 7 $+$ signs occurred in the last 7 trials, the validity of the runs might be suspect. To include *order* in our considerations, we incorporate the runs test in the probability alarm. A *run* is any set of differences having the same sign which occur in a row. For example, the sequence

$$- \quad - \quad - \quad + \quad + \quad + \quad + \quad + \quad + \quad +$$

has 2 runs, but

$$- \quad + \quad - \quad + \quad + \quad + \quad - \quad + \quad + \quad +$$

has 6 runs, although both sequences contain 7 $+$ signs out of 10 samples. If the median of the distribution of the output is known, we may denote the

[1] See Ref. 136 for an extension of the sign test.

Table 10.3 Critical Values of r in the Runs Test†

In the bodies of parts a and b are given various critical values of r for various values of n_1 and n_2. Any value of r which is equal to or smaller than that shown in part a or equal to or larger than that shown in part b is significant at the 0.05 level.

a

n_1 \ n_2	2	3	4	5	6	7	8	9	10	11	12	13	14	15	16	17	18	19	20
2											2	2	2	2	2	2	2	2	2
3					2	2	2	2	2	2	2	2	2	3	3	3	3	3	3
4				2	2	2	3	3	3	3	3	3	3	3	4	4	4	4	4
5			2	2	3	3	3	3	3	4	4	4	4	4	4	4	5	5	5
6		2	2	3	3	3	3	4	4	4	4	5	5	5	5	5	5	6	6
7		2	2	3	3	3	4	4	5	5	5	5	5	6	6	6	6	6	6
8		2	3	3	3	4	4	5	5	5	6	6	6	6	6	7	7	7	7
9		2	3	3	4	4	5	5	5	6	6	6	7	7	7	7	8	8	8
10		2	3	3	4	5	5	5	6	6	7	7	7	7	8	8	8	8	9
11		2	3	4	4	5	5	6	6	7	7	7	8	8	8	9	9	9	9
12	2	2	3	4	4	5	6	6	7	7	7	8	8	8	9	9	9	10	10
13	2	2	3	4	5	5	6	6	7	7	8	8	9	9	9	10	10	10	10
14	2	2	3	4	5	5	6	7	7	8	8	9	9	9	10	10	10	11	11
15	2	3	3	4	5	6	6	7	7	8	8	9	9	10	10	11	11	11	12
16	2	3	4	4	5	6	6	7	8	8	9	9	10	10	11	11	11	12	12
17	2	3	4	4	5	6	7	7	8	9	9	10	10	11	11	11	12	12	13
18	2	3	4	5	5	6	7	8	8	9	9	10	10	11	11	12	12	13	13
19	2	3	4	5	6	6	7	8	8	9	10	10	11	11	12	12	13	13	13
20	2	3	4	5	6	6	7	8	9	9	10	10	11	12	12	13	13	13	14

b

n_1 \ n_2	2	3	4	5	6	7	8	9	10	11	12	13	14	15	16	17	18	19	20
2																			
3																			
4				9	9														
5			9	10	10	11	11												
6			9	10	11	12	12	13	13	13	13								
7				11	12	13	13	14	14	14	14	15	15	15					
8				11	12	13	14	14	15	15	16	16	16	16	17	17	17	17	17
9					13	14	14	15	16	16	16	17	17	18	18	18	18	18	18
10					13	14	15	16	16	17	17	18	18	18	19	19	19	20	20
11					13	14	15	16	17	17	18	19	19	19	20	20	20	21	21
12					13	14	16	16	17	18	19	19	20	20	21	21	21	22	22
13						15	16	17	18	19	19	20	20	21	21	22	22	23	23
14						15	16	17	18	19	20	20	21	22	22	23	23	23	24
15						15	16	18	18	19	20	21	22	22	23	23	24	24	25
16							17	18	19	20	21	21	22	23	23	24	25	25	25
17							17	18	19	20	21	22	23	23	24	25	25	26	26
18							17	18	19	20	21	22	23	24	25	25	26	26	27
19							17	18	20	21	22	23	23	24	25	26	26	27	27
20							17	18	20	21	22	23	24	25	25	26	27	27	28

† Adapted from Frieda S. Swed and C. Eisenhart, Tables for Testing Randomness of Grouping in a Sequence of Alternatives, *Ann. Math. Statist.*, vol. 14, pp. 83–86, 1943, with the kind permission of the authors and publisher.

observations above the median by a + sign, those below by a − sign. If the number of runs of plus and minus signs is larger or smaller than expected by chance, we reject the hypothesis that the computer is working properly.

We can, with the aid of Table 10.3,[1] construct curves analogous to those developed for the other tests. The symbols n_1 and n_2 represent the number of plus and minus signs, respectively. The table gives the critical values of r, the number of runs for $\alpha = 0.05$. If r lies between these critical values, we assume that the computer is working properly. If not, we stop the simulation and seek the difficulty.

The xy plotter can be used to indicate the coordinate position (n_1, n_2). Critical values r_1' and r_2' are associated in Table 10.3 with each such coordinate position.

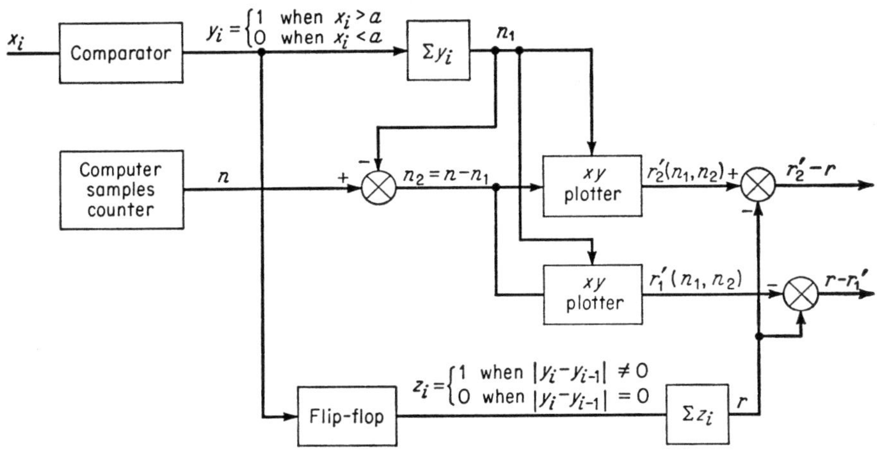

FIG. 10.10 Block diagram of the probability alarm using the runs test.

Mechanization of this test is shown in Fig. 10.10. The value of the ith sample, x_i, is compared with the median a in the comparator. The comparator output is 1 when $x_i > a$ and 0 otherwise. The sum of these outputs is n_1, the number of positive signs in the runs. The computer sample counter records the total number of samples n. Thus $n_2 = n - n_1$ is the number of negative signs. The terms n_1 and n_2 are fed to the x and y coordinates of the plotters, on whose conducting surfaces are plots of r_1' and r_2', respectively, as a function of the variables n_1 and n_2. (See Chap. 5 for methods of generating functions of two variables.)

The outputs of the xy plotters are then the values of r_1' and r_2' corresponding to the inputs n_1 and n_2. The quantity y_i is fed to a flip-flop (bistable multivibrator) whose output is unity when successive values of y_i differ. The output is zero otherwise. The sum of the z_i's is r, the total number

[1] Table 10.3 is for small samples. Tables for large samples are given in Ref. 136.

of runs. Finally the differences $r_2' - r$ and $r - r_1'$ are monitored. The computation stops when either $r_2' - r$ or $r - r_1'$ is negative.

10.5 Tolerance Assignment and Design of Experiment

Preceding sections have touched upon the problem of how much data should be taken in order to perform a meaningful statistical analysis. Such a consideration is extremely important but is only a part of the more general problem of determining the method of choosing a sample. For example, consider the problem of determining by simulation (experiment) the performance P of a system as a function of two parameters p_1 and p_2. P might be the steady-state response of a servo system with time constant τ and damping coefficient ζ. Then

$$p_1 = \tau \qquad p_2 = \zeta$$

If the input to the system has a random error imposed on it, then the output has a random error also. Hence, to determine P as a function of τ and ζ is not simply a matter of picking several values of τ and ζ and fitting a surface to the resulting values of P, for the values obtained as a result of the simulation are not those of P for given values of τ and ζ but rather values $P + \epsilon = P_s$, where ϵ is a random variable that represents the difference between the value of performance P_s obtained from the simulation and the "true" value P. Of importance is how much data should be taken for each choice of τ and ζ so that a reasonable estimate of P can be obtained at these values and *how many values of* τ *and* ζ *should be selected.* The latter question is extremely important, for if too many points are selected, then the data-reduction problem becomes cumbersome; if too few points are taken, the analysis may be meaningless. Of importance, also, are the selected values of τ and ζ. As a simple example that illustrates the point, suppose P were a function of one variable τ and suppose it were known that the relationship was of the form

$$P = m\tau + b$$

where m and b were unknown constants (i.e., the relationship is linear). Suppose that the purpose of the simulation is to determine the slope m and the intercept b which describe the straight line. If we chose only one value of τ, it is clear that we could not estimate either m or b. It is necessary, of course, to choose two values of τ, but which two? In this case, to obtain the best estimate of m and b, it is necessary to select extreme values of τ, that is, take half of the data at the smallest permissible value of τ and the remaining half at the largest permissible value of τ. In this case, it is *not* the most efficient use of the data to take samples at points evenly spaced between the largest and smallest values of τ.

The above example stresses the idea that not only is the amount of data an important consideration, but the entire method of choosing data points and also other considerations, such as when to stop testing and look at the data before performing additional tests (sequential testing),[1] are all grouped under the heading of "design of experiments."

We shall not here discuss this field in detail. This subject should be studied in specialized courses after a good general background in statistics is obtained. However, so important is this area that an attempt is made here to acquaint the reader with at least some of the elementary notions involved in this discipline. To achieve this end we treat an important practical problem, that of determining optimum production tolerances by analog simulation. This problem illustrates how some techniques of experimental design are employed and also utilizes several concepts already discussed in this chapter and the preceding one. Furthermore, some general statistical methods not previously touched upon are presented which play an important role in the solution of this problem.[2]

10.5.1 Introduction

The procedure usually followed in determining by analog simulation the optimum parameter values for various types of control systems is to determine a control-system configuration by simplified analysis and then to program a computer in a realistic way, including all significant nonlinearities and appropriate random input disturbances. Next, a performance criterion of some sort is selected for the control system under study, and system parameters (gains, time constants, etc.) are varied until an optimum system is obtained. At this point, simulation activity usually ceases and is not resumed again unless the actual control system either cannot be constructed to obtain the desired parameter values or else fails to attain the level of performance indicated by the earlier simulation.

Although this approach has produced many successful systems, it is seriously deficient in one respect: the essentially statistical nature of an actual system resulting from a manufacturing process is completely ignored. The representation of a control system as a differential equation with known coefficients is not realistic. An accurate simulation must effectively consider an ensemble of control systems (i.e., a random differential equation) characterized by parameters with common statistical properties. These statistical properties ordinarily take the form of production tolerances. Such tolerances must be considered concurrently with the search for optimum parameters if the simulator is to provide a reliable evaluation of the system performance.

[1] In previous discussions we have considered designs only where the sampling is random and the sample size is fixed. It should be emphasized that this method of taking data is not always the best and that for many applications other techniques are superior.

[2] This work was performed by R. B. McGhee and A. Levine at the Hughes Aircraft Company and is discussed in detail in Ref. 149.

The addition of random parameter values to a control-system simulation not only allows a more realistic determination of performance but also permits a more systematic and organized assignment of production tolerances. Conventionally, whereas the mean values for parameters are selected primarily on the basis of performance, the tolerances are selected primarily on the basis of cost. These two indices are somewhat at odds. For example, if cost were no object, but the desire to maximize performance dominated, then the ideal choice for each tolerance would be zero. On the other hand, if one desired simply to minimize cost per unit, infinite tolerances (100 per cent acceptance of production) would be the proper policy. In the practical situation, a compromise between performance and cost must be effected. This compromise is too often accomplished on the sometimes dubious basis of intuition and "engineering judgment." It is unfortunate that such methods are used at a decision level so vital to the ultimate performance and cost of a product—more so, since the trade-off can be made more objective through analog simulation. By using a simulator to obtain performance as a function of the tolerance assignment, one can choose tolerances so as to maximize performance for a fixed cost or minimize cost for a fixed performance. In statistical terms, one can use the simulator to obtain optimum values for both the mean (nominal value) and standard deviation (production tolerances) for each important parameter.[1]

10.5.2 Proposed Tolerance Optimization Technique

Determination of the Performance Function. In order to arrive at an optimum set of tolerances, it is necessary first to determine the dependence of the control-system performance index $P(Q)$ on the tolerance set Q. It is proposed that $P(Q)$ be determined by simulation according to the following procedure:

1. Optimum nominal parameters are determined (see Chap. 8).
2. A tolerance set Q, whose components are the standard deviations assigned to each parameter, is selected. This vector, together with the mean values determined in step 1, is used to adjust filters operating on a stationary noise source so as to obtain voltages having the same mean and variance as the parameters under investigation.
3. These voltages are fed to integrators as initial conditions while the computer is in RESET. Each integrator in turn provides the input to a multiplier representing the value of the random coefficients in the simulation. When the computer is placed in OPERATE, the integrators have no input. Thus, the value of the coefficient is fixed during the solution of the differential equation but random when the computer is in RESET. This effectively simulates a new production sample each time the computer is placed in OPERATE.

[1] A first step in this direction has been taken by Davis [130].

4. A large number of runs are made with a fixed tolerance set. An average performance index is obtained for the whole set of runs (e.g., probability of hit for a missile system).

5. The components of the tolerance set are varied in a systematic way, with a large number of computer runs being made for each new set of tolerances.

6. With the aid of regression analysis [125] the results are used to fit a polynomial which represents the solution as a function of the problem parameters. An arbitrarily high order of polynomial fit may be used if sufficient data are taken on the analog computer. For instance, let $Q = (\sigma_1, \ldots, \sigma_n)$ represent the standard deviations of n parameters on which tolerances are to be set. Then the performance function $P(Q)$ can be written

$$P(Q) = a_0 + \sum_{i=1}^{n} b_i\sigma_i + \sum_{i=1}^{n}\sum_{j=1}^{n} c_{ij}\sigma_i\sigma_j + \sum_{i=1}^{n}\sum_{j=1}^{n}\sum_{k=1}^{n} d_{ijk}\sigma_i\sigma_j\sigma_k + \text{higher-order terms}$$

where a_0, b_i, c_{ij}, d_{ijk}, . . . are coefficients evaluated by regression analysis. The order of terms in the equation depends upon the nature of the surface of $P(Q)$ and the amount of computer data.

The enumerated steps lead to a representation of a system performance function $P(Q)$ without an a priori assumption regarding the form of the functional dependence of the system performance on parameter values.

It should be noted that step 5 may easily involve hundreds of thousands of runs. For this reason, it appears that the proposed simulation technique is economically feasible only when high-speed analog computers such as the GPS (General Purpose Simulator Instruments Company) are employed (see Sec. 11.4). These computers are capable of providing solutions at rates up to 50 cps and can compute average performance indices simultaneously with solution of the differential equations of the control system.

Determination of the Cost Function. Once $P(Q)$ is obtained, it is necessary to determine the dependence of the unit cost of the system, $C(Q)$, on the tolerance assignment. This is a difficult problem which has long concerned manufacturing industries. It does not appear that analog simulation can aid in the determination of $C(Q)$. Nevertheless, $C(Q)$ is required in order to optimize Q; it is therefore necessary to discuss the estimation of costs sufficiently to indicate the difficulties to be expected and the general nature of the cost function.

Among the factors to be taken into account in the determination of $C(Q)$ are the raw cost per unit, salvage value, maintenance costs resulting from tolerances too wide or too narrow, lifetime, etc. Initial estimation of $C(Q)$ based on such considerations can probably best be accomplished by reference to manufacturing experience with previous products similar to the one under consideration. As production proceeds, the original estimate may be refined on the basis of actual costs. In addition to the experience accumulated by various manufacturing organizations, considerable literature exists on the subject of cost estimation [128, 129].

Although the general problem is much too extensive to treat here, one can observe from experience that $C(Q)$ is generally hyperbolic in nature. That is, reduction of any element of $Q = (\sigma_1, \sigma_2, \ldots, \sigma_n)$ to zero will result in infinite cost, whereas minimum cost is achieved by making all tolerances infinite.

Optimization of the Tolerance Set. After $P(Q)$ has been determined, the optimization of Q may be considered. The first step required is to select a criterion for optimization. An obvious criterion is to optimize performance for a fixed cost. That is, on a surface of constant cost in the Q space, one seeks the point of maximum performance. On the other hand, it might be more desirable to minimize cost for a fixed performance. More complicated optimization criteria are possible by specifying the cost one is willing to accept as a function of performance, etc.

This optimization can be carried out on the computer by the gradient methods described in Chap. 8. The procedure of optimization of either the performance function or the cost function while the other is held constant is an optimization with equality constraints and is described in detail in Sec. 8.5.

As more data become available, the estimate of the cost function will change, thereby shifting the optimum. However, *the original performance function $P(Q)$ remains valid; so the apportionment of tolerances may be altered to a new optimum without further simulation.*

10.5.3 Example[1]

Description of Simulation. In order to establish the feasibility of the proposed technique, let us undertake a simulation of a hypothetical radar-homing missile. Besides random parameters, this simulation includes random radar-tracking noise and a random missile-heading error at launch. The performance index is selected probability of hit P_h as measured against an idealized strip target. To limit the amount of data to be taken, only two parameters are to be varied randomly. These parameters are the navigation gain K and the principal missile filtering-time constant τ. The missile being simulated derives steering information from the angular rate of the line of sight from missile to target; so K is given by

$$K = \frac{\text{steady-state missile turning rate}}{\text{line-of-sight angular rate}} \tag{10.15}$$

Figure 10.11 is a functional block diagram of the missile simulated.

[1] For a description of other simulation aspects of this missile problem, see Chap. 7, Examples 4 and 5.

To measure P_h with sufficient precision, it is necessary to simulate 500 attacks for each value of the tolerance set $Q = (\sigma_K, \sigma_\tau)$. By considering 4 values for each tolerance, a total of 16 cases are generated, resulting in a grand total of 8,000 simulated attacks. The total running time for a real-time computer is about one week. Thus, this example shows quite clearly that a problem of realistic complexity will require high-speed simulation.

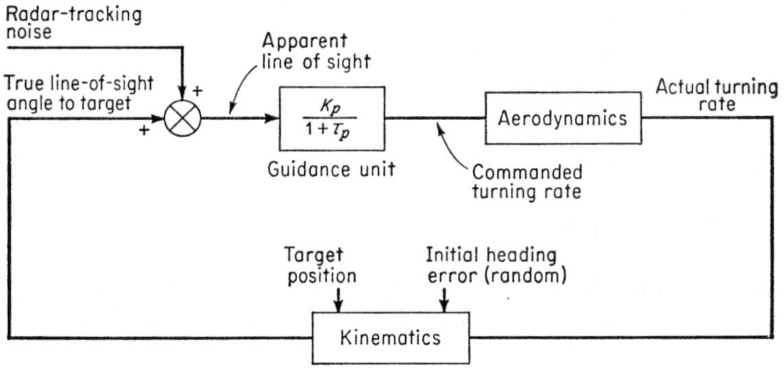

FIG. 10.11 Radar-homing missile navigating in a plane.

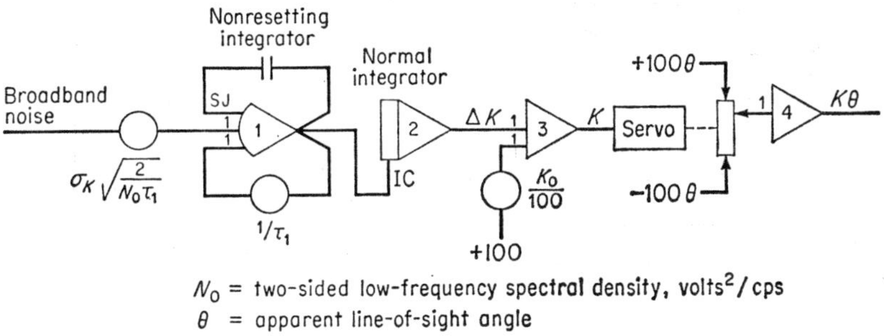

FIG. 10.12 Random variation of navigation gain.

In this simulation we have assumed that the population statistics for both K and τ are gaussian, with mean values K_0 and τ_0 and standard deviation σ_K and σ_τ, respectively. The sample values are therefore obtained by continuous low-pass filtering of noise generators, using integrators which do not reset. Figure 10.12 shows the computer program used to vary randomly the navigation gain K from run to run. (See Chap. 11 for a detailed description of the simulation of noise processes.) Table 10.4 shows the results obtained in the simulation. This table presents the *measured* P_h as a function of the percentage standard deviation assigned to the randomized parameters K and τ. The raw data of this table have been subjected to quadratic regression analysis on σ_K and σ_τ using a digital computer;

TABLE 10.4 MEASURED PROBABILITY OF HIT AS A FUNCTION OF PARAMETER STANDARD DEVIATIONS

σ_τ, per cent	σ_K, per cent			
	0	10	20	30
0	0.34	0.30	0.31	0.29
10	0.30	0.31	0.31	0.29
20	0.30	0.26	0.30	0.30
30	0.24	0.22	0.28	0.27

i.e., the data points have been fitted in a least-square sense by [see Ref. 125, chap. 7]

$$P_h = P_0 + a_1\sigma_K + a_2\sigma_\tau + a_{11}\sigma_K^2 + a_{12}\sigma_K\sigma_\tau + a_{22}\sigma_\tau^2 \tag{10.16}$$

The results are plotted in Fig. 10.13. Figure 10.14 is a contour map of Fig. 10.13 with hypothetical cost curves included. Examples of minimum-cost and maximum-performance optima are indicated.

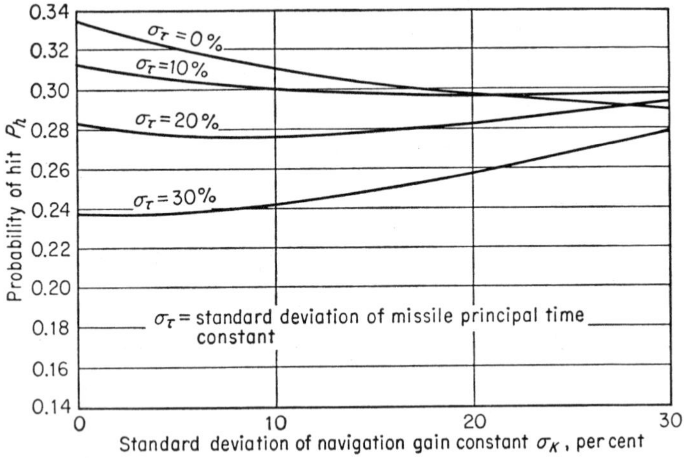

FIG. 10.13 Dependence of probability of hit on tolerance assignment for a homing missile.

Significance of the Experimental Results. The first and most obvious observation that can be made from Fig. 10.13 is that linearization of the dependence of P_h on σ_K and σ_τ is impossible. Linear dependence would result in equally spaced, parallel straight lines in Fig. 10.13. Replotting with σ_K^2 as the abscissa would show that linearization of the dependence of P_h on σ_K^2 and σ_τ^2 is also impossible. *Further examination of the figure reveals the remarkable fact that the performance of this system can actually be degraded by tightening tolerances.* For example, when $\sigma_\tau = 30$ per cent, the probability of hit is decreased from 0.28 to 0.24 as the tolerance on σ_K is reduced from 20 to 0 per cent. This is quite at odds with the widespread opinion that, when tolerances are relaxed in one part of a system, they must be tightened elsewhere to maintain the system performance. In the present example, this apparently anomalous behavior results from a cancellation of effects when K and τ both vary widely from their nominal values.

A more general and far-reaching conclusion arising from this example is concerned with the form of the mathematical model selected to represent the dependence of the system performance index on parameter *values*. In

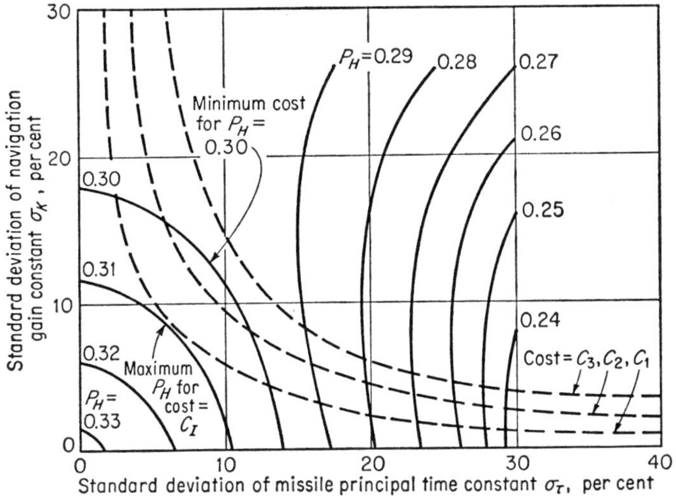

FIG. 10.14 Optimization of tolerances for minimum cost or maximum performance.

the more conventional approach to the study of tolerances, one begins by assuming a specific model for this dependence [128]. For example, on the basis of experience gained in the optimization of nominal values, the system designer might decide that an adequate model for the relationship between P_h and the values of the parameters K and τ is given by the quadratic expression

$$P_h = \gamma_0 + \gamma_1(K - K_0) + \gamma_2(\tau - \tau_0) + \gamma_{11}(K - K_0)^2 + \gamma_{12}(\tau - \tau_0)(K - K_0) + \gamma_{22}(\tau - \tau_0)^2 \quad (10.17)$$

Now since the values of K and τ were independent in this experiment (and very often in practice), the average or expected value of P_h measured over the ensemble of missiles is given by

$$P_h^{\text{av}} = E(P_h) = \gamma_0 + \gamma_{11}E[(K - K_0)^2] + \gamma_{22}E[(\tau - \tau_0)^2] \quad (10.18)$$

where the symbol E denotes the expected value or theoretical average resulting from repeated estimation of P_h. The other terms vanish because they have zero expected value. Now, by definition, the ensemble averages in the above expression are just the variances; i.e.,

$$E[(K - K_0)^2] = \sigma_K{}^2 = \text{variance of } K \tag{10.19}$$

$$E[(\tau - \tau_0)^2] = \sigma_\tau{}^2 = \text{variance of } \tau \tag{10.20}$$

so that

$$P_h{}^{\text{av}} = \gamma_0 + \gamma_{11}\sigma_K{}^2 + \gamma_{22}\sigma_\tau{}^2 \tag{10.21}$$

This is the analytic origin of the common assumption that performance depends linearly on the variances. If one accepts this assumption, then all that is required is to perform an experiment, selecting many values of the variables and recording the performance index along with the values of the variables. Regression analysis is then accomplished on Eq. (10.17), yielding the desired coefficients γ_0, γ_{11}, and γ_{22}.

Although the above approach will succeed if the assumed model, Eq. (10.17), is in fact correct, it has already been shown that the resulting conclusion of linear dependence of performance on parameter variances cannot possibly describe the behavior of the example at hand. The reason is that the effects of the terms σ_1, σ_2, and $\sigma_1\sigma_2$ are absent from Eq. (10.21), whereas a regression analysis of the experimental results based on Eq. (10.16) shows that $\sigma_1\sigma_2$ is in fact the most significant term. This term represents interaction effects which are of prime importance in this system. The actual financial cost of ignoring this term could be considerable since utilization of Eq. (10.17) would lead both to unnecessarily tight tolerances on K and to degraded performance when σ_τ is large.

The process of deriving the effects of tolerances from a model for performance dependence on parameter values can be turned around. That is, if one knows the coefficients in an equation like (10.16),

$$P_h = a_0 + a_1\sigma_K + a_2\sigma_\tau + a_{11}\sigma_K{}^2 + a_{12}\sigma_K\sigma_\tau + a_{22}\sigma_\tau{}^2 \tag{10.16}$$

then the effect of parameter values can be inferred (though not uniquely). For example, if it were true that

$$P_h = a_0 + a_1'|K - K_0| + a_2'|\tau - \tau_0| + a_{11}(K - K_0)^2 + a_{12}'|K - K_0|\,|\tau - \tau_0| + a_{22}(\tau - \tau_0)^2 \tag{10.22}$$

then Eq. (10.16) would follow and the coefficients in Eq. (10.22) could be obtained from a regression on tolerances such as was performed in the present example.

To summarize, in the light of the preceding discussion, it is maintained that one should in general reverse the usual approach to tolerance analysis by:

1. Performing a simulation with randomized coefficients as described in the section above on tolerance optimization
2. Deriving the mathematical model for the system from a regression analysis of the dependence of system performance on the tolerance vector

In this way, new insight can be obtained into the relationship between performance and parameter values, and serious errors in tolerance assignment can be avoided.

10.6 Design of Experiments

10.6.1 The Need for Experimental Design

The experiment just described involved only two parameters. In a more realistic problem it is likely that a much larger number of system parameters would be considered in the assignment of optimum tolerances. Utilization of the method presented above leads to an alarming increase in the total number of runs as the number of parameters increases, unless the experiment is carefully designed. The problem of determining just how to take data so as to obtain the most information with a minimum of effort falls within the realm of statistical design of experiment. This field is so extensive that it is neither possible nor desirable to consider it in detail here. Fortunately, there are several excellent texts on the subject which treat a variety of types of experiment [131].

Although a general discussion of design of experiment is not appropriate, there are two aspects of this field which are particularly useful in experiments of the type discussed above. In the first place, it is necessary to be able to estimate the total amount of data required to define $P(Q)$ before beginning the experiment. Secondly, in the event this number is too large, it is very desirable to have a means for reducing this total by a judicious choice of data points in the tolerance vector space. Both of these problems are considered below in connection with the present experiment.

10.6.2 Estimation of the Amount of Data

In the missile simulated, the measured values of the performance index P_h possess a binomial distribution because they result from repeated trials of the same experiment. One can therefore estimate the variability of successive measurements of P_h [125]. It turns out that, if P_h is the true or theoretical probability of hit and $\hat{P}_h$ is the measured ratio of hits to total firings, then

$$\hat{P}_h - \frac{1}{\sqrt{n}} \leq P_h \leq \hat{P}_h + \frac{1}{\sqrt{n}} \tag{10.23}$$

where n = total number of simulated firings

with probability 0.95 or better. Consequently, since the data in Table 10.4 were obtained by making 500 simulated firings per data point,

$$\hat{P}_h - \frac{1}{\sqrt{500}} \leq P_h \leq \hat{P}_h + \frac{1}{\sqrt{500}} \tag{10.24}$$
$$\hat{P}_h - 0.045 \leq P_h \leq \hat{P}_h + 0.045$$

That is, we can have at least 95 per cent confidence that the true P_h lies within ± 0.045 of the values listed in Table 10.4.[1] Utilization of quadratic regression [Eq. (10.16)] considerably improves the confidence we may have in the final result of Table 10.4.

If one were to decide that a precision of ± 0.045 with 95 per cent confidence was required of every point and, furthermore, that m values of each of k parameters must be considered, then for this example the total number of runs would be given by

$$n = 500m^k \tag{10.25}$$

This number may be so large as to make a "brute-force" determination of $P(Q)$ impractical, even with a high-speed computer. In such an event, some means such as the experimental design described below must be found for reducing this total.

10.6.3 Factorial Design

In the example treated in Sec. 10.5.3 the data points were taken at uniformly spaced points in a σ_K, σ_τ space (Table 10.4). Fortunately, it is not necessary to take data at all of these points to determine a quadratic relationship between $P(Q)$ and Q. One type of design of experiment which allows fewer data points to be considered is called *fractional factorial design* [132]. This design involves a sequential determination of the points at which observations should be made, and is particularly well suited to the determination of $P(Q)$. The total number of computer runs may be drastically reduced, amounting possibly to several orders of magnitude in the total effort required to define $P(Q)$. Although the complete theory is somewhat involved, Table 10.4 can be used to illustrate the basic idea of fractional factorial design.

Referring to the experimental example, if one knew the relative importance of the terms σ_K, σ_τ, and $\sigma_K\sigma_\tau$ in determining $P(Q)$, then a more intelligent selection of sampling points could be made. Now an *estimate* of the relative effects can be obtained by evaluating $P(Q)$ only at the vertices of Table 10.4. The relative effect of each of the three terms above is obtained as follows:

1. Estimated effect of σ_K = average of P_h over column 2 − average P_h over column 1 = $(0.29 + 0.27)/2 - (0.34 + 0.24)/2 = -0.01$.

2. Estimated effect of σ_τ = average of P_h over row 2 − average P_h over row 1 = $(0.24 + 0.27)/2 - (0.34 + 0.29)/2 = -0.06$.

3. Estimated effect (interaction)[2] of $\sigma_K\sigma_\tau$ = average difference of row 2 − average difference of row 1 = $(0.27 - 0.24)/2 - (0.29 - 0.34)/2 = 0.04$.

[1] Methods also exist for the estimation of confidence intervals in the event that the variables do not possess a binomial distribution. See Ref. 125, chap. 10.

[2] An interaction between two variables implies that the total effect is not just the sum of the two effects when they are taken separately.

Thus it can be seen that the "interaction effect" of $\sigma_K\sigma_\tau$ is greater than the "main effect" due to σ_K and nearly as great as the "main effect" due to σ_τ. Although this result is obtained from only four data points, it is (qualitatively) quite in accord with the final result obtained by regression analysis (Fig. 10.13). The significance of this test is simply that the $\sigma_K\sigma_\tau$ term cannot be dropped from the regression equation (10.16). There is, however, a possibility of dropping the linear term σ_K. If this were done, then less data would be required to estimate the remaining regression coefficients. A more sophisticated approach to fractional factorial design must consider confidence intervals and other effects to determine whether or not a given term can be dropped from Eq. (10.16).

When a large number of parameter tolerances are involved in Q, the dropping of even one term from the regression equation may provide a considerable saving in the simulation effort required to obtain $P(Q)$. Even if no terms can be dropped from the regression equation, uniformly spaced data points do not provide the best grid for measurement of $P(Q)$. Further development of fractional factorial design provides a solution to the problem of determining the optimum location for data points in the parameter tolerance space.

11 Application of Statistics to Computer Operations [1]

11.1 Introduction

The previous two chapters, which included estimation, testing of hypotheses and some discussion of design of experiments, developed some procedures which can be used for analyzing data obtained from the simulation of noise problems. In addition, these chapters emphatically show that statistical problems cannot be solved by the computer in the same fashion as deterministic problems. In this chapter we grapple with the problems of simulating noise processes on the analog computer.

We define a noise problem as one whose solution can be specified only in a probabilistic sense. Here we differentiate between the noise that is a part of the problem and the noise generated by computer elements, such as operational amplifiers. We shall assume in this chapter that we can scale the problem so that noise in computer elements has a negligible effect on the solution.

The three major phases of successful computer solutions of noise problems are:

1. Accurate simulation of the problem statistics.
2. Measurement and evaluation of solution data.
3. Detection and elimination of inaccurate results caused by computer errors. (Detection techniques are discussed in Chap. 10.)

We begin the chapter by discussing the essential aspects of the correct simulation of noise processes. Proper simulation is attained when the variables in the computer have the same statistical characteristics as the problem variables. We must, therefore, make certain that every noise source used in the computer has its analogous counterpart in the problem. These noise sources may be required to generate continuously varying forcing functions, random initial conditions, and system parameters either fixed or varying during the computer run.

[1] See Appendix 4 for a discussion of some of the basic concepts of probability and statistics which are used in this chapter.

Two of the more easily constructed noise sources which generate continuous waveforms have either a normal[1] amplitude or Poisson zero-crossing distribution. We shall include a brief explanation of how one might build one of these sources. Then, having dispensed with the mechanism for generating noise, we consider the several pitfalls that are encountered in its proper application. The noise generator must on occasion simulate white noise, which has an infinite power output. It is impossible to meet this condition in a physical system, but we shall show that an adequate approximation to white noise can be made in most cases.

Because gaussian noise so predominates in nature, a computer facility may have only gaussian noise generators with known spectral densities. We shall describe how any given noise source may be transformed to (1) gaussian noise with a different spectral density, (2) nongaussian noise with a given one-dimensional probability distribution, having no restriction on the spectral density, and (3) a nongaussian noise with a given spectral density and one-dimensional probability density. Incidentally, technique 2 may be applied to the transformation of one arbitrary distribution to another.

Time and amplitude scaling occur in noise problems. We shall give the scaling of spectral density special attention because it is somewhat different from the scaling that we have already encountered in deterministic problems.

Our next subject covers the taking of computer data. Any computer solution to a stochastic problem usually involves the use of an average of a finite number of samples to estimate a statistical parameter. The determination of what constitutes a sufficient amount of data to meet the accuracy requirements of the problem is an important topic that is covered in Chap. 10. Here we shall concentrate upon the setting of confidence limits on our estimate. First we examine continuous waveforms with ergodic properties. We describe how to measure and estimate such time averages as the mean, variance, spectral density, correlation function, and one-dimensional probability density. Note that the computer noise sources must be regularly calibrated by the measurement of some of these quantities.

The variance, spectral density, and correlation function are all associated with the second moment. Hence their determination is made in a similar manner in that we average a product of two variables. In fact, if certain assumptions are made, the variance and spectral density are computed precisely the same way. In the special case where gaussian noise is involved, the variance may be computed in the way that we compute the mean of other signals. These subjects and the several techniques for making the calculations will be discussed subsequently in some detail. Then we shall consider the evaluation of confidence intervals when ensemble averages are taken.

[1] The terms *normal* and *gaussian* will be used interchangeably in this chapter.

A frequently useful characteristic of data taking is that in a problem with several random variables the solution may be obtained under certain conditions when one or more variables are changed in a nonrandom fashion. This property allows us to obtain the constituent parts of a seemingly unobtainable solution. Suppose the statistical properties of one of the parameters are unknown and await some physical experiment before they can be evaluated. In anticipation of this experiment we can vary this one parameter deterministically and incorporate the statistical characteristics *when they become known* to produce the solution.

After expounding upon this important subject we next discuss procedures for measuring multidimensional probabilities when a conditional probability is known. Our two alternative courses are to use a function generator or a comparator. A final and often overlooked consideration in data taking is the bias introduced into results by the operator. We describe how the operator may be an unwitting source of error and how these pitfalls may be avoided.

In most statistical problems the computations are time-consuming and burdensome. Confidence limits for large samples vary as $1/\sqrt{n}$, where n is the number of computer runs. These limits are related to the accuracy with which we can estimate a statistical parameter. Hence we observe, for example, that we must take four times as many runs (and also increase computer running time by a factor of 4) to double our accuracy. As a result the operator is continually assessing whether the accuracy improvement bought at the cost of time is worth the price. We shall show that in the special case of a linear problem with random forcing functions and initial conditions we can obtain statistical parameters like the mean and mean-square value of a quantity without having to use noise sources at all. With integrals involving the weighting functions, generated by the adjoint technique, we are able to obtain an exact measure of the statistical parameter with a few computer runs instead of an approximate measure with several runs. Where we can apply this method, the saving in effort is enormous.

11.2 Simulation of Noise Processes

Before discussing the numerous ramifications of programming statistical problems, we consider the nature of these problems as they are solved on the computer. In most computer problems in which statistics find application, the ith parameter takes on a value $\hat{f}_i = f_i + \epsilon_i$, where f_i is its nominal or desired value and ϵ_i is some random error term; i.e.,

$$\hat{f}_i - f_i = \epsilon_i$$

is the deviation of the value of the parameter from its desired value. Hence, the solution to the problem will be $S + \delta$, where S is the solution to the problem if each $\epsilon_i = 0$, and δ is a random value. The application of statistics comes in estimation of S knowing $S + \delta$. The quality S is estimated by a *time* or *ensemble* average.[1]

If a time average is taken, as in the case of the correlation function, a single computer run of sufficient time duration to estimate the parameter with the prerequisite accuracy is all that is required. An ensemble average, which is used whenever the solution is specified at a particular time of a random process, is the average of the results of several *like* computer runs. We use *like* in the sense that the system parameters in each computer run are the same except for the random variables which change according to their statistical properties. Usually it is assumed that each sample is statistically independent of all the others.[2] In other words, the value obtained during any computer run in no way affects the values obtained in succeeding runs. Thus, the computer must simulate a set of independent experiments. This point merits further comment.

Suppose we want to use n experiments to estimate a parameter. If this parameter is the solution of a system of equations, one method of estimating it would be to program n identical systems on the computer and assign a noise generator to each random variable in every system. This ensures that all n samples of a random variable will be statistically independent. With this program we can determine the n samples of the solution in the length of time it takes to get one sample. Although this parallel operation does speed the acquisition of data, it is manifestly unsatisfactory because of its equipment requirements.

A typical number for n is 100, and no computer facility would ever stock this astronomical number of noise generators required for each random variable. Instead, the computer obtains the n solution samples by running one system n successive times with the random variables changing from one run to the next. In contrast to obtaining all n samples in one computer run, we reduce the amount of equipment by n but run the computer about n times as long to get the n samples. For such an operation the computer is designed to cycle automatically through RESET, OPERATE, and DATA READOUT for the number of required runs. We term this *the repetitive mode of operation.*

Now the same noise generator changes the random variable from one run to the next. As we have already mentioned, all the samples of each random variable must be statistically independent.

However, only white-noise sources have the property that all samples of a random variable are independent no matter what their proximity in time—a point we discuss from another standpoint in Sec. 11.2.2. Since white noise is an idealized fiction, the best we can hope for is that these parameter samples are weakly correlated. If we assume that the noise generator has no periodicities, the correlation coefficient decreases as the time separation

[1] See Appendix 4 for a definition of time and ensemble averages.

[2] One sample is the data obtained during a *single* computer run.

increases. Thus the highest correlation will come between parameter samples in adjacent computer runs. The programmer should verify that the highest correlation is low enough to satisfy the problem requirements for approximate statistical independence.

We can illustrate these points about statistical independence by considering two ways the noise generator can be used to simulate random variables. A quantity can be varied continuously during the problem run, or it may change from one run to the next but remain constant during the run. In repetitive operation each cycle consists of a problem run followed by a period of RESET, when the data are processed and the computer prepared for the next run. Suppose in a hypothetical case that a noise generator emits the signal shown in Fig. 11.1, where succeeding intervals of OPERATE

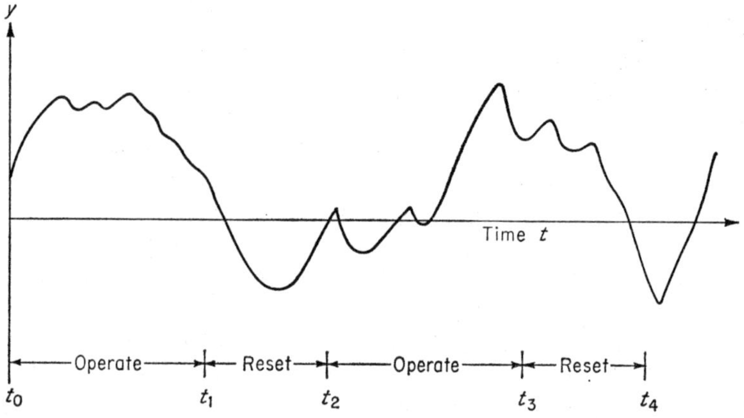

FIG. 11.1 A waveform.

and RESET are indicated. If the noise generator is used to produce a continuous sample, the highest correlation of successive runs occurs between the variables at the end of one run and the beginning of the next, for instance, the correlation between $y(t_1)$ and $y(t_2)$. We can reduce this correlation by increasing the RESET time. Where the noise generator produces a random variable, constant during each run, the noise source voltage is sampled once each cycle. Thus the time interval between samples is the period of one cycle. We can reduce the correlation between samples by lengthening the time of a cycle. Here we have the option of lengthening the RESET time or of time scaling to increase the OPERATE time.

When two or more random variables which are constant during a run have the same form of probability density and at most differ by a potentiometer scale factor, they may with proper care be generated by the same noise source. Now the added wrinkle of complication is that we must verify not only that the samples of a random variable are approximately uncorrelated but that the samples of the two random variables are uncorrelated with each other as well. Thus, to minimize the correlation we must space the samples as far apart in time as possible. If only two variables are involved, we can make them independent by the transformation described in Sec. 11.2.3.

11.2.1 Noise Sources[1]

A statistical problem can be simulated only if noise sources of known characteristics which satisfy the ergodic theorem[2] are available to generate the random variables. The continuous output of a noise source, as typified by Fig. 11.1, is divided into time strips, each of which represents the behavior of a random process during a computer run. The ergodic property ensures that each of these time strips will have the same statistical characteristics. We shall specify such sources by their one-dimensional probability density and spectral density. The use of these properties in simulating noise problems will be discussed in various sections of this chapter.

The gaussian distribution, which is more common than any other in physical problems, can be generated by a grid-controlled inert-gas vacuum tube in a transverse magnetic field, e.g., a 6D4 or 2D21. The transverse magnetic field smooths out the resonance peaks that arise from oscillations of the gas at higher frequencies in the spectral density. These tubes produce an approximately flat spectral density over a large range of frequency. However, the spectral density falls off sharply at low frequencies, a region usually important in analog-computer simulation.

A more accurate low-frequency noise generator uses a 5727 or 2D21 thyratron operating as a diode.[3] The output of the vacuum tube is demodulated and filtered to produce a flat spectral density in the frequency band approximately ± 40 cps around zero frequency. The mean value of the noise output will be almost zero, and the waveform will be almost stationary because of automatic gain control. The exact nature of its spectral density will be determined by the manner in which the noise is filtered.

Another useful noise generator is the Geiger-Müller counter, which in the vicinity of a radioactive source gives a Poisson probability density of impulses whose form is

$$p(\xi = n) = \frac{(\lambda t)^n e^{-\lambda t}}{(n)!} \tag{11.1}$$

where t = duration of time of observation
n = number of pulses produced by counter
λ = average rate of occurrence of pulses in unit time

[1] A more detailed discussion of noise sources is found in Ref. 113, pp. 26-01 to 26-23, by Henry Low, which also contains an excellent bibliography.

[2] See Appendix 4 for a discussion of the ergodic theorem.

[3] This device was designed by R. R. Bennett and H. Low at Hughes Aircraft Company during the early 1950s and is now being marketed by ELGENCO, Santa Monica, Calif., for the major computer manufacturers.

$p(\xi = n)$ = probability that *exactly* n pulses will occur in time t. If each pulse from the counter triggers a bistable multivibrator (a flip-flop) whose output is alternately $+A$ and $-A$ volts, the spectral density of the output of the multivibrator is

$$S(f) = \frac{A^2}{\lambda} \frac{1}{1 + (\pi f/\lambda)^2} \tag{11.2}$$

Of course, the number of zero crossings of the multivibrator output in a time duration t will have the same probability density as Eq. (11.1).

11.2.2 White Noise

A white-noise source is defined as noise having a constant (sometimes called flat) spectral density over the entire frequency range ($-\infty < f < \infty$). The one-dimensional probability density of such a source must be specified separately since it cannot be inferred from the statement that the noise is

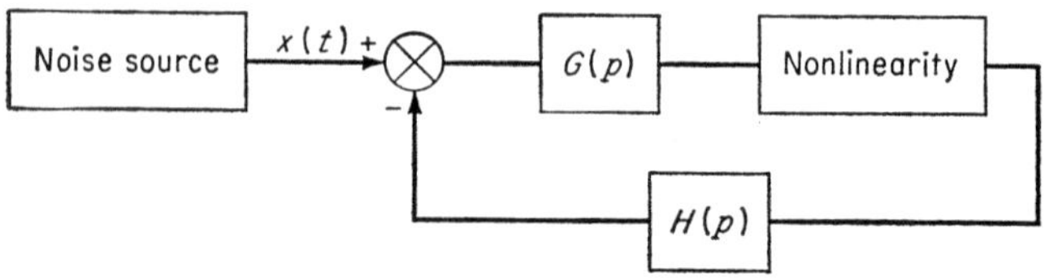

FIG. 11.2 A circuit used to demonstrate how the relative-bandwidth rule operates with a nonlinearity.

white. Such a noise source has an infinite power output and hence does not actually exist in physical problems. Yet, because it lends itself most readily to analysis, white noise is an extremely useful idealization which arises in mathematical problems[1] and is also applicable in certain physical situations. We have an empirical criterion, called the relative-bandwidth rule, which tells us when a noise source approximates white noise.

Relative-bandwidth rule: *Whether in a physical problem or a computer simulation, whenever the bandwidth*[2] *of the flat noise source is large compared with the bandwidth of the system into which it is inserted, e.g., about* 10:1, *the noise source is equivalent to a white-noise source.* Equivalence is used in the sense that first and second moments of the solution will be approximately the same for the two noise sources.

This criterion may be applied to nonlinear, as well as linear, systems if we exercise the proper precaution. Suppose the simulation of Fig. 11.2 calls for $x(t)$ to be white noise. The relative-bandwidth rule will apply if

[1] The correlation function is a delta function, and the spectral density is a constant, both of which are easy to work with in statistical problems.

[2] Bandwidth is defined as the frequency interval between half-power points, occurring at the positive and negative frequency.

$G(p)$ is a constant-coefficient differential operator having a much smaller bandwidth than $x(t)$.

To show that a close approximation to a white source exists under these conditions, let us prove that the correlation functions of white noise and filtered white noise passing through a system represented by a first-order filter are approximately the same. In Fig. 11.3, $x(t)$ is the white noise, $y(t)$ its filtered output, and $z_1(t)$ the output of the system.

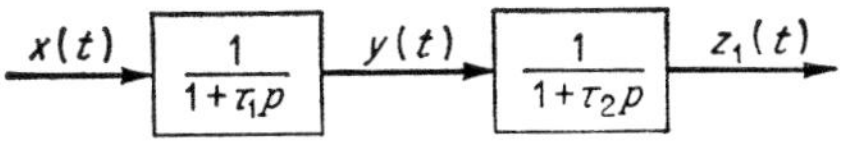

FIG. 11.3 Two first-order lags in cascade.

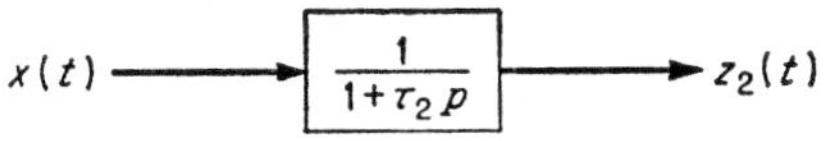

FIG. 11.4 A first-order lag.

If $x(t)$ has a flat spectral density of N_0, the spectral density of y is (see Sec. 11.2.5)

$$S_y(f) = \frac{N_0}{|1 + \tau_1 j\omega|^2} = \frac{N_0}{1 + \tau_1^2\omega^2} \tag{11.3}$$

and the spectral density of z_1 is

$$S_{z_1} = \frac{S_y}{|1 + \tau_2 j\omega|^2} = \frac{N_0}{(1 + \tau_1^2\omega^2)(1 + \tau_2^2\omega^2)} \tag{11.4}$$

For z_1 the correlation function is

$$R_{z_1} = \int_{-\infty}^{\infty} S_{z_1} e^{j\omega\tau} \frac{dw}{2\pi} \tag{11.5}$$

By substituting Eq. (11.4) into Eq. (11.5) and summing residues,

$$R_{z_1}(\tau) = \frac{N_0}{2(\tau_2^2 - \tau_1^2)} \left(\tau_2 e^{-|\tau|/\tau_2} - \tau_1 e^{-|\tau|/\tau_1} \right) \tag{11.6}$$

For comparison, suppose white noise is inserted directly at the input of the system, as shown in Fig. 11.4. Then the spectral density of z_2 becomes

$$S_{z_2}(f) = \frac{N_0}{1 + \tau_2^2\omega^2} \tag{11.7}$$

and the correlation function becomes

$$R_{z_2} = \frac{1}{2\pi} \int_{-\infty}^{\infty} S_{z_2} e^{j\omega\tau}\, dw \tag{11.8a}$$

$$R_{z_2} = \frac{N_0}{2\tau_2} e^{|\tau|/\tau_2} \tag{11.8b}$$

Comparing Eqs. (11.6) and (11.8*b*), we see that in the limit, as τ_2 becomes much larger than τ_1, R_{z_1} and R_{z_2} become equal:

$$\lim R_{z_1} = R_{z_2} = \frac{N_0}{2\tau_2} e^{-|\tau|/\tau_2} \qquad \frac{\tau_1}{\tau_2} \to 0 \tag{11.9}$$

We have shown that white noise and first-order filtered white noise produce similar correlation functions at the output of a system provided the relative-bandwidth rule applies. In a similar manner but with much more effort we could prove that the rule holds for white noise and higher-order filtered white noise through a higher-order system.

Example 1. In a system under study, white noise of spectral density $N_0 = 10$ volts²/cps is to be passed through a filter in the manner shown in Fig. 11.5 and the output is plotted as a function of time. The computer noise generator is first-order filtered white noise with a bandwidth of 40 rad/sec.

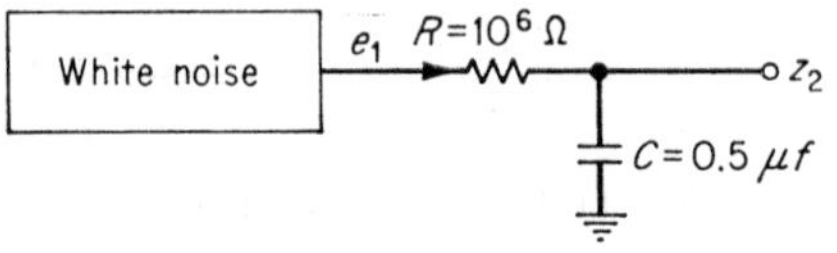

FIG. 11.5 A white-noise source feeding into an RC circuit.

The system transfer function is

$$\frac{z_2(\omega)}{e_1(\omega)} = \frac{1}{1 + RCj\omega} \tag{11.10}$$

To find the difference in correlation function at z_2 when the input is white noise or filtered white noise, we use Eqs. (11.6) and (11.8*b*). For white-noise input the correlation function R_{z_2} is, by Eq. (11.8*b*),

$$R_{z_2} = 10e^{-|\tau|/0.5} \qquad \text{volts}^2 \tag{11.11}$$

where $N_0 = 10$ volts²/cps, $\tau_2 = RC = 0.5$ sec.

For the computer noise-generator input the correlation function R_{z_1} is, by Eq. (11.6),

$$R_{z_1} = \frac{10}{2(0.2475)}(0.5e^{-|\tau|/0.5} - 0.05e^{-|\tau|/0.05}) \tag{11.12}$$

where

$$\frac{\tau_1}{2} = \frac{1}{\text{noise-generator bandwidth}} = \frac{1}{40 \text{ rad/sec}}$$

$$\tau_1 = \tfrac{1}{20} \text{ sec}$$

11.2.3 Transformation of the Characteristics of Noise Sources

A typical computer facility will solve problems requiring noise sources that have a wide variety of statistical characteristics. We could, if we were so inclined, have a different source to supply each characteristic. We would then use the source directly without modification when such a statistical property was called for. This minor convenience would be secured at the heavy cost of having several noise generators which were infrequently used.

In general, we expect a computer facility to have some noise sources of known statistical properties, which can be modified *with standard computer elements,* to fit the requirements of the problem. Furthermore, we shall assume that these noise sources are gaussian, because this distribution is easiest to generate and modify and also is the most likely of all the distributions to occur in physical problems. As we indicated in the introduction, the three transformations we shall consider are from the gaussian source with known spectral density to:

1. An arbitrary one-dimensional probability where spectral density is immaterial

2. A normal distribution whose spectral density differs from the known source
3. An arbitrary one-dimensional distribution with specified spectral density

A function generator is all that is required to make the transformation from one probability density to another. In the general case, illustrated by Fig. 11.6, $x(t)$ is a noise source with a known probability density $p(x)$, and $y(t)$ is the voltage with a desired probability density $p(y)$. What must the characteristic of the function generator be to make the proper transformation from $p(x)$ to $p(y)$? Or, to state it in another way, we wish to find the equation

$$y = f(x) \tag{11.13}$$

where y is a single-valued function of x. To obtain Eq. (11.13) we use the basic relationship for the transformation of a probability distribution of one random variable:[1]

$$\Pr(X < x) = \Pr(Y < y) \tag{11.14}$$

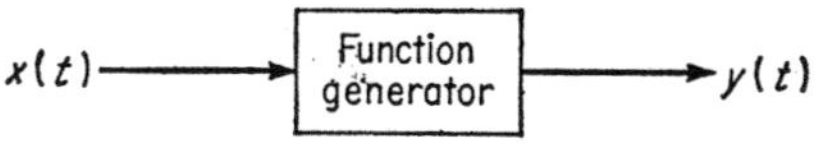

FIG. 11.6 A functional transformation which alters a probability density.

In Eq. (11.14) X is the random variable associated with the source of $x(t)$, and similarly Y is associated with the source of $y(t)$.† Equation (11.14) may also be written in terms of the probability-density functions of x and y, respectively—$p_X(x)$, $p_Y(y)$:

$$\int_{-\infty}^{y=f(x)} p_Y(\alpha)\,d\alpha = \int_{-\infty}^{x} p_X(\beta)\,d\beta \tag{11.15}$$

To obtain Eq. (11.13) we find the value of y which, for each given value of x, will satisfy Eq. (11.15).

Example 2. The voltage $x(t)$ of a noise generator has a gaussian density function

$$p_X(x) = \frac{1}{\sigma\sqrt{2\pi}} e^{-x^2/2\sigma^2} \tag{11.16}$$

which we want to change to a voltage $y(t)$ having a uniform density function

$$p_Y(y) = \begin{cases} \dfrac{1}{2a} & -a \le y \le a \\ 0 & \text{elsewhere} \end{cases} \tag{11.17}$$

[1] Equation (11.14) reads: "The probability that the random variable X is less than the quantity x equals the probability that the random variable Y is less than the quantity y." We shall use the symbol Pr ( ) for probability distribution and $p($ $)$ for probability density.

† There is a certain amount of confusion about why, for example, $x(t)$ is not a random variable. It is true that x does vary with time and in that sense is a variable. However, at any instant of time x will be a particular value. The random variable X assumes values of which x is one; and each one of these values of X has associated with it a certain probability of occurrence.

Using Eq. (11.15),

$$\frac{1}{\sigma\sqrt{2\pi}}\int_{-\infty}^{x} e^{-\alpha^2/2\sigma^2}\,d\alpha = \frac{1}{2a}\int_{-a}^{y} d\beta \tag{11.18}$$

$$\frac{1}{\sigma\sqrt{2\pi}}\int_{-\infty}^{x} e^{-\alpha^2/2\sigma^2}\,d\alpha = \frac{1}{2a}(y+a) \tag{11.19}$$

$$y = 2a\int_{0}^{x} \frac{1}{\sigma\sqrt{2\pi}} e^{-\alpha^2/2\sigma^2}\,d\alpha \tag{11.20}$$

The functional relationship between x and y is shown in Fig. 11.7.[1] The inverse transformation from a uniform to a gaussian density may also be made. Figure 11.8 shows the functional relationship where u is the uniform and v is the gaussian density. Figure 11.8 is much more difficult to implement on the computer than Fig. 11.7 because of the large slopes in the vicinity of $y = \pm a$.

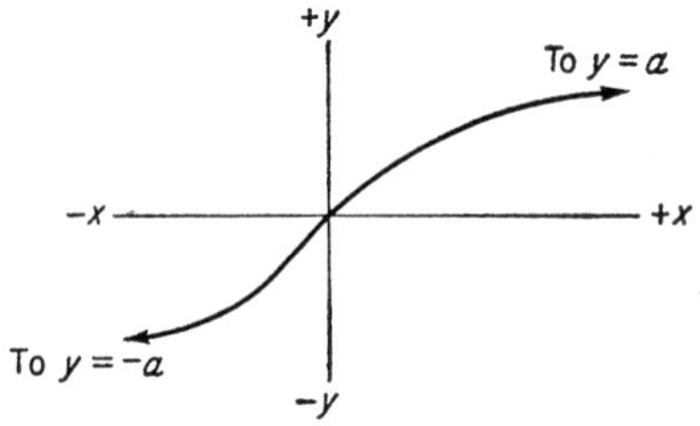

FIG. 11.7 A functional transformation to change from a uniform to a gaussian probability density.

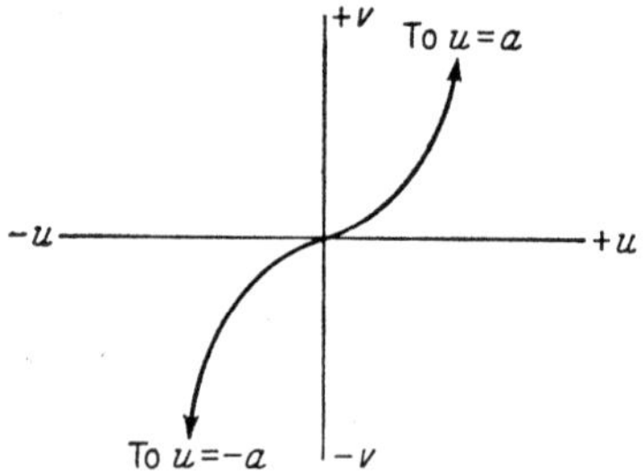

FIG. 11.8 A functional transformation to change from a gaussian to a uniform probability density.

There is another way to get a uniform probability density which requires considerably more computer equipment. If the voltage outputs $x(t)$ and $y(t)$ of a pair of independent gaussian noise sources are used to represent the rectangular coordinates (x, y) of a point P, the polar coordinates of the same point have a Rayleigh distribution for the magnitude r and a uniform distribution for the angle θ. An additional requirement of gaussian noise sources is that they have zero mean and the same standard deviation σ. The transformation from an (x, y) to an (r, θ) coordinate system may readily be made with a polar resolver.[2]

Thus if the two gaussian noise sources have probability-density functions of the form

$$p(x) = \frac{1}{\sigma\sqrt{2\pi}}\exp\left(-\frac{x^2}{2\sigma^2}\right) \tag{11.21}$$

$$p(y) = \frac{1}{\sigma\sqrt{2\pi}}\exp\left(-\frac{y^2}{2\sigma^2}\right) \tag{11.22}$$

[1] The integral $\int_0^x e^{-\alpha^2/2\sigma^2}\,d\alpha$ is called the error function (erf), and values of it may be found in Appendix 6.

[2] See Sec. 6.6 for a discussion of resolvers.

the polar coordinates have, respectively, a Rayleigh distribution,

$$p(r) = \begin{cases} \dfrac{1}{r\sigma^2}\exp\left(-\dfrac{r^2}{2\sigma^2}\right) & 0 \le r \le \infty \\ 0 & \text{elsewhere} \end{cases} \tag{11.23}$$

and a uniform distribution,

$$p(\theta) = \begin{cases} \dfrac{1}{2\pi} & -\pi \le \theta \le \pi \\ 0 & \text{elsewhere} \end{cases} \tag{11.24}$$

where $$r = \sqrt{x^2 + y^2} \qquad \theta = \tan^{-1}\frac{y}{x}$$

Parenthetically, we could start with two noise sources having Rayleigh and

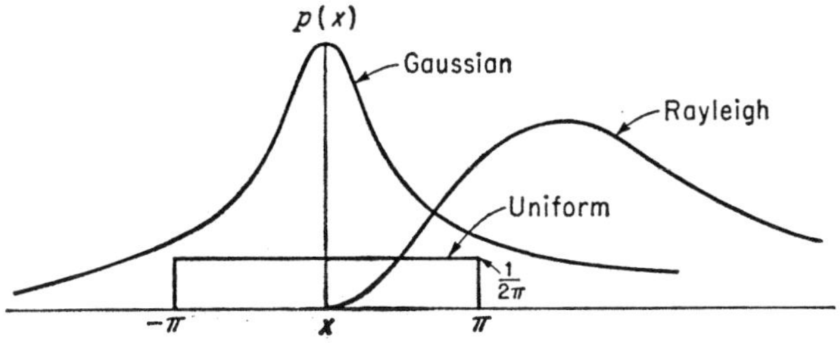

FIG. 11.9 Characteristics of gaussian, uniform, and Rayleigh probability densities.

uniform distributions and through a polar-to-rectangular coordinate transformation obtain two independent gaussian noise sources. Figure 11.9 shows the shapes of these three dissimilar probability densities.

Should the two gaussian noise sources be correlated, we can make a transformation which results in two gaussian random variables that are uncorrelated for a fixed time separation and hence independent. Taking the general relationship where the two variables x_1 and x_2 have different variances, we have the probability density

$$p(x_1, x_2) = \frac{1}{2\pi\sqrt{1-\rho^2}} \exp\left[\frac{1}{2(1-\rho^2)}\left(\frac{x_1{}^2}{\sigma_1{}^2} - \frac{2\rho x_1 x_2}{\sigma_1\sigma_2} + \frac{x_2{}^2}{\sigma_2{}^2}\right)\right] \tag{11.25}$$

We can make two different linear transformations[1] which result in independent random variables, distributed $N(0, 1)$.†

In the first set the transformation which makes y_1 and y_2 independent is

$$\begin{aligned} y_1 &= \frac{x_1}{\sigma_1} \\ y_2 &= \frac{1}{(1-\rho^2)^{\frac{1}{2}}}\left(\frac{x_2}{\sigma_2} - \frac{\rho x_1}{\sigma_1}\right) \end{aligned} \tag{11.26}$$

[1] See Ref. 115, pp. 589–600, for a further discussion of these transformations.

† By convention $N(u, \sigma^2) = (1/\sqrt{2\pi}\,\sigma)e^{-(x-u)^2/2\sigma^2}$. Thus $N(0, 1)$ is a gaussian distribution with zero mean and unit variance.

In the second set the transformation which makes z_1 and z_2 independent is

$$
\begin{aligned}
z_1 &= \frac{1}{\sigma_{z_1}} (x_1 \cos \alpha + x_2 \sin \alpha) \\
z_2 &= \frac{1}{\sigma_{z_2}} (-x_1 \sin \alpha + x_2 \cos \alpha)
\end{aligned}
\tag{11.27}
$$

where

$$\tan \alpha = \frac{1}{2\rho\sigma_{x_1}\sigma_{x_2}} [\sigma_{x_2}{}^2 - \sigma_{x_1}{}^2 \pm \sqrt{2\rho\sigma_{x_1}\sigma_{x_2} + (\sigma_{x_1} - \sigma_{x_2})^2}]$$

$$
\begin{aligned}
\sigma_{z_1}\sigma_{z_2} &= \sigma_{z_1}\sigma_{x_2} \sqrt{1 - \rho^2} \\
\sigma_{z_1}{}^2 + \sigma_{z_2}{}^2 &= \sigma_{x_1}{}^2 + \sigma_{x_2}{}^2
\end{aligned}
$$

11.2.4 Discrete Probabilities

If we have random initial conditions or constant coefficients which may assume only a discrete number of values, we can use a continuous noise

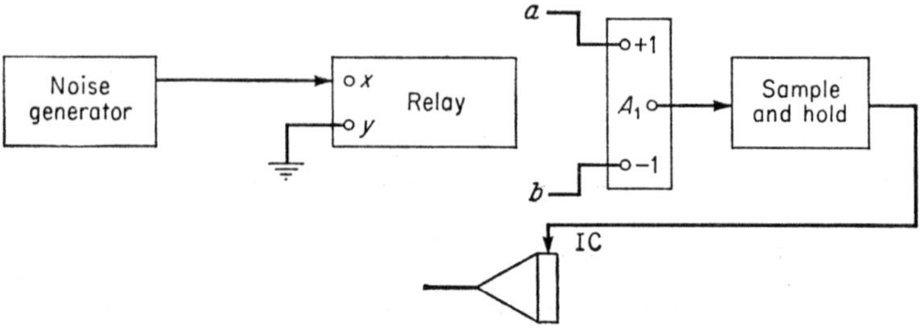

FIG. 11.10 A circuit for generating a random initial condition.

generator with a relay or some other suitable voltage comparator to assign probabilities for these values. Assume that we have an initial condition which has two values, a and b, that can occur with equal probability. Our noise source is gaussian with zero mean. A circuit we might use to set this initial condition is shown in Fig. 11.10. The noise generator causes the relay arm to connect first to $+1$ and then to -1. At some time during RESET the sample-and-hold circuit samples and holds either a or b, whichever the arm A_1 is connected to at the time of sampling. This then becomes the initial condition during the next computer run.

With a little ingenuity we can generate equally probable discrete events *even when we do not know the characteristics of the noise generator.* We require only that during RESET we can take independent samples of the position of the relay. Now we assign to the position $+1$ a probability p and to -1 a probability q. Then it follows that the occurrence first of the sample $+1$ and then of the sample -1 has the probability pq. This is the same as the occurrence first of -1 and then of $+1$. Thus, in this case we can obtain equal probability of two parameters by using compound events.

To reiterate this point we take a sequence of samples during RESET and consider them in ordered pairs. These pairs are mutually exclusive, i.e., nonoverlapping. The first pair, which is either (+, −) or (−, +), sets the initial condition at a or b, respectively. Before we get such a pair we might get others of the form (+, +) and (−, −), which we discard. Hence, we may need several samples to set the initial condition where, by comparison, we need only one when we know the noise-generator characteristics.

11.2.5 Spectral-density Transformation for Gaussian Noise

A simple relationship tells us that, if $S_x(f)$ is the spectral density of the input noise $x(t)$, the spectral density at the output of the system is

$$S_y(f) = S_x(f)|Y(j\omega)|^2 = S_x(f)[Y(j\omega)][Y(-j\omega)] \tag{11.28}$$

where $Y(j\omega)$ is the Fourier transform of the system impulse response.

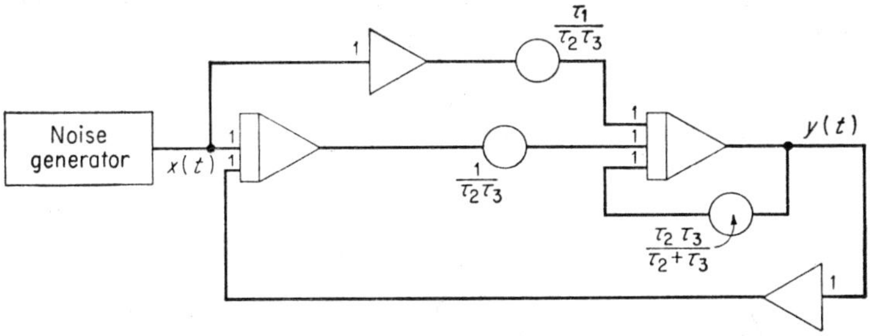

FIG. 11.11 A computer circuit for generating a particular spectral density.

Example 3. We have a noise generator with a voltage $x(t)$ having a spectral density

$$S_x(f) = \frac{N_0}{1 + \tau_1^2\omega^2} \tag{11.29}$$

which must be modified to a voltage $y(t)$ having a spectral density

$$S_y(f) = \frac{N_0}{(1 + \tau_2^2\omega^2)(1 + \tau_3^2\omega^2)} \tag{11.30}$$

By Eq. (11.28)

$$|Y(j\omega)|^2 = \frac{S_y(f)}{S_x(f)} = \frac{1 + \tau_1^2\omega^2}{(1 + \tau_2^2\omega^2)(1 + \tau_3^2\omega^2)} \tag{11.31}$$

Therefore the transfer function is

$$Y(j\omega) = \frac{1 + \tau_1 j\omega}{(1 + \tau_2 j\omega)(1 + \tau_3 j\omega)} \tag{11.32}$$

The computer system function is[1]

$$\frac{y(t)}{x(t)} = \frac{1 + \tau_1 p}{(1 + \tau_2 p)(1 + \tau_3 p)} \tag{11.33}$$

The computer program is shown in Fig. 11.11.

[1] The relation between $j\omega$ and $p = d/dt$ as operators is discussed in Chap. 4.

11.2.6 Transformation to an Arbitrary Probability and Spectrum

The probability and spectral characteristics of a noise generator are interrelated. Usually we cannot transform one of these characteristics without modifying the form of the other. The most important exception is gaussian noise passed through a linear filter. In this one case the spectral density will be altered at the filter output but the noise will remain gaussian. Thus, in the transformation of Sec. 11.2.5, if the input $x(t)$ is gaussian, so is the output $y(t)$.

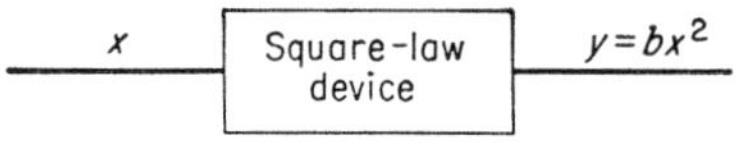

FIG. 11.12 A square-law device.

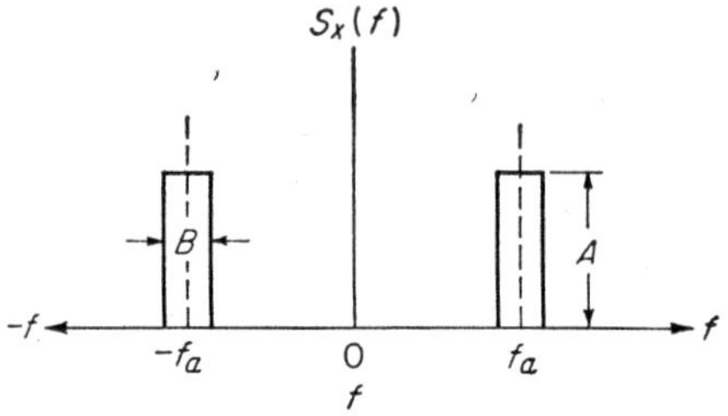

FIG. 11.13 A particular spectral density.

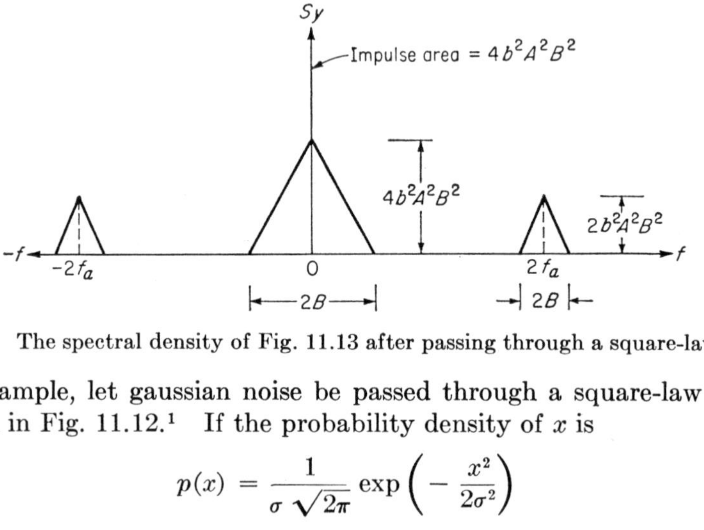

FIG. 11.14 The spectral density of Fig. 11.13 after passing through a square-law device.

For example, let gaussian noise be passed through a square-law device, as shown in Fig. 11.12.[1] If the probability density of x is

$$p(x) = \frac{1}{\sigma\sqrt{2\pi}} \exp\left(-\frac{x^2}{2\sigma^2}\right) \tag{11.34}$$

and the spectral density is as shown in Fig. 11.13, then the probability density of y is

$$p(y) = \begin{cases} \dfrac{1}{2by} \exp\left(-\dfrac{y}{2b\sigma^2}\right) & y \geq 0 \\ 0 & y < 0 \end{cases} \tag{11.35}$$

and its spectral density is as shown in Fig. 11.14.

This example and the preceding section serve to demonstrate the two types of calculations we can make:

[1] This example is taken from Ref. 116, which has an extensive discussion of noise through nonlinear devices on pages 250–311.

1. If the input probability and spectral density are known, the output probability and spectral density of a zero-memory nonlinear device can be calculated.

2. If the input to a linear filter is gaussian noise, its output probability, which is also gaussian, and spectral density can be calculated.

These two properties suggest that transformation from a gaussian noise source to one with a different probability and specified spectral density be carried out in two stages, as we show in Fig. 11.15.

The probability and spectral characteristics of $x(t)$ and $z(t)$ are specified. We begin by assuming that $y(t)$ has a gaussian distribution with a given variance. We then determine the nature of the nonlinear device which will transform $y(t)$ to $z(t)$. Knowing $z(t)$ and the functional transformation, we next evaluate the spectral density of $y(t)$. Finally, with both the spectral density of $x(t)$ and $y(t)$ known, we can calculate the properties of the linear filter by using Eq. (11.28).

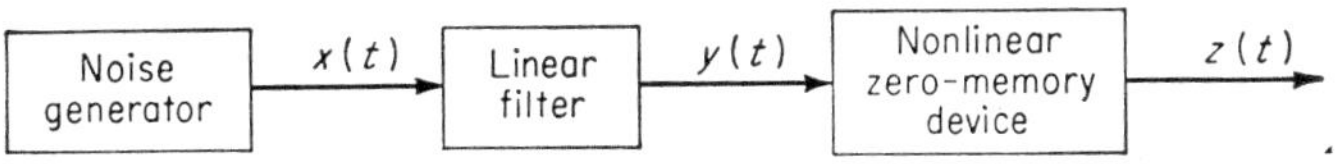

FIG. 11.15 A circuit for converting $x(t)$ to $z(t)$, which has an arbitrary probability density and spectral density.

Although the procedure which transforms $x(t)$ to $z(t)$ is simple conceptually, its implementation may be difficult to carry out. The characteristics of $z(t)$ may be such that there is no analytic closed-form solution for the functional transformation from $y(t)$ to $z(t)$. In that event we must use the digital computer to obtain an approximate numerical solution, which would then form the basis for programming the linear filter and nonlinear device on the analog computer which will generate an approximate $z(t)$.

11.2.7 Time and Amplitude Scaling

In this section when we speak of scaling it is in reference to the setting of a potentiometer which adjusts the output amplitude of a noise generator. Obviously when the noise characteristics are not expressed in terms of a voltage amplitude, this type of scaling is ineffective. For instance, if we wish to generate Poisson-distributed noise pulses by using a Geiger-Müller counter and radioactive source, we can vary the mean time between pulses by changing either the level of radioactivity or the proximity of the source to the counter. However, we regard such an adjustment as a modification of the equipment and therefore outside the scope of this book.

The most interesting scaling is associated with the adjustment of spectral-density levels. We shall treat this subject in detail, but first let us consider the scaling of moments. Most practical problems will involve either the first or second moment, but for generality we shall study the case of the

nth moment, i.e., the expectation value of x^n. Our simple circuit in Fig. 11.16 illustrates the computer diagram for which the gain k must be set.

Let the nth moments of x and y be

$$E(x^n) = \overline{x^n} = a \qquad E(y^n) = \overline{y^n} = b \tag{11.36}$$

Then

$$\overline{y^n} = \overline{(kx)^n} = k^n\overline{x^n} \tag{11.37}$$

Hence

$$k = \left(\frac{b}{a}\right)^{1/n} \tag{11.38}$$

As we shall see presently, spectral-density scaling is akin to scaling of the second moment.

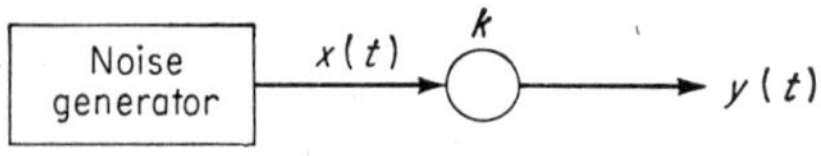

FIG. 11.16 The output of a noise generator whose amplitude is scaled by a potentiometer.

Before considering spectral-density scaling, we emphasize some important properties of noise. Where a two-dimensional distribution is entirely specified by its second moment, it can be derived from its spectral density. The prime example of this is the normal distribution. *However, the one-dimensional normal probability density and the spectral density are not uniquely related.* This very important point may be demonstrated by a simple illustration. Consider two gaussian noise sources having the same variance and zero mean, as shown in Fig. 11.17. Thus, by our assumptions

$$\sigma^2 = N_1B_1 = N_2B_2$$

If the two noise sources have different bandwidths, their spectral densities will not be equal. Let us feed the noise of each source into a rectangular

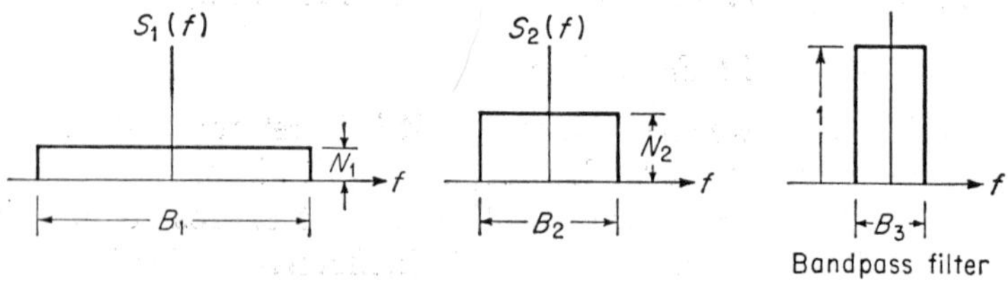

FIG. 11.17 Two spectral densities and a bandpass filter.

low-pass filter having a bandwidth B_3 and unity gain in the passband, as shown in Fig. 11.17, where $B_3 < B_2 < B_1$. From this it follows that $N_1 < N_2$. Since this is a linear filter, the output in each case will be a gaussian distribution. If source 1 is the input, the output will have a variance $\sigma_1{}^2 = N_1B_3$. For source 2 the output variance will be $\sigma_2{}^2 = N_2B_3$. Hence, two sources with the same probability distribution but different spectral densities have been affected differently by a filter.

Amplitude Scaling of Spectral Density. Amplitude and time scaling involving the spectral density are most easily explained by a typical example illustrated by Fig. 11.18. *We assume that the shape of the spectral density*

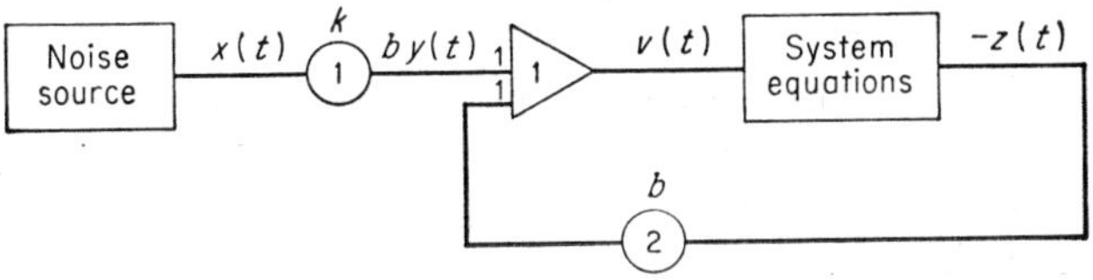

FIG. 11.18 A circuit which demonstrates noise scaling.

of $x(t)$ and $y(t)$ is the same, with the ratio of the two densities a constant at all frequencies.

On this diagram $y(t)$ is an angle measured in units of *degrees.* Therefore its spectral density is $N_0(f)$ deg²/cps. The scaling on $y(t)$ is

$$b \text{ volts} \equiv 1° \tag{11.39}$$

Hence the amplitude scaling on the spectral density of $y(t)$ is

$$M_0(f) = b^2 N_0(f) \qquad \text{volts}^2/\text{cps} \tag{11.40}$$

The noise-generator output spectral density is measured in $W_0(f)$ volts²/cps. By the Wiener-Khintchine equations the spectral density of $x(t)$ is

$$\begin{aligned} W_0(f) &= \int_{-\infty}^{\infty} R_x(\tau) \cos \omega\tau \, d\tau \\ &= \frac{1}{2T} \int_{-T}^{T} dt \int_{-\infty}^{\infty} x(t)x(t+\tau) \cos \omega\tau \, d\tau \qquad T \to \infty \end{aligned} \tag{11.41}$$

For $y(t)$ the spectral density is

$$M_0(f) = \frac{1}{2T} \int_{-T}^{T} dt \int_{-\infty}^{\infty} y(t)y(t+\tau) \cos \omega\tau \, d\tau \tag{11.42}$$

Substituting Eq. (11.43) in Eq. (11.42),

$$y = kx \tag{11.43}$$

where k is the setting of potentiometer 1 in Fig. 11.18, we get

$$\begin{aligned} M_0(f) &= k^2 \left[\frac{1}{2T} \int_{-T}^{T} dt \int_{-\infty}^{\infty} x(t)x(t+\tau) \cos \omega\tau \, d\tau \right] \\ &= k^2 W_0(f) \end{aligned} \tag{11.44}$$

Using Eq. (11.40),

$$k^2 W_0(f) = b^2 N_0(f) \tag{11.45}$$

and the potentiometer setting becomes

$$k = b \sqrt{\frac{N_0}{W_0}} \tag{11.46}$$

where $N_0(f)$ is measured in problem dimensions, in this case deg²/cps, and $W_0(f)$ is measured in volts²/cps.

Time Scaling. If the problem has a time scale different from the one used to specify the spectral density of the noise source, the potentiometer setting must account for this factor. The spectral density of the noise generator has the dimensions of time or, more precisely, the product of volts² and time. The noise in the problem has the dimensions of the product of (problem variable)² and time, or, in this example, deg² × time. If the time variable of the noise source is t_s and that of the problem is t_p, Eq. (11.46) may be rewritten to show dimensions of

$$k = b \frac{\text{volts}}{\text{deg}} \sqrt{\frac{N_0 \times \text{deg}^2 \times t_p}{W_0 \times \text{volts}^2 \times t_s}} \tag{11.47}$$

If the time scaling is

$$t_p = \alpha t_s \tag{11.48}$$

the setting of the potentiometer becomes

$$k = b \sqrt{\frac{N_0 \alpha}{W_0}} \tag{11.49}$$

Equation (11.49) states that, when the problem is slowed down ($\alpha > 1$), the potentiometer setting is increased, and vice versa.

Example 4. In Fig. 11.18 the spectral densities measured at zero frequency are W_0 and N_0 for x and y, respectively, and have the values

$$W_0 = 4.21 \text{ volts}^2/\text{cps} \qquad N_0 = 0.001 \text{ deg}^2/\text{cps}$$

We note that from Eq. (11.49) we need know only the ratio of spectral densities for scaling. Therefore, the value of each spectral density at any frequency is all that is required.

The problem is to be sped up by a factor of 5 ($\alpha = 0.2$), and the scaling at amplifier 1 is

$$25 \text{ volts} \equiv 1°$$

Using these quantities in Eq. (11.46), we obtain as our potentiometer setting

$$k = 25 \sqrt{\frac{0.001 \times 0.2}{4.21}} \approx 0.17 \tag{11.50}$$

11.3 Measurement and Evaluation of Data

In this section we shall be concerned with some of the methods by which we can extract the statistical properties of the solution from the data at our disposal. These properties, like the mean and variance, are called population parameters and are specific numbers. The ever-present restrictions on time and financial cost which we can expend on any problem limit us to time averages of finite duration and ensemble averages of finite number. Therefore, the sample average will change value from one sample to the next and will itself be a random variable with a certain probability density. We begin by discussing the procedures for measuring several quantities and the confidence limits associated with each measurement. For convenience we separate those measurements into the two categories of time averages and ensemble averages.

11.3.1 Parameter Estimates Using Time Averages

These measurements may be made on any waveform which is a stationary random process. Although on occasion this waveform is the solution of a problem, most likely it will be the noise-generator output. We may therefore regard the ensuing discussion on time averages as a means of monitoring our noise sources to verify that they are performing properly.

In each case the time average will be the estimate of the designated parameter. We shall assume that the time average is sufficiently long that the central limit theorem holds, which means that the sample average tends to be normally distributed.[1]

Thus we need know only the variance of the time average (a population parameter) to establish the confidence limits[2] on the parameter. Let the time average of one sample be designated as $\hat{x}_1(T)$, where T is the time duration of $\hat{x}_1(T)$, and let $\sigma^2_{\hat{x}_1}$ be the variance of $\hat{x}_1(T)$. Then the 95 per cent confidence level, alternatively called the significance level, is in the interval $\hat{x}_1(T) \pm 1.96\sigma_{\hat{x}_1}$, and the 99 per cent confidence level is $\hat{x}_1(t) \pm 2.58\sigma_{\hat{x}_1}$. Closely related to these are the so-called 2σ and 3σ intervals, which are $\hat{x}_1 \pm 2\sigma_{\hat{x}_1}$ and $\hat{x}_1 \pm 3\sigma_{\hat{x}_1}$. These are respectively the 95.5 and 99.7 per cent confidence levels.

The principal problem that we face in using these significance levels is that the variance of the time average is generally not known. We can surmount this difficulty in one of two ways. We proceed by taking several samples of the time average, with each sample measured for the same time interval. We calculate the variance of these samples, which is called the *sample variance.* Our first alternative is to use the *sample variance* instead of the *variance* in determining the confidence intervals. However, the *sample variance* will in general be different from the *variance* of the time average, with the difference on the average decreasing as the number of samples increases.

A second alternative is to use the Student t distribution, which is related to the gaussian distribution. It makes use of the sample variance rather than the true (population) variance. Thus with the t distribution we can get precise confidence intervals with only the sample variance of the time average. This distribution is discussed more fully in Sec. 11.3.2.

[1] The central limit theorem states that under certain conditions a sum of n random variables approaches a normal distribution as n becomes large. For a further discussion of this theorem, see Ref. 114, pp. 362–366, and Ref. 116, pp. 81–84.

[2] See Chap. 10 for an explanation of confidence limits.

For the reader confused by the terms *sample variance* and *variance* we digress briefly from our main narrative for an explanation. If we have n samples of the time average, the sample variance is the estimate of the variance of the population, calculated from the n samples. The variance (also called the population variance) of a single time average is the number we would get if we took an infinite number of sets of n samples each and found the average of the sample variances. Observe also that the variance is a *number* and the sample variance is a *random variable.*

Estimation of the Mean.[1] The mean of a stochastic variable may be estimated by a time average. This estimate is called the sample mean and may be obtained by the two simple circuits shown in Fig. 11.19. In Fig. 11.19a the switch S_1 is closed at $t = 0$ and opened at $t = T$. The output of the integrator at this latter time is the estimate of the mean. By comparison in Fig. 11.19b the signal is fed into a low-pass filter when the switch S is closed at $t = 0$, and remains closed thereafter. Following a transient settling time the term $\hat{x}'(t)$ is a continuous estimate of the mean. Although this circuit has the attractive feature of permitting continuous monitoring of the noise generator, we must reject it for the following reason.

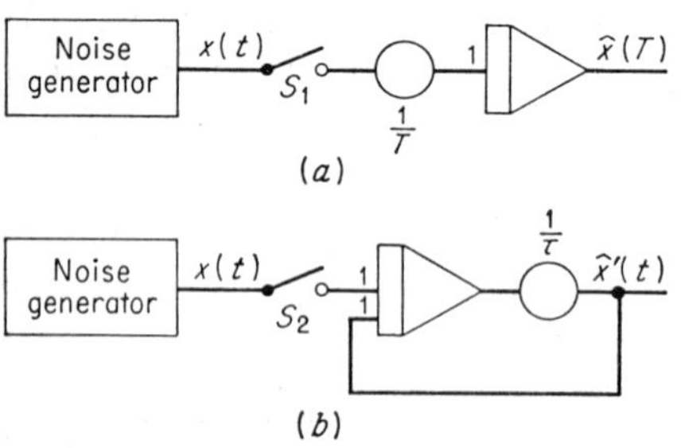

FIG. 11.19 Two averaging circuits.

A good noise source has a very small mean. This forces us to a low-pass filter with a very long time constant, or, equivalently, a very narrow frequency band (τ large), to keep the variance of $\hat{x}'(t)$ reasonably small. With this large time constant is associated a long transient settling time. Thus if a change occurs in the noise-generator output mean, we must wait a long time before we can observe its effect accurately. For this reason the circuit of Fig. 11.19a is better for the measurement of the mean. However, the circuit which continuously measures the average is a good one when the average is sufficiently large. We shall therefore study it in some detail when we consider methods for estimating the variance.

The variance of the sample mean $\sigma_{\hat{x}}^2$, derived from Fig. 11.19a, is

$$\sigma_{\hat{x}}^2 = \int_{-T}^{T} \frac{T - |\tau|}{T^2} [R_x(\tau) - m^2] \, d\tau \tag{11.51}$$

where $R_x(\tau)$ = correlation function of $x(t)$

m = mean of $x(t)$

As Eq. (11.51) shows, the variance of the sample mean can be calculated if we know the correlation function and the mean. We clearly do not know

[1] The equations we use for the variance of the estimates of the mean and variance are derived in Ref. 114, pp. 679–695.

the mean because we are using the sample mean to estimate it! Let us assume that the mean is small compared with the correlation function. What then?

If we do not know the correlation function precisely—and we may assume that this will almost always be the case—we can use the computer to obtain an estimate of it. We point out in a succeeding paragraph that it is difficult for the computer to obtain a measurement of $R(\tau)$ for any arbitrary τ. To determine $R(\tau)$ for a sufficient number of delay times to calculate Eq. (11.51) imposes the added burden of a large amount of data processing—a burden which is unnecessary because we have a better way to get the confidence limits.

When the correlation function is unknown, there is only one practical method for finding the confidence limits on the sample mean. We integrate $x(t)$ for n separate intervals of time T and obtain n integrator outputs designated $\hat{x}_i(T)$, where $i = 1, 2, \ldots, n$. Utilizing *all* our data, we form a new estimate of the sample mean by taking the ensemble average:

$$\hat{x}(T) = \sum_{i=1}^{n} \frac{\hat{x}_i(T)}{n} \tag{11.52}$$

The sample variance is

$$S_{\hat{x}}^2 = \sum_{i=1}^{n} \frac{x_i^2(T) - [\hat{x}(T)]^2}{(n-1)n} \qquad (11.53)^1$$

We reduce our confidence interval by using all the samples in the ensemble to determine the sample mean. In contrast to the variance of a single sample $\sigma_{\hat{x}_1}{}^2$ [see Eq. (11.51)], the variance of n averaged samples $\sigma_n{}^2$ is

$$\sigma_n{}^2 = \frac{\sigma_{\hat{x}_1}{}^2}{n} \tag{11.54}$$

This variance is approximated by its sample in Eq. (11.53). When $\sigma_{\hat{x}}{}^2$ is small, a situation stemming from large T or n, the confidence intervals will be about the same as those using the correct variance. In all other instances we must use the Student t distribution to find the confidence intervals, for reasons we delineate in the section covering the ensemble mean.

Example 5. For the sake of demonstrating the procedure by which confidence intervals are determined, let us assume that the x_i shown in Table 11.1 represent the integrator outputs for five different computer runs.

This sample mean is

$$\hat{x} = \sum_{i=1}^{n} \frac{x_i}{n} = \frac{9}{5} = 1.8 \tag{11.55}$$

[1] Observe that the term $n - 1$ appears in the denominator instead of n. This is necessary to make $S_{\hat{x}}{}^2$ an unbiased estimate of the variance. If the sample variance were calculated using the population mean instead of the sample mean $\hat{x}(T)$, the denominator would be n^2 instead of $n(n-1)$.

The sample variance is

$$\sigma_{\hat{x}}^2 = \sum_{i=1}^{n} \frac{(x_i)^2 - (\hat{x})^2}{n(n-1)} = \frac{43 - (5)(3.24)}{(5)(4)} \tag{11.56}$$

and

$$\sigma_{\hat{x}} = 1.17$$

When we have only five samples, we use the t distribution to find the confidence intervals. However, in this example we have kept the sample size small to keep the computations simple. We shall substitute the sample variance as an approximation to the true variance in the gaussian-distribution confidence limits and caution the reader that this technique is reserved for sample sizes 30 or larger.

TABLE 11.1 COMPUTER DATA

Computer run i	x_i	x_i^2
1	1	1
2	3	9
3	−2	4
4	2	4
5	5	25
Total........	9	43

By using the results of Eq. (11.56), our confidence interval on the population mean $\bar{x}$ at the 95 per cent significance level is

$$\hat{x} - 1.96\sigma_{\hat{x}} < \bar{x} < \hat{x} + 1.96\sigma_{\hat{x}} \tag{11.57}$$

or

$$-0.49 < \bar{x} < 4.09$$

At the 99 per cent significance level our interval is

$$\hat{x} - 2.58\sigma_{\hat{x}} < \bar{x} < \hat{x} + 2.58\sigma_{\hat{x}} \tag{11.58}$$

or

$$-1.22 < \bar{x} < 4.82$$

If instead we take n discrete periodic samples of the random process, $\hat{x}_n$ is the sample mean defined as

$$\hat{x}_n = \frac{\sum_{k=1}^{n} x(kL)}{n} \tag{11.59}$$

where L is the time interval between samples. The variance of $\hat{x}_n$ is

$$\sigma_{\hat{x}_n}^2 = \frac{\sigma_x^2}{n} + \sum_{k=-(n-1)}^{n-1} \left[\frac{n-k}{n^2} R_x(\tau) - m^2 \right] \tag{11.60}$$

where σ_x^2 is the variance of the random process $x(t)$.

In Eq. (11.51) the standard deviation σ_{x_1} is proportional to $1/T$ for large T. Similarly, in Eq. (11.60), $\sigma_{\hat{x}_n}^2$ is proportional to $1/n$. This is an inherent and exasperating property of the variance of all large-sample estimates, whether they are of the mean or any other parameter. This property shows us the effort we must expend to improve the accuracy of the measurement. As we have already pointed out, the confidence interval is proportional to the standard deviation. Thus if we want to change this interval by a factor α, we must change the measurement time by a factor $1/\alpha^2$. To reduce the interval by 10 we must increase the time of measurement one hundredfold. In colloquial terms, the cussedness principle of statistics is that we *pay* in time more than we *receive* in accuracy. *It is, therefore, axiomatic that we always estimate a parameter with the least accuracy commensurate with the problem requirements.*

Estimation of the Variance and Spectral Density. The estimate of the spectral density of a noise source in the neighborhood of a particular frequency is equivalent to the estimate of the variance of a noise signal. To show this equivalence we consider the procedure for estimating the spectral density. We pass the noise-generator signal $x(t)$ through a bandpass filter whose output is $y(t)$. The bandwidth of the filter is sufficiently narrow that $x(t)$ has an approximately constant spectral density N_0 in the region where the filter attenuation is small. If $\overline{x(t)}$, the mean of $x(t)$, is zero, then $\overline{y(t)} = 0$ and the average measurement of $y^2(t)$ is an estimate of the variance of $y(t)$ as well as the spectral density of $x(t)$.

If we are examining the spectral density of $x(t)$ in the neighborhood of zero frequency and our first-order filter time constant is τ_1, then the spectral density of $y(t)$ is

$$S_y(f) = \frac{N_0}{1 + \tau_1{}^2 w^2} \tag{11.61}$$

and its variance is

$$E[y^2(t)] = \frac{1}{2\pi} \int_{-\infty}^{\infty} \frac{N_0}{1 + \tau_1{}^2 w^2}\, dw = \frac{N_0}{2\tau_1} \tag{11.62}$$

Thus, we see that the estimations of the variance of $y(t)$ and spectral density of $x(t)$ are the same in this case except for a scale factor.[1]

When the noise is normally distributed, $N(0, \lambda^2)$, we have two distinct ways of estimating the variance λ^2, as we show in Fig. 11.20. Thus, in this figure

$$E(y^2) = \lambda^2 \qquad E(|y|) = \sqrt{\frac{2}{\pi}}\,\lambda \tag{11.63}$$

We observe that the absolute-value device estimates the standard deviation instead of the variance.

Now let us begin our comparison of these two methods by assuming that

[1] The topic of spectral-density estimation is more extensive than our discussion would indicate. Reference 117 treats the subject in all its important essentials.

$y(t)$ has the correlation function

$$R_y(t) = N_0 e^{-|t|/\tau_f} \tag{11.64}$$

Or, stated equivalently, it is white noise of spectral density N_0 which has been passed through a first-order filter with time constant τ_f. Taking the averaging device as an integrator and integrating for a time T, we set potentiometer 1 at $1/T$ and potentiometer 2 at $(1/T)\sqrt{\pi/2}$. The sample means $\hat{z}_1(T)$ and $\hat{w}_1(T)$ are unbiased estimates of λ^2 and λ, respectively. The standard deviations of these sample means are[1]

$$\begin{aligned} \sigma_{\hat{z}_1}(T) &= \sqrt{\frac{2\tau_f}{T}}\,\lambda^2 \\ \sigma_{\hat{w}_1}(T) &= \sqrt{\frac{0.531\tau_f}{T}}\,\lambda \end{aligned} \tag{11.65}$$

We compare these two methods by first assuming that $\hat{z}_1(T)$ and $\hat{w}_1(T)$ are normally distributed and then comparing their confidence intervals at

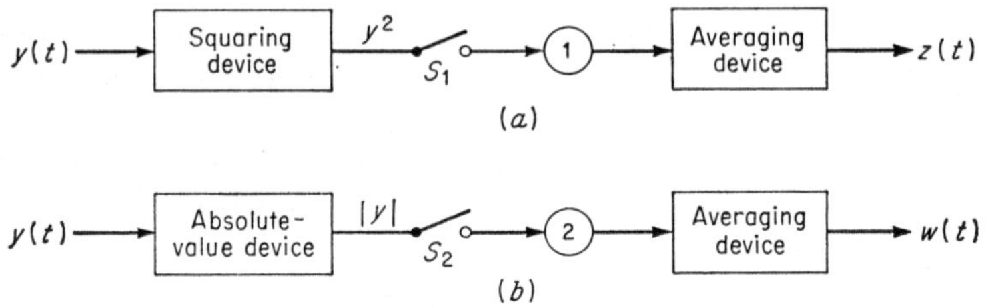

FIG. 11.20 Two averaging circuits which determine the variance of a gaussian distribution.

a given significance level to see which is smaller. Let us use the 95.5 per cent level to indicate the relative size of the confidence intervals.

Thus the confidence intervals on λ^2 and λ derived from their estimates are

$$\hat{z}_1(T) - 2\lambda^2\sqrt{\frac{2\tau_f}{T}} < \lambda^2 < \hat{z}_1(T) + 2\lambda^2\sqrt{\frac{2\tau_f}{T}} \tag{11.66}$$

and

$$\hat{w}_1(T) - 2\lambda\sqrt{\frac{0.531\tau_f}{T}} < \lambda < \hat{w}_1(T) + 2\lambda\sqrt{\frac{0.531\tau_f}{T}} \tag{11.67}$$

To facilitate the comparison we assume that the time interval for averaging is sufficiently long to make the estimate approximately equal to the parameter being measured. Under this condition we have

$$\hat{z}_1(T) \approx \lambda^2 \tag{11.68}$$
$$\hat{w}_1(T) \approx \lambda \tag{11.69}$$

[1] The results of the estimation of standard deviation are based on unpublished work by R. B. McGhee and A. Levine at Hughes Aircraft Company in 1959.

Substituting these results into Eqs. (11.66) and (11.67), we obtain

$$\lambda^2\left(1 - 2\sqrt{\frac{2\tau_f}{T}}\right) < \lambda^2 < \lambda^2\left(1 + 2\sqrt{\frac{2\tau_f}{T}}\right) \qquad (11.70)$$

and

$$\lambda\left(1 - 2\sqrt{\frac{0.531\tau_f}{T}}\right) < \lambda < \lambda\left(1 + 2\sqrt{\frac{0.531\tau_f}{T}}\right) \qquad (11.71)$$

In order to make both sets of confidence intervals apply to the same parameter, we square the terms of Eq. (11.71) to get

$$\lambda^2\left(1 - 2\sqrt{\frac{0.531\tau_f}{T}}\right)^2 < \lambda^2 < \lambda^2\left(1 + 2\sqrt{\frac{0.531\tau_f}{T}}\right)^2 \qquad (11.72)$$

Expanding the upper interval in Eq. (11.72),

$$I_2 = \lambda^2\left(1 + 2\sqrt{\frac{0.531\tau_f}{T}}\right)^2 - \lambda^2 = \lambda^2\left(1 + 4\sqrt{\frac{0.531\tau_f}{T}} + 4\,\frac{0.531\tau_f}{T}\right) - \lambda^2$$

For T large the smallest term, which is $0(1/T)$, can be neglected and we get for the interval I_2

$$I_2 = 2.92\lambda^2\sqrt{\frac{\tau_f}{T}} \qquad (11.73)$$

This compares in Eq. (11.66) with an upper interval on λ^2 of $2.83\lambda^2\sqrt{\tau_f/T}$. Thus we see that the squaring method has somewhat smaller confidence intervals than the absolute-value technique.

If we are using the variance to estimate the spectral density, then τ_f is of special interest in the confidence limits. We recall that τ_f is the time constant of the spectral-density filter characteristics and is inversely proportional to the spectral-density bandwidth. Thus the smaller the bandwidth over which the spectral density is measured, the larger the relative confidence limits. We define the relative confidence limit as

$$\frac{2\hat{\sigma}_{z_1}}{E[z_1(T)]} = \frac{2\hat{\sigma}_{z_1}}{\lambda^2}$$

From this we conclude that, the smaller the bandwidth of the spectral-density estimate, the longer the time over which we must integrate for comparable accuracy.

We cite some numbers taken from Middleton [114, p. 687] to give the reader some idea of the integrating time involved. Suppose $f_f = 10^3$ cps, where $f_f = 1/2\pi\tau_f$ and $2\hat{\sigma}_{z_1}(T)/E[\hat{z}_1(T)] = 0.01$. Then from Eq. (11.66), $T \geq 12$ sec. If $f_f = 1$ cps, then $T = 200$ min. For an $f_f = 10^{-2}$ cps, $T = 1{,}600$ days, a time well beyond the accuracy of the computer and the patience of the computer operator.

The use of an integrator as an averaging device to estimate the variance has a serious drawback. It cannot monitor the noise continuously. As we have already pointed out, noise generators are our most likely candidate for time-average estimates. These noise sources are operating all the time during the running of a problem, and we would like to check their performance regularly. It is at least a nuisance to have to switch an integrator on and off. If, instead, our averaging device is a low-pass filter, whose output is read by a meter or digital voltmeter, the averaging is a continuous process which may be observed at any time. In this case the relative confidence limit is an important criterion, for it gives us a measure of the jitter of the meter needle or digital voltmeter numbers. The less the fluctuation around the average, the easier is our task in trying to read the average. In Fig. 11.20 let these outputs be $z(t) = z_2(t)$ and $w(t) = w_2(t)$. Then, for the variance estimate the relative confidence limit is

$$\frac{2\sigma_{\bar{z}_2}(t)}{\bar{z}_2(t)} = 2\sqrt{\frac{f_0}{f_f}} \tag{11.74}$$

and for the standard-deviation estimate we have

$$\frac{2\sigma_{\bar{w}_2}(t)}{\bar{w}_2(t)} = 1.03\sqrt{\frac{f_0}{f_f}} \tag{11.75}$$

where f_0 = bandwidth of low-pass filter
f_f = noise-generator spectral-density bandwidth
$\bar{z}_2(t)$ = steady-state mean of $z_2(t)$
$\sigma^2_{\bar{z}_2}(t)$ = variance of $\bar{z}_2(t)$
$\bar{w}_2(t)$ = steady-state mean of $w_2(t)$
$\sigma^2_{\bar{w}_2}(t)$ = variance of $\bar{w}_2(t)$

We observe that the meter wiggle is considerably less for the measurement of the standard deviation. However, because the confidence interval is somewhat greater, there will be a larger error when $w_2(t)$ is used instead of $z_2(t)$ to estimate the variance. To make the limits compatible, i.e., based on an estimate of the same parameter, we square the relative confidence limit of Eq. (11.75) and compare it with Eq. (11.74).

After the switch S_1 is closed, the mean and variance of $z_2(t)$, starting from zero, increase monotonically and approach their final values asymptotically. The mean and variance of $w_2(t)$ exhibit a similar pattern. These final values are called the steady-state values. For $t > 5/2\pi f_0$ the actual values of the mean and variance are within 1.0 per cent of the final values.

Equations (11.74) and (11.75) apply after the averaging has gone on for a long enough time to allow all transient effects to die out.[1] Thus, the steady-state output of this filter is stationary noise. Note that the relative confidence interval settles down to a constant value, unlike the integrator output, whose confidence interval decreases with time of integration. We can reduce this interval by decreasing the smoothing filter bandwidth f_0.

[1] Section 11.5 also discusses the transient behavior of the mean and variance.

However, if we have a change in the variance of a noise generator, it takes the filter approximately $t = 5/2\pi f_0$ sec to settle down to its new level. So we face our usual bedevilment in statistical measurements, the trade-off between accuracy and running time.

In the preceding discussion we have shown that one has the option of two different methods to measure the variance. The squaring circuit is the more accurate, but only by a few per cent. Because an absolute-value device is simpler and more accurate than a squaring device, it is the preferable circuit.

Estimation of the Correlation Function. If we restrict ourselves to the use of standard computer components, the correlation function is the statistical characteristic which requires the most effort to measure. Figure

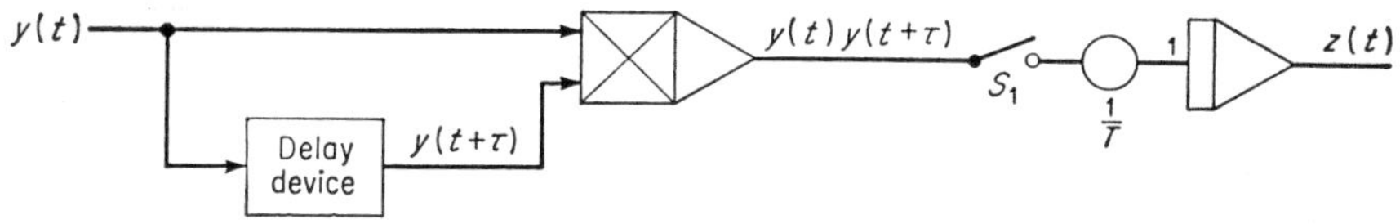

FIG. 11.21 A circuit which evaluates the autocorrelation function.

11.21 shows how $z(t)$, the estimate of $R_y(\tau)$ which is the correlation function of $y(t)$, would be programmed on the computer. The key element is the delay device. In Sec. 5.4 we described several techniques which will produce delays of electrical signals. Of these only the Padé approximation works with standard computer parts, and it is accurate only for smaller delay times. Where we take discrete samples, the sample-and-hold circuits will perform the job very well.

If $y(t)$ is gaussian white noise with variance λ^2 and is passed through a first-order low-pass filter of time constant $1/\alpha$, the variance of $z(t)$ is[1]

$$\sigma_z^2(\tau, T) = \frac{\{2\alpha T - 1 + 2e^{-2\alpha T} + [(2\alpha T + 1)(2\alpha T - 1) - 2(\alpha T)^2]e^{-2\alpha T}\}\,\lambda^4}{2(\alpha T)^2} \tag{11.76}$$

where τ is the delay time. For $\tau = 0$ and T large we have

$$\sigma_z^2(0, T) = \frac{2\lambda^4}{\alpha T} \tag{11.77}$$

which is the same as Eq. (11.66).

For $y(t)$ which is gaussian noise with any spectral density,

$$\sigma_z^2(\tau, T) = \frac{1}{T}\int_0^T (T - \beta)[R_y^2(\beta) + R_y(\tau + \beta)R_y(\tau - \beta)]\,d\beta \tag{11.78}$$

where $R_y(\tau) = E[y(t)y(t + \tau)]$.

[1] Equations (11.76) and (11.78) are derived in Ref. 118, pp. 225–299.

Suppose we use our sample-and-hold circuits to generate samples of the form $y(kL)y(kL + \tau)$ and estimate our correlation function as

$$\hat{R}_y(\tau) = \frac{1}{n} \sum_{k=0}^{n-1} y(kL)y(kL + \tau) \qquad (11.79)$$

where L is the time interval between successive samples. Then the variance of $\hat{R}$ is[1]

$$\sigma_{\hat{R}}^{\;2} = \frac{1}{N^2} \sum_{k=-(n-1)}^{n-1} (n - k)[R_y^2(kL) + R_y(kL + \tau)R_y(kL - \tau)] \qquad (11.80)$$

where $R_y(kL) = E[y(t)y(t + kL)]$.

Estimation of Probability Density. The probability density of a waveform like the one in Fig. 11.22 may be defined as the average length of time

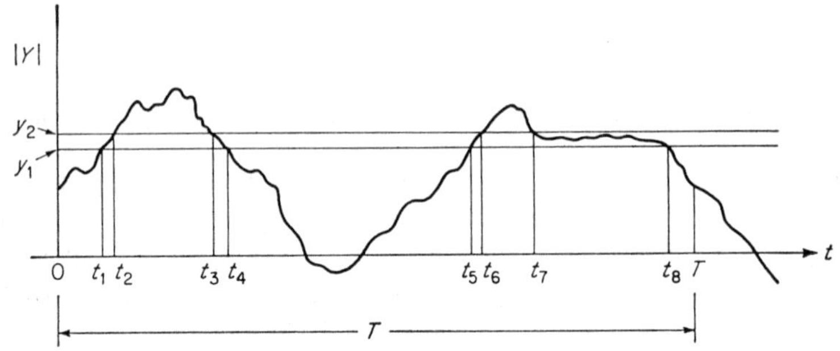

FIG. 11.22 A waveform.

the voltage is in some narrow range of values. Thus the probability density for the range $y_1 < Y < y_2$ may be estimated for a time strip of interval T in Fig. 11.22 as

$$p_i = \frac{1}{T}[(t_2 - t_1) + (t_4 - t_3) + (t_6 - t_5) + (t_8 - t_7)] \qquad (11.81)$$

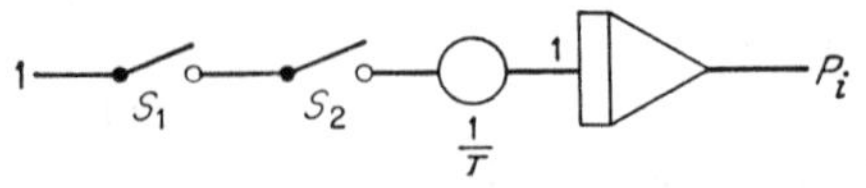

FIG. 11.23 A circuit which determines the probability density of the amplitude of a continuous waveform.

The result of Eq. (11.81) may be obtained on the computer by using the circuit in Fig. 11.23. Switch S_1 is closed by a relay when $Y > y_1$, and switch S_2 is opened by another relay when $Y > y_2$. If we take n samples of time interval T and then estimate our probability p_{av} as

$$p_{av} = \sum_{i=1}^{n} \frac{p_i}{n} \qquad (11.82)$$

[1] This result is derived in Ref. 119, pp. 342–344.

then, for large T, p_i will be normally distributed. The estimate of the variance of p_{av} is

$$\hat{\sigma}_p{}^2 = \sum_{i=1}^{n} \frac{(p_i - p_{av})^2}{n(n-1)} \tag{11.83}$$

We can then use either the gaussian distribution of the sample mean p_{av} or the more precise t distribution to fix the confidence intervals.

11.3.2 Parameter Estimates Using Ensemble Averages

In Sec. 11.3.1 we showed how one may estimate the mean and probability density by a combination of time and ensemble averages. However, in most problems the solution is a nonstationary random variable for which time averages cannot be applied. The solution can be specified only in terms of an ensemble average at each value of the independent variable. Ensemble averages can be obtained only from repeated computer runs simulating the same problem.

For example, suppose we want to determine the mean of a nonstationary random variable $x(t)$ as a function of time. For simplicity we assume that we are interested in m discrete times, $t_1, t_2, \ldots, t_m$, and that we estimate our mean with n samples. The data we accumulate are shown in the matrix array of Table 11.2. The sample mean at any time t_i is

$$x(t_i) = \sum_{k=1}^{n} \frac{x_k(t_i)}{n}$$

This illustration shows that the data storage and processing are significantly greater for ensemble averages than for time averages. In most cases

TABLE 11.2 MATRIX ARRAY OF DATA FOR THE ENSEMBLE MEANS

Computer run	*Time* t_1	t_2	$\cdots$	t_m
1	$x_1(t_1)$	$x_1(t_2)$	$\cdots$	$x_1(t_m)$
2	$x_2(t_1)$	$x_2(t_2)$	$\cdots$	$x_2(t_m)$
$\cdots$	$\cdots$	$\cdots$	$\cdots$	$\cdots$
n	$x_n(t_1)$	$x_n(t_2)$	$\cdots$	$x_n(t_m)$

we must store every point in the matrix, which means that it is necessary to retain $n \times m$ separate pieces of data. Our data handling is markedly reduced in a category of problem, termed the final-value problem, in which we are interested in the behavior of the random variable at a specific time. Some examples of random variables in final-value problems are:

1. The thickness of the finished steel in a rolling mill
2. The characteristics of the gasoline produced in a fractionation process
3. The terminal miss in a ballistic missile

Our discussion relates to ensemble averages at a single time, which makes it directly applicable to final-value problems. However, ensemble averages taken at several problem times use the same statistical concepts, with the estimate at each time treated as a different random variable.

Estimation of the Mean. If for n computer runs the samples at a particular time are $x_i(T)$, where $i = 1, 2, \ldots, n$, then the sample mean is

$$\hat{x}(T) = \sum_{i=1}^{n} \frac{x_i(T)}{n} \tag{11.84}$$

The sample variance is

$$S^2(T) = \sum_{i=1}^{n} \frac{[x_i(T) - \hat{x}(T)]^2}{n - 1} \tag{11.85}$$

If the parent population from which the samples are drawn is normally distributed, so is the sample mean for any number of samples. As we have already pointed out, the confidence interval for the population mean is based on the sample mean and the population variance. In the usual situation we know only the sample variance. Hence the confidence interval based upon this estimate is approximate. At this juncture we employ the t distribution, which gives the precise confidence interval when the true variance is unknown. The t-interval test is exact when the parent population is normal. Empirical results show that the t test works well even for nonnormal distributions which do not depart significantly from normality.

For nonnormal distributions the t distribution becomes better as the number of samples increases. Assuming that the samples are either independent or weakly correlated, we know from the central limit theorem that the distribution of a sum of samples more closely approaches the normal as the number of samples increases. Suppose we have n samples, $x_1, \ldots, x_n$, which we divide into average sums y_j, of m samples each, where $km = n$. Thus

$$y_j = \sum_{i=1+(j-1)m}^{jm} \frac{x_i}{m} \qquad j = 1, 2, \ldots, k$$

As the number of samples n increases, so will the number of samples m comprising each y_j. Thus, y_j will tend toward a normal distribution and a confidence interval derived from a t distribution of the y_j's will become more precise.

We use Appendix 11 to calculate the confidence intervals for a given significance level and a specific number of samples. The relationships between the degrees of freedom v and the number of samples n is

$$v = n - 1$$

Let β_v be the percentile for a given number of samples n and a specified significance level. Then the symmetrical limits on the mean $\bar{x}$ are, using Eqs. (11.84) and (11.85),

$$\hat{x}(t) - \frac{\beta_v S}{\sqrt{n}} < \bar{x} < \hat{x}(T) + \frac{\beta_v S}{\sqrt{n}} \tag{11.86}$$

Had we used the sample variance to estimate the population variance directly in the normal distribution, our β_v would be, respectively, 1.96 and 2.58 for the 95 and 99 per cent confidence levels. These values are approached for large values of n when we use the precise t distribution.

The symmetrical limits always produce the minimum interval for the t distribution. Since we usually want to minimize the region in which we specify our population mean, the symmetrical limits are most commonly used.

Appendix 11, in indicating the symmetrical limits, specifies that, if $1 - \alpha$ is the significance level, $\alpha/2$ per cent of the time the mean will be below the minimum bound and the same per cent above the maximum bound. If we want the 95 per cent significance level, we obtain the $t_{0.975}$ $(-t_{0.025})$ percentile. For example, if we want a 95 per cent significance level on the mean being below a given value, we read the proper quantity β_v on the $t_{0.95}$ $(-t_{0.05})$ column and obtain the interval

$$-\infty < \bar{x} < \hat{x}(T) + \beta_v \frac{S}{\sqrt{n}}$$

Similarly we can get asymmetrical limits. Take β_{v_1} from the $t_{0.975}$ $(-t_{0.025})$ column and β_{v_2} from the $t_{0.99}$ $(-t_{0.01})$ column. Then

$$\hat{x}(T) - \beta_{v_1} \frac{S}{\sqrt{n}} < \bar{x} < \hat{x}(T) + \beta_{v_2} \frac{S}{\sqrt{n}}$$

is the 96.5 $[1 - (0.025 + 0.01)]$ per cent confidence interval.

Example 6. In Example 5 we used the sample variance as an estimate of the variance of the sample mean. Then, with this approximation, we found the confidence intervals. To show how much in error we were, let us apply the same data to find the *exact* confidence intervals with the t distribution.

Using Appendix 11, we observe that the limits for the 95 per cent significance level are $\beta_v = \pm 2.78$ for five computer runs. Thus, referring to data of Example 5 and Eq. (11.86),

$$-2.12 < \bar{x} < 5.72$$

For the 99 per cent confidence interval we note that $\beta_v = \pm 4.60$ and the interval is

$$-4.69 < \bar{x} < 8.29$$

As we can see by a comparison of the results of Examples 5 and 6, the use of the sample variance with the gaussian distribution *always* yields smaller intervals than the correct t distribution.

Estimation of the Variance. When the sample variance of Eq. (11.85) is derived from a normal population, the variable $(n - 1)S^2/\sigma^2$ is χ^2-distributed. The confidence limits resulting in the minimum interval on the standard deviation σ, whose estimate is $\hat{\sigma}$, are of the form

$$\left(\hat{\sigma} - a\frac{S}{\sqrt{n}}\right) < \sigma < \left(\hat{\sigma} + b\frac{S}{\sqrt{n}}\right) \tag{11.87}$$

where, particularly for small samples, a and b are not equal and the calculation of a and b requires some effort.[1] A close approximation to the minimum confidence interval is the one where, for a significance level $1 - \alpha_1$, on the average σ will fall below the lower limit $\alpha_1/2$ per cent of the time and will fall above the upper limit the remaining $\alpha_1/2$ per cent of the time.

We use Appendix 7 to provide these limits, which are written

$$\frac{S}{\chi_{v,1-\alpha_1/2}} < \sigma < \frac{S}{\chi_{v,\alpha_1/2}} \tag{11.88}$$

For example, for 11 samples and a 95 per cent confidence interval we have

$$\begin{aligned} v &= n - 1 = 10 \\ 1 - \alpha_1 &= 0.95 \\ \alpha_1 &= 0.05 \\ 1 - \frac{\alpha_1}{2} &= 0.975 \\ \frac{\alpha_1}{2} &= 0.025 \end{aligned}$$

Therefore, for $v = 10$ in the column, $100\alpha = 97.5$, we read $\chi^2_{10,0.975} = 2.05$; and in the column $100\alpha = 2.5$ we read $\chi^2_{10,0.025} = 0.325$. These quantities substituted in Eq. (11.88) provide a two-sided confidence interval substantially close to the minimum.

For sample sizes $n > 30$, the χ^2 distribution approaches the normal distribution. We can then use the symmetrical interval

$$\left(1 - \frac{1.41}{\sqrt{n}}\right)S < \sigma < \left(1 + \frac{1.41}{\sqrt{n}}\right)S \tag{11.89}$$

for the 95 per cent confidence level and

$$\left(1 - \frac{1.83}{\sqrt{n}}\right)S < \sigma < \left(1 + \frac{1.83}{\sqrt{n}}\right)S \tag{11.90}$$

for the 99 per cent level. We observe that for large samples the confidence interval is always less on the standard deviation than it is on the mean by a factor of $\sqrt{2}$.

[1] The procedure for calculating a and b is described in Ref. 120, pp. 260–263.

The χ^2 test works well even for nonnormal distributions which depart moderately from normality. For precisely the reasons used to justify the t test for nonnormal population, the χ^2 test improves as the sample size increases.

Estimation of Probability. The probability that a random variable will be within a closed region is estimated by a set of computer runs which consist in counting the number of times the sample falls inside the region. If we designate each sample lying within the region as a success, then the probability estimate $\hat{p}$ is

$$\hat{p} = \frac{\text{number of successes}}{\text{number of computer runs}}$$

The quantity $\hat{p}$ has a binomial distribution whose confidence intervals for a given significance level are dependent only upon the measured probability and number of computer runs.[1] The reader should note that the binomial distribution arises from the counting procedure in the estimate. Thus we have the same distribution for $\hat{p}$ regardless of the shape or dimension of the probability distribution we are estimating.

In happy contrast to some other measured parameters we have discussed, p has precise confidence intervals because the $\hat{p}$ distribution is the same for all random variables. In Appendix 8 are graphs of the confidence intervals for the 95 and 99 per cent significance levels. For each sample size n, there are two curves which give the upper and lower value of the confidence interval. Having the measured probability $\hat{p}$, we obtain the corresponding bounds of the theoretical probability p on the upper and lower curves, which gives us directly the confidence interval. For example, at the 95 per cent significance level, if $\hat{p} = 0.35$, then the confidence interval on p for $n = 100$ is $0.23 < p < 0.45$.

By observing the line $p = \hat{p}$ we see that the confidence intervals are not symmetrical about p except in the neighborhood of $p = 0.50$. The graph shows the confidence intervals for the region $0 \leq p \leq 0.50$ on the left ordinate, and for the region $0.50 \leq p \leq 1.0$ on the right ordinate. The reason that we can present the intervals in this form is that in measuring $\hat{p}$ we also measure its complement, $1 - \hat{p}$. Therefore, the parameters p and $1 - p$ must have the same confidence intervals. Thus, if $p > 0.50$, we can find the equivalent confidence limits on $1 - p$ associated with $1 - \hat{p}$. Assume that these are α and β, so that the interval is

$$\alpha \leq 1 - p \leq \beta$$

Then the confidence limits on p are

$$1 - \beta \leq p \leq 1 - \alpha$$

[1] For an additional discussion of confidence limits associated with the binomial distribution, see Chap. 9.

11.3.3 Taking Data

Content of the Data. The mean, variance, probability distribution, spectral density, and correlation function are some of the many characteristics of a random variable which can be measured. In deterministic problems we select the variables of interest and record their behavior. In statistical problems we have an additional stratum of selection by being compelled to choose as well the important statistical characteristics to be measured. Furthermore, because we estimate these parameters, we must also choose the confidence intervals which meet our accuracy specifications. As we have shown, choosing the confidence intervals has as a direct consequence the choosing of the computer running time for the estimation of a given parameter.

When we are running the problem, we usually have a good idea of our needs and can pick the statistical characteristics and confidence intervals we wish to record. Besides, if we err in our judgment, we can easily return to the particular case and rerun it to suit our requirements while the problem is still programmed on the computer. The difficulty arises when our problem solutions have usefulness beyond the time that the computer program has been dismantled. Very often, as a natural evolution of thinking about the problem, we find that our area of interest changes. Thus the characteristics which command our attention now may be different from the ones we want to know about several months hence. To set up and rerun a large problem takes considerably more effort than would have been required to accumulate the additional data when the problem was first programmed.

The determination of adequate confidence intervals can be a vexing task. Unless we are clairvoyant, a gift denied to most of us, we rarely know that our future accuracy of estimate must be greater than our present one. Why then should we make more computer runs than we apparently need? Particularly if other problems are waiting in the wings, the operator will have to do some soul searching if he delays them by running the present problem to get an accuracy that seems excessive.

Our choice is a simpler one when we consider which characteristics to record. If the equipment is available, we should record as much of the data as possible and in the most usable form. Since this does not affect the overall computer time, we sacrifice nothing but convenience by this course of action.

A case in point is the measurement of probability in a closed region. Suppose our problem is the landing of a vehicle on the moon, whose surface in this discussion is assumed flat. The region has some irregular boundary, of interest at the moment. If we want to find the probability of falling within the region, we simply run the problem several times and count the number of times the solution falls within the boundary.

But let several months pass, with the problem removed from the computer and the solution filed in a suitable niche. The landing is now to be made in a different region. Inevitably we shall find that our previous results cannot be extended to cover this situation. Our way out of this embarrassment is to record the probability distribution at many points on the surface instead of the single probability of falling in a particular region. Let us represent the position on the moon's surface in terms of the rectangular coordinates of Fig. 11.24. The vehicle-landing probability distribution at x_1, y_1 is

$$P_1(x_1, y_1) = \Pr(X < x_1, Y < y_1) \tag{11.91}$$

Thus, the probability P that the vehicle will fall in the shaded region is

$$P = P_4 + P_1 - P_2 - P_3 \tag{11.92}$$

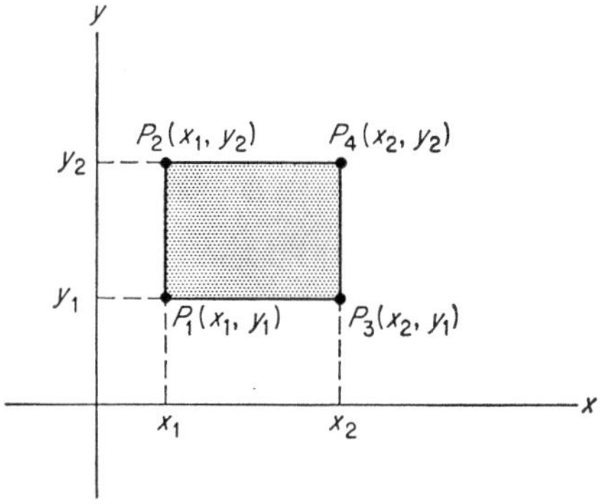

FIG. 11.24 Illustration of a technique to determine the probability of being in a rectangular region when the probability distributions of the corners are known.

If we measure the distribution over a grid of values of x and y, we can then divide a region into a set of rectangular areas like the one shown in Fig. 11.24. The distribution is a well-behaved function which increases monotonically in the direction of increasing values of x and y. We can therefore use some simple interpolation scheme to find corners of the rectangles that lie on the boundary but are not points of the measured grid of probability distributions.

Elimination of One or More Random Parameters. When we have two or more random parameters which are not random processes, e.g., initial conditions, coefficients of a differential equation, starting phase angle of a forcing function, one or more of them may be treated as a nonrandom variable whose values are selected by the computer operator. This manner of solving a statistical problem is useful when we do not know the precise properties of these parameters. On occasion this situation can arise in physical problems that are being studied while experimental data are being compiled. The computer study can be run in anticipation of the receipt of these data. We shall consider the case where the statistical properties of one random variable are unknown. The procedure can be extended to two or more random variables, but is usually too unwieldy to be practical.

Suppose we want to measure an expectation value of the solution $f(X, Y, Z)$, which is a function of the random parameters X, Y, Z, where the parameters have the probability-density function

$$p_1(x, y, x) = p_2(x)p_3(y, z|x)$$

The term $p_3(y, z|x)$ is a conditional-probability density, because it is the probability density of y and z given $X = x$. When y and z are independent of x, $p_3(y, z|x) = p_3(y, z)$. When y and z are also independent of each other, $p_3(y, z) = p_4(y)p_5(z)$. Since all expectation values can be treated the same way, we use the mean as an illustration:

$$E[f(X, Y, Z)] = \overline{f(X, Y, Z)} = \int_{-\infty}^{\infty} \int_{-\infty}^{\infty} \int_{-\infty}^{\infty} f(x, y, z)p_1(x, y, z)\, dx\, dy\, dz$$

$$= \int_{-\infty}^{\infty} E[f(X, Y, Z)|X = x]p_1(x)\, dx \qquad (11.93)$$

where $$E[f(X, Y, Z)|X = x] = \int_{-\infty}^{\infty} \int_{-\infty}^{\infty} f(x, y, z)p_3(y, z|x)\, dy\, dz \qquad (11.94)$$

Equation (11.94) is a conditional expectation and is the mean we would get as a solution when $X = x$. In practice we obtain the conditional mean for several values of the random parameter, $X = x_i$, $i = 1, 2, \ldots, n$. Then we assign to each of the x_i a probability measure $p(x_i)$ and obtain a discrete approximation to Eq. (11.93):

$$\hat{f}(X, Y, Z) = \sum_{i=1}^{n} \hat{E}[f(X, Y, Z)|X = x_i]p(x_i) \qquad (11.95)$$

where (1) $\hat{E}[f(X, Y, Z)|X = x_i]$ is an estimate of $E[f(X, Y, Z)|X = x_i]$ based on the average of a finite number of computer samples, and (2) $\hat{f}(X, Y, Z)$ is the estimate of $\overline{f(X, Y, Z)}$.

For example, suppose that in the differential equation

$$\frac{d^2y}{dt^2} + \frac{dy}{dt} + a_0 y = 0$$

the two initial conditions and the coefficient a_0 are random variables which are statistically independent. Our solution is the mean value of y at time t_2, that is, $\overline{y(t_2)}$.

Assume that the statistical characteristics of a_0 are unknown. We choose a particular value of a_0, say $a_0 = 1$, and make n runs. For each run the values of the initial conditions are set by noise generators having the proper probability densities. We average the sample values $y_1, y_2, \ldots, y_3$ and obtain

$$\hat{E}[y(t_2)|a_0 = 1] = \sum_{i=1}^{n} \frac{y_i}{n}$$

We repeat this process for several values of a_0. Note that the probability densities for the initial conditions do not change with each different value of a_0 because they are statistically independent. If they are not independent, then their probability densities are a function of a_0, and must be changed with each different value of this parameter. Finally, after assigning a probability measure for each value of a_0 run on the computer, we can use Eq. (11-95) to obtain $\hat{E}[y(t_2)]$.

The procedure we have outlined here can be extended to the measurement of $E[g(X, Y, Z)]$, where $g(X, Y, Z)$ is some function of $f(X, Y, Z)$, for instance $g(X, Y, Z) = [f(X, Y, Z)]^n$.

When the output is a probability measure, we have similar justification for eliminating one of the random variables. In a particular problem let X, Y, and Z be input random variables and let U and V be output, i.e., solution, random variables. Then the output two-dimensional probability $\Pr(u_1 < U < u_2; v_1 < V < v_2)$ is the probability that the random variables U and V will fall in the region $(u_1 < U < u_2, v_1 < V < v_2)$. The two output random variables are, of course, functions of the input random variables.

Now if we choose to eliminate Z as a random variable, the output probability is then a function of X and Y for each given value z. If we designate this probability

$$p_z(U, V) = \Pr[(u_1 < U < u_2, v_1 < V < v_2)|z] \tag{11.96}$$

it follows that

$$\Pr(u_1 < U < u_2, v_1 < V < v_2) = \int_{-\infty}^{\infty} p_z(U, V)p(z)\,dz \tag{11.97}$$

where $p(z)$ is the probability density of Z.

If we obtain $p_{z_i}(U, V)$ for several values of z_i and give a probability measure for each $p(z_i)$, we can approximate Eq. (11.97) as

$$\hat{\Pr}(u_1 < U < u_2, v_1 < V < v_2) = \sum_{i=1}^{n} \hat{p}_{z_i}(U, V)p(z_i) \tag{11.98}$$

where (1) $\hat{p}_{z_i}(U, V)$ is the estimate of $p_{z_i}(U, V)$ based on a finite number of computer samples and (2) $\hat{\Pr}(u_1 < U < u_2, v_1 < V < v_2)$ is the estimate of $\Pr(u_1 < U < u_2, v_1 < V < v_2)$.

Multidimensional Probability Measurement with Known Conditional Probabilities. We have already shown that a multidimensional probability can be estimated by counting the relative frequency with which the computer solutions fall within a pertinent region. If we know a priori a conditional probability associated with this multidimensional probability, our procedure can be modified.

As an example, let our problem be the determination of probability of kill for a hit-to-kill missile fired against an aircraft. In this illustration we shall be concerned with a two-dimensional representation of the target. Thus the shaded area of Fig. 11.25 is the region within which the missile path must fall to constitute a successful flight. However, hitting the aircraft is not sufficient. The probability of killing the target is not the same for hits on different parts of the aircraft. This problem, therefore, has a three-dimensional probability of kill. In addition to the two spatial random variables X and Y, we have K, a random variable representing the two alternatives of kill or no kill. Thus K can have two values which by our convention are unity for a kill and zero for no kill. Our probability of kill density, $p(K = 1, x, y)$, is

$$p(K = 1, x, y) = p_1(K = 1|x, y)p_2(x, y) \tag{11.99}$$

where:

1. $p_2(x, y)$ is the joint spatial probability density; $\int_R p_2(x, y)\, dx\, dy$, where R is the shaded region, is the probability of R hitting the target.

2. $p_1(K = 1|x, y)$ is the conditional-probability density of kill, given $X = x$, $Y = y$. When a point (x_i, y_i) falls outside the shaded region, then $p(K = 1|x_i, y_i) = 0$.

The probability of kill is

$$p(K = 1) = \iint_{-\infty}^{\infty} p(K = 1, x, y)\, dx\, dy \tag{11.100}$$

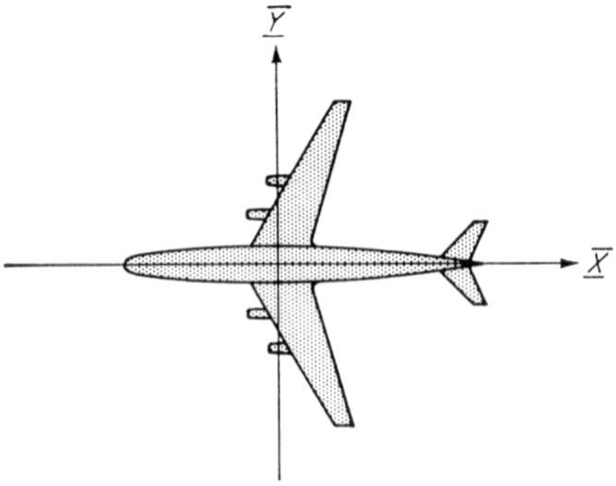

FIG. 11.25 A planar view of an aircraft target.

The coordinate point (x, y) is fed into a function generator which is programmed to produce the function of two variables, $f(x, y) = p(K = 1|x, y)$. Hence, for the rth run we record a number, $f(x_r, y_r)$, which is the conditional probability of kill, $p(K = 1|x_r, y_r)$. For a sample of n computer runs, our estimate of probability of kill is

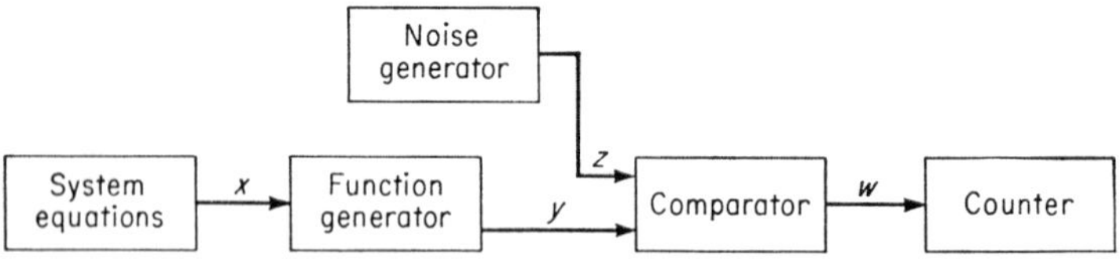

FIG. 11.26 A circuit which evaluates a two-dimensional probability when a conditional probability is known.

$$\hat{p}(K = 1) = \frac{1}{n} \sum_{r=1}^{n} f(x_r, y_r) \tag{11.101}$$

An alternative scheme of some merit involves the use of a comparator and an additional noise generator shown in Fig. 11.26.[1] Although this pro-

[1] This technique, including Fig. 11.27, is taken from Ref. 121.

cedure can work for higher-dimensional distributions, it is practical only for the two-dimensional probability. In this method we need know only the conditional density function $p(K = 1|x)$ associated with the system equations output x and the probability density of the noise-generator output.

The comparator works so that $w = 1$ when $z - y \geq 0$, and $w = 0$ otherwise. If the probability density of the noise-generator output is

$$p(z) = \frac{1}{\sqrt{2\pi}\,\sigma} \exp\left[-\frac{(z - \bar{z})^2}{2\sigma^2}\right] \tag{11.102}$$

then $p(w = 1|y)$, the conditional probability of a count $w = 1$, given y, is

$$p(w = 1|y) = \frac{1}{\sqrt{2\pi}\,\sigma} \int_{-\infty}^{y} \exp\left(-\frac{z - z}{2\sigma^2}\right) dz \tag{11.103}$$

We note that, because y is a function of x, the probability of a count $p(w = 1)$ is

$$p(w = 1) = \int_{-\infty}^{\infty} p(w = 1|y)p(x)\,dx \tag{11.104}$$

and the probability of kill, $p(K = 1)$, is

$$p(K = 1) = \int_{-\infty}^{\infty} p(K = 1|x)p(x)\,dx \tag{11.105}$$

Since we want to use the count as a measure of the probability of kill, we equate Eqs. (11.104) and (11.105):

$$p(w = 1) = p(K = 1) \tag{11.106}$$

To satisfy Eq. (11.106) we must find the functional relationship

$$y = f(x) \tag{11.107}$$

where for any given value of x we obtain a y which satisfies the condition

$$p(w = 1|y) = p(K = 1|x)$$

because for each value of x there is a *single* value of y. However, it is not necessary to have only one value of x for each value of y (see Fig. 11.27*b*).

We can do this by construction when we draw the two conditional probabilities on the same graph. We illustrate this procedure using Fig. 11.27*a*, where we have chosen $p(K = 1|x)$ to be a triangular density function. Now we take a value x_1 and find the value y_1 which makes

$$p(w = 1|y_1) = p(K = 1|x_1)$$

We plot this point in Fig. 12.27*b* and repeat the procedure until we have the functional relationship between x and y.

Thus we see that every x_j produces a y_j such that

$$p(w = 1|y_j) = p(K = 1|x_j)$$

Note that we perform a functional transformation on x to obtain y.

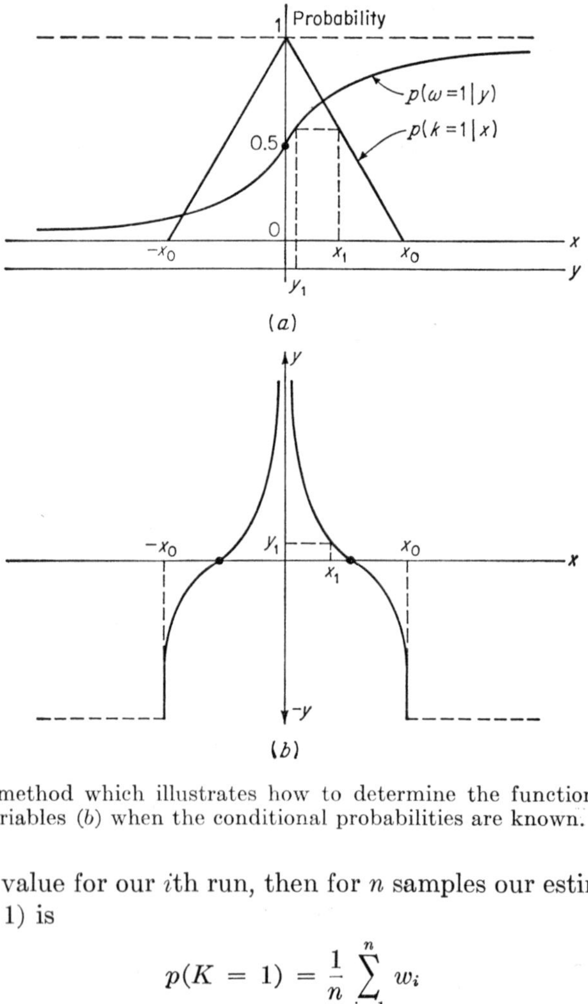

FIG. 11.27 A method which illustrates how to determine the functional relationship between two variables (*b*) when the conditional probabilities are known.

If w_i is the value for our ith run, then for n samples our estimated probability $\hat{p}(K = 1)$ is

$$p(K = 1) = \frac{1}{n} \sum_{i=1}^{n} w_i \tag{11.108}$$

The advantage of the second method rests in the summation which gives the measured probability. As Eq. (11.108) shows, because w_i is either 0 or 1, we need only the simplest counting circuit composed of binary flip-flops. On the other hand, to perform the operation in Eq. (11.105) we must for the two-dimensional case add a set of numbers that can assume several values.

We could also generate $p(K = 1)$ by feeding x directly into the comparator and transforming the output of the noise source z to a new variable u so that for all x

$$\int_{-\infty}^{x} p(u)\,du = p(K = 1|x) \tag{11.109}$$

Having determined the $p(u)$ which satisfies Eq. (11.109), we could make the transformation from z to u by the method outlined in Sec. 11.2.3.

Interpretation of Statistical Measurements. A person unskilled in the ways of statistics is apt to attribute to confidence intervals a significance which they do not have and, as a result, bias his results.

On occasion we have a good idea of the value of the quantity we are measuring. In fact, the measurement may be made as a check on the performance of the computer. What do we do when such a population parameter falls outside the confidence interval for a set of n samples? We know that, if the significance level is 95 per cent, we can expect such an occurrence on the average 5 per cent of the times we estimate a parameter. Yet an event which occurs only one time in twenty is infrequent enough to make us wonder whether it is in the statistical nature of the problem or in a computer malfunction. Faced with this uncertainty, the best we can do is to repeat the experiment with another set of n runs.

If this second set of runs has a confidence interval which encloses the anticipated value of the parameter, we have obtained evidence that decreases the likelihood of computer error. At this juncture some people erroneously discard the first set of data and keep only the second set of data. They use the fallacious argument that an estimate which places the parameter value outside the confidence limits is suspect. The only correct procedure without additional information is to combine the two sets of data into a single set of $2n$ samples and base the confidence intervals on all the data. To do otherwise would mean that we are arbitrarily discarding 5 per cent of the results. We would then be allowing our personal prejudices to bias the problem statistics.

Statistical results can never be discarded unless they are inaccurate or are discarded in such a way that no bias is introduced into the solution. Nothing prevents us from deciding to ignore every second or third computer run. However, we cannot single out data with *specific* statistical properties and eliminate these data while using only the rest of the data. Because the accuracy of our estimate increases with the number of samples, this means that we should always use *all* the computer solutions we obtain.

Another common problem involves the plotting of data, illustrated by a typical example shown in Fig. 11.28. The variable x is a problem parameter, whereas y is a solution parameter. For five values of x ($x_1, \ldots, x_5$) we have secured five measured values of y designated by the circles. At each value of x we can determine a confidence interval for y which we show with a symbol Φ. Our problem comes in attempting to describe the behavior of y for values of x other than those for which computer runs were made.

First, we must recognize that a continuous curve of confidence intervals is a difficult one to draw. In general, the most we can say is that the confidence intervals increase the farther removed we are from one of the values of x for which we have data. Figure 11.28 reflects this phenomenon in a qualitative way. The general subject of confidence intervals of a line is beyond the scope of this book. The reader may find additional illuminating material in Ref. 122. Within the region bounded by the two confidence-limit curves we have a band describing the estimation of y as a function of x.

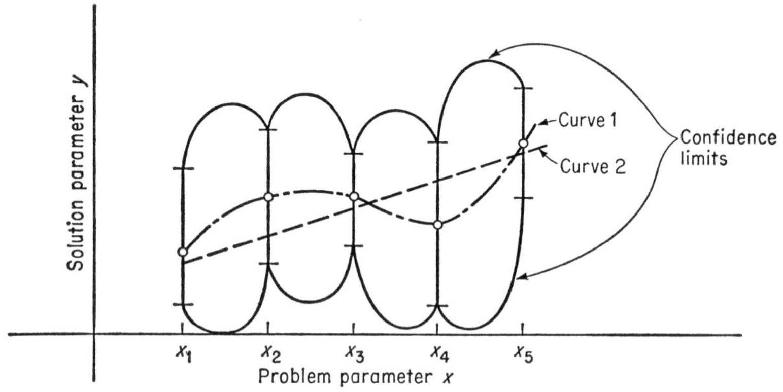

FIG. 11.28 Continuous confidence intervals determined from discrete data point.

This is anathema to most engineers who want a single-valued relationship between the two variables.

Sometimes the measured value is also most likely the parameter value, but this depends upon the probability distribution of the parameter and is not always true. With an intuitive feeling that even if this is not true it should be, some engineers will draw the curve so that it passes through all the measured points, as we show in curve 1. If we do not know what the correct curve should be, who can say that any other is better? On the other hand, we might have some additional knowledge which indicates that the curve is monotonically increasing. In that event curve 2 might be a better one.

11.4 Repetitive Operation

We define repetitive operation as the cycling of computer modes without stopping. Usually this involves only the RESET and OPERATE modes, but others, particularly HOLD, may also be programmed. During the time that the computer is in repetitive operation it functions without any assistance from the operator.

The repetitive operation is ideally suited for the taking of an ensemble average, where we must obtain several samples in which only the statistical parameters differ from one run to the next. In one typical cycle we begin in RESET with all quantities set to their correct values. We then switch to OPERATE and run the problem until its termination. Then we return to RESET, allowing enough time to restore the problem to its initial state. Statistical quantities, like initial conditions and coefficients, are set by noise generators. The remaining quantities return to where they were at the start of the previous run. We may, at our option, record the sample values, which are stored as numbers, either during the RESET mode or following the OPERATE mode. Since printing takes some time, we shorten the length of a cycle by using OPERATE for data printout. Where the storage of information is done electronically, as in the case of a probability count, nothing is gained by a delay in the recording of the sample.

The newer models of the EASE, PACE, and DYSTAC[1] computers all have repetitive operations mode capability. So does the GPS computer,[2] which, from its inception 15 years ago, has always been a repetitive computer. The GPS computer has a time-base range from 1 sec to 0.3 msec, whereas the others have a range from 1 sec to 5.0 msec.

The time base is the interval of time required by the computer to solve the problem for one unit interval of problem independent variable, provided the computer has not been time-scaled.

We mention in passing that all four of these computers also have an iterative operation mode, whereby some of the integrators can be programmed to be in OPERATE while others are in RESET. This mode[3] can be used to evaluate multiple integrals, optimize parameters, and solve partial differential equations. Also it may find application in solving statistical decision problems.

An important measure of the performance of a computer in the repetitive mode is the minimum length of time required for one cycle. In a problem which has several different cases, part of the problem running time is consumed in setting up each case. However, the greatest amount of time is absorbed in obtaining the samples for each case. *The minimum time per cycle is fixed not by the maximum repetition rate, but rather by the minimum time base of the computer.*

A cycle of time duration t consists of two parts: t_1, the time in OPERATE, and t_2, the time in RESET. Although a smaller time base usually requires less reset time, we shall assume that t_2 is a constant. However, the mini-

[1] These computers are built respectively by Beckman Instruments, Inc., Richmond, Calif.; Electronic Associates, Inc., Long Beach, Calif.; and Computer Systems, Inc., Monmouth Junction, N.J.

[2] This computer is built by the GPS Instrument Co., Needham, Mass. It was designed originally at the MIT Instrumentation Laboratory.

[3] For a brief description of the iterative computer, see Ref. 123.

mum value of t_1 is directly affected by the time base and always satisfies Eq. (11.110):

$$\min t_1 = k\alpha \tag{11.110}$$

where α is the time base of the computer and k is the maximum range of the independent variable in the problem. We can see from this equation that the GPS computer will have an OPERATE cycle considerably less than the other three machines.

Since the other computers are about five to ten times more accurate, the faster speed of the GPS computer gives it no clear-cut superiority. However, if the computers use the same running time to generate data, the GPS can give more accurate statistical results under certain circumstances. Suppose that we are measuring the standard deviation and that we want the confidence interval to be such that the limits deviate no more than 15 per cent from the measured variance. For the EASE, PACE, and DYSTAC we must obtain 100 samples to achieve this accuracy.

The GPS can run about ten times as fast as the other computers and can therefore get 1,000 samples in the time that the others get the necessary 100. To translate this into some meaningful figures, let us assume that the correct sample value of the standard deviation is $\hat{\sigma}$ and has a 5 per cent error on the GPS and 1 per cent error on the other computers. Therefore, for the GPS with 1,000 samples the calculated confidence interval at the 95 per cent significance level is

$$(\hat{\sigma} + 0.05\hat{\sigma})\left(1 - \frac{1.41}{\sqrt{1{,}000}}\right) < \sigma < (\hat{\sigma} + 0.05\hat{\sigma})\left(1 + \frac{1.41}{\sqrt{1{,}000}}\right) \tag{11.111}$$

or

$$1.003\hat{\sigma} < \sigma < 1.097\hat{\sigma}$$

The other computers with 100 samples have the confidence interval

$$(\hat{\sigma} + 0.01\hat{\sigma})\left(1 - \frac{1.41}{\sqrt{100}}\right) < \sigma < (\hat{\sigma} + 0.01\hat{\sigma})\left(1 + \frac{1.41}{\sqrt{100}}\right) \tag{11.112}$$

or

$$0.896\hat{\sigma} < \sigma < 1.151\hat{\sigma}$$

If it were without error, the GPS would have the confidence interval

$$0.953\hat{\sigma} < \sigma < 1.047\hat{\sigma} \tag{11.113}$$

and the other computers,

$$0.859\hat{\sigma} < \sigma < 1.141\hat{\sigma} \tag{11.114}$$

When we compare the actual computer results with their errorless counterparts, we note from Eqs. (11.111) and (11.113) that the maximum difference between the measured and errorless GPS data is $0.144\hat{\sigma} = 1.097\hat{\sigma} - 0.953\hat{\sigma}$. For the other computers this difference is $0.292\hat{\sigma} = 1.151\hat{\sigma} - 0.859\hat{\sigma}$.

Thus we see that, in running the less accurate GPS computer the same length of time as the others, we can nevertheless get more accurate results because the larger number of samples allows us to estimate the parameter with less statistical uncertainty. This advantage of the GPS diminishes and ultimately disappears as the number of samples increases, because the confidence interval decreases and the computer error becomes increasingly more important.

11.5 Ensemble Averaging without Sampling

In the one special case of random initial conditions and forcing function in a linear system, ensemble averaging can be accomplished without recourse to sampling. Furthermore, at most, only $n + 1$ computer runs are needed to measure *exactly* the mean, variance, and higher-order moments for an nth-order differential equation when the initial conditions are statistically independent.

The linear equation

$$\frac{d^n y}{dt^n} + a_{n-1} \frac{d^{n-1} y}{dt^{n-1}} + \cdots + a_1 \frac{dy}{dt} + a_0 y = g(t) \qquad (11.115)$$

where

$$g(t) = 0 \qquad \text{for } t \leq 0$$

has as a solution

$$y(t) = c_0 f_0(t) + c_1 f_1(t) + \cdots + c_i f_i(t) + \cdots + c_{n-1} f_{n-1}(t) + \int_0^t h(t, \tau) g(\tau)\, d\tau \qquad (11.116)$$

The symbol $f_i(t)$ represents the solution of the homogeneous differential equation with unity initial condition on the ith derivative and zero initial conditions on all other derivatives.[1] The coefficient c_i represents the magnitude of the initial conditions on $d^i y/dt^i$ and is, therefore, a random variable. The term $h(t, \tau)$ is the weighting function.

11.5.1 Mean

The ensemble mean $E[y(t)]$ is, from Eq. (11.116),

$$E[y(t)] = f_0(t) E(c_0) + f_1(t) E(c_1) + \cdots + f_i(t) E(c_i) + \cdots + f_{n-1}(t) E(c_{m-1}) + \int_0^t h(t, \tau) E[g(\tau)]\, d\tau \qquad (11.117)$$

Thus we can find $E[y(t)]$ in a single computer run by using the mean values of the forcing function and each initial condition. Note that the forcing function does not need to be stationary.

11.5.2 Variance

The ensemble mean can be computed without any restrictions on the random variables. For the variance and higher-order moments we impose the following conditions:

[1] See Sec. 3.2 for a more detailed discussion of this point. Because we are dealing only with y and none of its derivatives here, we have eliminated the superscripts and used a simpler notation.

1. The forcing function is white noise.
2. The forcing function and all initial conditions have zero mean.

When all initial conditions are independent random variables, our variance $E[y^2(t)]$ is

$$E[y^2(t)] = \sum_{i=0}^{n-1} E(c_i^2)f_i^2(t) + N_0 \int_0^t h^2(t, \tau)\, d\tau \tag{11.118}$$

where N_0 is the two-sided spectral density of the forcing-function white noise.

When the initial conditions are correlated, our variance is

$$E[y^2(t)] = \sum_{i=0}^{n-1} E(c_i^2)f_i^2(t) + \sum_{i=0}^{n-1} \sum_{\substack{j=0 \\ i \neq j}}^{n-1} E(c_ic_j)f_i(t)f_j(t) + N_0 \int_0^t h^2(t, \tau)\, d\tau \tag{11.119}$$

That part of Eq. (11.118) contributed by the random initial conditions can be generated if we use Eq. (11.115). We can generate a particular $f_i(t)$ and, with additional computer elements, $E(c_i^2)f_i^2(t)$ by using the associated initial conditions. In n computer runs we can have all the terms. If we want the computer to combine these terms, we can go to the other extreme in computer equipment used and program n identical system equations. Each equation would have a different nonzero unity initial condition. Thus one computer run would suffice, as we show in Fig. 11.29*a*.

The generation of random initial-condition terms in Eq. (11.119) involves more effort because of the cross-product terms $f_i(t)f_j(t)$. If we want the computer to plot these terms for us, we must program at least two system equations, as we show in Fig. 11.29*b*. If we are willing to compute the $f_i(t)f_j(t)$ terms ourselves, we can use the technique outlined for Eq. (11.118). This gives us $f_i(t)$ and $f_i^2(t)$ in each computer run. Returning to our cross-product terms, we use one equation to produce $f_i(t)$ and the other to produce $f_j(t)$. In the process we shall also get $f_i^2(t)$ and $f_j^2(t)$. Therefore, we need $(n^2 - n)/2$ computer runs to obtain all the individual terms when we generate the $f_i(t)$ two at a time.

The term arising from the forcing function involves the weighting function, which we can generate by using the adjoint of Eq. (11.115).[1] Because the weighting function is a function of two variables, we can evaluate the integral only for discrete values of t. This means that we have only discrete values of $E[y^2(t)]$ even though the contribution from the initial conditions is computed continuously for the entire range of t.

Although we may need several computer runs to determine the variance

[1] See Sec. 3.4 and Appendix 2 for a more detailed discussion of the adjoint method for generating weighting functions.

for a range of values of t when we have a random forcing function, these runs are few in comparison with those required to estimate the variance by sampling. Remember also that when we are finished we have an *exact* measure of the variance at the values of t for which we have made the runs. The contrast between the two methods is even more vivid if we have either only random initial conditions or random functions and, in the case of the

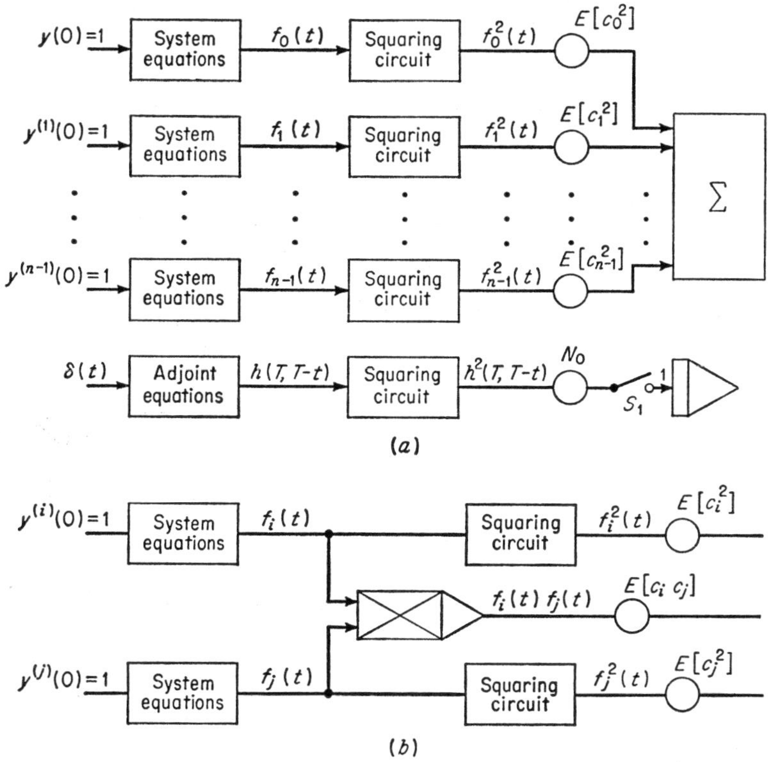

FIG. 11.29 (*a*) A circuit for generating the variance of the solution when all initial conditions are independent; (*b*) a circuit for generating the variance of the solution when initial conditions are correlated and $i \neq j$.

latter, are dealing with final-value systems. Now one run will suffice for an exact measure of the variance.

This is easy enough to see for the initial conditions, but there are some interesting ramifications worth considering in the case of the random forcing functions. As we have previously stated, the final-value system is one in which the solution is of interest at a particular value of t, say $t = T$. The weighting function of such a system can be obtained in one computer run of the adjoint system. Furthermore, if we have several random forcing functions, all of which are independent white-noise sources inserted at several system inputs, their combined effect on the variance can also be determined in a single computer run.

Let Fig. 11.30*a* represent the system with $g_1(t)$, $g_2(t)$, and $g_3(t)$, the random noise sources, having the spectral densities N_1, N_2, and N_3. The variance

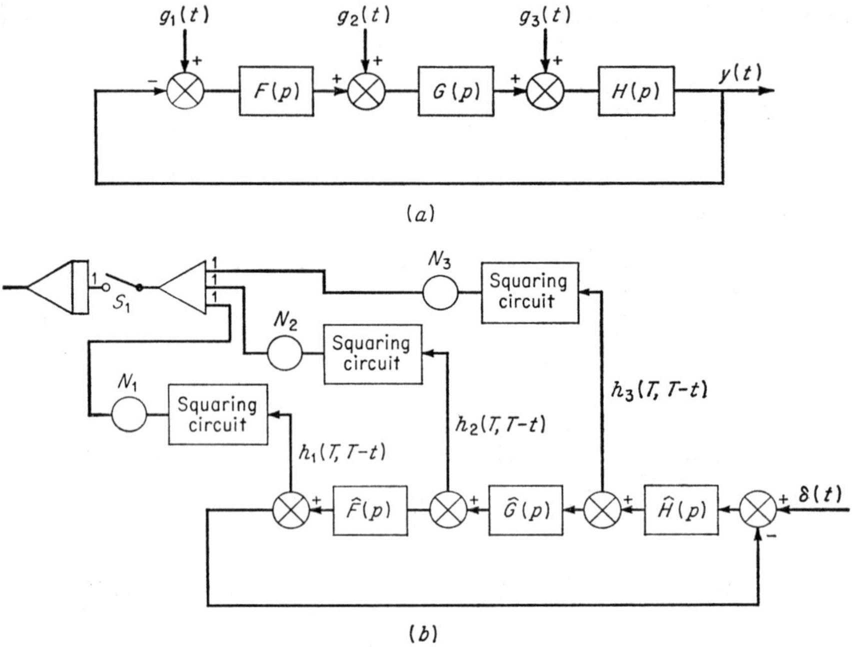

FIG. 11.30 (*a*) A circuit with three noise inputs; (*b*) a circuit for generating the variance of the output of (*a*).

of $y(t)$, $E[y^2(t)]$, is

$$E[y^2(t)] = N_1 \int_0^t h_1^2(t, \alpha)\, d\alpha + N_2 \int_0^t h_2^2(t, \alpha)\, d\alpha + N_3 \int_0^t h_3^2(t, \alpha)\, d\alpha \tag{11.120}$$

where $h_1(t, \alpha)$, $h_2(t, \alpha)$, and $h_3(t, \alpha)$ are the weighting functions, with $g_1(t)$, $g_2(t)$, and $g_3(t)$, respectively, as inputs and $y(t)$ as the output.

Figure 11.30*b* is the adjoint system. If the switch S_1 opens at $t = T$, the output of integrator 1 will read $E[y^2(T)]$ for a problem which starts at $t = 0$ and has a final value at $t = T$.

11.5.3 Whitening the Noise

When the input noise is not white, we can still obtain $E[y^2(t)]$ from Eq. (11.118), provided we modify the system in a certain way. We must make the system input white noise. How we can do this can best be illustrated by

Fig. 11.31. There are no difficulties if we wish to determine $E[y^2(t)]$ for white noise $w(t)$ applied at point A. From the diagram it is evident that this is precisely the same as the $E[y^2(t)]$ for filtered white noise, with $x(t)$ applied at point B. This, then, is the clue to using the weighting-function method when our input noise is not white. We add an additional transfer function at the point where the noise is applied such that, if the input is white noise, the output has the problem noise spectral density. We then determine the weighting function for the system plus the additional transfer function. This is the weighting function used in Eqs. (11.118) and (11.119).

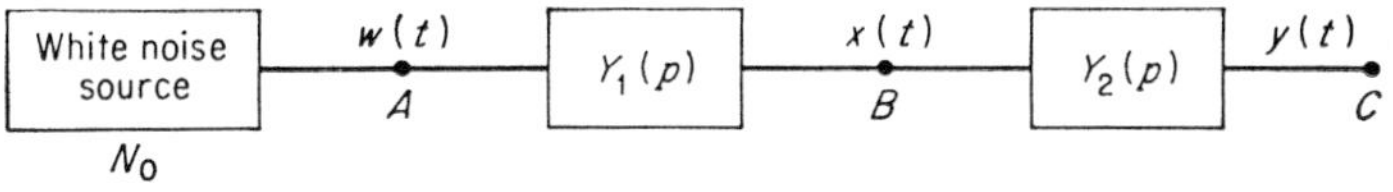

FIG. 11.31 A block diagram demonstrating noise whitening.

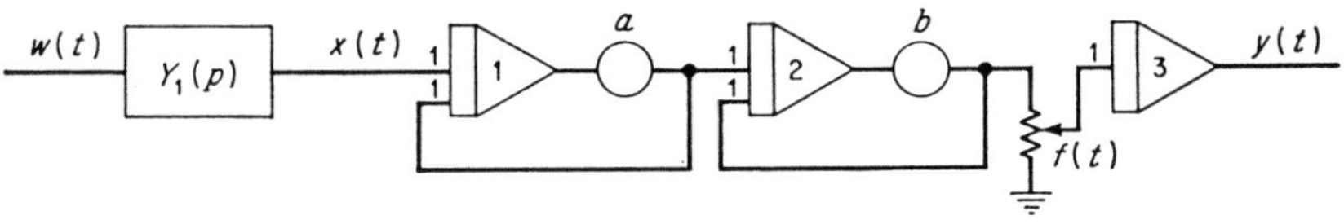

FIG. 11.32 A computer circuit.

Example 7. For the system in Fig. 11.32 we wish to solve for $E[y^2(t)]$, where $x(t)$ has a spectral density

$$S_x(f) = \frac{M_0}{1 + \tau_1{}^2 w^2} \tag{11.121}$$

To use the weighting-function method our first step is to find the filter $Y_1(p)$ through which we must pass white noise $w(t)$ to obtain $x(t)$.

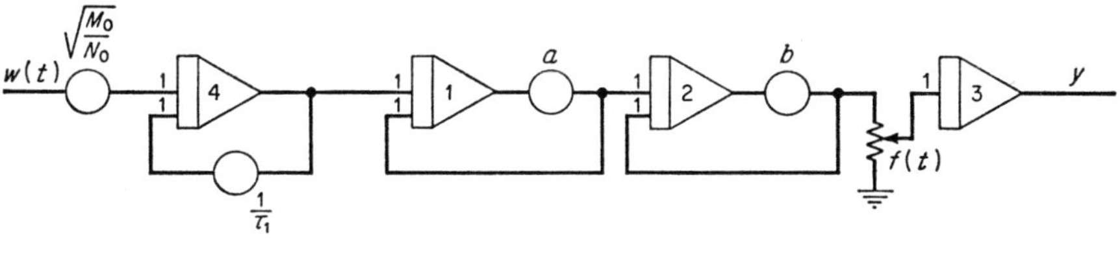

FIG. 11.33 A computer circuit with noise whitening.

If $w(t)$ has a spectral density of N_0, then by Eq. (11.28)

$$\begin{aligned} N_0|Y_1(jw)|^2 &= \frac{M_0}{1 + \tau_1{}^2 w^2} \\ |Y_1(jw)|^2 &= \frac{M_0}{N_0} \frac{1}{1 + \tau^2 w^2} \\ Y_1(jw) &= \frac{M_0}{N_0} \frac{1}{1 + \tau_1 jw} \end{aligned} \tag{11.122}$$

or in computer notation

$$Y_1(p) = \frac{M_0}{N_0} \frac{1}{1 + \tau_1 p}$$

The modified system whose weighting function is the proper one to substitute in Eq. (11.118) is shown in Fig. 11.33. Note again that $w(t)$ has a spectral density of N_0.

The modified-system weighting function with $w(t)$ as an input and $y(t)$ as an output is $h_1(t_1\tau)$. Thus with $x(t)$ having the spectral density indicated in Eq. (11.121), we get

$$E[y^2(t)] = N_0 \int_0^t h_1^2(t, \tau)\, d\tau \tag{11.123}$$

which can be evaluated after we generate $h_1(t, \tau)$ by using the adjoint of Fig. 11.33

APPENDIX

1 Properties of the Delta Function[1]

Background and Definition. The delta function (δ function) plays an important role in operational calculus as well as boundary-value and point-source problems (e.g., impulse forces and point electric charges). The function also arises in noise problems and in the evaluation of the weighting function. Because of their wide applicability, problems involving the δ function will be programmed on the computer. This function is so unusual that the computer operator must thoroughly understand its properties in order to program the computer correctly.

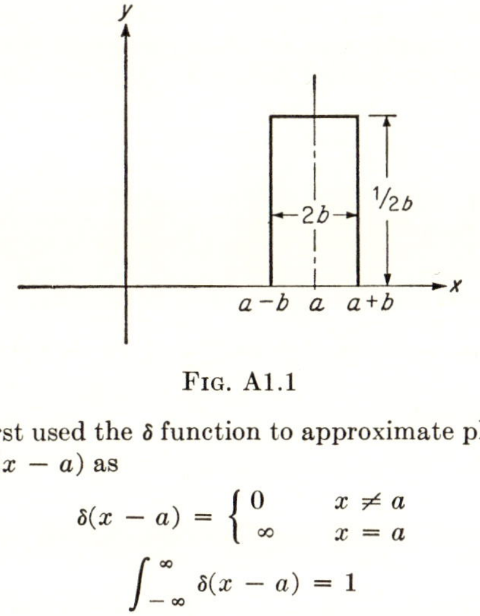

Fig. A1.1

The physicist Dirac first used the δ function to approximate physical phenomena. He defined the δ function $\delta(x - a)$ as

$$\delta(x - a) = \begin{cases} 0 & x \neq a \\ \infty & x = a \end{cases} \tag{A1.1}$$

and

$$\int_{-\infty}^{\infty} \delta(x - a) = 1 \tag{A1.2}$$

We see from this definition that the δ function is zero anywhere but at one point, where it is infinite. Nonetheless the area of the δ function is unity. The most common representation of this function is shown as the rectangular pulse in Fig. A1.1. In the limit as $b \to 0$, we satisfy the conditions of Eq. (A1.1) while holding the area fixed at unity.

For all its beguiling simplicity the δ function has been anathema to the classical mathematician because of its discontinuity and lack of differentiability in the conventional sense. The δ function is always ultimately associated with integration and indeed has no meaning except as it relates to this operation. The mathematician, confronted with a δ-function integration, preferred to bypass it by using an equivalent Stieltjes integral.

[1] For further discussion of this subject, see Ref. 45, pp. 130–145.

Recently, the δ-function manipulations have been given a rigorous mathematical foundation by Schwartz.[1]

Basic Characteristics. Before enumerating some of the formal operations of the δ function, let us consider some of its properties:

1. $\delta(x - a)$ is a function of two variables, x and a. When $a = 0$, the function is often written $\delta(x)$.

2. When written as $c\delta(x - a)$, the quantity c refers to the area of the δ function. It makes no sense to say that this is the magnitude of the δ function, since it is infinite at $x = a$ and zero elsewhere for any finite c. It is customary when plotting δ functions to represent them as shown in Fig. A1.2, with the height corresponding to the area of the function.

3. The quantity $[\delta(x - a)]^n$ is undefined for $n \neq 0, 1$.

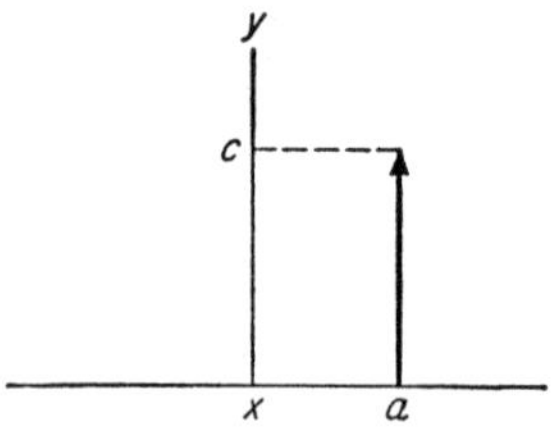

FIG. A1.2

Sifting Property. The major reason for the existence of the δ function is the sifting operation which it performs on another function. This characteristic is

$$\int_{-\infty}^{\infty} \delta(t - a) f(t)\, dt = f(a) \tag{A1.3}$$

In this integration a single value of $f(t)$ is chosen at the point where $\delta(t - a)$ is not zero. Assuming that $f(t)$ is continuous in the neighborhood of a, we may prove Eq. (A1.3) by referring to Fig. A1.1. Thus

$$\int_{-\infty}^{\infty} f(t)\delta(t - a)\, dt = \frac{1}{2b}\int_{a-b}^{a+b} f(t)\, dt \tag{A1.4}$$

By the mean-value theorem

$$\int_{a-b}^{a+b} f(t)\, dt = 2bf(a + \theta b) \qquad -1 \leq \theta \leq 1 \tag{A1.5}$$

Substituting Eq. (A1.5) into (A1.4) and performing the necessary limiting operation, we have

$$\int_{-\infty}^{\infty} f(t)\delta(t - a)\, dt = \begin{cases} \lim\limits_{b\to 0} f(a + \theta b) & \text{(A1.6a)} \\ f(a) & \text{(A1.6b)} \end{cases}$$

Identities. Several useful identities of the δ function are

$$\delta(x) = \delta(-x) \tag{A1.7}$$

$$\delta(ax) = \frac{1}{a}\,\delta(x) \tag{A1.8}$$

$$\delta(a - x)\delta(x - b) = \delta(a - b) \tag{A1.9}$$

$$\int_{-\infty}^{\infty} f^{(n)}(x)\,\delta(x - a)\, dx = \frac{d^n f(a)}{da^n} \qquad f^{(n)}(x) = \frac{d^n f}{dx^n} \tag{A1.10}$$

[1] See Ref. 138. In addition, Ref. 45, pp. 135–145, comments on Schwartz's work.

$$\int_{-\infty}^{\infty} f(x)\delta^{(n)}(x - a)\,dx = (-1)^n \frac{d^n f(a)}{da^n} \qquad \delta^{(n)}(x - a) = \frac{d^n\delta}{dx^n} \tag{A1.11}$$

$$\frac{d^n\delta(x)}{dx^n} = \frac{(-1)^n n!\delta(x)}{x^n} \tag{A1.12}$$

$$\delta(x) = \int_{-\infty}^{\infty} e^{j\omega x}\frac{d\omega}{2\pi} \tag{A1.13}$$

$$\delta\left[\prod_{i=1}^{n}(x - a_i)\right] = \sum_{j=1}^{n}\frac{1}{\prod_{\substack{i=1\\ i\neq j}}^{n}(a_j - a_i)}\,\delta(x - a_j) \tag{A1.14}$$

Equation (A1.7) is proved by making a change of variable from x to $-x$ in the integral $\int_{-\infty}^{\infty}\delta(x)\,dx$, and Eq. (A1.8) is proved by a change of variable from x to ax in the same integral. Equation (A1.8) has special importance in analog computing because it shows how to handle time scaling on δ functions. We shall return to this point.

Equation (A1.11) is proved by using integration by parts. For the case $n = 1$,

$$\int_{-\infty}^{\infty} f(x)\frac{d\delta(x - a)}{dx}\,dx = f(x)\delta(x - a)\Big|_{-\infty}^{\infty} - \int_{\infty}^{\infty}\frac{df}{dx}(x)\delta(x - a)\,dx \tag{A1.15}$$

Since $\delta(x - a) = 0$ at $x = \pm\infty$,

$$f(x)\delta(x - a)\Big|_{-\infty}^{\infty} = 0 \tag{A1.16}$$

Hence

$$\int_{-\infty}^{\infty} f(x)\delta^{(1)}(x - a)\,dx = (-1)\frac{df(a)}{da} \tag{A1.17}$$

By a similar procedure we may show that Eq. (A1.11) holds for all n. We can also prove the identity (A1.12) by performing the integration

$$\int_{-\infty}^{\infty} x^n\delta^{(n)}(x)\,dx$$

by parts and observing that

$$\int_{-\infty}^{\infty}\delta^{(n)}(x)\,dx = \delta^{(n-1)}(x)$$

Equation (A1.13) arises from the Fourier transform pair,

$$f(x) = \frac{1}{2\pi}\int_{-\infty}^{\infty} F(\infty)e^{j\omega x}\,d\omega \tag{A1.18}$$

$$F(\omega) = \int_{-\infty}^{\infty} f(t)e^{-j\omega x}\,dx \tag{A1.19}$$

If we assume $f(x) = \delta(x)$ and substitute in Eq. (A1.19), then

$$F(\omega) = 1 \tag{A1.20}$$

Substitution of Eq. (A1.20) into (A1.18) gives the identity (A1.13). This relation is very important in the analysis of noise problems. The technique of proving Eq. (A1.14) may be demonstrated by showing that

$$\delta[(x - a_1)(x - a_2)(x - a_3)] = \frac{1}{(a_1 - a_2)(a_1 - a_3)}\delta(x - a_1) + \frac{1}{(a_2 - a_1)(a_2 - a_3)}\delta(x - a_2) + \frac{1}{(a_3 - a_1)(a_3 - a_2)}\delta(x - a_3) \tag{A1.21}$$

As we have seen, the δ function is nonzero only when its argument is zero. In the case of Eq. (A1.21) this occurs at the values $x = a_1$, $x_2 = a_2$, $x = a_3$. Consider what happens in the region $a_1 - \epsilon/2 \leq x \leq a_1 + \epsilon/2$. In this case we see that $x - a_2 \rightarrow a_1 - a_2$ for ϵ arbitrarily small. Similarly, $x - a_3 \rightarrow a_1 - a_3$ in this region. Therefore we may write

$$\delta[(x - a_1)(x - a_2)(x - a_3)] \rightarrow \delta[(x - a_1)(a_2 - a_1)(a_3 - a_1)]$$

$$a_1 - \frac{\epsilon}{2} < x < a_1 + \frac{\epsilon}{2} \rightarrow 0 \qquad \text{(A1.22)}$$

By Eq. (A1.8)

$$\delta[(x - a_1)(a_2 - a_1)(a_3 - a_1)] = \frac{1}{(a_2 - a_1)(a_3 - a_1)} \delta(x - a_1) \qquad \text{(A1.23)}$$

Similarly we may obtain the other two terms on the right side of Eq. (A1.21). Likewise one may show that Eq. (A1.14) applies for the general case.

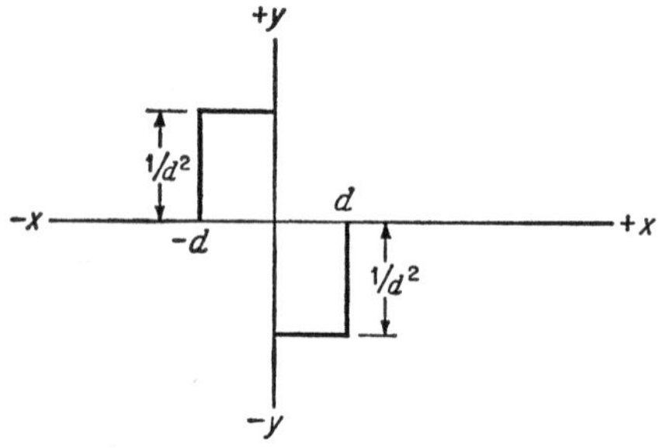

FIG. A1.3

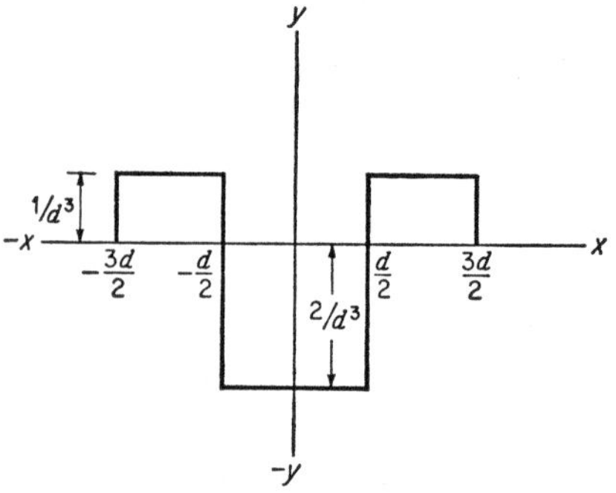

FIG. A1.4

Derivatives. In some of the equations (A1.8) to (A1.14) we have referred to derivatives of the δ function. Although the derivatives are somewhat less important than the δ function, they do have some usefulness. As we shall see presently, there is an equivalence between initial conditions and the δ function and its derivatives. We must be careful to define precisely the derivatives of the discontinuous δ function.

If we define $\delta^{(1)}(x)$ as

$$\delta^{(1)}(x) = \frac{1}{d}\left[\delta\left(x + \frac{d}{2}\right) - \delta\left(x - \frac{d}{2}\right)\right] \qquad d \rightarrow 0 \qquad \text{(A1.24)}$$

where $d = 2b$, then with Fig. A1.1 as a model for $\delta(x)$, Fig. A1.3 is the pictorial representation for $\delta^{(1)}(x)$. $\delta^{(1)}(x)$ is called a doublet, and the next higher derivative, called a triplet, is

$$\delta^{(2)}(x) = \frac{1}{d}\left[\delta^{(1)}\left(x + \frac{d}{2}\right) - \delta^{(1)}\left(x - \frac{d}{2}\right)\right] \qquad d \rightarrow 0 \qquad \text{(A1.25)}$$

as is shown in Fig. A1.4.

In general we can show that

$$\delta^{(n)}(x) = \frac{1}{d^{n+1}} \sum_{k=0}^{n+1} (-1)^k u\left[x + \frac{d}{2}(n + 1 - 2k)\right] \qquad d \rightarrow 0 \qquad \text{(A1.26)}$$

where $u(x - k)$ is defined in Eq. (A1.27).

The Unit Step. The most important of the functions related to the δ function is its integral, called a unit step and defined as

$$u(x - a) = \int_{-\infty}^{t} \delta(t - a)\, dt = \begin{cases} 1 & \text{for } t > a \\ 0 & \text{for } t < a \end{cases} \tag{A1.27}$$

The key to the behavior of the unit step is the sign of the argument. When $t - a$ is positive, the unit step is unity. Elsewhere it is zero except at $t - a = 0$. For this reason the unit step is not an even function; that is, $u(t - a) \neq u(a - t)$. Rather,

$$u(t - a) = 1 - u(a - t)$$

Also

$$u(t - a)u(t - b) = u(t - b) \qquad \text{when } b > a$$

The unit step is sometimes written $\mathbf{1}(x - a)$ and is shown in Fig. A1.5. This function also has a sifting property. Suppose in Fig. A1.6 that the switch closes at $t = a$ and opens at $t = b$.

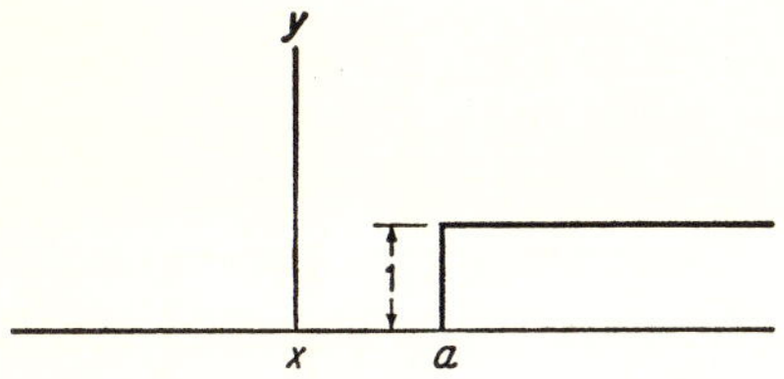

FIG. A1.5

FIG. A1.6

We may write

$$g(t) = \begin{cases} f(t) & a \leq t \leq b \\ 0 & \text{elsewhere} \end{cases} \tag{A1.28a}$$

or

$$\begin{aligned} g(t) &= f(t)[u(t - a) - u(t - b)] \\ &= f(t)[u(t - a)u(b - t)] \end{aligned} \tag{A1.28b}$$

Correspondingly we may write an integral as

$$\int_a^{\infty} f(x)\, dx = \int_{-\infty}^{\infty} f(x)u(x - a)\, dx \tag{A1.29}$$

Now let us show how with time scaling initial conditions remain invariant. If Fig. A1.7*a* is our circuit, the δ-function input can by Eq. (A1.27) be replaced by the initial

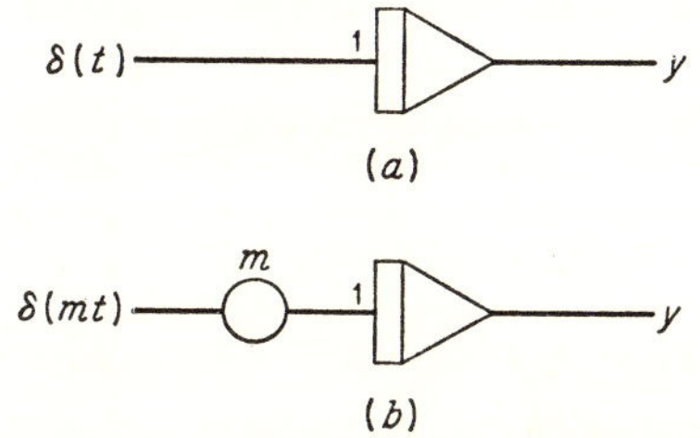

FIG. A1.7

condition $y = -u(t)$. If, however, we time-scale in Fig. A1.7*b* by putting a gain of m at the input to integrator 1, we must change the δ function to $\delta(mt)$. By Eq. (A1.8), y remains unchanged.

Relation between Initial Conditions and the δ Functions. Of some interest is the equivalence between initial conditions and various order derivatives of δ functions in linear differential equations. Zadeh solved the case for the constant-coefficient equation [139].[1] His work was later extended to the variable-coefficient differential equation by Solodov [141], whose results are enumerated below.

Assume that we have the differential equation

$$\sum_{i=0}^{n} a_i p^i y = 0 \tag{A1.30}$$

with initial conditions at $t = \alpha$, $y_0, y_1, \ldots, y_{n-1}$, where

$$y_j = \left(\frac{d^j y}{dt^j}\right)_{t=\alpha}$$

Then its alternative form, with all initial conditions zero, is

$$\sum_{i=0}^{n} a_i p^i y = G_0\delta(t - \alpha) + G_1\delta^{(1)}(t - \alpha) + \cdots + G_{n-1}\delta^{(n-1)}(t - \alpha) \tag{A1.31}$$

where

$$\delta^{(k)}(t - \alpha) = \frac{d^k\delta}{dt^k}$$

We can then show that

$$\begin{aligned} G_0 = {} & [a_1(\alpha) - a_2^{(1)}(\alpha) + \cdots + (-1)^{n-1}a_n^{(n-1)}(\alpha)]y_0 \\ & + [a_2(\alpha) - a_3^{(1)}(\alpha) + \cdots + (-1)^{n-2}a_n^{(n-2)}(\alpha)]y_1 \\ & + \cdots + a_n(\alpha)y_{n-1} \end{aligned} \tag{A1.32a}$$

$$\begin{aligned} G_1 = {} & -\{[a_2(\alpha) - 2a_3^{(1)}(\alpha)\, 3a_4^{(2)}(\alpha) - \cdots + (-1)^{n-1}(n - 1)a_n^{(n-2)}(\alpha)]y_0 \\ & + [a_3(\alpha) - 2a_4^{(1)}(\alpha) + \cdots + (-1)^{n-2}(n - 2)a_n^{(n-3)}(\alpha)]y_1 \\ & + \cdots + a_n(\alpha)y_{n-2}\} \end{aligned} \tag{A1.32b}$$

$$G_{n-1} = (-1)^{n-1}a_n(\alpha)y_0 \tag{A1.32c}$$

For the general coefficient the form is

$$G_m = \sum_{i=0}^{n-m-1} \sum_{k=0}^{n-m-i-1} (-1)^k a_{k+m+i+1}^{(k)} y_i \tag{A1.32d}$$

where

$$a_j^{(k)}(\alpha) = \frac{d^k a_j(\alpha)}{d\alpha^k}$$

In the constant-coefficient equation the relationship simplifies to

$$G_k = \sum_{i=k}^{n-1} a_{i+1} y_{i-k} \tag{A1.32e}$$

The Unit Impulse and the Laplace Transform. We can use the δ function in conjunction with the Laplace transform to find the weighting function for a constant-coefficient equation. The weighting function is the solution of the equation

$$\sum_{k=0}^{n} a_k p^k y = 0 \tag{A1.33}$$

with all initial conditions zero except

$$y_{n-1} = \frac{1}{a_n}$$

[1] For a simple derivation of this case see Ref. 140, p. 26.

The self-same weighting function can be found if we solve the equation

$$\sum_{k=0}^{n} a_k p^k y = \delta(t) \tag{A1.34}$$

However, if we take the Laplace transform of Eq. (A1.34), we find that we are solving Eq. (A1.33) with an initial condition $1/2a_{n-1}$.

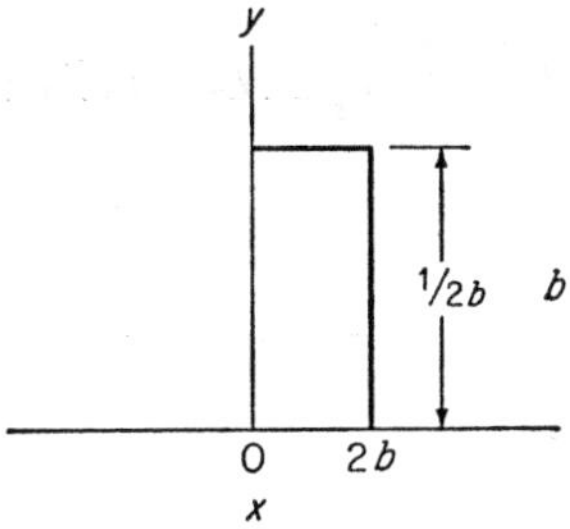

FIG. A1.8

The reason for this is that the Laplace transform applies for the region $0 < t < \infty$. Therefore, only half the area of the δ function is used (see Fig. A1.1). Engineers overcome this difficulty by redefining the δ function and calling it a unit impulse, as shown in Fig. A1.8.

APPENDIX

2 Properties of the Weighting and Green's Functions and the Adjoint Equation [1,2]

In Sec. 2.4.1 we showed that the weighting function can be used in a convolution integral to obtain the solution of the inhomogeneous linear differential equation. These are three similar functions: *the single-sided Green's function, the weighting function,* also called *the impulse-response function,* and *the classical Green's function,* which are used in the convolution integral. Because all these functions arise from the same differential equation, there is an inevitable confusion about the extent of applicability of each function. This appendix is written principally to dispel that uncertainty.

Single-sided Green's Function. The single-sided Green's function is the solution of the inhomogeneous equation

$$a_n \frac{d^n y}{dx^n} + a_{n-1} \frac{d^{n-1} y}{dx^{n-1}} + \cdots + a_1 \frac{dy}{dx} + a_0 y = \delta(x - \alpha) \tag{A2.1}$$

with all initial conditions, $y^{(0)}(\alpha)$, $y^{(1)}(\alpha)$, . . . , $y^{(n-1)}(\alpha)$, zero, or equivalently is the solution of the homogeneous equation

$$a_n \frac{d^n y}{dx^n} + a_{n-1} \frac{d^{n-1} y}{dx^{n-1}} + \cdots + a_1 \frac{dy}{dx} + a_0 y = 0 \tag{A2.2}$$

with initial conditions on y and its first $n - 2$ derivatives all zero and with

$$y^{(n-1)}(\alpha) = \frac{1}{a_n(\alpha)}$$

The solution

$$y = K(x, \alpha)$$

is a function of two variables α and x, where:

1. α is the value of the independent variable at which initial conditions are applied (or at which the δ function is not zero).
2. x is the value of the independent variable at which y is evaluated.

[1] There is much more to the Green's function than we discuss in this appendix. A recent article by Zadeh [145] describes the major advances since 1950.

[2] We use the same nomenclature as Miller in Ref. 144. We have also utilized some of the ideas in his article.

The important properties of the single-sided Green's function are:

1. At $x = \alpha$

   a. $\frac{\partial^r}{\partial^r x} K(x, \alpha) = 0$ for $r = 0, 1, \ldots, n - 2$

   b. $\frac{\partial^{n-1}}{\partial x^{n-1}} K(x, \alpha) = \frac{1}{a_n(\alpha)}$

2. $\frac{\partial^r}{\partial x^r} [K(x, \alpha)]$ is continuous for all $r = 0, 1, \ldots, n - 1$.
3. $K(x, \alpha)$ is unique.
4. There are no restrictions on the values of α relative to x. The function $K(x, \alpha)$ can be nonzero for $\alpha > x$ as well as $\alpha < 0$.

Property 4 distinguishes the single-sided Green's function from the impulsive response.

Impulsive Response (Weighting Function). We can best describe the impulsive response by showing how it differs from the Green's function. In Eq. (A2.1) we obtain the Green's functions $K(x, \alpha)$ by applying the δ function at $x = \alpha$ and solving the problem for *both* increasing and decreasing values of x. By comparison we obtain the impulsive response $W(x, \alpha)$ by running the problem *only* for increasing values of x.

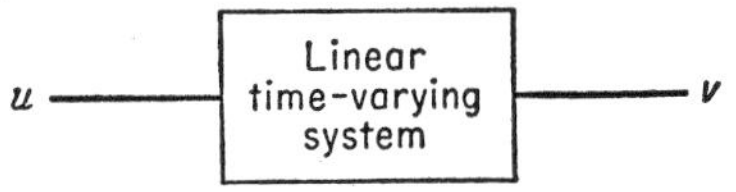

FIG. A2.1

When they both come from the same differential equation, the relation between the single-sided Green's function and the impulsive response is

$$W(x, \alpha) = K(x, \alpha) \qquad x \geq \alpha \tag{A2.3a}$$

$$W(x, \alpha) = 0 \qquad x < \alpha \tag{A2.3b}$$

Therefore, with the exception of property 4, $W(x, \alpha)$ has the same properties as those enumerated for the Green's function.

Property 1 is of particular interest because it may be used to determine the order of the differential equation from which either $W(x, \alpha)$ or $K(x, \alpha)$ came.[1]

The impulsive response may also be defined by the behavior of the linear time-varying system in Fig. A2.1. When the input u is $\delta(t - \alpha)$, the output v is the impulse response $W(t, \alpha)$. From the physical nature of things, when our independent variable is time, $V = 0$ for $\alpha > t$, because we cannot have an output before we have an input. Such a system is said to be nonanticipatory, and as a result the impulsive response is sometimes called the nonanticipatory Green's function.

A linear time-varying system may have the form

$$a_n \frac{d^n v_1}{dt^n} + a_{n-1} \frac{d^{n-1} v_1}{dt^{n-1}} + \cdots + a_1 \frac{dv_1}{dt} + a_0 v_1 = b_m \frac{d^m u}{dt^m} + b_{m-1} \frac{d^{m-1} u}{dt^{m-1}} + \cdots + b_1 \frac{du}{dt} + b_0 u \tag{A2.4a}$$

or, in operator notation,

$$L(v_1) = M(u) \tag{A2.4b}$$

[1] For additional work on the synthesis of differential equations from the impulsive response, see Ref. 145, pp. 1492–1494, and Ref. 149.

Taking $u = \delta(t - \alpha)$, then only for the equations

$$L(v_2) = \delta(t - \alpha) \tag{A2.5}$$

i.e., for $m = 0$, is the Green's function defined for Eq. (A2.4*b*). However, the impulsive response is not so restricted. There is a relationship between the impulsive response of Eqs. (A2.4*b*) and (A2.5), which we show in our discussion of adjoints.

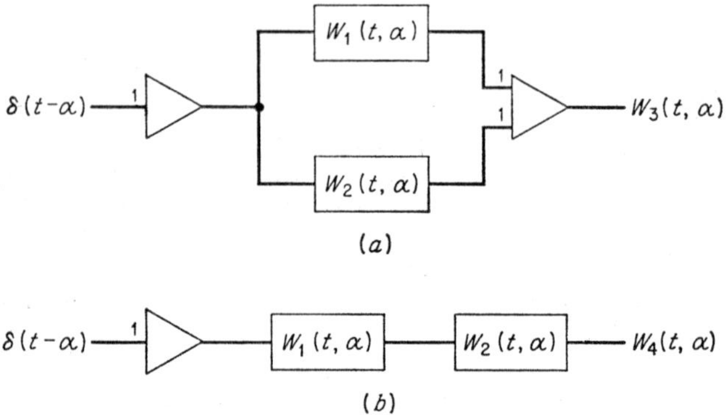

FIG. A2.2

Before going to that subject let us consider how weighting functions combine in parallel and cascade. Figure A2.2*a* and *b* show symbolically these two situations. Then

$$W_3(t, \alpha) = W_1(t, \alpha) + W_2(t, \alpha) \tag{A2.5a}$$

and

$$W_4(t, \alpha) = \int_\alpha^t W_2(t, \beta) W_1(\beta, \alpha)\, d\beta \tag{A2.5b}$$

Adjoint Operators and Their Properties.[1] Given the linear differential operator

$$L = a_n \frac{d^n}{dt^n} + a_{n-1} \frac{d^{n-1}}{dt^{n-1}} + \cdots + a_1 \frac{d}{dt} + a_0 \tag{A2.6a}$$

the adjoint operator is[2]

$$L^* = (-1)^n \frac{d^n}{dt^n} a_n + (-1)^{n-1} \frac{d^{n-1}}{dt^{n-1}} a_{n-1} + \cdots + (-1) \frac{d}{dt} a_1 + a_0 \tag{A2.6b}$$

Our principal interest in adjoint operators, as we showed in Chap. 2, is that the computer more conveniently generates an impulse response of a differential equation using the adjoint of the equation instead of the equation itself. However, adjoints have some additional properties worth relating.

If in Eqs. (A2.5) the impulse response is

$$v_2 = W(t, \alpha)$$

[1] For additional information about adjoint operators and the classical Green's function, see Ref. 24, pp. 123–125, 210–218, 254–270, and Ref. 45, pp. 42–44, 134–194.

[2] The asterisk will be used to indicate the adjoint of an operator.

then in Eq. (A2.4*b*) for an arbitrary u we may write

$$v_1(t) = \int_{-\infty}^{t} W(t, \alpha) M[u]\, d\alpha \tag{A2.7}$$

After some manipulations we obtain[1]

$$v_1(t) = \int_{\infty}^{t} M^*[W(t, \alpha)] u(\alpha)\, d\alpha \tag{A2.8}$$

Thus we see that the integral is the convolution of the input $u(\alpha)$ and an equivalent impulse response $M^*[W(t, \alpha)]$, where

$$M^*[W(t, \alpha)] = \sum_{r=0}^{m} (-1)^r \frac{\partial^r}{\partial t^r} [b_r(t) W(t, \alpha)] \tag{A2.9}$$

We also note that if the single-sided Green's function in Eqs. (A2.5) is

$$v_2 = K(t, \alpha)$$

then the solution of Eq. (A2.4*b*) is

$$v_1(t) = \int_{-\infty}^{\infty} K(t, \alpha) M[u]\, d\alpha = \int_{-\infty}^{\infty} M^*[K(t, \alpha)] u(\alpha)\, d\alpha \tag{A2.10}$$

However, $M^*[K(t, \alpha)]$ is not a Green's function because it does not satisfy property 1.

Another interesting property of adjoints stems from Green's equation. If u and v are any two differentiable functions and L and L^* are a differential operator and its adjoint, then by Green's equation

$$vL(u) - uL^*(v) = \frac{d}{dt} [P(u, v)] \tag{A2.11}$$

The term $P(u, v)$ is called the *bilinear concomitant*. With L defined by Eq. (A2.6*a*),

$$\begin{aligned} P(u, v) = {} & u \left[a_1 v - \frac{d}{dt} (a_2 v) + \cdots + (-1)^{n-1} \frac{d^{n-1}}{dt^{n-1}} (a_n v) \right] \\ & + \frac{du}{dt} \left[a_2 v - \frac{d}{dt} (a_3 v) + \cdots + (-1)^{n-2} \frac{d^{n-2}}{dt^2} (a_n v) \right] + \cdots + a_n v \frac{d^{n-1} u}{dt^{n-1}} \end{aligned} \tag{A2.12}$$

Equation (A2.11) can be used to show that any solution of the homogeneous adjoint equation

$$L^*(v) = 0$$

is an integrating factor for the inhomogeneous equation

$$L(u) = f(t)$$

Take one solution as $v_1(t)$. Then, since $Lv_1 = 0$, we may write Eq. (A2.11) as

$$v_1 L(u) = \frac{d}{dt} [P(u, v_1)] = v_1 f(t) \tag{A2.13}$$

At $t = 0$ we take all initial conditions $u^{(0)}, u^{(1)}, \ldots, u^{(n-1)}$ as zero. Then, integrating (A2.13),

$$\int_0^t \frac{d}{d\tau} [P(u, v_1)]\, d\tau = \int_0^t v_1 f(\tau)\, d\tau \tag{A2.14a}$$

or

$$P(u, v_1) \Big|_0^t = \int_0^t v_1 f(\tau)\, d\tau \tag{A2.14b}$$

[1] This result comes from Ref. 143. Borskii also proves several other important properties in this same article.

Since, at $t = 0$, $P(u, v_1) = 0$, we may write

$$P(u, v_1) = \int_0^t v_1 f(\tau)\, d\tau \tag{A2.14c}$$

As we see in Eq. (A2.12), $P(u, v)$ contains the first $n - 1$ derivatives of u. Thus in Eq. (A2.14c) we have an equation of order $n - 1$ to solve for u. If we had all n solutions, $v_1, v_2, \ldots, v_n$, of the homogeneous adjoint equations, we would be able to get n equations:

$$\begin{aligned} P(u, v_1) &= \int_0^t v_1 f(\tau)\, d\tau \\ P(u, v_2) &= \int_0^t v_2 f(\tau)\, d\tau \\ &\cdots\cdots\cdots\cdots \\ P(u, v_n) &= \int_0^t v_n f(\tau)\, d\tau \end{aligned} \tag{A2.15}$$

We could then solve for u in this set of equations. By a similar technique we may also solve the homogeneous variable-coefficient equation [24, p. 124].

Lest the reader look to this procedure as a means of solving variable-coefficient equations, we point out that it is no easier to get solutions of the homogeneous adjoint equation than solutions of the original homogeneous equation.

Classical Green's Function. The classical Green's function differs from the other two functions already described in that it is related to the two-point boundary-value problem, whereas the others are associated with initial-value problems. For Eq. (A2.1) the classical Green's function $H(x, \alpha)$ must satisfy boundary conditions at a and b of the form

$$U_i(a) = \alpha_{0i} u^{(0)}(a) + \alpha_{1i} u^{(1)}(a) + \cdots + \alpha_{n-1,i} u^{(n-1)}(a) = 0 \qquad i = 0, 1, 2, \ldots, r$$
$$U_j(b) = \alpha_{0j} u^{(0)}(b) + \alpha_{1j} u^{(1)}(b) + \cdots + \alpha_{n-1,j} u^{(n-1)}(b) = 0 \qquad j = r + 1, r + 2, \ldots, n - 1$$

The properties of the classical Green's function are:

1. $\dfrac{\partial^r}{\partial x^r} H(x, \alpha)$ is continuous for $r = 0, 1, \ldots, n - 2$.
2. $\dfrac{\partial^{n-1}}{\partial x^{n-1}} H(x, \alpha) = \dfrac{1}{a_n(\alpha)}$.
3. $H(x, \alpha)$ is unique.
4. The only restriction on x and α is $a \le x; \alpha \le b$.

APPENDIX

3 Short-circuit Transfer Impedances[†]

Transfer impedance function	Network	Relations	Inverse relations
A	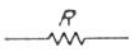	$A = R$	$R = A$
$\dfrac{A}{1 + pT}$	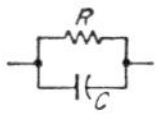	$A = R$ $T = RC$	$R = A$ $C = \dfrac{T}{A}$
$A(1 + pT)$	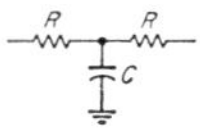	$A = 2R$ $T = \dfrac{RC}{2}$	$R = \dfrac{A}{2}$ $C = \dfrac{4T}{A}$
$A\,\dfrac{1 + p\theta T}{1 + pT}$ $\theta < 1$	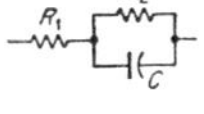	$A = R_1 + R_2$ $T = R_2C$ $\theta = \dfrac{R_1}{R_1 + R_2}$	$R_1 = A\theta$ $R_2 = A(1 - \theta)$ $C = \dfrac{T}{A(1 - \theta)}$
	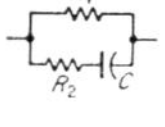	$A = R_1$ $T = (R_1 + R_2)C$ $\theta = \dfrac{R_2}{R_1 + R_2}$	$R_1 = A$ $R_2 = \dfrac{A\theta}{1 - \theta}$ $C = \dfrac{T(1 - \theta)}{A}$

† This table was compiled by S. Godet of Reeves Instrument Company, New York. It is reproduced by permission of McGraw-Hill Book Company, Inc., from the article Driftless D-C Amplifier, by F. R. Bradley and R. McCoy, *Electronics*, April, 1952.

Transfer impedance function	Network	Relations	Inverse relations
	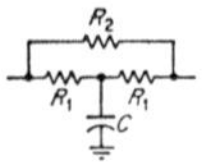	$A = \dfrac{2R_1R_2}{2R_1 + R_2}$ $T = \dfrac{R_1C}{2}$ $\theta = \dfrac{2R_1}{2R_1 - R_2}$	$R_1 = \dfrac{A}{2(1 - \theta)}$ $R_2 = \dfrac{A}{\theta}$ $C = \dfrac{4T(1 - \theta)}{A}$
$A\,\dfrac{1 + pT}{1 + p\theta T}$ $\theta < 1$	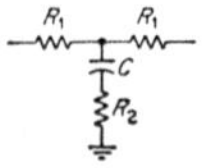	$A = 2R_1$ $T = \left(R_2 + \dfrac{R_1}{2}\right) C$ $\theta = \dfrac{2R_2}{2R_2 + R_1}$	$R_1 = \dfrac{A}{2}$ $R_2 = \dfrac{A\theta}{4(1 - \theta)}$ $C = \dfrac{4T(1 - \theta)}{A}$
	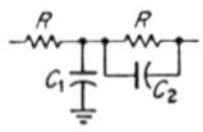	$A = 2R$ $T = \dfrac{R}{2}\,(C_1 + C_2)$ $\theta = \dfrac{2C_2}{C_1 + C_2}$	$R = \dfrac{A}{2}$ $C_1 = \dfrac{2T(2 - \theta)}{A}$ $C_2 = \dfrac{2T\theta}{A}$
	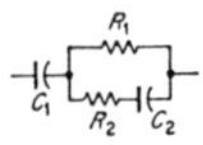	$B = C_1$ $T_2 = (R_1 + R_2)C_2$ $T_1T_3 = R_1R_2C_1C_2$ $T_1 + T_3 = R_1C_1 + R_2C_2 + R_1C_2$	$R_1 = \dfrac{T_1 + T_3 - T_2}{B}$ $R_2 = \dfrac{T_1T_3(T_1 + T_3 - T_2)}{B(T_3 - T_2)(T_2 - T_1)}$ $C_1 = B$ $C_2 = \dfrac{B(T_3 - T_2)(T_2 - T_1)}{(T_1 + T_3 - T_2)^2}$
$\dfrac{1}{pB} \times \dfrac{(1+pT_1)(1+pT_3)}{1 + pT_2}$ $T_1 < T_2 < T_3$	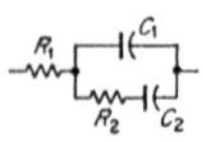	$B = C_1 + C_2$ $T_2 = R_2\,\dfrac{C_1C_2}{C_1 + C_2}$ $T_1T_3 = R_1R_2C_1C_2$ $T_1 + T_3 = R_1C_1 + R_2C_2 + R_1C_2$	$R_1 = \dfrac{T_1T_3}{BT_2}$ $R_2 = \dfrac{(T_1T_2 + T_2T_3 - T_1T_3)^2}{BT_2(T_3 - T_2)(T_2 - T_1)}$ $C_1 = \dfrac{BT_2^2}{T_1T_2 + T_2T_3 - T_1T_3}$ $C_2 = \dfrac{B(T_3 - T_2)(T_2 - T_1)}{T_1T_2 + T_2T_3 - T_1T_3}$
	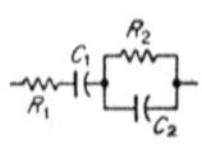	$B = C_1$ $T_2 = R_2C_2$ $T_1T_3 = R_1R_2C_1C_2$ $T_1 + T_3 = R_1C_1 + R_2C_2 + R_2C_1$	$R_1 = \dfrac{T_1T_3}{BT_2}$ $R_2 = \dfrac{(T_3 - T_2)(T_2 - T_1)}{BT_2}$ $C_1 = B$ $C_2 = \dfrac{BT_2^2}{(T_3 - T_2)(T_2 - T_1)}$

Transfer impedance function	Network	Relations	Inverse relations
$\frac{1}{pB} \times \frac{(1+pT_1)(1+pT_3)}{1+pT_2}$ $T_1 < T_2 < T_3$	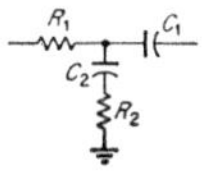	$B = C_1$ $T_2 = R_2C_2$ $T_1T_3 = R_1R_2C_1C_2$ $T_1 + T_3 = R_1C_1 + R_2C_2 + R_1C_2$	$R_1 = \frac{T_1T_3}{BT_2}$ $R_2 = \frac{T_1T_2T_3}{B(T_3 - T_2)(T_2 - T_1)}$ $C_1 = B$ $C_1 = \frac{B(T_3 - T_2)(T_2 - T_1)}{T_1T_3}$
$\frac{1}{pB} \times (1+pT_1)(1+pT_2)$ $T_1 \neq T_2$	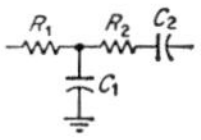	$B = C_2$ $T_1T_2 = R_1R_2C_1C_2$ $T_1 + T_2 = R_1C_1 + R_2C_2 + R_1C_2$	$R_1 = \frac{(\sqrt{T_1} - \sqrt{T_2})^2}{B}$ $R_2 = \frac{\sqrt{T_1T_2}}{B}$ $C_1 = \frac{B\sqrt{T_1T_2}}{(\sqrt{T_1} - \sqrt{T_2})^2}$ $C_2 = B$
$\frac{1}{pB} \times \frac{(1+pT_1)(1+pT_2)}{p\sqrt{T_1T_2}}$ $T_1 \neq T_2$	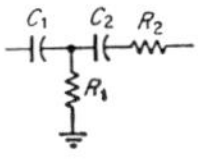	$B = C_2$ $T_1T_2 = R_1R_2C_1C_2$ $T_1 + T_2 = R_1C_1 + R_2C_2 + R_1C_2$	$R = \frac{(\sqrt{T_1} - \sqrt{T_2})^2}{B}$ $R_2 = \frac{\sqrt{T_1T_2}}{B}$ $C_1 = \frac{B\sqrt{T_1T_2}}{(\sqrt{T_1} - \sqrt{T_2})^2}$ $C_2 = B$
$\frac{1}{pB} \times \frac{(1+pT_1)(1+pT_2)}{p^2T_1T_2}$ $T_1 < T_2$	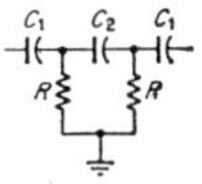	$B = \frac{C_1C_2}{C_1 + 2C_2}$ $T_1 = RC_1$ $T_2 = R(C_1 + 2C_2)$	$R = \frac{T_1(T_2 - T_1)}{2BT_2}$ $C_1 = \frac{2BT_2}{T_2 - T_1}$ $C_2 = \frac{BT_2}{T_1}$
$A\frac{1 + pT_1}{1 + p^2T_1T_2}$	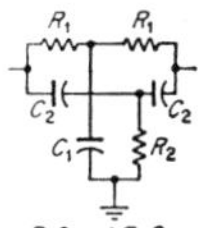 $R_1C_1 = 4R_2C_2$	$A = 2R_1$ $T_1 = \frac{R_1C_1}{2} = 2R_2C_2$ $T_2 = R_1C_2$	$R_1 = \frac{A}{2}$ $R_2 = \frac{AT_1}{4T_2}$ $C_1 = \frac{4T_1}{A}$ $C_2 = \frac{2T_2}{A}$
$\frac{1}{pB}$	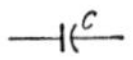	$B = C$	$C = B$

Transfer impedance function	Network	Relations	Inverse relations
$\frac{1}{pB}(1 + pT)$	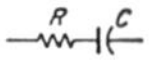	$B = C$ $T = RC$	$R = \frac{T}{B}$ $C = B$
$\frac{1}{pB}\frac{1 + pT}{pT}$	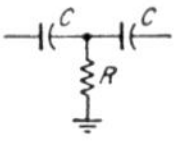	$B = \frac{C}{2}$ $T = 2RC$	$R = \frac{T}{4B}$ $C = 2B$
$\frac{1}{pB}\frac{1 + pT}{1 + p\theta T}$ $\theta < 1$	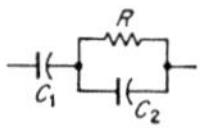	$B = C_1$ $T = R(C_1 + C_2)$ $\theta = \frac{C_2}{C_1 + C_2}$	$R = \frac{T(1 - \theta)}{B}$ $C_1 = B$ $C_2 = \frac{B\theta}{1 - \theta}$
	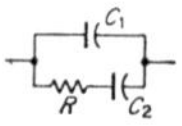	$B = C_1 + C_2$ $T = RC_2$ $\theta = \frac{C_1}{C_1 + C_2}$	$B = \frac{T}{B(1 - \theta)}$ $C_1 = B\theta$ $C_2 = B(1 - \theta)$
$\frac{1}{pB} \times \frac{(1+pT_1)(1+pT_3)}{1 + pT_2}$ $T_1 < T_2 < T_3$	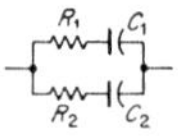	$B = C_1 + C_2$ $T_1 = R_1C_1$ $T_2 = (R_1 + R_2)\frac{C_1C_2}{C_1 + C_2}$ $T_3 = R_2C_2$	$R_1 = \frac{T_1(T_3 - T_1)}{B(T_2 - T_1)}$ $R_2 = \frac{T_3(T_3 - T_1)}{B(T_3 - T_2)}$ $C_1 = \frac{B(T_2 - T_1)}{T_3 - T_1}$ $C_2 = \frac{B(T_3 - T_2)}{T_3 - T_1}$
$A \times \frac{1 + pT_1}{1 + pT_1 + p^2T_1T_2}$	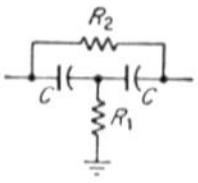	$A = R_2$ $T_1 = 2R_1C$ $T_2 = \frac{R_2C}{2}$	$R_1 = \frac{AT_1}{4T_2}$ $R_2 = A$ $C = \frac{2T_2}{A}$
$A \times \frac{1 + pT_2}{1 + pT_1 + p^2T_1T_2}$	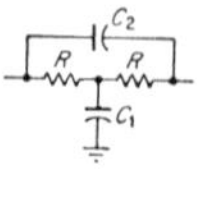	$A = 2R$ $T_1 = 2RC_2$ $T_2 = \frac{RC_1}{2}$	$R = \frac{A}{2}$ $C_1 = \frac{4T_2}{A}$ $C_2 = \frac{T_1}{A}$

Transfer impedance function	Network	Relations	Inverse relations
$A \times \dfrac{1 + pT_3}{1 + pT_1 + p^2T_1T_2}$ $T_2 > \dfrac{T_1}{4}$ (complex roots) $T_3 > T_2$	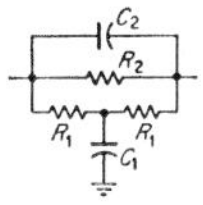	$A = \dfrac{2R_1R_2}{(2R_1 + R_2)}$ $T_1 = \dfrac{R_1(R_1C_1 + 2R_2C_2)}{2R_1 + R_2}$ $T_2 = \dfrac{R_1R_2C_1C_2}{R_1C_1 + 2R_2C_2}$ $T_3 = \dfrac{R_1C_1}{2}$	$R_1 = \dfrac{AT_3^2}{2[T_3^2 - T_1(T_3 - T_2)]}$ $R_2 = \dfrac{AT_3^2}{T_1(T_3 - T_2)}$ $C_1 = \dfrac{4[T_3^2 - T_1(T_3 - T_2)]}{AT_3}$ $C_2 = \dfrac{T_1T_2}{AT_3}$
	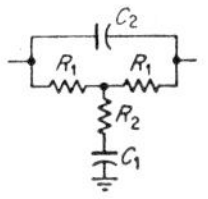	$A = 2R_1$ $T_1 = R_2C_1 + 2R_1C_2$ $T_2 = \dfrac{R_1(R_1 + 2R_2)C_1C_2}{R_2C_1 + 2R_1C_2}$ $T_3 = \left(R_2 + \dfrac{R_1}{2}\right)C_1$	$R_1 = \dfrac{A}{2}$ $R_2 = \dfrac{AT_1(T_3 - T_2)}{4[T_3^2 - T_1(T_3 - T_2)]}$ $C_1 = \dfrac{4[T_3^2 - T_1(T_3 - T_2)]}{AT_3}$ $C_2 = \dfrac{T_1T_2}{AT_3}$
	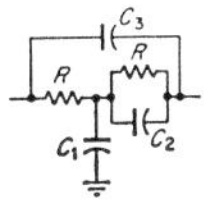	$A = 2R$ $T_1 = R(C_2 + 2C_3)$ $T_2 = \dfrac{RC_3(C_1 + C_2)}{C_2 + 2C_3}$ $T_3 = \dfrac{R}{2}(C_1 + C_2)$	$R = \dfrac{A}{2}$ $C_1 = \dfrac{2[2T_3^2 - T_1(T_3 - T_2)]}{AT_3}$ $C_2 = \dfrac{2T_1(T_3 - T_2)}{AT_3}$ $C_3 = \dfrac{T_1T_2}{AT_3}$
$A \times \dfrac{1 + pT_3}{1 + pT_1 + p^2T_1T_2}$ $T_2 > \dfrac{T_1}{4}$ (complex roots) $T_3 < T_1$	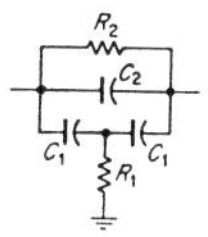	$A = R_2$ $T_1 = 2R_1C_1 + R_2C_2$ $T_2 = \dfrac{R_1R_2C_1(C_1 + 2C_2)}{2R_1C_1 + R_2C_2}$ $T_3 = 2R_1C_1$	$R_1 = \dfrac{AT_3^2}{4[T_1T_2 - T_3(T_1 - T_3)]}$ $R_2 = A$ $C_1 = \dfrac{2[T_1T_2 - T_3(T_1 - T_3)]}{AT_3}$ $C_2 = \dfrac{T_1 - T_3}{A}$
	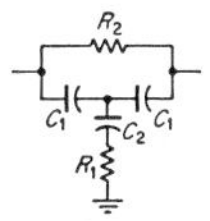	$A = R_2$ $T_1 = \dfrac{C_1(2R_1C_2 + R_2C_1)}{2C_1 + C_2}$ $T_2 = \dfrac{R_1R_2C_1C_2}{2R_1C_2 + R_2C_1}$ $T_3 = \dfrac{2R_1C_1C_2}{2C_1 + C_2}$	$R_1 = \dfrac{AT_3^2}{4[T_1T_2 - T_3(T_1 - T_3)]}$ $R_2 = A$ $C_1 = \dfrac{2T_1T_2}{AT_3}$ $C_2 = \dfrac{4T_1T_2[T_1T_2 - T_3(T_1 - T_3)]}{AT_3 2(T_1 - T_3)}$

Transfer impedance function	Network	Relations	Inverse relations
$A \times \dfrac{1 + pT_3}{1 + pT_1 + p^2T_1T_2}$ $T_2 > \dfrac{T_1}{4}$ (complex roots) $T_3 < T_1$	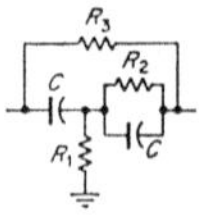	$A = R_3$ $T_1 = \dfrac{R_1(2R_2 + R_3)C}{R_1 + R_2}$ $T_2 = \dfrac{R_2R_3C}{2R_2 + R_3}$ $T_3 = \dfrac{2R_1R_2C}{R_1 + R_2}$	$R_1 = \dfrac{AT_3^2}{2[2T_1T_2 - T_3(T_1 - T_3)]}$ $R_2 = \dfrac{AT_3}{2(T_1 - T_3)}$ $R_3 = A$ $C = \dfrac{2T_1T_2}{AT_3}$
$A(1 + pT_1) \times (1 + pT_2)$ $T_1 < T_2$	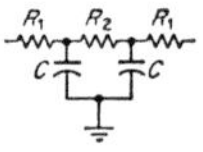	$A = 2R_1 + R_2$ $T_1 = \dfrac{R_1R_2}{2R_1 + R_2} C$ $T_2 = R_1C$	$R_1 = A \dfrac{T_2 - T_1}{2T_2}$ $R_2 = A \dfrac{T_1}{T_2}$ $C = \dfrac{2T_2^2}{A(T_2 - T_1)}$
$\dfrac{1}{pB} \dfrac{1 + p\theta T}{1 + pT}$ $\theta < 1$	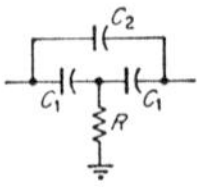	$B = C_2$ $T = RC_1 \dfrac{2C_2 + C_1}{C_2}$ $\theta = \dfrac{2C_2}{2C_2 + C_1}$	$R = \dfrac{T\theta^2}{4B(1 - \theta)}$ $C_1 = \dfrac{2B(1 - \theta)}{\theta}$ $C_2 = B$
	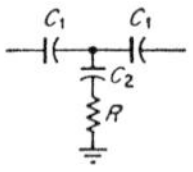	$B = \dfrac{C_1^2}{2C_1 + C_2}$ $T = RC_2$ $\theta = \dfrac{2C_1}{2C_1 + C_2}$	$R = \dfrac{T\theta^2}{4B(1 - \theta)}$ $C_1 = \dfrac{2B}{\theta}$ $C_2 = \dfrac{4B(1 - \theta)}{\theta^2}$
	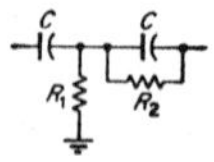	$B = \dfrac{R_1}{R_1 + R_2} C$ $T = R_2C$ $\theta = \dfrac{2R_1}{R_1 + R_2}$	$R_1 = \dfrac{T\theta^2}{2B(2 - \theta)}$ $R_2 = \dfrac{T\theta}{2B}$ $C = \dfrac{2B}{\theta}$
$A \times \dfrac{1 + pT_2}{(1+pT_1)(1+pT_3)}$ $T_1 < T_2 < T_3$	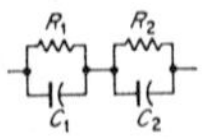	$A = R_1 + R_2$ $T_1 = R_1C_1$ $T_2 = \dfrac{R_1R_2}{R_1 + R_2} (C_1 + C_2)$ $T_3 = R_2C_2$	$R_1 = \dfrac{A(T_2 - T_1)}{T_3 - T_1}$ $R_2 = \dfrac{A(T_3 - T_2)}{T_3 - T_1}$ $C_1 = \dfrac{T_1(T_3 - T_1)}{A(T_2 - T_1)}$ $C_2 = \dfrac{T_3(T_3 - T_1)}{A(T_3 - T_2)}$

Transfer impedance function	Network	Relations	Inverse relations
$A \times \dfrac{1 + pT_2}{(1+pT_1)(1+pT_3)}$ $T_1 < T_2 < T_3$	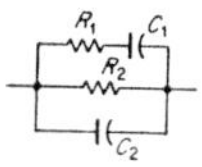	$A = R_2$ $T_2 = R_1C_1$ $T_1T_3 = R_1R_2C_1C_2$ $T_1 + T_3 = R_1C_1 + R_2C_2 + R_2C_1$	$R_1 = \dfrac{AT_2^2}{(T_3 - T_2)(T_2 - T_1)}$ $R_2 = A$ $C_1 = \dfrac{(T_3 - T_2)(T_2 - T_1)}{AT_2}$ $C_2 = \dfrac{T_1T_3}{AT_2}$
	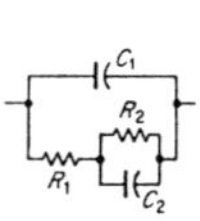	$A = R_1 + R_2$ $T_2 = \dfrac{R_1R_2}{R_1 + R_2} C_2$ $T_1T_3 = R_1R_2C_1C_2$ $T_1 + T_3 = R_1C_1 + R_2C_2 + R_2C_1$	$R_1 = \dfrac{T_2^2}{T_1T_2 + T_2T_3 - T_1T_3}$ $R_2 = \dfrac{A(T_3 - T_2)(T_2 - T_1)}{T_1T_2 + T_2T_3 - T_1T_3}$ $C_1 = \dfrac{T_1T_3}{AT_2}$ $C_2 = \dfrac{(T_1T_2 + T_2T_3 - T_1T_3)^2}{AT_2(T_3 - T_2)(T_2 - T_1)}$
	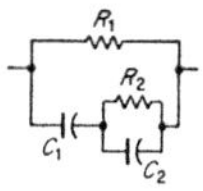	$A = R_1$ $T_2 = R_2(C_1 + C_2)$ $T_1T_3 = R_1R_2C_1C_2$ $T_1 + T_3 = R_1C_1 + R_2C_2 + R_2C_1$	$R_1 = A$ $R_2 = \dfrac{A(T_3 - T_2)(T_2 - T_1)}{(T_1 + T_3 - T_2)^2}$ $C_1 = \dfrac{T_1 + T_3 - T_2}{A}$ $C_2 = \dfrac{T_1T_3(T_1 + T_3 - T_2)}{A(T_3 - T_2)(T_2 - T_1)}$
$A \times \dfrac{1 + pT_2}{(1+pT_1)(1+pT_3)}$ $T_2 \leq T_1 \leq T_3$	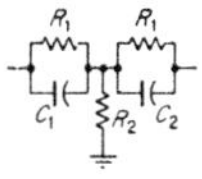	$A = 2R_1 + \dfrac{R_1^2}{R_2}$ $T_1 = R_1C_1$ $T_2 = \dfrac{R_1R_2}{R_1 + 2R_2} (C_1 + C_2)$ $T_3 = R_1C_2$	$R_1 = \dfrac{AT_2}{T_1 + T_3}$ $R_2 = \dfrac{AT_2^2}{(T_1 + T_3)(T_1 + T_3 - 2T_2)}$ $C_1 = \dfrac{T_1(T_1 + T_3)}{AT_2}$ $C_2 = \dfrac{T_3(T_1 + T_3)}{AT_2}$
$A \times \dfrac{1 + pT_2}{(1+pT_1)(1+pT_3)}$ $T_1 \leq T_3 \leq T_2$	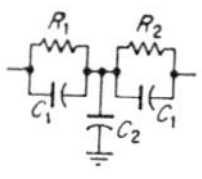	$A = R_1 + R_2$ $T_1 = R_1C_1$ $T_2 = \dfrac{R_1R_2}{R_1 + R_2} (2C_1 + C_2)$ $T_3 = R_2C_1$	$R_1 = \dfrac{AT_1}{T_1 + T_3}$ $R_2 = \dfrac{AT_3}{T_1 + T_3}$ $C_1 = \dfrac{T_1 + T_3}{A}$ $C_2 = \dfrac{T_1 + T_3}{A}\left(\dfrac{T_2}{T_3} + \dfrac{T_2}{T_1} - 2\right)$

APPENDIX

4 A Background in Probability and Random Processes[1]

Two events, A and B, are said to be mutually exclusive if the occurrence of one of the events precludes the occurrence of the other. For example, the events "heads on a toss of a coin" and "tails on the same toss of the coin" are mutually exclusive events. A function P of event A is called a probability if:

Axiom 1. $0 \leq P(A) \leq 1$. $P(A)$ = probability of A occurring.

Axiom 2. $P(A \text{ or } B) = P(A) + P(B)$ if A and B are mutually exclusive.

Axiom 3. $P(\Omega)$ = probability of any possible event = 1.

Example 1. Let S be the 36 possible outcomes of a throw of dice. Let $P(B)$ for any event $B = n_B/36$. n_B = number of ways of obtaining B. For example, take the specific event: A = event "two same faces." Then $P(A) = \frac{6}{36}$.

Note that the probability function has the same characteristics as does the function representing the mass of a physical body with the exception that the probability function is normalized with respect to 1. We can then interpret the probability function as being a "mass measure" of events. This interpretation will aid us in the understanding of the mean and variance of a function. For the moment, however, let us consider another interpretation.

Concept of an Experiment. The idea underlying the word *probability* is that of uncertainty. This concept of uncertainty is so broad that, in order to develop some mathematical theory or theories of probability, it is necessary to distinguish between various kinds of uncertainty. To do so in this text would lead us astray from our main theme. Hence, we shall discuss briefly only two kinds of expressions of uncertainty. These two kinds can be typified by the statements:

1. There is a 50-50 chance that, if I wait on the corner of Gauss and Euclid Avenues on May 26, 1984, John will arrive by 2 P.M.
2. There is a 50-50 chance that, if I toss a coin, it will come up heads.

Statements 1 and 2 both express uncertainty about the outcome of events A and B, respectively, where

$A \equiv$ event that John will arrive at G and E Avenues on May 26, 1984, by 2 P.M.
$B \equiv$ event that outcome of toss of a coin will be heads

The characteristic which distinguishes the events A and B is that the question as to whether or not A will occur can be cleared up once and for all just by waiting until 2 P.M., May 26, 1984, whereas the question concerning the outcome of a coin toss can never be

[1] This appendix was written by A. Levine.

finally cleared up, for once the coin is tossed and the outcome decided, the outcome of the next toss is still in question. To rephrase the distinction, one can think of a sequence of events such as B; that is, one can consider conceptually the sequence

$$B = B_1, B_2, B_3, \ldots$$

where

$$B_i = \text{outcome of } i\text{th toss}$$

On the other hand, one cannot consider a sequence of events such as A, for once 2 P.M. is past on the appointed day, one of the events "John arrived," "John did not arrive," has occurred. It is not possible again to stand on the corner and wait on May 26, 1984, until 2 P.M. Succinctly, the difference between the events A and B is that A cannot be repeated, whereas B can. To understand why this difference is so important, we must determine what we mean when we say that an event has a 50-50 chance of occurring, or that the probability of an event is $\frac{1}{2}$. In the case of coin tossing, this statement means that in a large number of tosses the event "heads" should occur in "approximately" one-half of the tosses. If the probability of heads were $\frac{2}{3}$, then we would interpret this statement to mean that "heads" is more likely than "tails" in the sense that in a large number of tosses approximately two-thirds of the tosses will be heads.

Certainly we cannot interpret the probability statement 1 in the same way. To say that John's arrival has probability $\frac{1}{2}$ cannot mean that in a large number of tries in which I wait for John on May 26, 1984, until 2 P.M. on G and E corners, John makes it only half the time; once 2 P.M., May 26, 1984, has passed, the issue of John's arrival has been decided. As another example, consider the statement "the probability that the moon has another side is $\frac{2}{3}$." This statement cannot mean that, in a long series of trials in which we travel to the other side of the moon, approximately two-thirds of the time the moon has another side. We need travel to the moon only once to determine whether or not it has another side. Once determined, the uncertainty disappears. Thus, the "frequency" interpretation of such statements as (1) makes no sense. We assume that the number called "probability of an event" represents the frequency of occurrence of the event in a long sequence of trials.

Some important probability functions are the following:

Binomial:

$$P(S_n = x) = \binom{n}{x} p^x q^{n-x}$$

$$S_n = x = \text{event ``}x\text{ successes in }n\text{ trials''}$$

$$p = \text{probability of success on a single trial}$$

$$q = 1 - p$$

$$\binom{n}{x} = \frac{n!}{x!(n-x)!}$$

Poisson:

$$P(X_t = k) = \frac{e^{-\lambda t}(\lambda t)^k}{k!}$$

$$X_t = k \equiv \text{event } k \text{ successes in time } t$$

$$\lambda = \text{const}$$

Normal:

$$P(a < X < b) = \int_a^b \frac{e^{-(x-u)^2/2\sigma^2}}{\sqrt{2\pi}\,\sigma}\,dx$$

$$a < X < b \equiv \text{event the quantity } X \text{ lies between two numbers } a \text{ and } b$$

$$\mu, \sigma^2 = \text{const}$$

χ^2 (chi-squared):

$$P(a < \chi^2 < b) = \int_a^b \frac{(\chi^2)^{n/2-1}e^{-\chi^2/2}}{2^{n/2}\Gamma(n/2)}\,d\chi^2 \qquad \chi^2 \geq 0$$

Definition of *conditional probability:*

$P(R|H) \equiv$ probability of event R occurring, given that event H has occurred

Definition of *joint probability:*

$P(RH) \equiv$ probability that *both* event R and event H have occurred

Definition of *statistical independence:*

If $P(R|H) \equiv P(R)$, then R is said to be independent of H.

The relationship between joint and conditional probability is

$$P(RH) = P(R|H)P(H)$$

Example 2. Let

n_W = total number of women in population
n_R = total number of rich women
n_H = total number of red-headed women
n_{RH} = total number of rich, red-headed women

Then

$$P(R) = \frac{n_R}{n_W}$$

$$P(H) = \frac{n_H}{n_W}$$

$$P(RH) = \frac{n_{RH}}{n_W}$$

$$P(R|H) = \frac{n_{RH}}{n_H} = \frac{n_{RH}}{n_W}\frac{n_W}{n_H} = \frac{P(RH)}{P(H)}$$

Theorem. Two events A and B are independent *if and only if*

$$P(AB) = P(A)P(B)$$

Two important theorems using conditional probabilities are:

1. *Total-probability rule.* Let $H_1, H_2, \ldots$ be mutually exclusive events such that they include every possible outcome of the experiment. Then

$$P(B) = \sum_i P(H_i)P(B|H_i)$$

where B is any event.

2. *Bayes' rule*

$$P(A|B) = \frac{P(AB)}{P(B)} = \frac{P(AB)}{\sum_i P(H_i)P(B|H_i)} = \frac{P(A)P(B|A)}{\sum_i P(H_i)P(B|H_i)}$$

Random Variables. To each random event $S_1, S_2, \ldots$ it is possible to assign a number. Hence, if the number x is assigned to the two events S_1, S_2, then

$$P(X = x) = P(S_1 \text{ or } S_2)$$

The variable X is called a random variable henceforth and can assume all the numbers assigned to the random events. We shall discuss probabilities only in terms of random variables.

Expected Value. Interpreting $P(X < x)$ as the probability mass of the event $(X < x)$, we define

$$p(x) = \frac{d}{dx}[P(X < x)]$$

(when it exists) as the *probability density*. We note that (Axiom 3)

$$P(X < \infty) = 1$$

if

$$\int_{-\infty}^{\infty} p(x)\, dx = 1$$

and

$$\int_{a}^{b} p(x)\, dx = P(a < X < b)$$

We define the *center of mass* of events in the same way as we define the center of mass of a physical body:

$$\text{CM} = \frac{\int_{-\infty}^{\infty} xp(x)\, dx}{\int_{-\infty}^{\infty} p(x)\, dx}$$

In this case,

$$\text{CM} = \int_{-\infty}^{\infty} xp(x)\, dx$$

is called the *expected value* or *mean* of the random variable and is indicated by the symbol $E(X)$.

The term *expected value* arises from our "frequency interpretation" of the quantity $p(x)$. If $p(x)$ represents or is proportional to the "frequency" of occurrence of the value of x in a long sequence of trials, then $E(X)$ represents the average value of x over the sequence of trials. Averages of important distributions are given below:

Binomial:
$$E(X) = \sum_{x=0}^{n} x \binom{n}{x} p^x q^{n-x} = np$$

Poisson:
$$E(X) = \sum_{k=0}^{\infty} k \frac{e^{-\lambda t}(\lambda t)^k}{k!} = \lambda t$$

Normal:
$$E(X) = \int_{-\infty}^{\infty} x \frac{e^{-(x-u)^2/2\sigma^2}}{\sqrt{2\pi}\,\sigma} = \mu$$

Variance. If again we interpret $p(x)$ as the probability mass density of x_1, we can define the "moment of inertia" (MI) of events as we do with physical bodies:

$$\text{MI} = \int_{-\infty}^{\infty} \frac{(x - \theta)^2 p(x)\, dx}{\int_{-\infty}^{\infty} p(x)\, dx} = \int_{-\infty}^{\infty} (x - \theta)^2 p(x)\, dx \qquad \theta = E(x)$$

The moment of inertia represents the *spread* of the mass about the center of mass. Using the frequency interpretation, we say that it is the average spread about the expected value. Hence in probability theory we give it the name of *variance*, or $V(x)$. Some variances of important distributions are:

Binomial:
$$V(X) = \sum_{x=0}^{n} (x - np)^2 \binom{n}{x} p^x q^{n-x} = npq$$

Poisson:
$$V(X) = \sum_{k=0}^{n} \frac{(k - \lambda t)^2 e^{-\lambda t}(\lambda t)^k}{k!} = \lambda t$$

Normal:
$$V(X) = \int_{-\infty}^{\infty} \frac{(x - \mu)^2 e^{-(x-u)^2/2\sigma^2}}{\sqrt{2\pi}\,\sigma} = \sigma^2$$

Some important theorems concerning expected values and variances are:

THEOREM 1. Let $S_n = \sum_{i=1}^{n} X_i$ be the sum of a random variable. Then

$$E(S_n) = \sum_{i=1}^{n} E(X_i)$$

THEOREM 2. $E(a) = a$, $a = \text{const.}$

THEOREM 3. $E(X + a) = E(X) + a$, $a = \text{const.}$

The following theorems concerning the variance follow as direct consequences of Theorems 1 to 3:

THEOREM 4*a*. $V(X) = E(X^2) - \theta^2$, $\theta = E(X)$.

THEOREM 4*b*. $V(aX) = a^2V(X)$, $a = \text{const.}$

THEOREM 4*c*. $V(aX + b) = a^2V(X)$; $a, b = \text{const.}$

THEOREM 4*d*. If $X_1, X_2, \ldots, X_n$ are *independent* random variables, then

$$V(S_n) = \sum_{i=1}^{n} V(X_i)$$

THEOREM 5. If Y is normally distributed with mean μ and variance σ^2, then

$$\frac{Y - \mu}{\sigma}$$

is distributed $N(0, 1)$, that is, is normally distributed with zero mean and unit variance.

An extremely important and useful theorem is Theorem 6, the central limit theorem.

THEOREM 6. If $X_1, X_2, \ldots, X_n$ are independent random variables, each variable having finite mean and finite variance, then

$$S_n = \sum_{i=1}^{n} X_i$$

is approximately normally distributed with expected value (mean) $E(S_n)$ and variance $V(S_n)$, the approximation becoming better as n increases.

THEOREM 7. If X_1 has the Poisson distribution

$$\frac{e^{-\lambda t}(\lambda t)^k}{k!}$$

and if X_2 has the same distribution, then $X_1 + X_2$ has the distribution

$$\frac{e^{-2\lambda t}(2\lambda t)^k}{k!}$$

Moment-generating Function. An important function of use in calculating expected values is $E(e^{\theta X})$, the moment-generating function. Since

$$e^{\theta X} = \sum_{i=0}^{\infty} \frac{(\theta X)^i}{i!}$$

we note that

$$E(e^{\theta X}) = \sum_{i=0}^{\infty} \frac{\theta^i E(X^i)}{i!}$$

Hence

$$\frac{\partial^i}{\partial \theta^i} E(e^{\theta X}) \bigg|_{\theta=0} = E(X^i)$$

Moment-generating functions of some important distributions are

Binomial: $$\sum_{x=0}^{\infty} e^{\theta x} \binom{n}{x} p^x q^{n-x} = (q + pe^{\theta})^n$$

Poisson: $$\sum_{k=0}^{\infty} \frac{e^{\theta k}(\lambda t)^k e^{-\lambda t}}{k!} = e^{-\lambda t(1-e^{\theta})}$$

Normal: $$\int_{-\infty}^{\infty} \frac{e^{\theta x} e^{-(x-u)^2/2\sigma^2}}{\sqrt{2\pi}\,\sigma}\,dx = e^{ut+\sigma^2 t^2/2}$$

Random Processes. A random process is a time sequence of random variables. An n-dimensional joint-probability density of the random process which is a function only of the time difference between the variables is said to be *stationary* in the *strict sense.* For example, if $(y_1, t_1; \ldots)$ denotes $y(t)$ evaluated at t_1, then strict stationarity implies

$$p(y_1, t_1; y_2, t_2; \ldots ; y_n, t_n) = p(y_1, t_1 + t; y_2, t_2 + t; \ldots ; y_n, t_n + t) \qquad \text{for all } t$$

If
$$E[y(t_1)y(t_2)] = E[y(t_1 + t)y(t_2 + t)]$$

for all t, then second moments are invariant under translation and the process is said to be *stationary* in the *wide sense.* For the gaussian distribution which is completely defined by its second moments, strict-sense stationarity and wide-sense stationarity are equivalent.

Time and Ensemble Averages. A *time average* is an average of a function of a single event taken over all time. An *ensemble average* occurs where the average is taken at specified times over all events in the random process. The *ergodic theorem* states that time and ensemble averages are equal when the process is ergodic. Stationarity is a necessary condition of ergodicity.[1]

For example, the second moment for an ergodic process has the equivalence

$$\int_{-\infty}^{\infty}\int_{-\infty}^{\infty} y(t_2)y(t_2)p[y(t_1)y(t_2)]\,dy(t_1)\,dy(t_2) = \lim_{T\to\infty} \frac{1}{2T}\int_{-T}^{T} y(t_1 + t)y(t_2 + t)\,dt$$

The ergodic property is very important for computer operators. Where it holds, the data may be taken in the time average of a single computer run instead of in the ensemble averaging of several systems programmed to run simultaneously.

Correlation Function and Spectral Density. The *correlation function* $R_y(\tau)$ for a stationary process is a real-valued function defined as

$$R_y(\tau) = \lim_{T\to\infty} \frac{1}{2T}\int_{-T}^{T} y(t)y(t + \tau)\,dt$$

It is related to the *spectral density* $S_y(f)$ in the Wiener-Khintchine relationship by the Fourier transform pair

$$S_y(f) = \int_{-\infty}^{\infty} R_y(\tau)e^{-j\omega\tau}\,d\tau$$
$$R_y(\tau) = \int_{-\infty}^{\infty} S_y(f)e^{j\omega\tau}\,df$$

[1] For a detailed discussion of the ergodic theorem and its implications, see Ref. 114, pp. 56–62.

Properties of the Correlation Function (Stationary Processes)

1. $R(\tau) = R(-\tau)$
2. $R(0) \geq |R(\tau)|$
3. $R(0) = \lim_{T\to\infty} \frac{1}{2T} \int_{-T}^{T} y^2(t)\,dt = \overline{y^2}$
4. $\lim_{\tau\to\infty} R_y(\tau) = E(y^2)$
5. Relation between the correlation coefficient $\rho_y(\tau)$ and the correlation function

$$\rho_y(\tau) = \frac{R_y(\tau) - E^2(y)}{\sigma_y^{\,2}}$$

Spectral Density

1. $S_y(f) = S_y(-f)$
2. $\overline{y^2} = \int_{-\infty}^{\infty} S_y(f)\,df$

APPENDIX

5 Median and Mid-range Estimates of the Mean[1]

N	Median		Mid-range	
	Variance	E	Variance	E
2	0.500	1.000	0.500	1.000
3	0.449	0.743	0.362	0.920
4	0.298	0.838	0.298	0.838
5	0.287	0.697	0.261	0.767
6	0.215	0.776	0.236	0.706
7	0.210	0.679	0.218	0.654
8	0.168	0.743	0.205	0.610
9	0.166	0.669	0.194	0.572
10	0.138	0.723	0.186	0.539
11	0.137	0.663	0.178	0.510
12	0.118	0.709	0.172	0.484
13	0.117	0.659	0.167	0.461
14	0.102	0.699	0.162	0.440
15	0.102	0.656	0.158	0.422
16	0.0904	0.692	0.154	0.392
17	0.0901	0.653	0.151	0.389
18	0.0810	0.686	0.148	0.375
19	0.0808	0.651	0.145	0.362
20	0.0734	0.681	0.143	0.350
∞	$1.57/N$	0.637	.....	0.000

[1] From "Introduction to Statistical Analysis," 2d ed., p. 406, by W. J. Dixon and F. J. Massey, Jr. Copyright © 1957. McGraw-Hill Book Company, Inc., New York. Used by permission.

6 Cumulative

$$\text{Area} = \frac{1}{\sqrt{2\pi}\,\sigma}\int_{-\infty}^{x} \exp\left[\frac{-(y-\mu)^2}{2\sigma^2}\right] dy \qquad \text{where } x = \mu + z\sigma$$

z	x	Area	z	x	Area
−3.25	$\mu - 3.25\sigma$	.0006	−1.00	$\mu - 1.00\sigma$	.1587
−3.20	$\mu - 3.20\sigma$	.0007	− .95	$\mu - .95\sigma$	.1711
−3.15	$\mu - 3.15\sigma$	.0008	− .90	$\mu - .90\sigma$	.1841
−3.10	$\mu - 3.10\sigma$	.0010	− .85	$\mu - .85\sigma$	.1977
−3.05	$\mu - 3.05\sigma$	.0011	− .80	$\mu - .80\sigma$	.2119
−3.00	$\mu - 3.00\sigma$	.0013	− .75	$\mu - .75\sigma$	.2266
−2.95	$\mu - 2.95\sigma$	.0016	− .70	$\mu - .70\sigma$	.2420
−2.90	$\mu - 2.90\sigma$	.0019	− .65	$\mu - .65\sigma$	.2578
−2.85	$\mu - 2.85\sigma$	.0022	− .60	$\mu - .60\sigma$	.2743
−2.80	$\mu - 2.80\sigma$	.0026	− .55	$\mu - .55\sigma$	.2912
−2.75	$\mu - 2.75\sigma$	.0030	− .50	$\mu - .50\sigma$	.3085
−2.70	$\mu - 2.70\sigma$	.0035	− .45	$\mu - .45\sigma$	.3264
−2.65	$\mu - 2.65\sigma$	.0040	− .40	$\mu - .40\sigma$	.3446
−2.60	$\mu - 2.60\sigma$	.0047	− .35	$\mu - .35\sigma$	.3632
−2.55	$\mu - 2.55\sigma$	.0054	− .30	$\mu - .30\sigma$	.3821
−2.50	$\mu - 2.50\sigma$	.0062	− .25	$\mu - .25\sigma$	.4013
−2.45	$\mu - 2.45\sigma$	.0071	− .20	$\mu - .20\sigma$	.4207
−2.40	$\mu - 2.40\sigma$	.0082	− .15	$\mu - .15\sigma$	.4404
−2.35	$\mu - 2.35\sigma$	.0094	− .10	$\mu - .10\sigma$	.4602
−2.30	$\mu - 2.30\sigma$	.0107	− .05	$\mu - .05\sigma$	.4801
−2.25	$\mu - 2.25\sigma$	.0122			
−2.20	$\mu - 2.20\sigma$	.0139			
−2.15	$\mu - 2.15\sigma$	.0158	.00	μ	.5000
−2.10	$\mu - 2.10\sigma$	.0179			
−2.05	$\mu - 2.05\sigma$	.0202			
−2.00	$\mu - 2.00\sigma$	.0228	.05	$\mu + .05\sigma$	.5199
−1.95	$\mu - 1.95\sigma$	.0256	.10	$\mu + .10\sigma$	.5398
−1.90	$\mu - 1.90\sigma$	.0287	.15	$\mu + .15\sigma$	.5596
−1.85	$\mu - 1.85\sigma$	.0322	.20	$\mu + .20\sigma$	.5793
−1.80	$\mu - 1.80\sigma$	.0359	.25	$\mu + .25\sigma$	.5987
−1.75	$\mu - 1.75\sigma$	.0401	.30	$\mu + .30\sigma$	.6179
−1.70	$\mu - 1.70\sigma$	.0446	.35	$\mu + .35\sigma$	.6368
−1.65	$\mu - 1.65\sigma$	.0495	.40	$\mu + .40\sigma$	.6554
−1.60	$\mu - 1.60\sigma$	.0548	.45	$\mu + .45\sigma$	.6736
−1.55	$\mu - 1.55\sigma$	.0606	.50	$\mu + .50\sigma$	.6915
−1.50	$\mu - 1.50\sigma$	.0668	.55	$\mu + .55\sigma$	.7088
−1.45	$\mu - 1.45\sigma$	.0735	.60	$\mu + .60\sigma$	.7257
−1.40	$\mu - 1.40\sigma$	.0808	.65	$\mu + .65\sigma$	.7422
−1.35	$\mu - 1.35\sigma$	.0885	.70	$\mu + .70\sigma$	.7580
−1.30	$\mu - 1.30\sigma$	.0968	.75	$\mu + .75\sigma$	.7734
−1.25	$\mu - 1.25\sigma$	.1056	.80	$\mu + .80\sigma$	.7881
−1.20	$\mu - 1.20\sigma$	.1151	.85	$\mu + .85\sigma$	.8023
−1.15	$\mu - 1.15\sigma$	.1251	.90	$\mu + .90\sigma$	.8159
−1.10	$\mu - 1.10\sigma$	.1357	.95	$\mu + .95\sigma$	.8289
−1.05	$\mu - 1.05\sigma$	.1469	1.00	$\mu + 1.00\sigma$	.8413

Normal Distribution[1]

z	x	Area	z	x	Area
1.05	$\mu + 1.05\sigma$	.8531	−4.265	$\mu - 4.265\sigma$	.00001
1.10	$\mu + 1.10\sigma$	.8643	−3.719	$\mu - 3.719\sigma$	.0001
1.15	$\mu + 1.15\sigma$	.8749	−3.090	$\mu - 3.090\sigma$	.001
1.20	$\mu + 1.20\sigma$	.8849	−2.576	$\mu - 2.576\sigma$	.005
1.25	$\mu + 1.25\sigma$	.8944	−2.326	$\mu - 2.326\sigma$	.01
1.30	$\mu + 1.30\sigma$	.9032	−2.054	$\mu - 2.054\sigma$	.02
1.35	$\mu + 1.35\sigma$	.9115	−1.960	$\mu - 1.960\sigma$	.025
1.40	$\mu + 1.40\sigma$	.9192	−1.881	$\mu - 1.881\sigma$	.03
1.45	$\mu + 1.45\sigma$	.9265	−1.751	$\mu - 1.751\sigma$	.04
1.50	$\mu + 1.50\sigma$	.9332	−1.645	$\mu - 1.645\sigma$	.05
1.55	$\mu + 1.55\sigma$	.9394	−1.555	$\mu - 1.555\sigma$	.06
1.60	$\mu + 1.60\sigma$	.9452	−1.476	$\mu - 1.476\sigma$	.07
1.65	$\mu + 1.65\sigma$	.9505	−1.405	$\mu - 1.405\sigma$	.08
1.70	$\mu + 1.70\sigma$	.9554	−1.341	$\mu - 1.341\sigma$	.09
1.75	$\mu + 1.75\sigma$	.9599	−1.282	$\mu - 1.282\sigma$	.10
1.80	$\mu + 1.80\sigma$	.9641	−1.036	$\mu - 1.036\sigma$	.15
1.85	$\mu + 1.85\sigma$	.9678	− .842	$\mu - .842\sigma$	.20
1.90	$\mu + 1.90\sigma$	.9713	− .674	$\mu - .674\sigma$	.25
1.95	$\mu + 1.95\sigma$	.9744	− .524	$\mu - .524\sigma$	.30
2.00	$\mu + 2.00\sigma$	.9772	− .385	$\mu - .385\sigma$	.35
2.05	$\mu + 2.05\sigma$	.9798	− .253	$\mu - .253\sigma$	.40
2.10	$\mu + 2.10\sigma$	.9821	− .126	$\mu - .126\sigma$	.45
2.15	$\mu + 2.15\sigma$	.9842	0	μ	.50
2.20	$\mu + 2.20\sigma$	.9861	.126	$\mu + .126\sigma$	.55
2.25	$\mu + 2.25\sigma$	.9878	.253	$\mu + .253\sigma$	.60
2.30	$\mu + 2.30\sigma$	.9893	.385	$\mu + .385\sigma$	.65
2.35	$\mu + 2.35\sigma$	.9906	.524	$\mu + .524\sigma$	.70
2.40	$\mu + 2.40\sigma$	.9918	.674	$\mu + .674\sigma$	.75
2.45	$\mu + 2.45\sigma$	.9929	.842	$\mu + .842\sigma$	.80
2.50	$\mu + 2.50\sigma$	.9938	1.036	$\mu + 1.036\sigma$	.85
2.55	$\mu + 2.55\sigma$	.9946	1.282	$\mu + 1.282\sigma$	.90
2.60	$\mu + 2.60\sigma$	.9953	1.341	$\mu + 1.341\sigma$	.91
2.65	$\mu + 2.65\sigma$	.9960	1.405	$\mu + 1.405\sigma$	.92
2.70	$\mu + 2.70\sigma$	.9965	1.476	$\mu + 1.476\sigma$	.93
2.75	$\mu + 2.75\sigma$	.9970	1.555	$\mu + 1.555\sigma$	.94
2.80	$\mu + 2.80\sigma$	.9974	1.645	$\mu + 1.645\sigma$	.95
2.85	$\mu + 2.85\sigma$	.9978	1.751	$\mu + 1.751\sigma$	.96
2.90	$\mu + 2.90\sigma$	.9981	1.881	$\mu + 1.881\sigma$	.97
2.95	$\mu + 2.95\sigma$	.9984	1.960	$\mu + 1.960\sigma$	.975
3.00	$\mu + 3.00\sigma$	.9987	2.054	$\mu + 2.054\sigma$	.98
3.05	$\mu + 3.05\sigma$	.9989	2.326	$\mu + 2.326\sigma$	.99
3.10	$\mu + 3.10\sigma$	.9990	2.576	$\mu + 2.576\sigma$	.995
3.15	$\mu + 3.15\sigma$	.9992	3.090	$\mu + 3.090\sigma$	.999
3.20	$\mu + 3.20\sigma$	.9993	3.719	$\mu + 3.719\sigma$	.9999
3.25	$\mu + 3.25\sigma$	.9994	4.265	$\mu + 4.265\sigma$	.99999

[1] From "Introduction to Statistical Analysis," 2d ed., pp. 382–383, by W. J. Dixon and F. J. Massey, Jr. Copyright © 1957. McGraw-Hill Book Company, Inc., New York. Used by permission.

APPENDIX

7 *Percentiles of*

df	100σ											
	.05	.1	.5	1.0	2.5	5.0	10	20	30	40	50	60
1	$.0^{6}39$	$.0^{5}157$	$.0^{4}39$	$.0^{3}16$	$.0^{3}98$	$.0^{2}39$	.016	.064	.148	.275	.455	.708
2	.001	.001	.005	.010	.025	.052	.106	.223	.356	.511	.693	.916
3	.005	.008	.024	.038	.072	.117	.195	.335	.475	.623	.789	.982
4	.016	.023	.052	.074	.121	.178	.266	.412	.549	.688	.839	1.011
5	.032	.042	.082	.111	.166	.229	.322	.469	.600	.731	.870	1.03
6	.050	.064	.113	.145	.206	.272	.367	.512	.638	.762	.891	1.04
7	.069	.085	.141	.177	.241	.310	.405	.546	.667	.785	.907	1.04
8	.089	.107	.168	.206	.272	.342	.436	.574	.691	.803	.918	1.04
9	.108	.128	.193	.232	.300	.369	.463	.598	.710	.817	.927	1.05
10	.126	.148	.216	.256	.325	.394	.487	.618	.727	.830	.934	1.05
11	.144	.167	.237	.278	.347	.416	.507	.635	.741	.840	.940	1.05
12	.161	.184	.256	.298	.367	.436	.525	.651	.753	.848	.945	1.05
13	.177	.201	.274	.316	.385	.453	.542	.664	.764	.856	.949	1.05
14	.193	.217	.291	.333	.402	.469	.556	.676	.773	.863	.953	1.05
15	.207	.232	.307	.349	.418	.484	.570	.687	.781	.869	.956	1.05
16	.221	.246	.321	.363	.432	.498	.582	.697	.789	.874	.959	1.05
17	.234	.260	.335	.377	.445	.510	.593	.706	.796	.879	.961	1.05
18	.247	.272	.348	.390	.457	.522	.604	.714	.802	.883	.963	1.05
19	.258	.285	.360	.402	.469	.532	.613	.722	.808	.887	.965	1.05
20	.270	.296	.372	.413	.480	.543	.622	.729	.813	.890	.967	1.05
22	.291	.317	.393	.434	.499	.561	.638	.742	.823	.897	.970	1.05
24	.310	.337	.412	.452	.517	.577	.652	.753	.831	.902	.972	1.05
26	.328	.355	.429	.469	.532	.592	.665	.762	.838	.907	.974	1.05
28	.345	.371	.445	.484	.547	.605	.676	.771	.845	.911	.976	1.04
30	.306	.386	.460	.498	.560	.616	.687	.779	.850	.915	.978	1.04
35	.394	.420	.491	.529	.588	.642	.708	.795	.862	.922	.981	1.04
40	.423	.448	.518	.554	.611	.663	.726	.809	.872	.928	.983	1.04
45	.448	.472	.540	.576	.630	.680	.741	.820	.880	.933	.985	1.04
50	.469	.494	.560	.594	.647	.695	.754	.829	.886	.937	.987	1.04
55	.488	.512	.577	.610	.662	.708	.765	.837	.892	.941	.988	1.04
60	.506	.529	.592	.625	.675	.720	.774	.844	.897	.944	.989	1.04
70	.535	.558	.618	.649	.697	.739	.790	.856	.905	.949	.990	1.03
80	.560	.582	.640	.669	.714	.755	.803	.865	.911	.952	.992	1.03
90	.581	.602	.658	.686	.729	.768	.814	.873	.917	.955	.993	1.03
100	.599	.619	.673	.701	.742	.779	.824	.879	.921	.958	.993	1.03
120	.629	.648	.699	.724	.763	.798	.839	.890	.929	.962	.994	1.03
140	.653	.671	.719	.743	.780	.812	.850	.898	.934	.965	.995	1.03
160	.673	.690	.736	.758	.793	.824	.860	.905	.939	.968	.996	1.02
180	.689	.706	.749	.771	.804	.833	.868	.910	.942	.970	.996	1.02
200	.703	.719	.761	.782	.814	.841	.874	.915	.945	.972	.997	1.02
250	.732	.746	.785	.804	.832	.858	.887	.924	.951	.975	.997	1.02
300	.753	.767	.802	.820	.846	.870	.897	.931	.956	.977	.998	1.02
350	.770	.783	.816	.833	.857	.879	.904	.936	.959	.979	.998	1.02
400	.784	.796	.827	.843	.866	.887	.911	.940	.962	.981	.998	1.02
450	.795	.807	.837	.852	.874	.893	.916	.944	.964	.982	.999	1.02
500	.805	.816	.845	.859	.880	.898	.920	.946	.966	.983	.999	1.01
750	.839	.848	.872	.884	.901	.917	.934	.956	.972	.986	.999	1.01
1,000	.859	.868	.889	.899	.914	.928	.943	.962	.976	.988	.999	1.01
5,000	.936	.939	.949	.954	.961	.967	.974	.983	.989	.995	1.00	1.00
∞	1	1	1	1	1	1	1	1	1	1	1	1

Read $.0^{3}16$ as .00016, etc.

the χ^2 */ df Distributions*[1]

100σ									df
70	80	90	95	97.5	99	99.5	99.9	99.95	
1.07	1.64	2.71	3.84	5.02	6.64	7.88	10.83	12.12	1
1.20	1.61	2.30	3.00	3.69	4.61	5.30	6.91	7.60	2
1.22	1.55	2.08	2.60	3.12	3.78	4.28	5.42	5.91	3
1.22	1.50	1.94	2.37	2.79	3.32	3.72	4.62	5.00	4
1.21	1.46	1.85	2.21	2.57	3.02	3.35	4.10	4.42	5
1.21	1.43	1.77	2.10	2.41	2.80	3.09	3.74	4.02	6
1.20	1.40	1.72	2.01	2.29	2.64	2.90	3.47	3.72	7
1.19	1.38	1.67	1.94	2.19	2.51	2.74	3.27	3.48	8
1.18	1.36	1.63	1.88	2.11	2.41	2.62	3.10	3.30	9
1.18	1.34	1.60	1.83	2.05	2.32	2.52	2.96	3.14	10
1.17	1.33	1.57	1.79	1.99	2.25	2.43	2.84	3.01	11
1.17	1.32	1.55	1.75	1.94	2.18	2.36	2.74	2.90	12
1.16	1.31	1.52	1.72	1.90	2.13	2.29	2.66	2.81	13
1.16	1.30	1.50	1.69	1.87	2.08	2.24	2.58	2.72	14
1.15	1.29	1.49	1.67	1.83	2.04	2.19	2.51	2.65	15
1.15	1.28	1.47	1.64	1.80	2.00	2.14	2.45	2.58	16
1.15	1.27	1.46	1.62	1.78	1.97	2.10	2.40	2.52	17
1.14	1.26	1.44	1.60	1.75	1.93	2.06	2.35	2.47	18
1.14	1.26	1.43	1.59	1.73	1.90	2.03	2.31	2.42	19
1.14	1.25	1.42	1.57	1.71	1.88	2.00	2.27	2.37	20
1.13	1.24	1.40	1.54	1.67	1.83	1.95	2.19	2.30	22
1.13	1.23	1.38	1.52	1.64	1.79	1.90	2.13	2.23	24
1.12	1.22	1.37	1.50	1.61	1.76	1.86	2.08	2.17	26
1.12	1.22	1.35	1.48	1.59	1.72	1.82	2.03	2.12	28
1.12	1.21	1.34	1.46	1.57	1.70	1.79	1.99	2.07	30
1.11	1.19	1.32	1.42	1.52	1.64	1.72	1.90	1.98	35
1.10	1.18	1.30	1.39	1.48	1.59	1.67	1.84	1.90	40
1.10	1.17	1.28	1.37	1.45	1.55	1.63	1.78	1.84	45
1.09	1.16	1.26	1.35	1.43	1.52	1.59	1.73	1.79	50
1.09	1.16	1.25	1.33	1.41	1.50	1.56	1.69	1.75	55
1.09	1.15	1.24	1.32	1.39	1.47	1.53	1.66	1.71	60
1.08	1.14	1.22	1.29	1.36	1.43	1.49	1.60	1.65	70
1.08	1.13	1.21	1.27	1.33	1.40	1.45	1.56	1.60	80
1.07	1.12	1.20	1.26	1.31	1.38	1.43	1.52	1.56	90
1.07	1.12	1.18	1.24	1.30	1.36	1.40	1.49	1.53	100
1.06	1.11	1.17	1.22	1.27	1.32	1.36	1.45	1.48	120
1.06	1.10	1.16	1.20	1.25	1.30	1.33	1.41	1.44	140
1.06	1.09	1.15	1.19	1.23	1.28	1.31	1.38	1.41	160
1.05	1.09	1.14	1.18	1.22	1.26	1.29	1.36	1.38	180
1.05	1.08	1.13	1.17	1.21	1.25	1.28	1.34	1.36	200
1.04	1.07	1.12	1.15	1.18	1.22	1.25	1.30	1.32	250
1.04	1.07	1.11	1.14	1.17	1.20	1.22	1.27	1.29	300
1.04	1.06	1.10	1.13	1.15	1.18	1.21	1.25	1.27	350
1.04	1.06	1.09	1.12	1.14	1.17	1.19	1.24	1.25	400
1.03	1.06	1.09	1.11	1.13	1.16	1.18	1.22	1.23	450
1.03	1.05	1.08	1.11	1.13	1.15	1.17	1.21	1.22	500
1.03	1.04	1.07	1.09	1.10	1.12	1.14	1.17	1.18	750
1.02	1.04	1.06	1.07	1.09	1.11	1.12	1.14	1.15	1,000
1.01	1.02	1.02	1.03	1.04	1.05	1.05	1.06	1.07	5,000
1	1	1	1	1	1	1	1	1	∞

[1] From "Introduction to Statistical Analysis," 2d ed., pp. 386–387, by W. J. Dixon and F. J. Massey, Jr. Copyright © 1957. McGraw-Hill Book Company, Inc., New York. Used by permission.

To use the χ^2 table for establishing confidence intervals on σ^2, we proceed as follows: Suppose $s^2 = 0.2$, n (the number of samples) $= 10$, and we wish to find a confidence interval on σ^2 with significance $1 - \alpha$. To illustrate, let $\alpha = 0.05$. Then we first find

$$\chi^2_{\alpha/2} = \chi^2_{0.025}$$

from the table by looking down the column marked $100\alpha/2 = 2.5$ until we come to the row indicated by $n - 1$ $(10 - 1)$. Hence,

$$\chi^2_{0.025} = 0.3$$

In addition, we find $\chi^2_{1-\alpha/2}$ by looking down the column marked $100(1 - \alpha/2) = 97.5$ until we come to the row marked $n - 1$. Hence

$$\chi^2_{1-\alpha/2} = \chi^2_{0.975} = 2.11$$

Then

$$\frac{s^2}{\chi^2_{1-\alpha/2}} < \sigma^2 < \frac{s^2}{\chi^2_{\alpha/2}}$$

constitutes the confidence interval on σ^2. In this case $s^2 = 0.2$; thus

$$\frac{0.2}{2.11} < \sigma^2 < \frac{0.2}{0.3}$$

with 95 per cent probability (or confidence).

APPENDIX

8 Confidence Intervals for Binomial Sampling at the 0.95 and 0.99 Confidence Levels[1]

Precise confidence intervals can be found by reading directly from Figs. A8.1 and A8.2, where c is the number of successes and n is the number of samples.

A Maximum Confidence Interval. All that is necessary to find maximum confidence intervals for the 0.95 confidence level is to use Fig. A8.1. Use the intersection of the line $c/n = 0.5$ with the curves marked by the number of samples you have. For example, let the number of samples equal 100. Then

$$\max \delta = \pm \frac{0.60 - 0.40}{2} = \pm 0.1$$

where 0.60 represents the intersection of $c/n = 0.5$ with the lower 100-sample curve and 0.40 represents the intersection with the upper 100-sample curve.

Use Fig. A8.2 to find maximum confidence intervals for the 0.99 level.

[1] Figures reprinted by permission from "Biometrica Tables for Statisticians," edited by E. S. Pearson and H. O. Hartley, 2d ed., vol. 1, pp. 204–205, Cambridge University Press, New York, 1958.

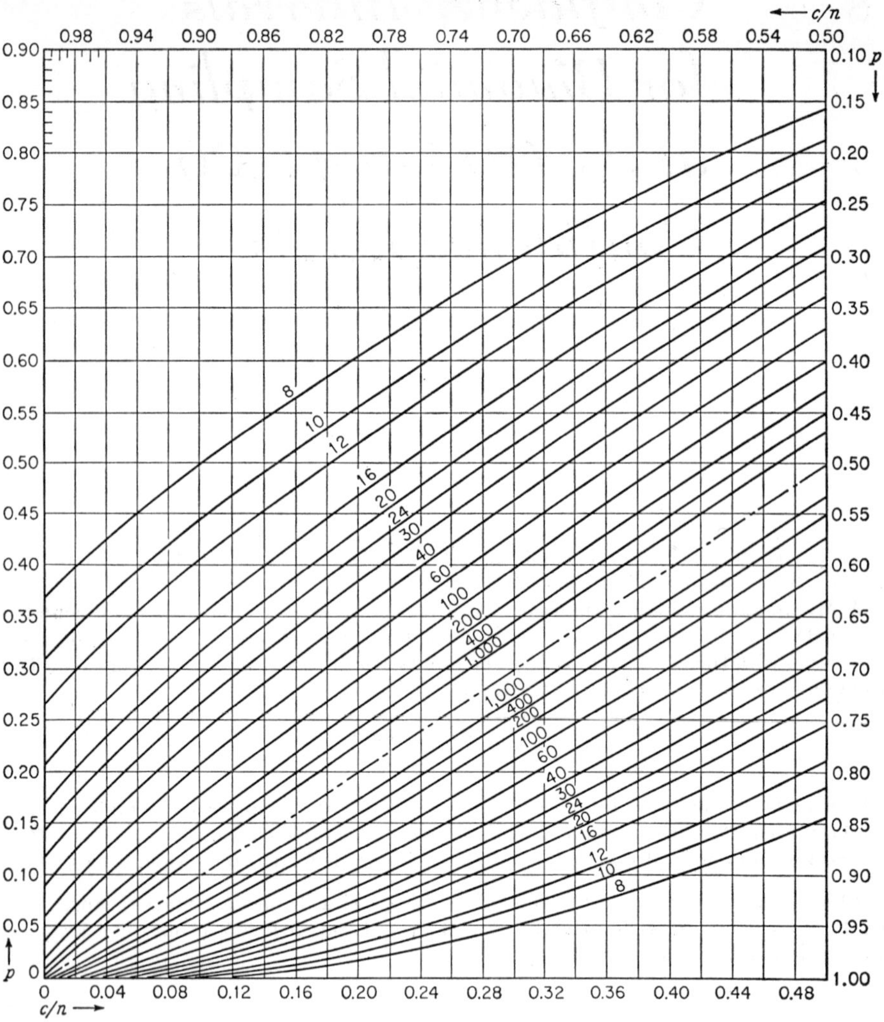

FIG. A8.1 Confidence coefficient, $1 - \alpha = 0.95$.

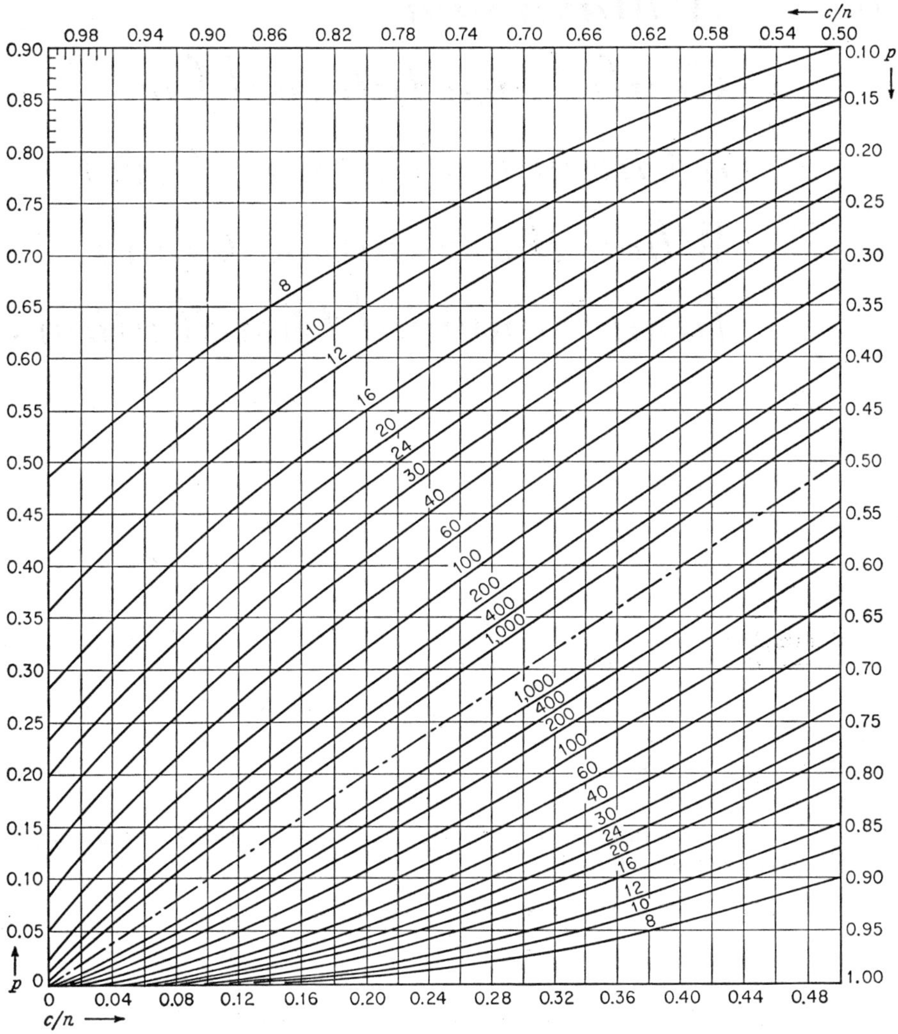

Fig. A8.2 Confidence coefficient, $1 - \alpha = 0.99$.

APPENDIX

9 A Maximum Confidence Interval for a Given Confidence Level When the Samples Come from Binomial Distributions[1]

Each sample x_{ij} comes from a population in which the probability of success is p_i (i represents the ith tactical situation). Let

$$\bar{x} = \frac{\sum_{ij} x_{ij}}{n}$$

where n represents the total number of runs. Since the samples are *independent*, the variance of $\bar{x}$, denoted by $V(\bar{x})$, is

$$V(\bar{x}) = \frac{\sum_{ij} V(x_{ij})}{n} = \frac{\sum_{i=1}^{n_T} p_i(1 - p_i)}{n_r n_T}$$

where n_r is the number of runs per case and n_T is the number of cases. Since

$$V(\bar{x}) = \frac{p_t(1 - p_t)}{n}$$

where p_t is the true average probability of hit, we note that the maximum value of $V(\bar{x})$ is

$$\max V(\bar{x}) = \frac{0.25}{n}$$

Since $n > 30$, we can use the central limit theorem to approximate the distribution of $\bar{x}$ by a normal distribution. Hence,

$$P = P[-\delta < \bar{x} - E(\bar{x}) < \delta]$$

[1] This appendix was written by A. Levine.

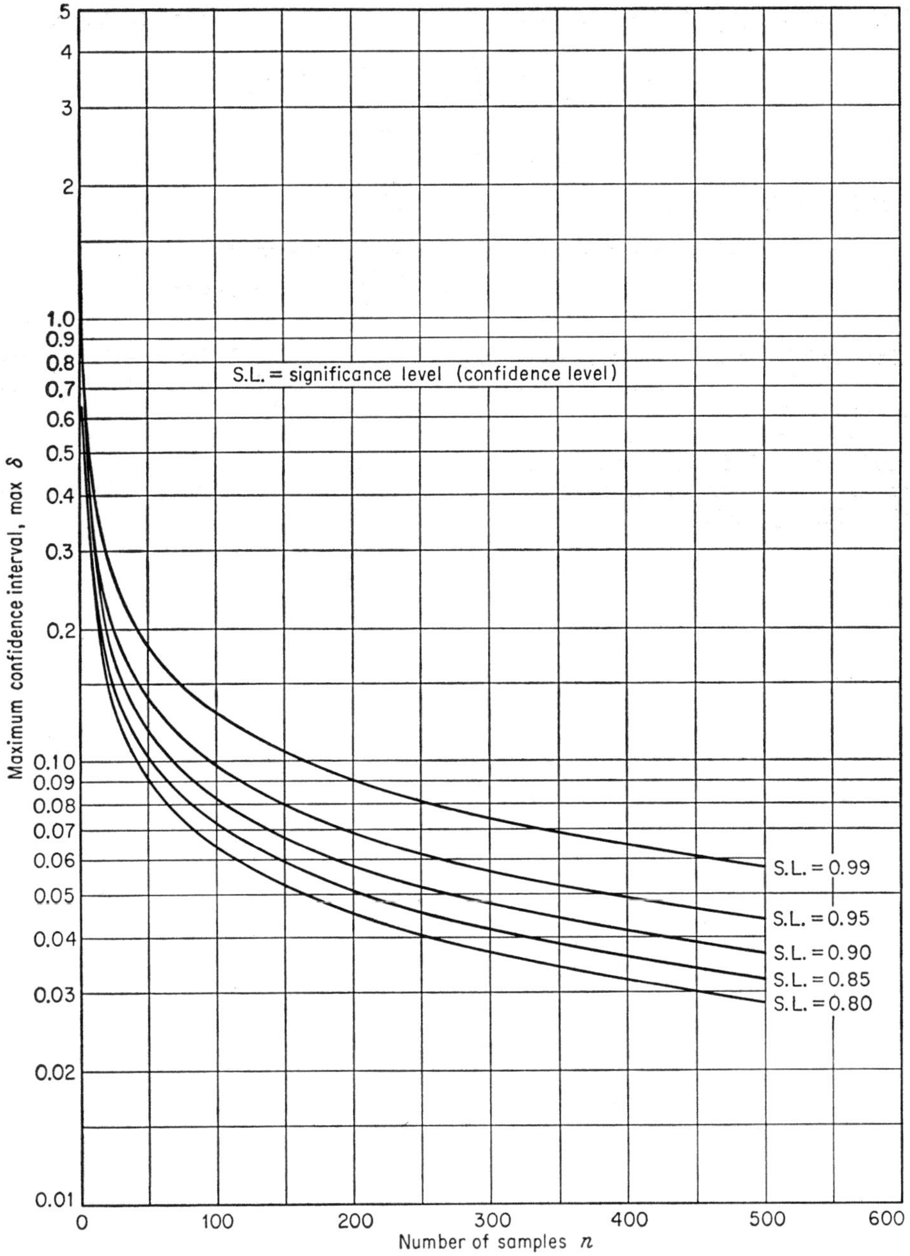
S.L. = significance level (confidence level)
S.L. = 0.99
S.L. = 0.95
S.L. = 0.90
S.L. = 0.85
S.L. = 0.80
Maximum confidence interval, max δ
Number of samples n
5
4
3
2
1.0
0.9
0.8
0.7
0.6
0.5
0.4
0.3
0.2
0.10
0.09
0.08
0.07
0.06
0.05
0.04
0.03
0.02
0.01
0
100
200
300
400
500
600

FIG. A9.1

is the probability that ($\bar{x}$ − mean of $\bar{x}$) is greater than $-\delta$ and less than δ. Thus,

$$P = \frac{1}{\sqrt{2\pi V(\bar{x})}} \int_{-\delta}^{\delta} e^{-[\bar{x}-E(\bar{x})]^2/2V(\bar{x})}\, d\bar{x}$$

$$= \frac{1}{\sqrt{2\pi}} \int_{-\delta/\sqrt{V(\bar{x})}}^{\delta/\sqrt{V(\bar{x})}} e^{-[\bar{x}-E(\bar{x})]^2/2}\, d\bar{x}$$

But $E(\bar{x})$ = true average probability of hit P_t. Hence

$$P(-\delta < \bar{x} - P_t < \delta) = P(\bar{x} - \delta < P_t < \bar{x} + \delta)$$

$$= \frac{1}{\sqrt{2\pi}} \int_{-\delta/\sqrt{V(\bar{x})}}^{\delta/\sqrt{V(\bar{x})}} e^{-(\bar{x}-P_t)^2/2}\, d\bar{x}$$

If we are interested in a 95 per cent confidence level, then we set the integral above equal to 0.95 and solve for the limits. However, we cannot solve for δ prior to experiment because $V(\bar{x})$ is unknown. Hence, we cannot get exact confidence limits by this method. But we *do* know that max $V(\bar{x}) = 0.25/n$. Hence, we can solve for the max δ for a confidence level of 0.95. To find the approximate amount of data necessary to attain a specified δ given the confidence level, use the procedure outlined in Sec. 9.4.4 in conjunction with Fig. A9.1.

APPENDIX

10 Evaluation of the Binomial Confidence Interval for Large Samples[1]

As has been pointed out in the text, when the number of samples is about 30, the normal distribution is a very good approximation to the binomial.

Using this fact so that $\bar{x}$ is approximately $N[p, (1 - p)p/n]$, where p is the overall true probability of hit and n is the number of samples ($n > 30$), we wish to find p such that

$$\int_{-\bar{x}_0}^{\bar{x}_0} \frac{1}{\sqrt{2\pi \frac{p(1-p)}{n}}} \exp \frac{-(\bar{x} - p)^2}{2 \frac{(1-p)p}{n}} \, d\bar{x} = Y = \text{significance level}$$

where $\bar{x}_0$ = measured sample mean

Normalizing,

$$\frac{1}{\sqrt{2\pi}} \int_{-(\bar{x}_0 - p)/\sqrt{p(1-p)/n}}^{(\bar{x}_0 - p)/\sqrt{p(1-p)/n}} e^{-u^2/2} \, du = Y$$

Equivalently,

$$\frac{1}{\sqrt{2\pi}} \int_{(\bar{x}_0 - p)/\sqrt{p(1-p)/n}}^{\infty} e^{-u^2/2} \, du = \frac{1 - Y}{2}$$

Using the tables of the error function to find the lower limit y for a given Y, we set

$$\frac{\bar{x}_0 - p}{\sqrt{\frac{p(1-p)}{n}}} = y$$

and solve for p. Substituting L for p and $\bar{x}$ for $\bar{x}_0$, we obtain Eq. (9.24).

The approximation obtained in the text is found by solving the quadratic and eliminating the terms y/n.

[1] This appendix was written by A. Levine.

APPENDIX

11 *Percentiles of the t Distribution*[1]

df	$t_{.60}$	$t_{.70}$	$t_{.80}$	$t_{.90}$	$t_{.95}$	$t_{.975}$	$t_{.99}$	$t_{.995}$
1	.325	.727	1.376	3.078	6.314	12.706	31.821	63.657
2	.289	.617	1.061	1.886	2.920	4.303	6.965	9.925
3	.277	.584	.978	1.638	2.353	3.182	4.541	5.841
4	.271	.569	.941	1.533	2.132	2.776	3.747	4.604
5	.267	.559	.920	1.476	2.015	2.571	3.365	4.032
6	.265	.553	.906	1.440	1.943	2.447	3.143	3.707
7	.263	.549	.896	1.415	1.895	2.365	2.998	3.499
8	.262	.546	.889	1.397	1.860	2.306	2.896	3.355
9	.261	.543	.883	1.383	1.833	2.262	2.821	3.250
10	.260	.542	.879	1.372	1.812	2.228	2.764	3.169
11	.260	.540	.876	1.363	1.796	2.201	2.718	3.106
12	.259	.539	.873	1.356	1.782	2.179	2.681	3.055
13	.259	.538	.870	1.350	1.771	2.160	2.650	3.012
14	.258	.537	.868	1.345	1.761	2.145	2.624	2.977
15	.258	.536	.866	1.341	1.753	2.131	2.602	2.947
16	.258	.535	.865	1.337	1.746	2.120	2.583	2.921
17	.257	.534	.863	1.333	1.740	2.110	2.567	2.898
18	.257	.534	.862	1.330	1.734	2.101	2.552	2.878
19	.257	.533	.861	1.328	1.729	2.093	2.539	2.861
20	.257	.533	.860	1.325	1.725	2.086	2.528	2.845
21	.257	.532	.859	1.323	1.721	2.080	2.518	2.831
22	.256	.532	.858	1.321	1.717	2.074	2.508	2.819
23	.256	.532	.858	1.319	1.714	2.069	2.500	2.807
24	.256	.531	.857	1.318	1.711	2.064	2.492	2.797
25	.256	.531	.856	1.316	1.708	2.060	2.485	2.787
26	.256	.531	.856	1.315	1.706	2.056	2.479	2.779
27	.256	.531	.855	1.314	1.703	2.052	2.473	2.771
28	.256	.530	.855	1.313	1.701	2.048	2.467	2.763
29	.256	.530	.854	1.311	1.699	2.045	2.462	2.756
30	.256	.530	.854	1.310	1.697	2.042	2.457	2.750
40	.255	.529	.851	1.303	1.684	2.021	2.423	2.704
60	.254	.527	.848	1.296	1.671	2.000	2.390	2.660
120	.254	.526	.845	1.289	1.658	1.980	2.358	2.617
∞	.253	.524	.842	1.282	1.645	1.960	2.326	2.576
df	$-t_{.40}$	$-t_{.30}$	$-t_{.20}$	$-t_{.10}$	$-t_{.05}$	$-t_{.025}$	$-t_{.01}$	$-t_{.005}$

When the table is read from the foot, the tabled values are to be prefixed with a negative sign. Interpolation should be performed using the reciprocals of the degrees of freedom.

[1] From "Introduction to Statistical Analysis," 2d ed., p. 384, by W. J. Dixon and F. J. Massey, Jr. Copyright © 1957. McGraw-Hill Book Company, Inc., New York. Used by permission.

APPENDIX

12 Conservative Confidence Intervals[1]

Consider the situation in which 10 independent simulation runs are made for each tactical situation (see Sec. 9.4.4). The outcome of each run is a random variable, having two outcomes, recorded only as a hit or a miss. Several comparisons can be made between experimental results concerning the "new" and "old" missile. The most obvious is the number of kills scored by each missile in a particular tactical situation. For example, using the data of Tables 9.1 and 9.2 in tactical situation 8, the experimental *estimate* of kill probability for the old missile was 0.65 and that for the new missile was 0.7. Using Fig. A8.1, we find that the 95 per cent confidence limits for 10 trials are

$$0.35 < \text{kill probability, new missile} < 0.93$$
$$0.3 < \text{kill probability, old missile} < 0.9$$

From the size of these intervals it is obvious that no decision can be made about the new missile with respect to the old missile.

A comparison of the relative effectiveness of the two missiles can also be made to cover a representative set of tactical situations. As an example, consider all *runs*. The resulting experimental estimates of overall kill probability are

$$\text{Experimental kill probability, new missile} = 0.85$$
$$\text{Experimental kill probability, old missile} = 0.78$$

The determination of the confidence limits of these data is complicated by the fact that the random variables involved have different expected values corresponding to different tactical situations. The following analysis leads to the most conservative estimate of confidence limits. Let

x_{ij} = a random variable denoting outcome of jth trial and ith tactical situation
n = number of tactical situations
m = number of trials per tactical situation

It is assumed that the tactical situations, equally weighted, represent a typical array. Then the experimental mean $\bar{x}$ of x_{ij} is an unbiased estimator of kill probability:

$$\bar{x} = \frac{\sum_{i=1}^{n} \sum_{j=1}^{m} x_{ij}}{mn}$$

The confidence limits imposed on kill probability are a function of the theoretical

[1] This appendix was written by A. Levine.

variance of $\bar{x}$ which is denoted $\sigma_{\bar{x}}^2$,

$$\sigma_{\bar{x}}^2 = V(\bar{x}) = V\left(\frac{1}{mn}\sum_{i=1}^{n}\sum_{j=1}^{m} x_{ij}\right) = \frac{1}{m^2n^2}\sum_{i=1}^{n} mV(x_i)$$

The variance of an individual random variable is

$$V(x_i) = p_i(1 - p_i)$$

where p_i is the expected value of x_i. The largest possible value of $V(x_i)$ occurs when $p_i = 0.5$. Thus, the most conservative estimate of $V(x_i)$ is 0.25. It is now obvious that the largest possible confidence interval for the composite estimate of the expected value of several random variables may be found by entering Fig. A8.1 at 0.5 and reading the ordinates off at the appropriate number of runs. The interval size, 2δ, may then be calculated and placed symmetrically about the unbiased estimate $\bar{x}$ of the kill probability. This yields the most conservative estimate of 95 per cent confidence limits. In the example stated above, $\delta \doteq 0.09$ for 100 samples and the confidence limits are

$$0.76 < \text{kill probability, new missile} < 0.94$$
$$0.69 < \text{kill probability, old missile} < 0.87$$

Since the estimate is conservative, it is possible to have limits greater than 1.00.

To extend this analysis to limits other than 0.95 and maintain a simple procedure, the overall number of runs must be large enough to allow approximation of the binomial distribution by a normal distribution.

A suitable estimator $s_{\bar{x}}^2$ of $\sigma_{\bar{x}}^2$ is

$$s_{\bar{x}}^2 = \frac{1}{m^2n^2}\sum_{i=1}^{n} m(0.25) = \frac{0.25}{mn}$$

By making the normal approximation, a δ can be found such that

$$\int_{-\delta}^{\delta} \frac{e^{-(\bar{x}-p)^2/2\sigma_{\bar{x}}^2}}{\sqrt{2\pi}\,\sigma_{\bar{x}}}\,d(\bar{x} - p) = P = \text{confidence-limit probability}$$

Table A12.1 Table for Use in Calculating Confidence Limits of Normally Distributed Random Variables

P	$f(P)$	P	$f(P)$
1.00	∞	0.65	0.935
0.99	2.58	0.60	0.842
0.98	2.33	0.55	0.755
0.95	1.96	0.50	0.674
0.90	1.64	0.40	0.524
0.85	1.44	0.30	0.385
0.80	1.28	0.20	0.253
0.75	1.15	0.10	0.126
0.70	1.04	0.00	0.000

P = confidence-interval probability

$\frac{\delta}{\sigma_{\bar{x}}} = f(P)$

δ = distance between confidence limit and experimental mean

$\sigma_{\bar{x}}$ = an estimator of standard deviation of $\bar{x}$

$\bar{x}$ = mean result of N trials performed on a normally distributed variable

If this integral is normalized, the deviation δ from the experimental mean may be expressed as a function of $\sigma_{\bar{x}}^2$ and P:

$$\delta = \sigma_{\bar{x}} f(P)$$

Using the estimator of $\sigma_{\bar{x}}$,

$$\delta = \frac{0.5}{\sqrt{mn}} f(P)$$

A listing of $f(P)$ versus P has been included here as Table A12.1 for convenience. This method can also be applied by starting with δ. Then a conservative estimate of the probability that the true kill probability lies within $\bar{x} \pm \delta$ may easily be found.

APPENDIX

13 Estimation of Sample Size[1,2]

A13.1 Estimation of Sample Size for Avoiding the Rejection of Differences When They Exist

The first step is to show that the probability of rejecting H_1 when it is true is a maximum when the quantities $n_x s_x^2$, $n_y s_y^2$ in Eq. (9.45) are a maximum. There are two cases to be treated, $n_r = 1$ and $n_r > 1$. We shall treat only $n_r > 1$. The other case is treated simply by redefining s_x^2 to be

$$s_x^2 = \sum_{i=1}^{n_T} \frac{(x_i - \bar{x})^2}{n_T}$$

where $\bar{x}$ is the overall mean.

The maximum value of s_x^2, s_y^2 is $0.25/n_T$. To see this we define

$$\bar{x}_i = \frac{\sum_{j=1}^{n_T} x_{ij}}{n_T}$$

It follows that

$$E(\bar{x}_i) = P_t = \text{true average probability of hit}$$

$$V(\bar{x}_i) = \frac{V_p}{n_T}$$

where

$$V_p = \frac{\sum_{j=1}^{n_T} p_j(1 - p_j)}{n_T}$$

$$p_j = \text{true probability of hit for } j\text{th tactical situation}$$

Note that

$$s_x^2 = \frac{\sum_{i=1}^{n_r} (\bar{x}_i - \bar{x})^2}{n_r}$$

where $\bar{x}$ is the overall sample mean. Since $[n_r/(V_p/n_T)]s_x^2$ is χ^2-distributed with $n_r - 1$

[1] See Sec. 9.5.3.

[2] This appendix was written by A. Levine.

degrees of freedom, it follows that

$$E\left(\frac{n_r n_T}{V_p} s_x^2\right) = n_r - 1$$

or

$$E(s_x^2) = \frac{V_p}{n_T}\left(1 - \frac{1}{n_r}\right) \qquad n_r > 1$$

and

$$V\left(\frac{n_T n_r}{V_p} s_x^2\right) = 2(n_r - 1) \qquad n_r > 1$$

or

$$V(s_x^2) = \frac{2V_p^2}{n_T^2 n_r^2}(n_r - 1) = \frac{2V_p^2}{n_r n_T}\left(1 - \frac{1}{n_r}\right)$$

Since the minimum value of n_r is 30, it follows that $V(s_x^2) \approx 0$. Hence we can use max $E(s_x^2)$ as the value of max s_x^2, that is,

$$\max s_x^2 = \frac{0.25}{n_T}\left(1 - \frac{1}{n_r}\right) < \frac{0.25}{n_T}$$

Similarly, max $s_y^2 < 0.25/n_T$.

It is clear that, the smaller the value of t in Eq. (9.45), the greater the probability of its *not* falling in the critical region. Since t is smallest when $n_x s_x^2$, $n_y s_y^2$ are largest, it follows that, for any value of θ, the true difference between P_{t_1} and P_{t_2}, the probability of not falling in the critical region, is largest when we set

$$s_x^2 = \frac{0.25}{n_T} \qquad s_y^2 = \frac{0.25}{n_T}$$

Hence the contention is proved.

If we assume that $n_x = n_y = n$, Eq. (9.45) becomes

$$t = \frac{(\bar{x} - \bar{y})\sqrt{n - 1}}{\sqrt{0.5/n_T}}$$

If θ is the true difference, then the mean of the t distribution is

$$\frac{\theta\sqrt{n - 1}}{\sqrt{0.5/n_T}}$$

By using the usual techniques, Figs. A13.1 to A13.3 and others like them for different significance levels can be drawn.

It is interesting to observe that for $n > 16$ we can use the normal distribution to approximate the t distribution. This normal distribution has mean $f(\theta) = \theta\sqrt{n - 1}/\sqrt{0.5/n_T}$ and unity variance. Hence, as an example, if one wishes to find the probability of accepting H_0 as a function of $\theta\sqrt{n_T}$ for the 0.025 significance level, one first establishes the critical region; that is, one finds the point along the *right* tail of the normal distribution such that the probability of falling beyond that point is $1 - 0.025$. Using the error-function tables, we find this point to be 1.96. We then evaluate

$$\frac{1}{\sqrt{2\pi}}\int_{-\infty}^{1.96} e^{-[x-f(\theta)]^2/2}\,dx$$

from the error-function tables. The value so obtained is the probability of accepting H_0 as a function of $\theta\sqrt{n_T}$ with n_r as a parameter.

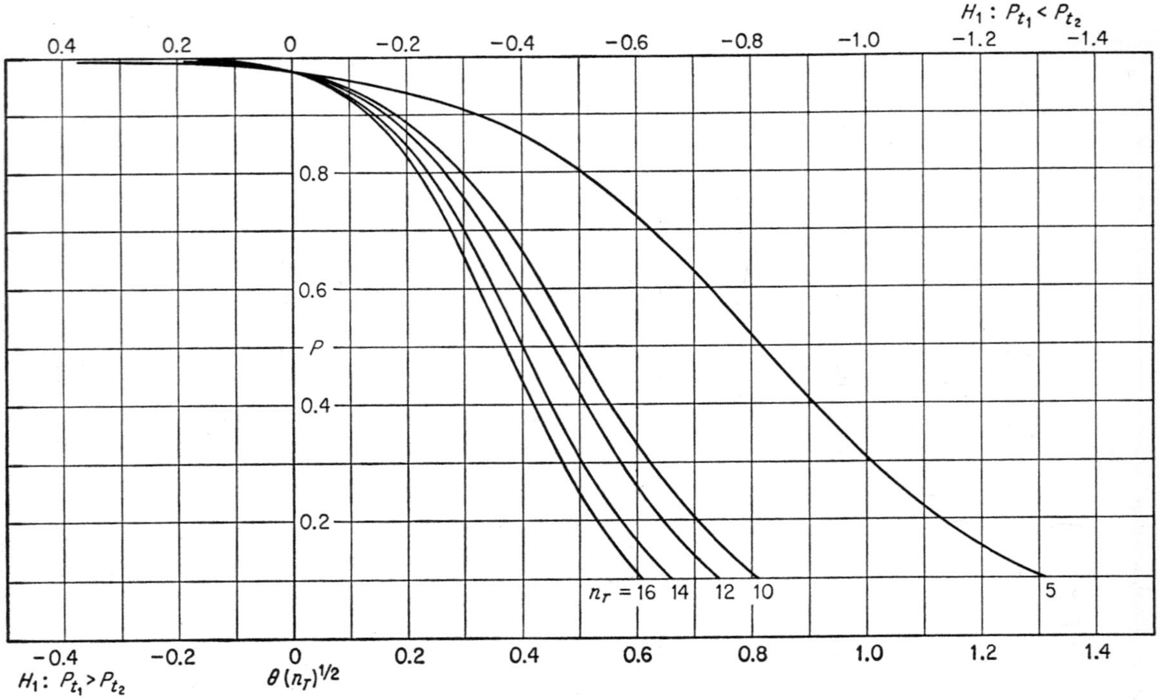

FIG. A13.1 Maximum probability of accepting H_0 as a function of θ. $\theta = P_{t_1} - P_{t_2}$; n_T = number of tactical runs; n_r = number of runs. Significance level = 0.025.

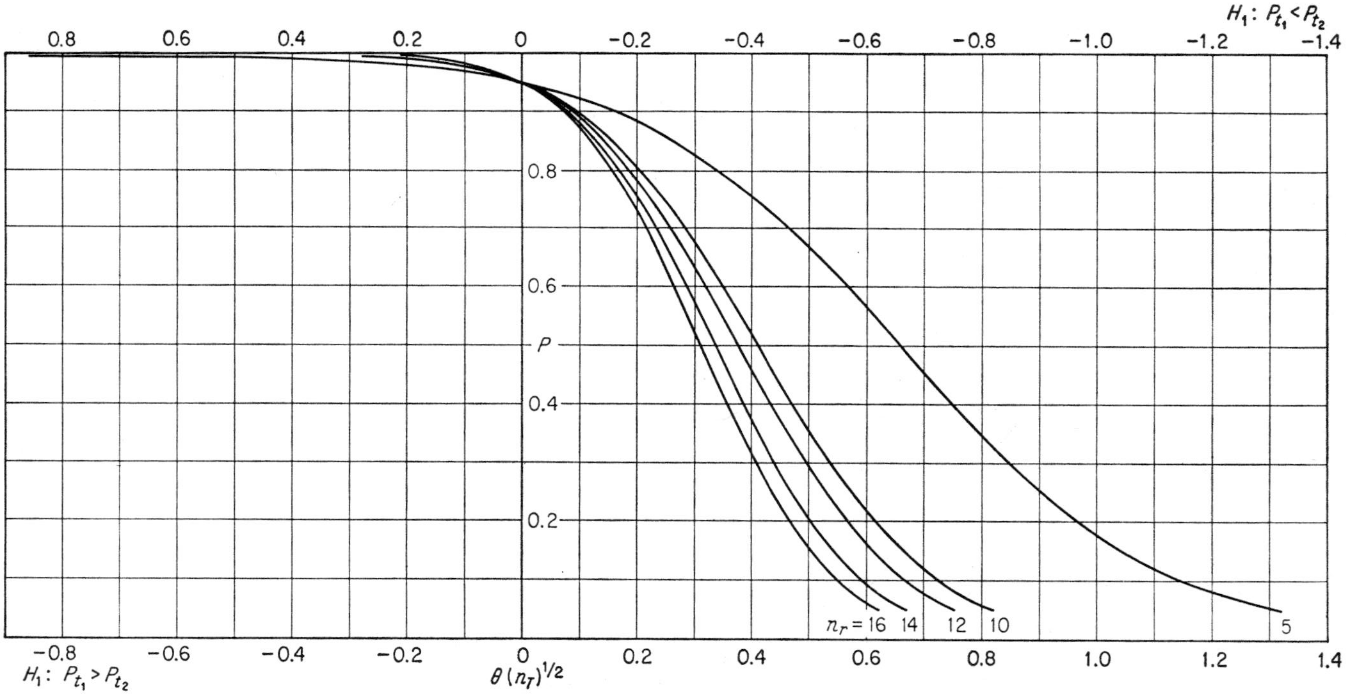

FIG. A13.2 Maximum probability of accepting H_0 as a function of θ. $\theta = P_{t_1} - P_{t_2}$; n_T = number of tactical runs; n_r = number of runs. Significance level = 0.05.

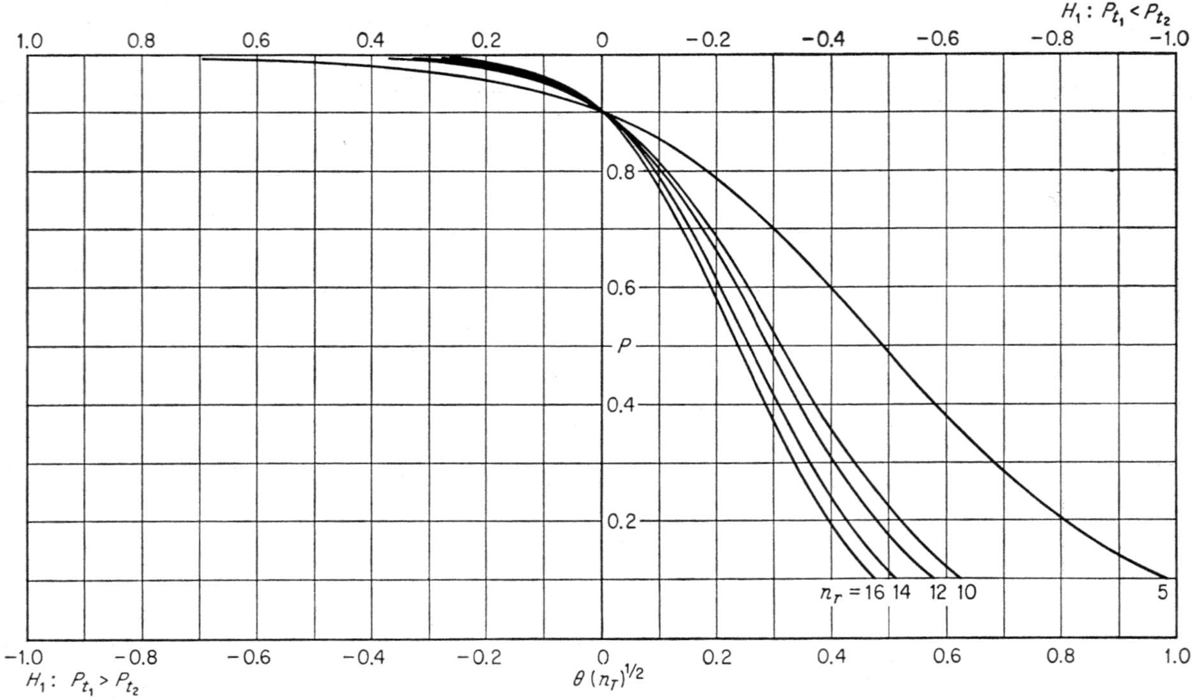

FIG. A13.3 Maximum probability of accepting H_0 as a function of θ. $\theta = P_{t_1} - P_{t_2}$; n_T = number of tactical runs; n_r = number of runs. Significance level = 0.1.

A13.2 Estimation of Sample Size for Avoiding Rejection of Differences When They Exist for the Case of Unweighted Hypotheses

We note from Sec. A13.1 that the mean of the t distribution is

$$q = \frac{\theta_0 \sqrt{n-1}}{\sqrt{0.5/n_T}}$$

That is, the frequency function $P(t)$ appears as shown in Fig. A13.4. Hence the probability of accepting $H_0: P_{t_1} < P_{t_2}$ when the mean of t is q is given by the area in the cross-hatched region. We can find this area from the Student t tables. By proper normalization, we obtain Fig. A13.5.

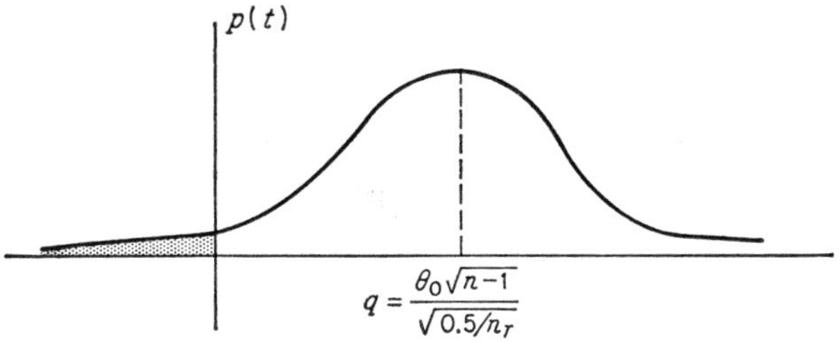

FIG. A13.4

To evaluate the probability of accepting H_0 as a function of $\theta \sqrt{n_T}$ when $n_r > 16$ we modify the technique outlined at the end of Sec. A13.1. We approximate the t distribution by the normal. Since the probability of accepting H_0 is just the probability of obtaining a negative value of t, it follows that

$$P(H_0|\theta_0) = \frac{1}{\sqrt{2\pi}} \int_{-\infty}^{0} e^{[x-f(\theta_0)]^2/2}\, dx$$

where

$$f(\theta_0) = \frac{\theta_0 \sqrt{n-1}}{\sqrt{0.5/n_T}}$$

By using the above relationships, the curves for $n_r = 26$, 50, and 100 were drawn in Fig. A13.5 (see page 432).

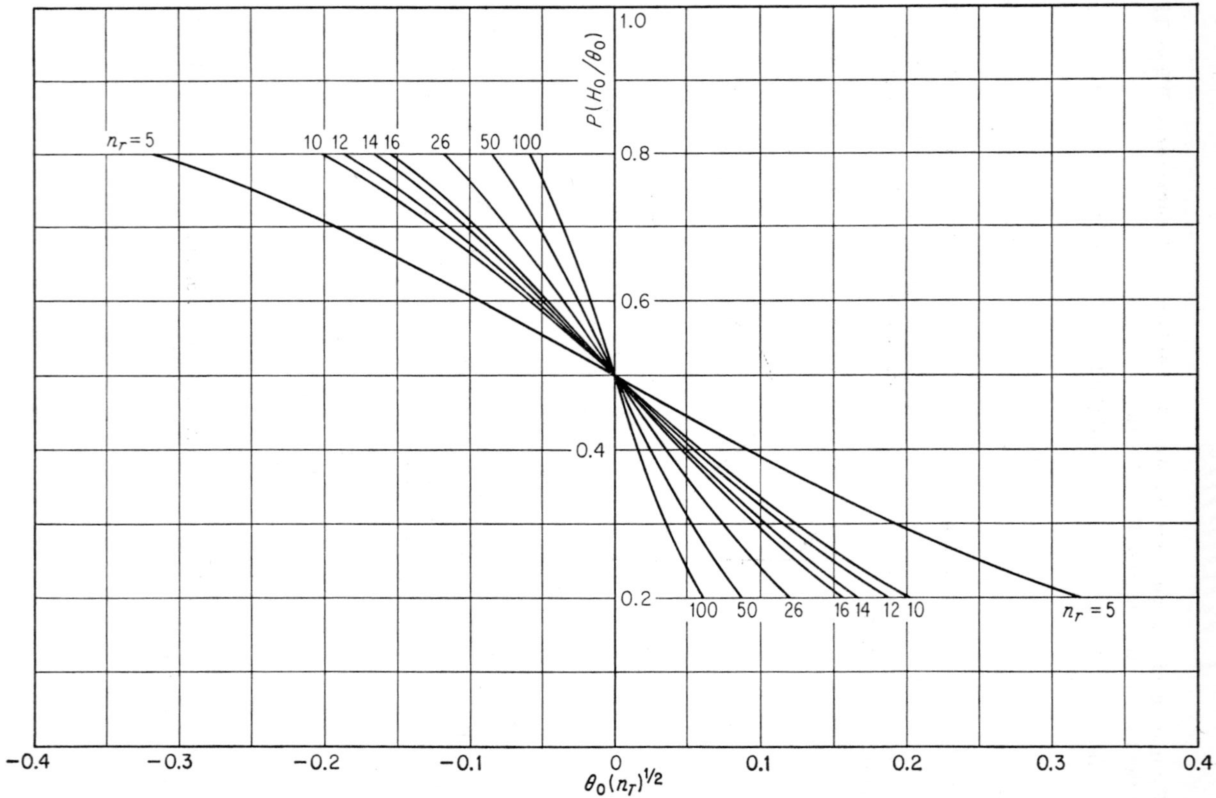

FIG. A13.5 Probability of accepting H_0 as a function of θ for unweighted hypotheses. $\theta = P_{t_1} - P_{t_2}$; n_T = number of tactical situations; n_r = number of runs.

Problems

Chapter 2

2.1. For the equation set

$$\frac{dy_1}{dt} = y_2 \qquad \frac{dy_2}{dt} = y_3 \qquad \frac{dy_3}{dt} = \ln y_2 \sin y_2 - e^{y_2}$$

find for which values (if any) of y_1, y_2, and y_3 the Lipschitz condition is not satisfied.

2.2. For the differential equation

$$\frac{dy}{dx} = f(x, y)$$

indicate for the following functions over what range of values of y the Lipschitz condition is satisfied:

(*a*) $f(x, y) = \begin{cases} 10 & y \geq 2 \\ -10 & y < 2 \end{cases}$ (*b*) $f(x, y) = \sin xy$

(*c*) $f(x, y) = xy^{\frac{2}{3}}$ (*d*) $f(x, y) = \begin{cases} \dfrac{4x^2y}{x^2 + y^2} & \text{when } x^2 + y^2 > 0 \\ 0 & \text{when } x^2 + y^2 = 0 \end{cases}$

2.3. Which of these equations is nonlinear? Explain your answer.

(*a*) $\dfrac{d^2y}{dx^2} + 10\dfrac{dy}{dx} + 3y = \sin y$ (*b*) $\left(\dfrac{d^2y}{dx^2} + y\right)^2 = \tan x$

(*c*) $\dfrac{d^2y}{dx^2} + e^x\dfrac{dy}{dx} + yf(y) = 0$

where $f(y)$ is shown in Fig. P 2.3.

(*d*) $\dfrac{dy_1}{dt} = y_1 - 3y_2 + y_1y_3 \qquad \dfrac{dy_2}{dt} = y_1 + y_2 \qquad \dfrac{dy_3}{dt} = 4y_2$

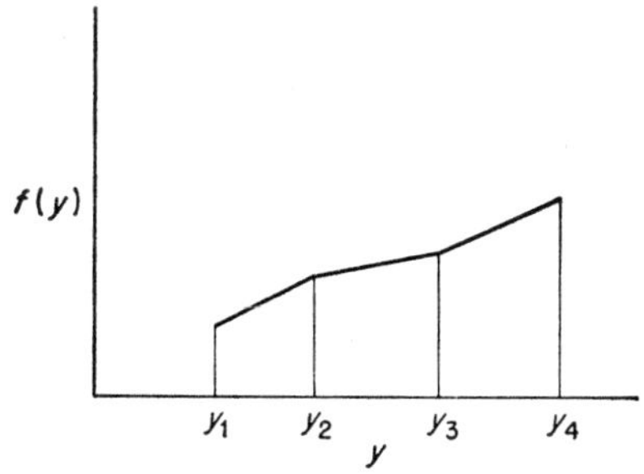

FIG. P 2.3

2.4. Given the set of equations

$$\frac{d^2\theta}{dt^2} + k_1\theta = -k_2\frac{d^2\gamma}{dt^2} \qquad \frac{d^2\gamma}{dt^2} + k_3\gamma = -k_2\frac{d^2\theta}{dt^2}$$

Eliminate θ and obtain a single equation in γ. Then rewrite as a set of first-order equations.

2.5. For the equation

$$\frac{d^2y}{dx^2} - 1 = 0$$

with the initial conditions, at $x = 0$,

$$y_1 = 1 \qquad \frac{dy_1}{dx} = 3$$

the solution is

$$y_a = e^{-x} - e^x$$

and for the initial conditions

$$y_2 = 4 \qquad \frac{dy_2}{dx} = -2$$

the solution is

$$y_b = e^x + 3e^{-x}$$

What is the solution y_c for the initial conditions

$$y_3 = 7 \qquad \frac{dy_3}{dx} = 9$$

HINT: Solve the set of equations

$$y_3 = \alpha_1 y_1 + \alpha_2 y_2 \qquad \frac{dy_3}{dt} = \alpha_1 \frac{dy_1}{dt} + \alpha_2 \frac{dy_2}{dt}$$

Then

$$y_c = \alpha_1 y_a + \alpha_2 y_b$$

2.6. In the equation

$$\frac{d^2y}{dt^2} + a_1(t)\frac{dy}{dt} + a_0 y = \sin t$$

the coefficient $a_1(t)$ is complex and is

$$a_1(t) = x(t) + iy(t)$$

If the initial conditions at $t = 0$ are

$$y(0) = 1 + i1 \qquad \frac{dy}{dt}(0) = -3$$

how would you alter this equation so that it can be programmed on the computer? Note that the computer can handle only real dependent variables.

2.7. The equation set

$$\frac{dx_1}{dt} = -1.1x_1 + \frac{5}{7t}x_2 + \frac{12}{7t}x_3$$

$$\frac{dx_2}{dt} = -0.33tx_1 + \left(\frac{1}{t} - 1.5\right)x_2 + 1.8x_3$$

$$\frac{dx_3}{dt} = -0.23tx_1 - 1.9x_2 + \left(1.8 + \frac{1}{t}\right)x_3$$

has as a solution

$$x_1 = c_1e^{-0.2t} + e^{-0.3t}(c_2 \cos 0.9t - c_3 \sin 0.9t)$$

$$x_2 = 0.3c_1te^{-0.2t} + te^{-0.3t}(0.4c_2 - 0.3c_3) \cos 0.9t - [(0.3c_2 + 0.4c_3) \sin 0.9t]$$

$$x_3 = 0.4c_1te^{-0.2t} + te^{-0.3t}[(0.3c_2 - 0.4c_3) \cos 0.9t - (0.4c_2 + 0.3c_3) \sin 0.9t]$$

Suppose we convert this equation set to a third-order equation in x_2 and add a forcing function, obtaining as our result

$$\frac{d^3x_1}{dt^3} + a_2(t)\frac{d^2x_1}{dt^2} + a_1(t)\frac{dx_1}{dt} + a_0(t)x_1 = f(t) \tag{a}$$

(*a*) Determine the values of a_1, a_2, and a_3.
(*b*) Obtain the weighting function.
HINT: x_1 is a solution of the homogeneous part of Eq. (*a*).

2.8. With the proper combination of initial conditions, two different differential equations can yield the same solution.

For the equation

$$\frac{d^2y_1}{dt^2} + y_1 = 0 \tag{a}$$

the solution is

$$y_1 = c_1 \sin t + c_2 \cos t$$

For the equation

$$\frac{d^3y_2}{dt^3} + \frac{d^2y_2}{dt^2} + \frac{dy_2}{dt} + y_2 = 0 \tag{b}$$

the solution is

$$y_2 = d_1 \sin t + d_2 \cos t + d_3e^{-t}$$

For what initial conditions in Eqs. (*a*) and (*b*) are y_1 and y_2 equal?

2.9. Given that, at $t = 0$, the initial conditions are

$$\left(\frac{d^3y}{dt^3}\right)_{t=0} = 5 \qquad \left(\frac{d^2y}{dt^2}\right)_{t=0} = -3$$

can you find a unique solution for the equation

$$\frac{d^2y}{dt^2} + 4y = 0$$

If your answer is yes, find the solution. If no, explain.

2.10. The equation

$$t\frac{d^3y}{dt^3} - \frac{d^2y}{dt^2} - t\frac{dy}{dt} + y = 0$$

has three solutions:

$$y_1 = \cosh t + 3t \qquad y_2 = e^{-t} - t \qquad y_3 = 3e^t + 7e^{-t} + 14t$$

Show that these do not constitute a linearly independent set of solutions. Find three linearly independent solutions and then obtain the solution for the initial conditions, at $t = 1$,

$$y(1) = 1 \qquad \frac{dy}{dt} = -1 \qquad \frac{d^2y}{dt^2} = 2$$

2.11. One solution of the equation

$$\frac{t^2}{2}\frac{d^2y}{dt^2} - t\frac{dy}{dt} + y = 0$$

is $y_1 = t$.

Find another solution so that the two solutions form an independent set. Show that they are independent. Then find the weighting function.

2.12. For the equation

$$\frac{d^2y}{dx^2} - x\frac{dy}{dx} + (x - 1)y = 0$$

one solution is

$$y_1 = e^x$$

Using the method of reduction of order, show that the other independent solution is

$$y_2 = e^x \int \exp\left(\tfrac{1}{2}x^2 - 2x\right) dx$$

2.13. The equation

$$\frac{d^3y}{dx^3} + \frac{d^2y}{dx^2} + \frac{dy}{dx} + y = e^x \qquad (a)$$

has as a solution to its homogeneous equation

$$y_H = c_1 \sin x + c_2 \cos x + c_3 e^{-x}$$

Find the weighting function and then solve for y in Eq. (a).

2.14. For the differential equation

$$\frac{d^2y}{dx^2} + w^2 y = f(x) \qquad (a)$$

when $f(x) = 0$, two independent solutions are

$$y_1 = \sin wx \qquad y_2 = \cos wx$$

Find the weighting function $K(x, \alpha)$ and then show how we can use the computer to find the coefficients of a Fourier series representation for an arbitrary $f(x)$.

HINT: If the Fourier series representation is

$$f(x) = \frac{a_0}{2} + \sum_{m=1}^{\infty} a_m \cos mw_i x + \sum_{m=1}^{\infty} b_n \sin mw_i x$$

then

$$a_m = \frac{w_i}{\pi} \int_0^{2\pi/w_i} f(x) \cos mw_i x \, dx \qquad (b)$$

$$b_m = \frac{w_i}{\pi} \int_0^{2\pi/w_i} f(x) \sin mw_i x \, dx \qquad (c)$$

Make the necessary transformations on the solution of Eq. (a),

$$y = \int_0^x K(x, \alpha) f(\alpha) \, d\alpha$$

to get it into the form of Eqs. (b) and (c).

2.15. Using Picard iteration, solve

$$\frac{dy}{dt} = t^2 - y$$

Begin the iteration with $y = 1$ at $t = 0$ and carry through four iterations.

Chapter 3

3.1. Figure P 3.1 represents a suspension system where M_1 and M_2 are masses, k_1 and k_2 are spring constants, b_2 is a shock-absorber damping constant, and x_1, x_2, x_3 are displacements.

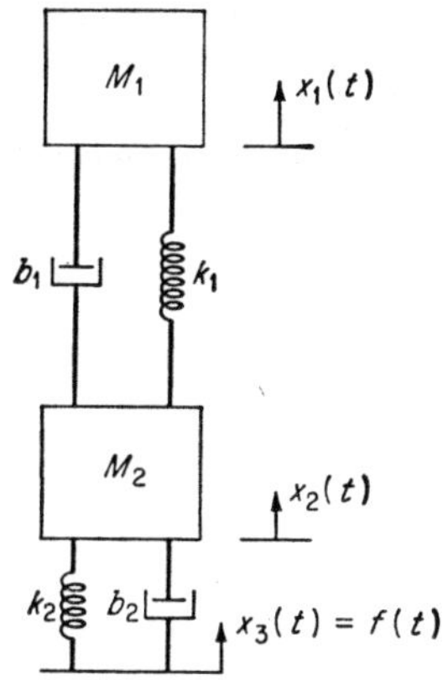

FIG. P 3.1

The equations of motion are

$$M_1 \frac{d^2x_1}{dt^2} + b_1 \left(\frac{dx_1}{dt} - \frac{dx_2}{dt} \right) + k_1(x_1 - x_2) = 0$$

$$M_2 \frac{d^2x_2}{dt^2} + b_2 \left(\frac{dx_2}{dt} - \frac{dx_1}{dt} \right) + k_1(x_2 - x_1) + k_2(x_2 - x_3) = 0$$

Program these equations on the computer. How many initial conditions must be specified?

3.2. (*a*) Program the computer to solve the equation

$$4 \frac{d^4y}{dt^4} - \frac{d^3y}{dt^3} \frac{d^2y}{dt^2} + \left(\frac{d^2y}{dt^2} \right)^3 = 0$$

where, at $t = 0$,

$$\left(\frac{d^4y}{dt^4} \right)_0 = 7 \qquad \left(\frac{d^2y}{dt^2} \right)_0 = -2$$

(*b*) At $t = 0$, what are the values of

$$\frac{d^5y}{dt^5} \quad \text{and} \quad \frac{d^6y}{dt^6}$$

3.3. Set up Mathieu's equation,

$$\frac{d^2x}{dt^2} + (a + b \cos wt)x = 0$$

assuming that $\cos wt$ is available. The quantities a and b are constant.

(*a*) How many initial conditions are needed to solve the problem?

(*b*) If the problem starts at $t = 0$, will the following three sets of initial conditions give the same solution?

1. $x(0) = 1 \qquad \left(\frac{d^2x}{dt^2} \right)_0 = -a(1 + b)$

2. $x(0) = 1 \qquad \left(\frac{dx}{dt} \right)_0 = 2$

3. $\left(\frac{dx}{dt} \right)_0 = 2 \qquad \frac{d^2x}{dt^2} = -a(1 + b)$

(*c*) If you wanted to find d^3x/dt^3, how would you obtain it on the computer without using differentiators? What would be the initial conditions of the differential equation you would have to solve?

3.4. The equations for neutron density Q in a nuclear pile containing uranium are

$$\tau \frac{dQ}{dt} = (k - 1 - \beta)Q + \sum_{i=1}^{6} \lambda_i P_i + S$$

$$\frac{dP_i}{dt} = \beta_i Q - \tau\lambda_i P_i \qquad i = 1, 2, \ldots, 6$$

where Q = neutron density
τ = average lifetime of a neutron
k = amount of new neutrons produced by a single neutron in time τ
S = amount of an additional source of neutrons
P_i = output share of delayed neutrons in a chain reaction
β_i, λ_i = const

Set up the computer equations. How many initial conditions are required to solve the problem?

3.5. Program the computer to make a phase-plane plot, that is, a plot of dy/dx versus y, of the equation

$$\frac{d^2y}{dx^2} + \left(\frac{dy}{dx}\right)^2 - y\frac{dy}{dx} = 0$$

Assume that we have an xy plotter (described in Chap. 5), a device which will plot any two variables in rectangular coordinates.

What initial conditions are required on the computer?

3.6. Given

$$\frac{d^2y}{dt^2} + f(t)\frac{dy}{dt} + ay = b\frac{dx}{dt} + cx$$

where a, b, c are constants.

Can the equation be set up on the computer using (*a*) the canonical technique, (*b*) the dummy-variable technique?

If your answer is yes, draw the computer diagram. If your answer is no, explain your answer.

3.7. Obtain the equation or set of equations represented by the computer diagram in Fig. P 3.7.

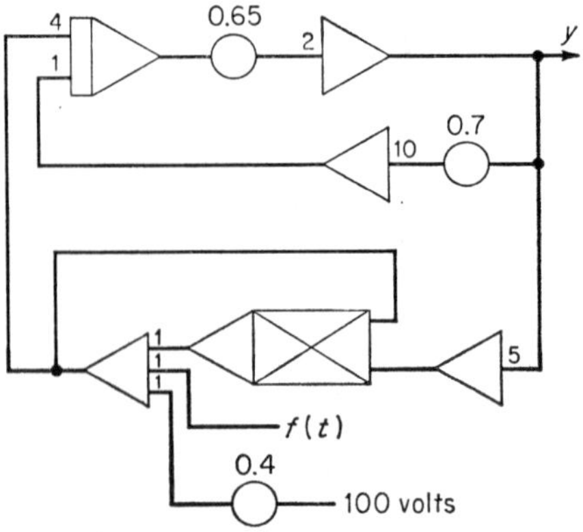

FIG. P 3.7

3.8. Suppose we want to solve a set of nonlinear equations of the form

$$\frac{dx_1}{dt} = f_1(x_1, x_2, \ldots, x_n)$$
$$\cdots\cdots\cdots\cdots\cdots$$
$$\frac{dx_n}{dt} = f_n(x_1, x_2, \ldots, x_n)$$

but have only m integrators, where $m < n$. Show how this problem can be solved on the computer.

HINT: Consult the Picard iteration technique.

3.9. Given $f(t)$, program the computer to solve the integral for any given t.

$$y(t) = \int_0^t f(\tau)e^{t-\tau}\, d\tau \tag{a}$$

Note that the solution of the differential equation

$$\frac{dy}{dt} + ay = 0 \tag{b}$$

is

$$y = e^{-at}$$

HINT: Find the weighting function of Eq. (b) and then form the integral in Eq. (a).

3.10. Obtain the adjoints of equations

(a) $a_2 \dfrac{d^2y}{dt^2} + a_1 \dfrac{dy}{dt} + a_0 y = 0 \qquad a_0 = e^{-\alpha t},\ a_1 = \dfrac{1}{t},\ a_2 = 1$

(b) $\dfrac{d^2y}{dt^2} + 3t^2 \dfrac{dy}{dt} + y^2 = 0$

(c) $x \dfrac{d^2y}{dx^2} + (\sin x - 1) \dfrac{dy}{dx} + 4x^3 y = 0$

Program the adjoint equations.

3.11. For equation

$$a_2(t)p^2y + a_1(t)py + a_0(t) = 0 \tag{a}$$

the adjoint is

$$\sum_{i=0}^{2} (-1)^i \frac{d^i}{dt^i} [a_i(t)u] = 0 \tag{b}$$

Show that the adjoint of Eq. (b) is Eq. (a).

3.12. The equation

$$\frac{d^2y}{dx^2} - y = 0 \tag{a}$$

has the boundary conditions, at $x = 0$,

$$y(0) + 2\left(\frac{dy}{dx}\right)_0 = 3$$

and, at $x = 1$,

$$y(1) + \left(\frac{dy}{dx}\right)_1 = 2$$

Can Eq. (a) satisfy these conditions? If it can, what must the initial conditions be at $x = 0$ to obtain the proper boundary conditions?

3.13. Suppose we use the dummy-variable method to solve the equation

$$\frac{d^3y}{dt^3} + a_2 \frac{d^2y}{dt^2} + a_1 \frac{dy}{dt} + a_0 y = b_2 \frac{dx^2}{dt^2} + b_1 \frac{dx}{dt} + b_0 \tag{a}$$

Assume that the weighting function for the dummy variable y_K is $K_1(t, \alpha)$. Program the computer by the adjoint method to find the weighting function $K_2(t, \alpha)$ so that for Eq. (*a*) you can write

$$y(t) = \int_0^t K_2(t, \alpha)x(\alpha)\, d\alpha \qquad (b)$$

HINT: Because

$$y_K(t) = \int_0^t K_1(t, \alpha)x(\alpha)\, d\alpha$$

it follows that

$$y(t) = \int_0^t K_1(t, \alpha)\left(b_2\frac{d^2x}{d\alpha^2} + b_1\frac{dx}{d\alpha} + b_0x\right) d\alpha \qquad (c)$$

Now integrate by parts until Eq. (*c*) is in the form of Eq. (*b*). Assume that x and its derivatives are zero at $\alpha = 0$, whereas $K_1(t, \alpha)$, $dK_1/d\alpha$, and $d^2K_1/d\alpha^2$ are zero at $\alpha = t$.

Chapter 4

4.1. For the operator

$$\frac{e_o}{e_i} = \frac{1}{1+p}$$

can we perform division and write

$$\frac{1}{1+p} = 1 - p + p^2 - p^3 + p^4 - \cdots + (-1)^n p^n \qquad n \to \infty$$

if

(*a*) $e_i = t^m$?
(*b*) $e_i = e^{-t}$?
(*c*) $e_i = f(t)$, where, at $t = 0$, $f(t)$ and all its derivatives are zero?

4.2. Show that

$$\frac{d^2y}{dt^2} = \frac{dx^2}{dt^2}$$

is not the same as

$$y = x$$

where there are initial conditions.

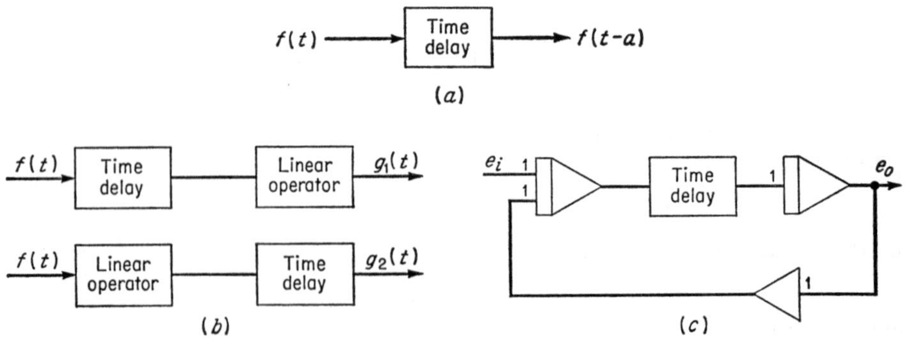

FIG. P 4.3

4.3. A time delay has the property shown in Fig. P 4.3*a*, where the delay time is a. In which of the following circumstances are the two operations of Fig. P 4.3*b* commutative, that is, $g_1(t) = g_2(t)$? The linear operator is (*a*) a constant-coefficient potentiometer, (*b*) an integrator, (*c*) a variable-coefficient potentiometer $h(t)$.

Using these results, if we multiply e_i by a factor k, how is e_o scaled in Fig. P 4.3*c*?

4.4. Show that in the two sets of block diagrams in Fig. P 4.4*a*, where the operator Q is nonlinear and the operators A, B, and C are commutative, there is an equivalence such that $y_1 = y_2$ and $y_3 = y_4$. Note that in Fig. P 4.4*a* the nonlinearity is in the forward loop and in Fig. P 4.4*b* the nonlinearity is in the feedback loop.

Does the constant-loop-gain rule apply? Explain.

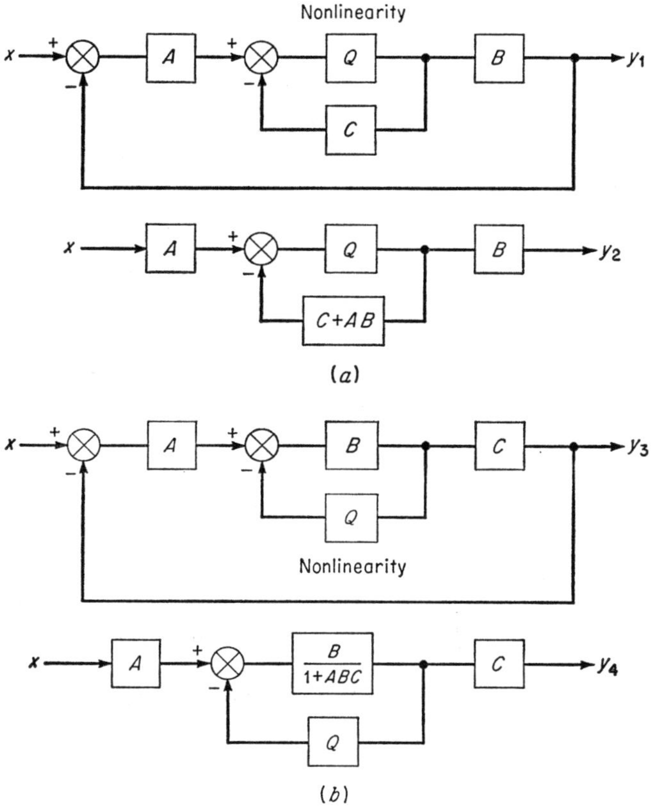

FIG. P 4.4

4.5. Given the circuit in Fig. P 4.5, what would the block diagram look like (to achieve the same result) if the forcing function is moved to (*a*) A or (*b*) B? All operators are commutative.

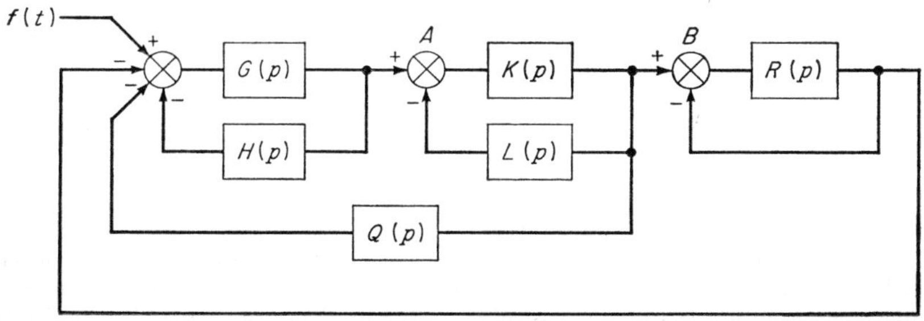

FIG. P 4.5

4.6. If we convert from Fig. P 4.6a to Fig. P 4.6b, find u_1, u_2, and u_3.

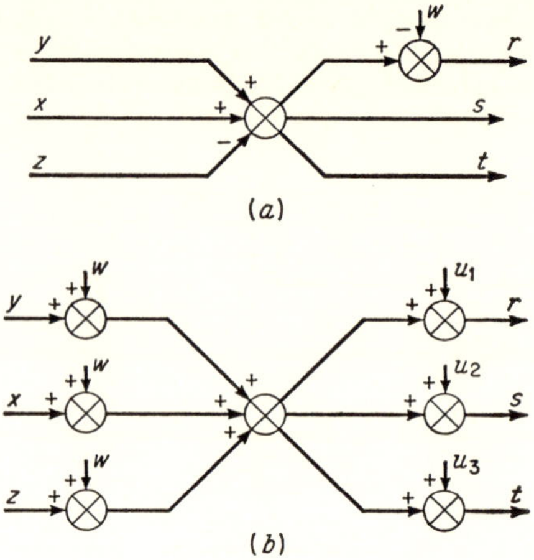

FIG. P 4.6

4.7. Given

$$\frac{d^2y}{dt^2} + f(t)\left(\frac{dy}{dt}\right)^2 + ay = \frac{dx}{dt} + bx$$

where a and b are constants.

If only x but not dx/dt is available, can the equation be set up on the computer by (a) canonical method, (b) dummy-variable method, (c) any other method?

At $t = 0$

$$y(0) = 1 \qquad \left(\frac{dy}{dt}\right)_0 = 0 \qquad \left(\frac{dx}{dt}\right)_0 = 2 \qquad x(0) = 0$$

If your answer is yes, program the computer. If no, explain.

4.8. Show the validity of the result of Example 4 by the following procedure: Take the equation

$$\sum_{i=0}^{n} a_n \frac{d^iy}{dt^i} = \sum_{j=0}^{n} b_j \frac{d^jx}{dt^j} \tag{a}$$

and reduce to the set of first-order equations

$$\begin{aligned} y &= y_1 + \alpha_n x \\ \frac{dy_1}{dt} &= y_2 + \alpha_{n-1}x \\ &\cdots\cdots\cdots\cdots \\ \frac{dy_{n-1}}{dt} &= y_n + \alpha_1 x \\ \frac{dy_n}{dt} &= -a_{n-1} - a_{n-2}y_{n-1} - a_1y_2 - a_0y_3 + \alpha_0 x \end{aligned} \tag{b}$$

Program equation set (b). Then by converting equation set (b) to an nth-order differential equation and comparing term by term with Eq. (a), find the α's in terms of the b's and show that this is the same result as we would obtain using the manipulation procedure of Example 4.

4.9. Assume that, in the circuit in Fig. P 4.9, $f(t)$ is a triangular wave and $y(t)$ is the solution. Suppose we have a square-wave generator, i.e., a device which generates $g(t) = df/dt$, and have no integrators available other than those which appear in the circuit. How do we insert $g(t)$ in the problem to generate $y(t)$? If necessary, rearrange computer elements and use additional amplifiers.

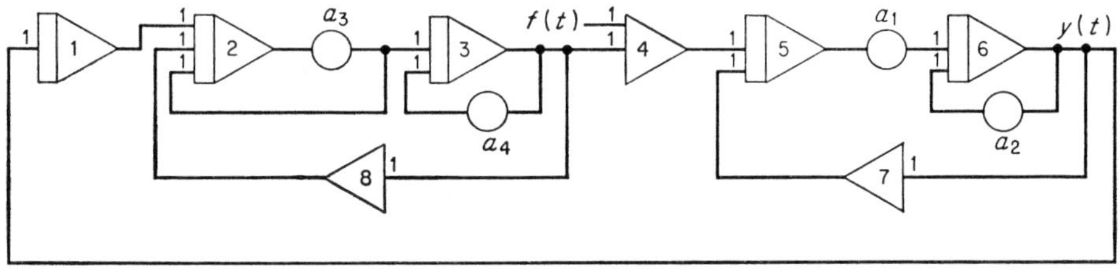

FIG. P 4.9

4.10. Using one high-gain amplifier, generate the following transfer functions with *RC* circuits:

(*a*) $\dfrac{e_o}{e_i} = -K_1 \dfrac{1}{1 + \tau p}$ (*b*) $\dfrac{e_o}{e_i} = -K_2 \dfrac{(1 + \tau_1 p)(1 + \tau_2 p)}{1 + \tau_3 p}$

(*c*) $\dfrac{e_o}{e_i} = -K_3 \dfrac{1 + \tau_1 p}{1 + \tau_2 p}$ (*d*) $\dfrac{e_o}{e_i} = -K_4(1 + \tau_1 p)$

4.11. Given that the initial conditions at $t = 0$ are

$$y = a_0 \qquad \frac{dx}{dt} = b_1$$

$$\frac{dy}{dt} = a_1 \qquad \frac{d^2x}{dt^2} = b_2$$

$$\frac{d^2y}{dt^2} = a_2$$

For the circuit shown in Fig. P 4.11, what are the initial conditions on the integrators?

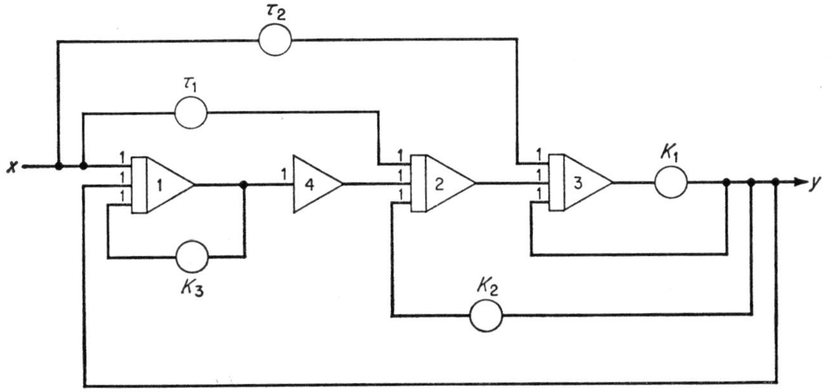

FIG. P 4.11

4.12. Given the equation

$$\sum_{i=0}^{n} a_i p^i y = \sum_{i=0}^{m} b_i p^i x$$

where the a_i's are constants but the b's are not and $a_n = 1$.
Figure P 4.12 is the computer program for the canonical form.

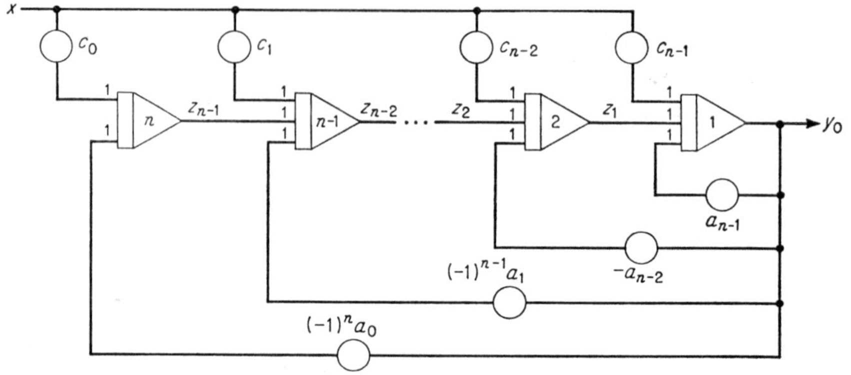

FIG. P 4.12

Find the values of the coefficients c_i in terms of the a's and b's. Also indicate the initial conditions on the integrators in terms of the initial conditions at $t = 0$ on y and its derivatives.

4.13. Given the differential equation

$$\frac{d^3y}{dt^3} + a_2\frac{d^2y}{dt^2} + a_1\frac{dy}{dt} + a_0 y = b_2\frac{d^2x}{dt^2} + b_1\frac{dx}{dt} + b_0 x$$

where all coefficients are constant. Find the values of the potentiometer coefficients in Fig. P 4.13 in terms of the a's and b's. Also indicate the variable which appears at the output of each integrator. NOTE: The c's and d's can be negative.

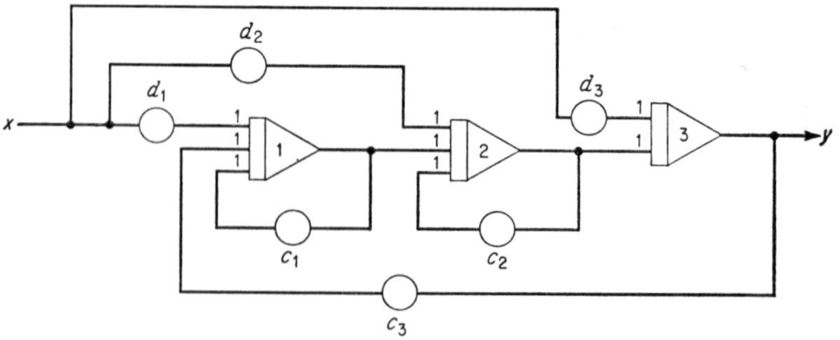

FIG. P 4.13

4.14. Given the equation

$$\frac{d^3y}{dt^3} + 3\frac{d^2y}{dt^2} - 4\frac{dy}{dt} + 2y = 7\frac{d^2x}{dt^2} + 0.5\frac{dx}{dt} - x$$

where the initial conditions at $t = 0$ are

$$\left(\frac{d^2y}{dt^2}\right)_0 = 6 \qquad \left(\frac{dy}{dt}\right)_0 = -2 \qquad y(0) = 1$$

$$\left(\frac{d^2x}{dt^2}\right)_0 = 1 \qquad \left(\frac{dx}{dt}\right)_0 = -3 \qquad x(0) = 4$$

Program the computer using the canonical and dummy-variable methods, and place the correct initial conditions on the integrators.

4.15. Find the transfer functions for the two circuits in Fig. P 4.15.

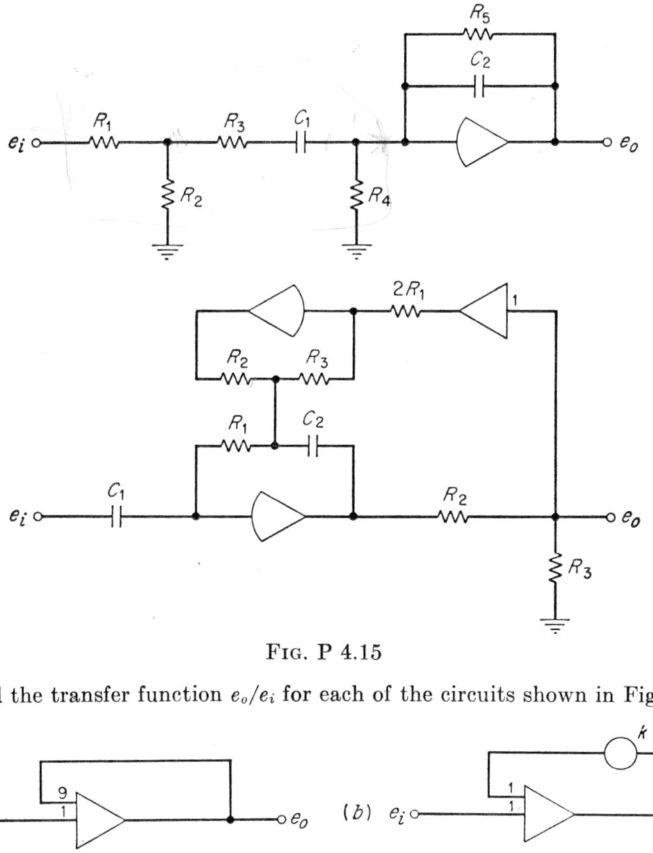

FIG. P 4.15

4.16. Find the transfer function e_o/e_i for each of the circuits shown in Fig. P 4.16.

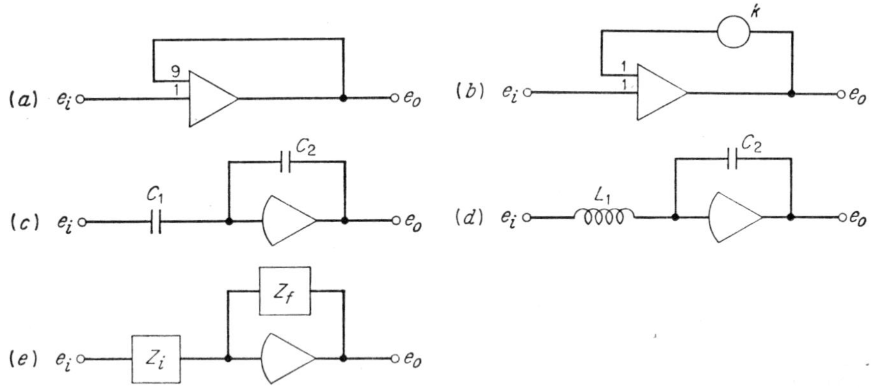

FIG. P 4.16

4.17. Table P 4.17 lists the approximate transfer functions and the error for the three circuits in Fig. P 4.17. The quantity $-A$ is the amplifier gain, and the z's are short-circuit transfer impedances. Derive these terms.

TABLE P 4.17

Circuit	Approximate transfer function e_2/e_1	Error
1	$-\frac{z_2}{z_1}$	$\frac{1}{A}\left(1+\frac{z_2}{z_1}+\frac{z_2}{z_3}\right)$
2	$-\frac{z_2}{z_1}$	$\frac{1}{A}\left(1+\frac{z_2}{z_1}+\frac{z_2}{z_4}+\frac{z_3}{z_4}+\frac{z_2z_3}{z_1z_4}\right)$
3	$-\left(\frac{z_2}{z_1}+\frac{1+\frac{z_2}{z_5}}{1+\frac{z_3}{z_4}+\frac{z_3}{z_5}}\right)$	$\frac{1}{A}\left(1+\frac{z_2}{z_1}+\frac{\frac{z_2}{z_3}+\frac{z_2}{z_4}}{1+\frac{z_5}{z_3}+\frac{z_5}{z_4}}\right)$

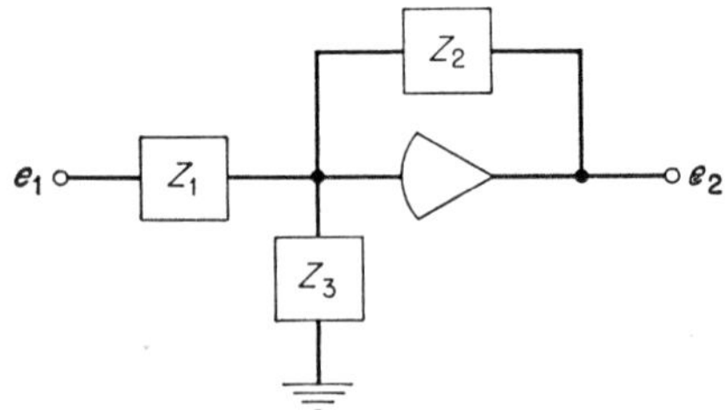

Circuit 1

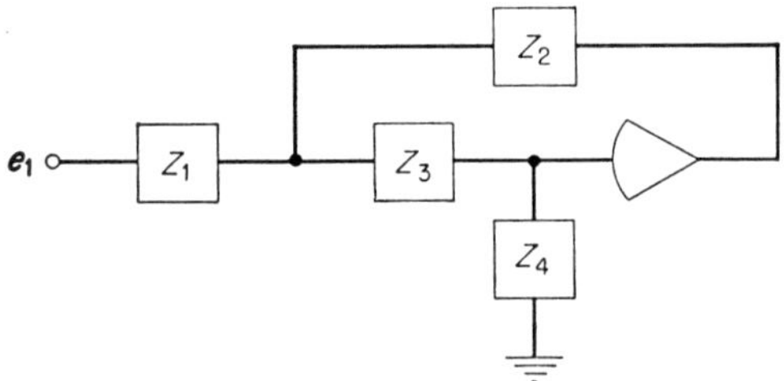

Circuit 2

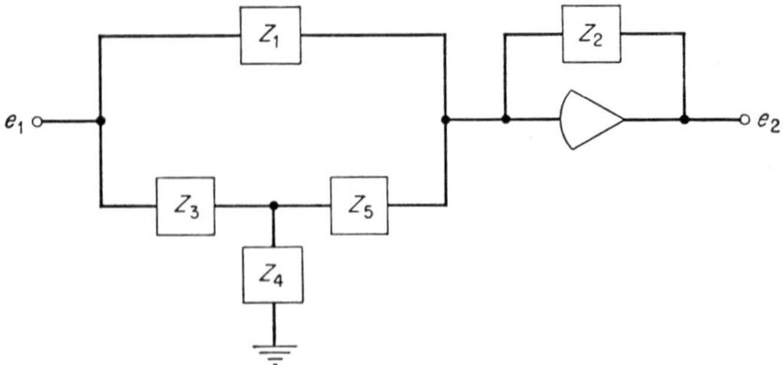

Circuit 3

FIG. P 4.17

4.18. For the circuit in Fig. P 4.18 are there any amplifiers or integrators which can be eliminated if all we want to obtain is e_o? HINT: To see if any elements can be eliminated, place as many elements as possible on their respective forward loops.

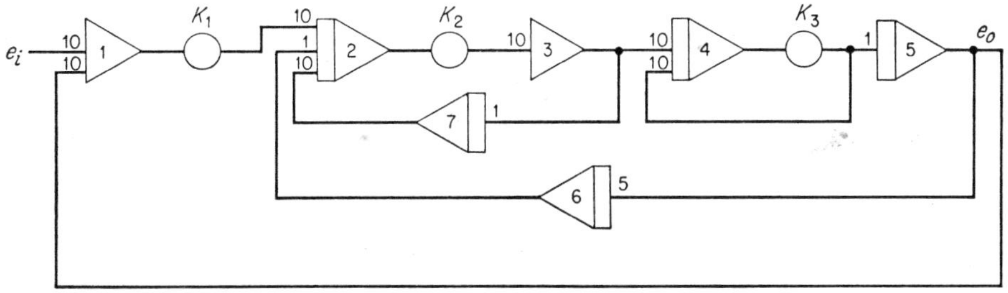

FIG. P 4.18

4.19. The circuit shown in Fig. P 4.19 has been programmed on the computer. Indicate how you would reprogram the computer to obtain dy/dt or $-dy/dt$ and d^2y/dt^2 or $-d^2y/dt^2$, using no more than one additional amplifier but any number of additional potentiometers.

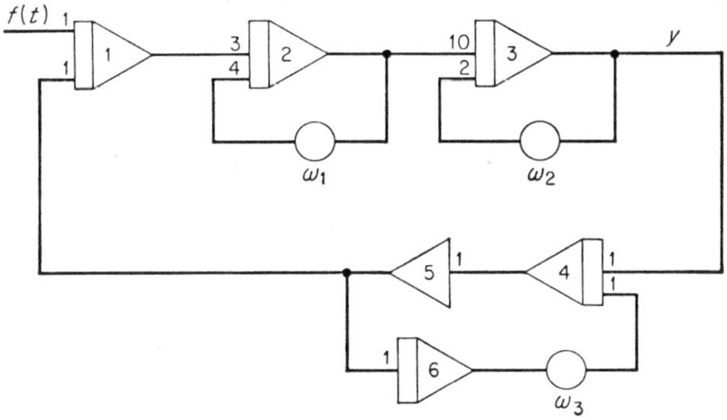

FIG. P 4.19

4.20. If we use the symbolic notation $e^{-\alpha p}$ to represent a time delay of α sec, where $e^{-\alpha p}f(t) = f(t - \alpha)$, manipulate the circuit of Fig. P 4.20 to eliminate the time delay from the forward path of loop 2. The quantities τ, K_1, K_2, . . . , K_5 are constants.

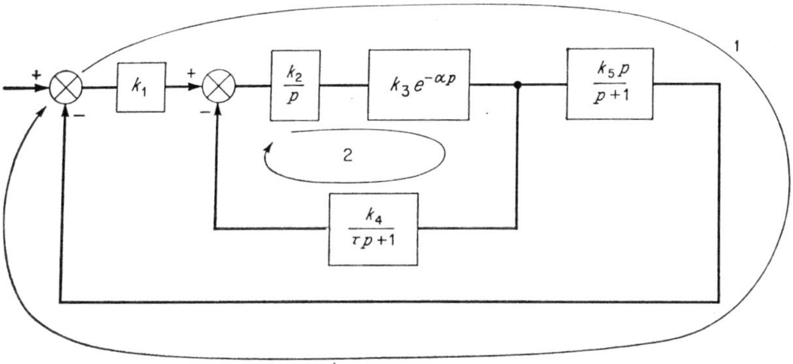

FIG. P 4.20

4.21. Figures P 4.21*a* and *b* represent the same set of differential equations. However, Fig. P 4.21*a* has one more integrator than Fig. P 4.21*b* and hence requires one more initial condition. Explain this apparent paradox.

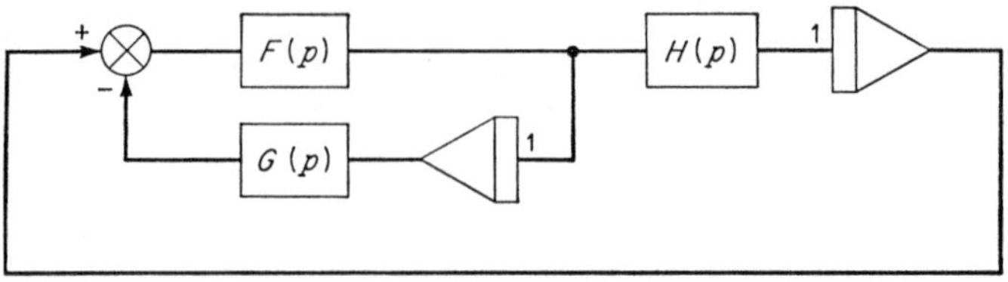

(*a*)

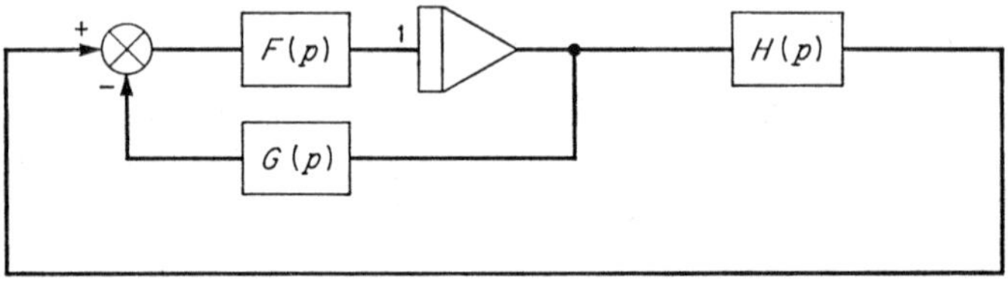

(*b*)

FIG. P 4.21

4.22. Given the equation

$$\frac{d^2y}{dt^2} + c_1 \frac{dy}{dt} + c_2 y = x$$

where, at $t = a$, the initial conditions are $y = 1$, $dy/dt = -5$.

Set up the computer to solve for y in the region

(*a*) $a \leq t \leq b$.

(*b*) $b \leq t \leq a$.

(*c*) If we time-scale at the input of each integrator for part *b*, are the integrator initial conditions the same or different for parts *a* and *b*?

(*d*) If we solve part *b* by first making the transformation $t' = -t$ and then programming the resultant equation with t' as the independent variable, are the integrator initial conditions the same or different for parts *a* and *b*?

4.23. For the equation

$$\frac{d^2y}{dt^2} + a_1 \frac{dy}{dt} + a_2 y = f(t)$$

where

$$f(t) = \sin wt$$

the computer operator obtains a solution at $t = 10$ for y and dy/dt. However, he neglects to record the initial conditions y and dy/dt at $t = 0$. Describe how he should program the computer with the information he has to find the initial conditions.

4.24. Given the equation

$$\frac{d^2y}{dt^2} + 3\frac{dy}{dt} + y = \frac{dx}{dt} + 7x$$

where at time $t = 0$

$$\frac{dy}{dt} = 1 \qquad y(0) = 2 \qquad x(0) = 3$$

The input x is available but not its derivative. Show the computer diagram when the computer is (*a*) sped up by a factor of 5, (*b*) slowed down by a factor of 5. Use both the canonical and dummy-variable forms.

Is x affected by the scaling? Explain your answer.

4.25. Figure P 4.25 represents a computer diagram before scaling. The computer operator slows down his problem by a factor of 10 but forgets to change the input gains of integrator 1. Which differential equation does he solve if he plots y on the reduced time scale? If the initial conditions are

$$IC_1 = A \qquad IC_2 = B \qquad IC_3 = C$$

what are the initial conditions of the differential equation?

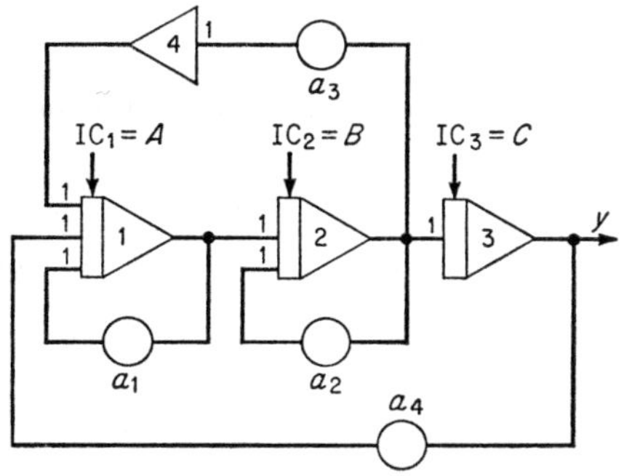

FIG. P 4.25

4.26. With a 10:1 slowdown of the computer solution, write the system of equations in terms of x and y and their derivatives represented by the computer diagram of Fig. P 4.26.

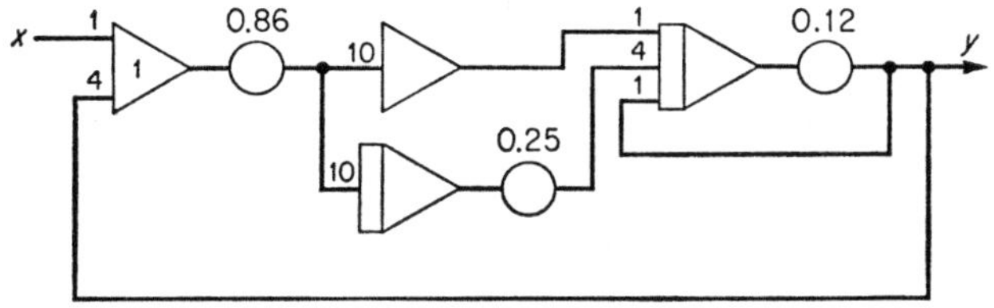

FIG. P 4.26

4.27. Time-scale the circuit in Fig. P 4.27 to run the problem backward.

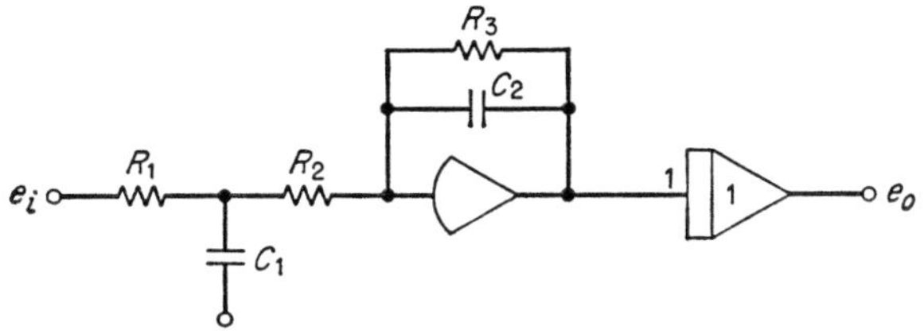

FIG. P 4.27

4.28. Given a system represented by the equations

$$\frac{d^2x_2}{dt^2} = 20x_2 + 40x_1 + x_0 + f(t)$$

$$\frac{d^2x_1}{dt^2} = -(75x_1 + 30x_2)$$

$$\frac{dx_0}{dt} = -5x_1$$

where $f(t)$ is a system input, i.e., a forcing function. The equations programmed on the computer give the following maxima during the run on a 1:1 scaling:

$$\max x_1 = 1 \text{ volt} \qquad \max x_2 = 250 \text{ volts}$$

Scale so that

(*a*) 10 volts $\leq \max x_1 \leq$ 100 volts
10 volts $\leq \max x_2 \leq$ 100 volts

(*b*) No loop gain exceeds 10.

4.29. For which of these nonlinearities can the problem be run backward as well as forward? See Chap. 5 for a description of phenomena *a* and *b* and Ref. 46 for the others.

(*a*) Rate limiting, (*b*) dead space, (*c*) backlash, (*d*) Coulomb friction, (*e*) hysteresis.

Chapter 5

5.1. Factorials may be determined by using a gamma function $\Gamma(\alpha)$, where

$$\Gamma(\alpha) = (\alpha - 1)!$$

$$\Gamma(\alpha) = \int_0^\infty x^{\alpha-1}e^{-x}\,dx \qquad (a)$$

Use Eq. (*a*) to determine the value of $n!$ when $n = 3, 4$. NOTE: This is intended only as an exercise and is not suggested as a practical way to evaluate factorials. When n is small, $n!$ is easily calculated directly. When n is large, one should use Stirling's approximation.

5.2. Given

$$y = \frac{1}{1 + \tau_1{}^2w^2}$$

Program the computer to plot log y versus log w, where $w > 0$.

5.3. Using the method of generalized integration, draw the computer diagram to plot the function

$$a(z) = \frac{4R[z(R + z)]^{\frac{1}{2}}}{R + 2z} \qquad -5 \leq z \leq 5$$

for $R = 8, 15$.

Indicate the values of the initial conditions.

5.4. Using no more than one electronic multiplier, generate

$$y = (t + a)^{-4}$$

HINT: Start with $z = (t + a)^{-2}$.

5.5. Using generalized integration (dx/dt is available), program the computer to solve

(*a*) $y = \tan x$ (*b*) $y = \ln x$ (*c*) $y = \left(\dfrac{dy}{dx}\right)^{dy/dx}$

5.6. Program and solve the integrals

(*a*) $\displaystyle\int_{-3}^{5} e^{-y^2/2}\,dy$ (*b*) $\displaystyle\int_1^5 \frac{e^{-3}}{z}\,dz$

5.7. Determine differential equations to generate the functions

(*a*) $y = \ln(1 + t)$ (*b*) $y = (t + a)^\alpha$

(*c*) $y = a^{-t^2}$ (*d*) $y = t \exp \dfrac{1}{t}$

(*e*) $y = c_1t^3 + c_2t^{-2}$ (*f*) $y = \dfrac{a}{3}t^2 + c_1t + \dfrac{c_2}{t}$

(*g*) $y = t(c_1 + c_2 \ln t) + \frac{3}{4}t^3$ (*h*) $y = c_1t \cos at + c_2t \sin at$

5.8. In Fig. P 5.8, where

$$e_i = \sin wt \qquad y = e^{-at}$$

what is the output e_o?

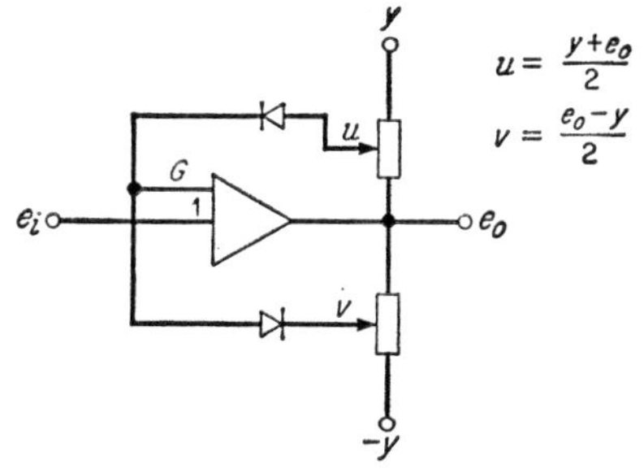

FIG. P 5.8

5.9. For the circuit in Fig. P 5.9 what is z?

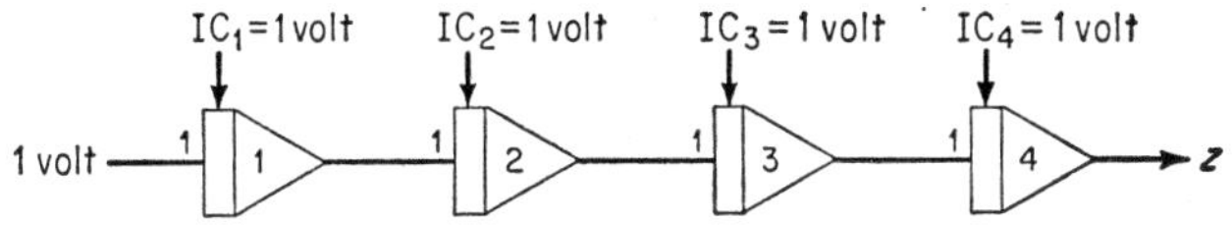

FIG. P 5.9

5.10. For the circuits in Fig. P 5.10 plot e_o versus e_i.

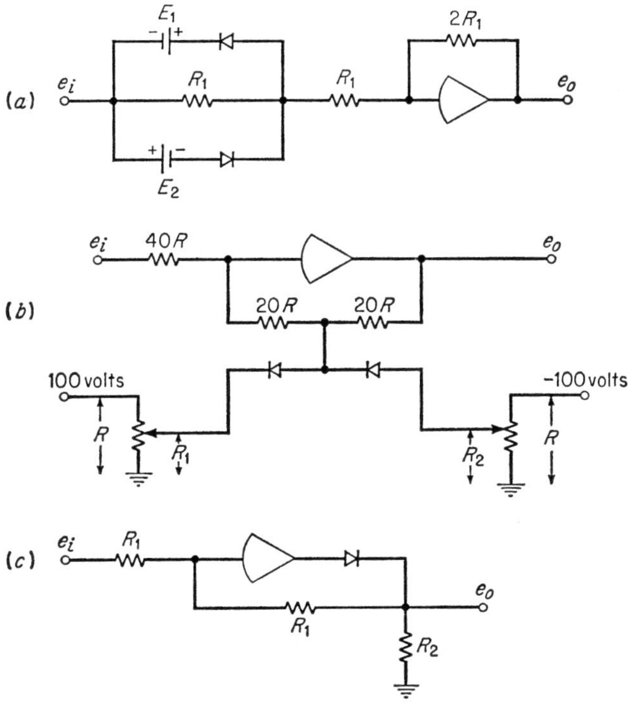

FIG. P 5.10

5.11. Explain the operation of the circuit in Fig. P 5.11.

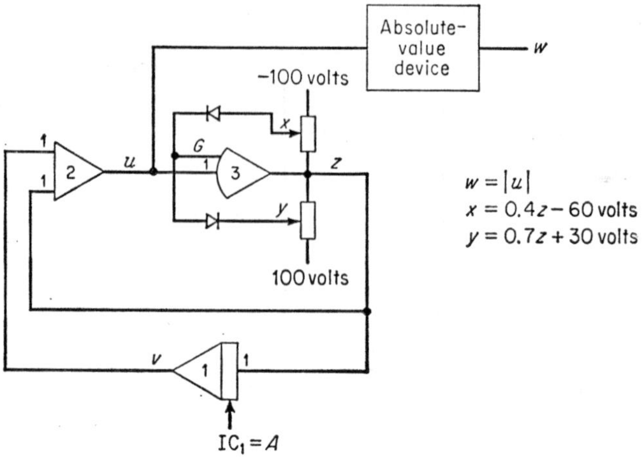

FIG. P 5.11

5.12. If we use a differentiator, we can obtain powers of x with the circuit shown in Fig. P 5.12. Explain how the circuit works.

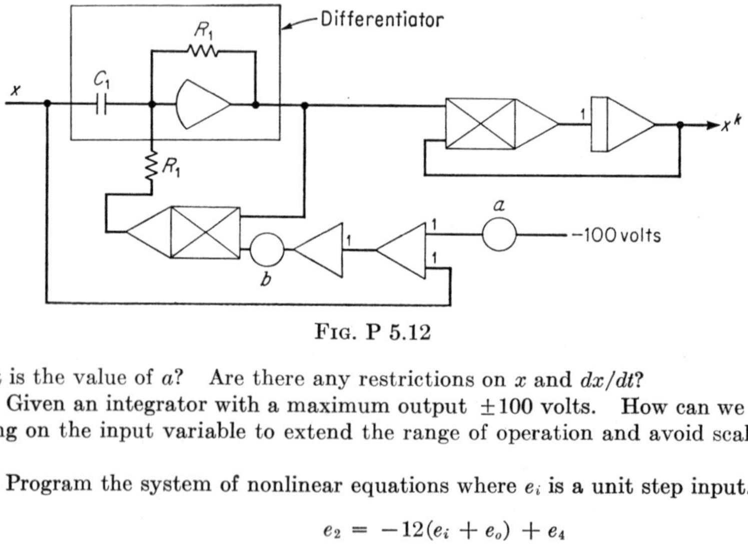

FIG. P 5.12

What is the value of a? Are there any restrictions on x and dx/dt?

5.13. Given an integrator with a maximum output ± 100 volts. How can we employ switching on the input variable to extend the range of operation and avoid scaling the input?

5.14. Program the system of nonlinear equations where e_i is a unit step input.

$$e_2 = -12(e_i + e_o) + e_4$$

$$e_3 = (e_2)^{\frac{1}{3}}$$

$$167e_3 + 234\frac{de_3}{dt} = \frac{de_4}{dt} + 0.2\frac{d^2e_4}{dt^2}$$

$$e_4 = \frac{de_0}{dt} + 0.025\frac{d^2e_0}{dt^2}$$

Draw a computer diagram using one servomultiplier with a scale factor of unity.

(*a*) No loop gain > 10.

(*b*) 10 volts $\leq$ max $|e_0, e_2, e_4| \leq$ 100 volts.

The expected voltages are

$$|e_2| < 12 \text{ volts} \qquad |e_4| < 10 \text{ volts} \qquad |e_0| < 2 \text{ volts}$$

Assume that the range of all amplifiers is ± 100 volts.

5.15. The antenna power spectrum can be represented as

$$P(\theta) = E(\theta)E^*(\theta)$$

$E^*(\theta)$ is the complex conjugate of $E(\theta)$.

$$E(\theta) = \int_{-L/2}^{L/2} f(y) \exp \{j[\beta y \sin \theta + \psi(y)]\} \, dy$$

where L = antenna aperture
$\beta = 2\pi/\tau$
$f(y)$ = amplitude distribution
$\psi(y)$ = phase

Program the computer to evaluate $E(\theta)$ for discrete values of θ.

5.16. The circuit shown in Fig. P 5.16 has been set up on the computer. Under ideal conditions the solution at the output of integrator 2 is

$$y = \sin wt$$

However, because of leakage in the program board the solution is

$$y = e^{at} \sin wt$$

By adding one or more connections between the elements, make the necessary compensations so that

$$y = \sin wt$$

Use additional potentiometers, integrators, and amplifiers if necessary.

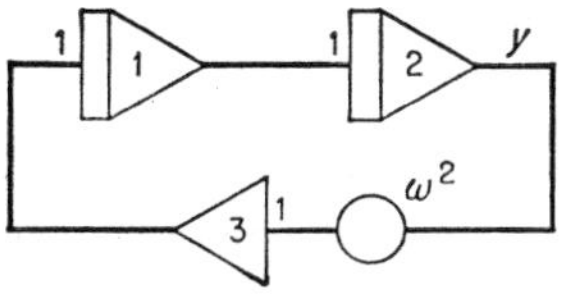

Fig. P 5.16

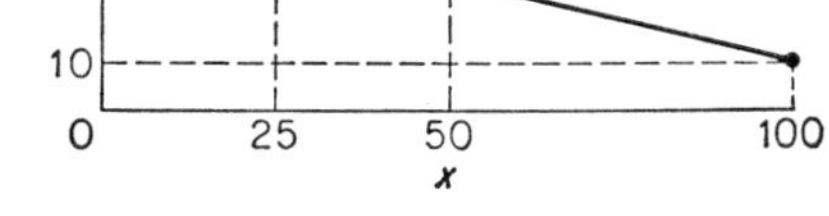

Fig. P 5.17

5.17. Suppose we want to generate the function in Fig. P 5.17 using a diode function generator. If the maximum slope on each diode is unity, we see that the required slope of the first segment is 2. How can we change the x variable to x', where $0 \leq x' \leq 100$ volts, so that no diode segment slope exceeds unity?

5.18. Assume that we have 10 diodes with break points spaced at equal intervals, 0°, 18°, 36°, . . . , 162°. If we want to minimize the least-squares error, i.e.,

$$\int_0^{180°} [\sin x - f(x)]^2 \, dx$$

formulate but do not solve the equations which will calculate the slopes of the individual diodes.

5.19. Let us generate

$$z = xe^{-y} \cos y$$

on a tapped potentiometer with 10 equally spaced taps and y as the input to the servomotor.

If we have available voltages ± 100 volts and the range $0 \leq y \leq 4\pi$, what is the error to three significant figures at $y = 3\pi/4,\ 2\pi/3$?

Adjust so that there is no error at the tap points.

5.20. Every straight-line segment function $f(x)$ can also be represented by the sum

$$f(x) = a + bx + \Sigma c_i|x - x_i|$$

where the x_i are the end points of the segments. Note that in this representation we use absolute-value devices instead of diode segments. Find the value of the coefficients if

$$\begin{aligned} f(x) &= 2x - 3 && 0 \leq x \leq 1 \\ f(x) &= x - 2 && 1 \leq x \leq 3 \\ f(x) &= 10 - 3x && 3 \leq x \leq 4 \\ f(x) &= 2 - x && 4 \leq x \leq 6 \\ f(x) &= -4 && x \leq 6 \end{aligned}$$

HINT: The term $g_i(x) = -\alpha|x - x_i| + \alpha|x - x_{i-1}|$ has a slope 2α in the region $x_{i-1} \leq x \leq x_i$ and zero slope elsewhere. Use proper combinations of the g_i to satisfy the conditions of $f(x)$ in each interval (x_{i-1}, x_i).

5.21. Let $f(x, y) = \sin xy$, a function defined in the region $0 \leq x \leq \pi/2$, $0 \leq y \leq \pi/2$. Using Eterman's method, represent the function by its approximation

$$f(x, y) = g_1(x) + g_2(y) + K$$

where K is a constant.

5.22. Suppose we are studying a system which contains a digital computer that solves the equation

$$y_n = K_1x_n - K_2 + x_{n-1}K_3y_{n-1} - K_4y_{n-2}$$

where

$$y_n = y(nT) \qquad x_n = x(nT)$$

T is a constant and n is an integer. The input $x(t)$ is continuous and is zero for $t < 0$.

Program the computer to solve the equation beginning from $t = 0$.

5.23. Given the function $f(t)$, generate the variable time delay $f[t - g(t)]$ using a Padé approximation $(n = 2)$.

HINT: Generate a fixed delay with respect to an appropriate variable. Then program the Padé circuit with respect to this variable by the method of generalized integration.

Chapter 6

6.1. An amplifier can be considered an implicit-function solver.

(*a*) What is the implicit function being solved?

(*b*) What is the error in the solution?

6.2. Set up the following equations and solve implicitly for y where the variables x and z are inputs. Calculate the error in the solution. Under what conditions does each error become large?

(*a*) $y^2 + \sin y + x = 0$ (*b*) $e^y + 2y + x + yz = 0$

(*c*) $y = \cosh^{-1} x$ (*d*) $y = x^{\frac{1}{2}}$

(*e*) $y = \ln x$

6.3. Set up the computer to solve the following differential equations:

$$\frac{dy}{dt} = \tfrac{1}{2}(1 + y^2)^{\frac{1}{2}} \qquad (a)$$

At $t = 0$, $y = 1$.

$$\frac{d^2y}{dt^2} + k\sqrt{t}\,\frac{dy}{dt} + \frac{1}{t}\,y = \sin wt \qquad t \geq 1 \qquad (b)$$

where k is a constant. At $t = 1$, $y = 0.5$, $dy/dt = 1.3$.

Assume that we have a function generator which gives $z = e^x$ when x is an input.

6.4. Using multipliers and implicit-function techniques, generate $y = x^{1.95}$ given x.

6.5. Conformal mapping is a transformation of variables from one surface to another whereby angles are preserved. Let

$$w = x + iy \qquad z = u + iv$$

Then some conformal transformations are

(a) $w = z^2$ (b) $w = a \ln z$

(c) $w = az + \dfrac{b}{cz} + d$, where $z \neq -\dfrac{d}{c}$ (d) $w = z + \dfrac{1}{z}$

If w is given, program the computer to obtain z in each of these transformations.

6.6. Solve for a and b as a function of u where a_1, a_2 are constants.

$$(1 - ja_1u)^{-\frac{1}{2}}(1 - ja_2u)^{-\frac{1}{2}} = e^{jub}(1 - jau)^{-1}$$

HINT: Break up into real and imaginary parts and note that the square root gives two solutions. Choose the one which gives the smallest positive angle θ, where for $(1 + j\alpha u)^{\frac{1}{2}}$

$$\theta = \tan^{-1}\frac{\alpha u}{2}$$

6.7. (a) What is the expression for y in Fig. P 6.7?

(b) If amplifier 1 has a capacitor C_1, placed from grid to output, what is the new expression for y?

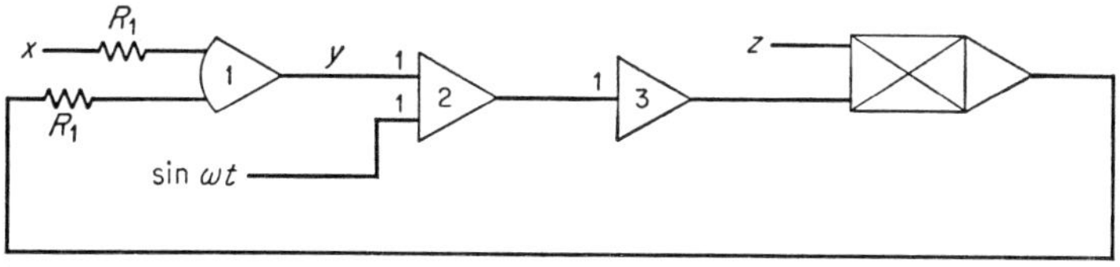

FIG. P 6.7

6.8. Show that $w = \sin^{-1}\theta$ in Fig. P 6.8.

HINT: This circuit uses generalized integrators (see Chap. 5).

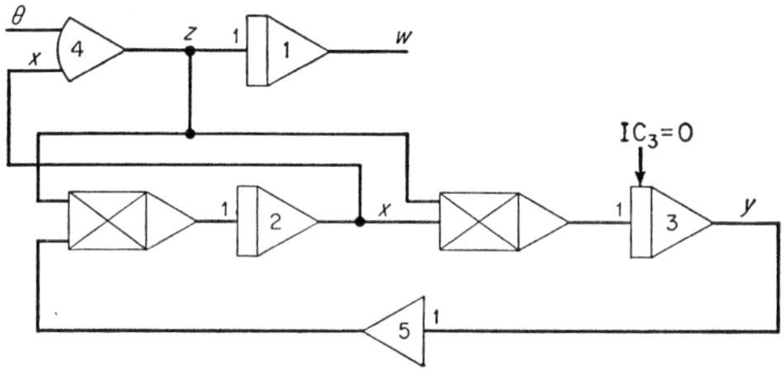

FIG. P 6.8

6.9. We are given the simultaneous set of linear algebraic equations

$$\begin{bmatrix} a_{11} & a_{12} & \cdots & a_{1n} \\ a_{21} & a_{22} & \cdots & a_{2n} \\ \cdots & \cdots & \cdots & \cdots \\ a_{n1} & a_{n2} & \cdots & a_{nn} \end{bmatrix} \begin{bmatrix} x_1 \\ x_2 \\ \cdots \\ x_n \end{bmatrix} = \begin{bmatrix} y_1 \\ y_2 \\ \cdots \\ y_n \end{bmatrix} \qquad (a)$$

where the a's are given.

Program the computer to find the coefficients of the inverse matrix b_{ij} in the equation

$$\begin{bmatrix} b_{11} & b_{12} & \cdots & b_{1n} \\ b_{21} & b_{22} & \cdots & b_{2n} \\ \cdots & \cdots & \cdots & \cdots \\ b_{n1} & b_{n2} & \cdots & b_{nn} \end{bmatrix} \begin{bmatrix} y_1 \\ y_2 \\ \cdots \\ y_n \end{bmatrix} = \begin{bmatrix} x_1 \\ x_2 \\ \cdots \\ x_n \end{bmatrix} \qquad (b)$$

Neglect computer stability problems.

HINT: In Eq. (*a*) choose a particular set of y's and solve from the x's. Substitute the x's and y's in Eq. (*b*) to solve for the b's. Observe that each set of y's in Eq. (*a*) will give us n equations in equation set (*b*). Choose enough independent sets of y's in Eq. (*a*) to enable us to get all the b's.

6.10. For the equation set

$$5 \cos \theta \cos \psi - 3 \sin \phi \sin \theta \sin \psi = 0$$
$$2 \sin \phi \sin \theta \cos \psi - 3 \cos \theta \sin \psi = 0$$

set up the computer in all the possible ways to solve for θ and ψ. Assume that ϕ is available as an input. Indicate the approximate errors in the solution.

6.11. Solve implicitly for z in the equation

$$\frac{z}{y} + \frac{y}{z} = k$$

Use no more than two multipliers to perform the necessary operations. y is available, and k is a positive constant. Are there any restrictions in the range of values of y and k?

6.12. Using implicit-function techniques with resolvers, solve for y in the equation

$$y = 1 - \frac{\phi}{(1 - k\phi^2) \tan \phi}$$

where ϕ is available and k is a constant.

For what values of ϕ does the solution error become large?

6.13. The equation set

$$y_1 = x_1 \cos \alpha - x_2 \sin \alpha \qquad y_2 = 2x_1 \sin \alpha - x_2 \cos \alpha$$

is not a rotation. Why?

6.14. Assume that a resolver operates in the range of angle $-\pi \leq \theta \leq \pi$. How can we use relays and other appropriate equipment to generate sines and cosines over the range $-3\pi \leq \theta \leq 3\pi$?

HINT

$$\sin (\pi - x) = -\sin (\pi + x) \qquad \cos (\pi - x) = \cos (\pi + x)$$

Chapter 7

7.1. Scale, with variable time z, the equation

$$\frac{d^2y}{dt^2} + a\frac{dy}{dt} + by = 0$$

so that $z = e^t$.

7.2. Given that we want to solve the equation

$$y = A \sin \theta$$

where $A = 100$ and $\theta = 1 + 2t$

for $0 \leq t \leq 5$

How do we scale, using a servomultiplier or electronic multipliers to minimize the solution error?

Refer to generalized integration in Chap. 5.

7.3. Indicate the expansion of the first three terms in the series around $x = x_a$:

(*a*) $\frac{d}{dx}(\tanh x)$ (*b*) $\frac{d}{dx}(\sinh^{-1} x)$ (*c*) $\cosh x$

(*d*) $\int_0^x ye^{-y^2}\,dy$ (*e*) $\ln(\sin x)$ (*f*) $\sin x$

7.4. Perturb the following quantities about their nominal values:

(*a*) $f(\theta, \psi, \phi) = \cos\theta \sin\phi \sin\psi$ (*b*) $g(x, \theta) = e^{x \cos\theta}$

(*c*) $f(x) = x^{x+1}$

7.5. Given

$$\frac{d^2y}{dt^2} + \exp\left(y\frac{dy}{dt}\right)\frac{dy}{dt} + y = g(t)$$

Perturb around $g(t)$ and obtain the equations.

7.6. Figure P 7.6 shows the dynamics of a rocket flying in the atmosphere where

$L \equiv$ lift force $D \equiv$ drag force $w \equiv$ wind velocity

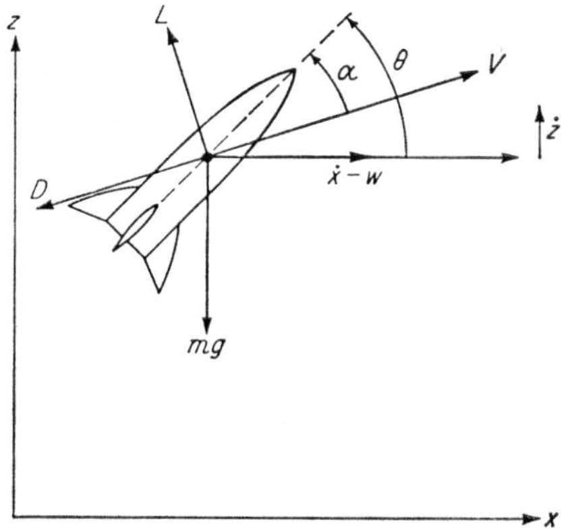

FIG. P 7.6

The rocket equations of motion are

$$\alpha = \theta - \tan^{-1}\frac{\dot{z}}{\dot{x} - w} \qquad \sin(\theta - \alpha) = \frac{\dot{z}}{V}$$

$$V = [\dot{z}^2 + (\dot{x} - w)^2]^{\frac{1}{2}} \qquad \cos(\theta - \alpha) = \frac{\dot{x} - w}{V}$$

$$\begin{aligned}
m\ddot{x} &= L\sin(\theta - \alpha) - D\cos(\theta - \alpha)\\
m\ddot{z} &= L\cos(\theta - \alpha) - D\sin(\theta - \alpha) - mg\\
I\ddot{\theta} &= M\\
L &= \tfrac{1}{2}\rho V^2 C_L S\\
D &= \tfrac{1}{2}\rho V^2 C_D S\\
M &= \tfrac{1}{2}\rho V^2 C_m S b\\
C_L &= f_1(\alpha)\\
C_D &= f_2(\alpha)\\
C_m &= f_3(\alpha)\\
\rho &= f_4(z)
\end{aligned}$$

I, b, S, m, and g are constants.

Perturb w and ρ and find the perturbation equation.

This problem is taken from Ref. 84.

7.7. In Fig. P 7.7*a* and *b* we have two circuits for obtaining e^x/x. Assume that the maximum output voltage of the function generator and the servomultiplier is 100 volts and that each has the same error, δ. For the range $1 \le x \le 6$, scale the two circuits to minimize the solution error and compare the results. The servomultiplier scale factor is taken as unity.

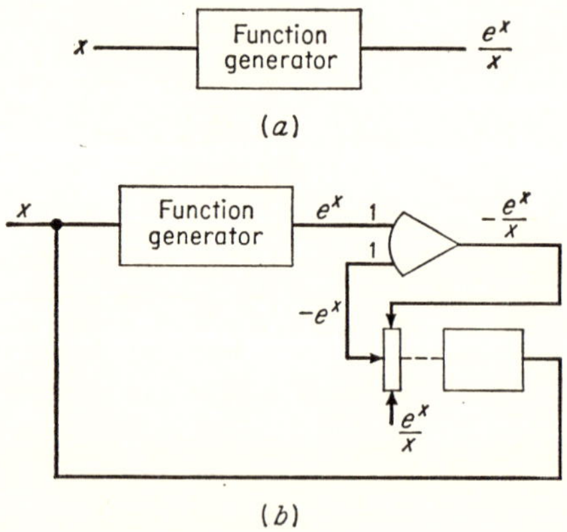

FIG. P 7.7

7.8. What do we do if we have to determine $\partial^2F/\partial x^2$, $\partial^2F/\partial y^2$, $\partial^2F/\partial x\,\partial y$ graphically at some nominal value x_0, y_0, where $F(x, y)$ is plotted as a continuous function of x and discrete function of y?

7.9. What would be the parameter-influence coefficients for variation of the initial conditions in Eq. (7.67)? Assume that n and λ are constants.

HINTS: See Appendix 1 for the equivalence of initial conditions and forcing functions.

7.10. For the equation

$$\frac{d^2x}{dt^2} + u\frac{dx}{dt} + \lambda x + vx^3 = f(t)$$

program the computer to obtain the parameter-influence coefficients for u, λ, and v.

7.11. Suppose we have x, dx/dt, and d^2x/dt^2 at $t = t_1$ and want to extrapolate to find x at $t = t_2$. Program the computer to accomplish this.

7.12. In a given problem we find that the relays controlling the change from RESET to OPERATE in the integrators switch at different times. To reduce the error due to this problem, would you speed up or slow down the problem solution by time scaling? Explain your answer.

7.13. Suppose we had the circuit in Fig. P 7.13, where, because of the servopotentiometer, we had excessive noise on z. Revise the circuit to reduce the noise. Do not rescale the problem.

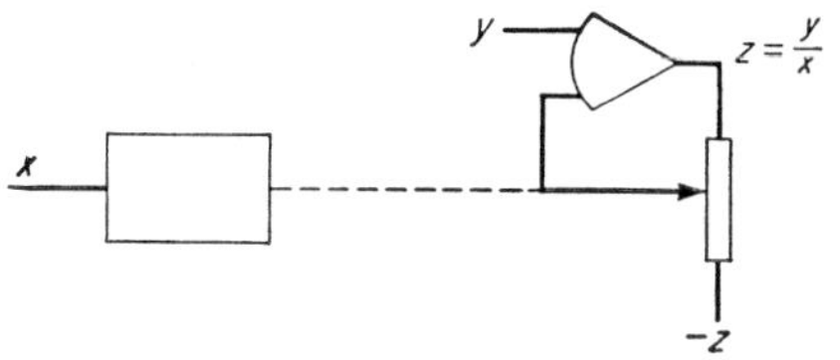

FIG. P 7.13

7.14. In simulating the beam rider geometry shown in Fig. P 7.14 we are interested in placing the missile and target on the same beam. We must therefore measure the distance ϵ, where

$$\epsilon = -x \sin \phi + y \cos \phi \qquad (a)$$

Usually, ϵ is very small and x and y, which must be scaled separately, are large. Since we cannot scale directly on ϵ, the computer solution may be considerably in error.

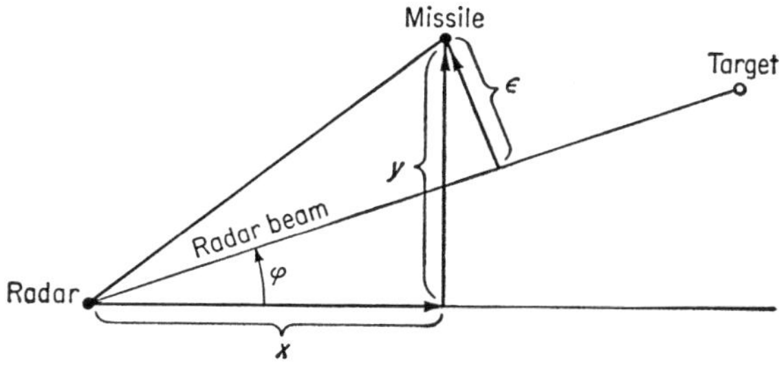

FIG. P 7.14

Suppose that the derivatives $\dot{x}$, $\dot{y}$, $\dot{\phi}$ are available and instead we generate

$$\dot{\epsilon} = -R\dot{\phi} - \dot{x} \sin \phi + \dot{y} \cos \phi$$

where

$$R = x \cos \phi + y \sin \phi$$

and integrate to get ϵ.

Under what conditions is this more accurate than the generation of ϵ directly by Eq. (a)? In your discussion assume that certain errors arise in the generation of sines and cosines and products and that these are of the same order of magnitude. Neglect effects of potentiometer wiper noise, which would, of course, be reduced by simulating $\dot{\epsilon}$ and integrating.

This problem was suggested by L. Bauer [137, pp. 5–60, where the reader will also find a partial discussion of the answer].

7.15. Suppose the function generator is set to generate Fig. P 7.15 and the reference voltage increases by 10 per cent. Show that the result is the same before and after the shift. Assume that the input scales are proportional to the reference voltage, and that the output is measured relative to the reference.

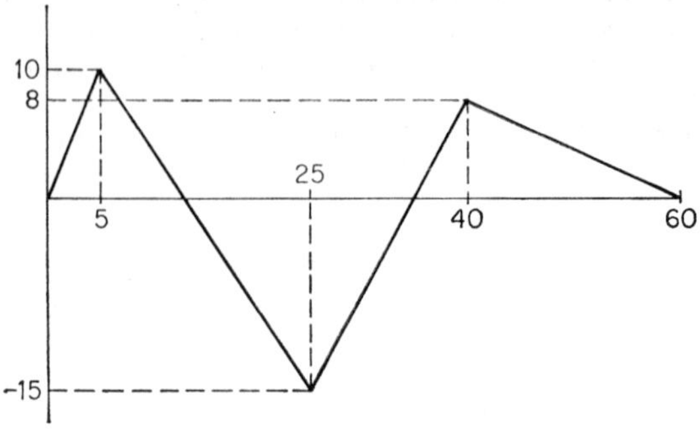

FIG. P 7.15

7.16. In a readout circuit suppose you resort to time scaling to reduce the relay-time switching error. Would you speed up or slow down the problem? Explain. (Compare with Prob. 7.12.)

Chapter 8

8.1. Which of the following functions are distance functions?

(*a*) $d_1(u, v) = \begin{cases} \sin(u - v) & \text{where } 0 \le u - v \le \pi \\ 0 & \text{elsewhere} \end{cases}$

(*b*) $d_2(u, v) = \sin|u - v|$ (*c*) $d_3(u, v) = e^{|u-v|}$

(*d*) $d_4(u, v) = 0$ (*e*) $d_5(u, v) = |u - v|^5$

8.2. For the polynomial

$$x^3 + a_1x^2 + a_2x + a_3 = 0$$

obtain the equations of steepest descent when

(*a*) $a_1 = 3$ $a_2 = 0$ $a_3 = -4$

(*b*) $a_1 = -7$ $a_2 = 3 - j5$ $a_3 = 10$

NOTE: a_2 is complex.

8.3. For the equation

$$\begin{aligned} f(x_1, x_2, x_3) &= 0 \\ &= x_1 - x_2 + 2x_3 + x_1^2 + 5x_1x_2 + 3x_1x_3 - 7x_2x_3 - 2x_2^3 + 6x_3^2 \end{aligned}$$

set up the equations of steepest descent to solve for x_1, x_2, and x_3. Assuming $x_3 = k$, set up the equations to plot the solution in the x_1, x_2 plane.

8.4. The circuit in Fig. P 8.4 when functioning properly has the solution

$$\dot{x} = wy \qquad \dot{y} = -wx$$

where $$x = \cos wt \qquad y = -\sin wt$$

However, because of drift, the amplitude of the signal is incorrect. To remedy this we can use an error term

$$\epsilon = 1 - \cos^2 wt - \sin^2 wt = 1 - \alpha^2 - y^2$$

and minimize it by using the distance function

$$d = \epsilon^2$$

How are the equations for $\dot{x}$ and $\dot{y}$ modified to minimize ϵ? (See also Fifer [46], pp. 407–411, for additional discussion of this technique.)

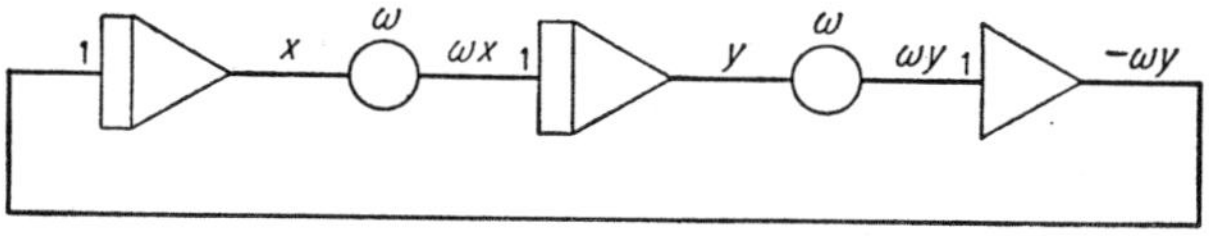

FIG. P 8.4

8.5. In the implicit equation

$$F(x, y, z) = 0 \qquad (a)$$

let us solve for z. If there are multiple solutions, a given computer program can obtain only half the solutions (see Chap. 6). How can we use optimization techniques to obtain all the solutions without modification of the circuit? For any given solution is this a better or worse method than the direct solution of Eq. (a)?

8.6. The matrix equation

$$[A]\,[x] = \lambda_i[x]$$

where

$$[A] = \begin{bmatrix} a_{11} & a_{12} & \cdots & a_{1n} \\ a_{21} & a_{22} & \cdots & a_{2n} \\ \cdots & \cdots & \cdots & \cdots \\ a_{n1} & a_{n2} & \cdots & a_{nn} \end{bmatrix}$$

$a_{ij} = a_{ji}$ (the matrix is symmetric and real), and

$$[x] = \begin{bmatrix} x_1 \\ x_2 \\ \cdot \\ \cdot \\ \cdot \\ x_n \end{bmatrix}$$

has n values of λ_i which will satisfy the equation.

These values are designated eigenvalues, and each vector $[x]$ associated with an eigenvalue is called an eigenvector.

The eigenvalues satisfy the equation

$$\lambda = [A]\,[x]\,[x] \qquad (a)$$

subject to the constraint

$$[x]\,[x] = x_1^2 + x_2^2 + \cdots + x_n^2 = 1$$

Thus the maximum of Eq. (a) yields the maximum eigenvalue, and the minimum yields the minimum eigenvalue.

Set up the steepest-descent equations which will determine the eigenvectors for the minimum and maximum eigenvalues when

$$A = \begin{bmatrix} 1 & 3 & -4 \\ 3 & 0 & 1 \\ -4 & 1 & 7 \end{bmatrix}$$

HINT: There must be three equations, one for each x_i.

See also Ref. 137 for a detailed discussion of this technique of determining eigenvectors.

8.7. Obtain a least-squares fit to

$$f(t) = \sin t$$

using straight-line segments, where

$$g(t) = b_0 + b_1t + b_2(t - a_2)u(t - a_2) + b_3(t - a_3)u(t - a_3)$$

for the range $0 \le t \le 3\pi/4$; $b_0, \ldots, b_3$ and a_2 and a_3 are the parameters to be varied; and $u(x - a_i)$ are unit step functions.

Thus, starting with the equation

$$E = \int_0^{3\pi/4} [f(t) - g(t)]^2 \, dt$$

we obtain the set of six equations

$$\frac{\partial E}{\partial \alpha_i} = 0 \qquad \alpha_i = b_0, \ldots, b_3, a_2, a_3$$

This set of simultaneous equations in the unknowns α_i can be solved by a method of steepest descent.

Set up the approximate distance function and then obtain the steepest-descent equations.

8.8. Suppose $g(\alpha_1, \alpha_2)$ is a distance function we want to minimize. Its relations to another function, $f(\alpha_1, \alpha_2)$, is shown in Fig. P 8.8. The analytic form of $f(\alpha_1, \alpha_2)$ is known. What are the steepest-descent equations? Give numerical quantities when they can be calculated.

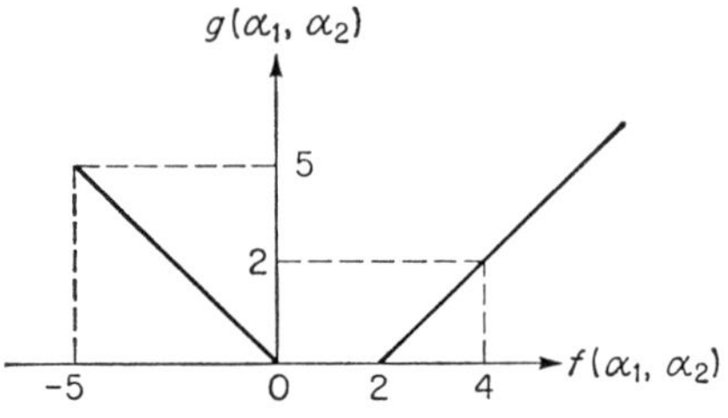

FIG. P 8.8

8.9. The distance between a line $Ax + By + C = 0$ and a point $P(x_1, y_1)$ is

$$r = \left| \frac{Ax_1 + By_1 + C}{(A^2 + B^2)^{\frac{1}{2}}} \right|$$

Assume that there are three lines which have the equations

(1) $\qquad Ax + By + C = 0$

(2) $\qquad Ex + Fy + G = 0$

(3) $\qquad Hx + Ky + L = 0$

and that no more than two of the lines are parallel.

We shall designate the distance between line 1 and a point in the plane as r_1. Also r_2 and r_3 will be the distance from this point to lines 2 and 3, respectively.

We wish to find the point in the plane that will minimize each of the functions d_1 and d_2, where

$$d_1 = r_1^2 + r_2^2 + r_3^2 \qquad d_2 = |r_1| + |r_2| + |r_3|$$

Set up the computer equations so that from an arbitrary starting point (x, y) the computer will by the method of steepest descent determine the point at which d_1 and d_2 are minimized. NOTE: The equations for minimizing d_1 and d_2 may be different.

8.10. The behavior of the electrical circuit in Fig. P 8.10 for S_1 closing at $t = 0$ is described by the equation

$$e(t) = i(t)R(t) + \frac{1}{C}\int_0^t i(\tau)\,d\tau$$

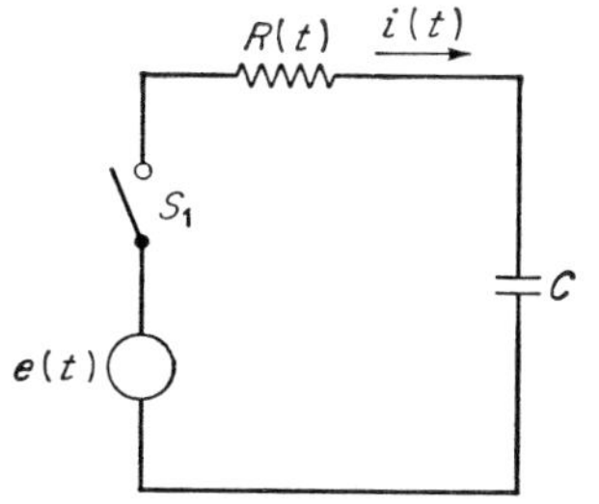

FIG. P 8.10

where C is a constant. Let

$$e(t) = K \qquad \text{for } t \geq 0$$

Suppose we want to change the capacitor C to a voltage $v(t)$, where

$$v(t) = \frac{1}{C}\int_0^t i(\tau)\,d\tau$$

and

$$v(t) < e(t)$$

The resistor has the behavior

$$R(t) = a + bt + ct^2$$

Set up the equations to minimize on the computer the energy designated as

$$\int_0^t i^2(\tau)R(\tau)\,d\tau$$

8.11. We want to approximate the function $f(x)$ by the polynomial

$$g(x) = a + bx + cx^2$$

The distance function d to be minimized has the form

$$d = \sum_{i=1}^{4} [f(x_i) - g(x_i)]^2 \qquad a^2 + b^2 + c^2 \leq 10$$

$$d = \sum_{i=1}^{4} |f(x_i) - g(x_i)| \qquad a^2 + b^2 + c^2 > 10$$

where x_1, x_2, x_3, x_4 are fixed values of x. Obtain the necessary equations and draw the computer diagram to find a, b, and c by a continuous steepest-descent method.

8.12. Program the computer to minimize the function

$$F(x_1, x_2, \ldots, x_n) = \sum_{i=1}^{n} x_i \ln x_i$$

subject to the conditions

(a) $\sum_{i=1}^{4} ix_i = 10$ (b) $x_i \geq 0$ for all i

8.13. Given a waveform $f(t)$ which is the solution of a fourth-order differential equation, we want to fit it with a second-order differential equation

$$\frac{d^2y}{dt^2} + a\frac{dy}{dt} + by = 0$$

where a, b, and the initial conditions are varied.

Program the computer to minimize the following two distance functions by the method of discrete steepest descent:

$$(a)\ d_1 = \int_0^{t_1} |f(t) - y|\, dt \qquad (b)\ d_2 = \int_0^{t_2} [f(t) - y]^2\, dt$$

NOTE: Include the equations which yield the parameter-influence coefficients.

8.14. A farmer has a rectangular plot of ground available for feeding a goat, with the perimeter shown in Fig. P 8.14. The goat is to be tethered at the point (0, 0) and allowed to move in a circular path of radius R. The farmer now has a problem in determining the length R of the rope. He would like to have the goat forage as much as possible in the rectangular plot. However, outside the plot there are shrubs which will affect the

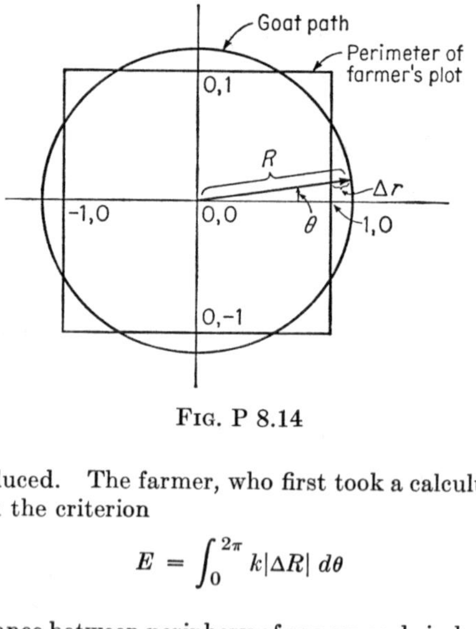

FIG. P 8.14

quality of the milk produced. The farmer, who first took a calculus course in an agricultural college, decides on the criterion

$$E = \int_0^{2\pi} k|\Delta R|\, d\theta$$

where ΔR = radial distance between periphery of square and circle as defined in the figure
$k = 1$ when $\Delta R < 0$
$k = 1 + \Delta R$ when $\Delta R \geq 0$

Set up the computer to minimize E and thereby find R.

8.15. In trying to solve a set of n linear simultaneous equations of the form

$$\sum_{j=1}^{n} -(a_{ij}x - b_i) = 0 \qquad i = 1, 2, \ldots, n$$

we face the problem of finding a stable computer configuration. The instabilities are caused by time lags in the computer elements. Rather than attempt to find which of the $n!$ computer setups is stable, we program instead the following set of equations:

$$\sum_{j=1}^{n} a_{ij}x_j - b_i = e_i$$

$$xj = - \sum_{c=1}^{n_i} a_{ij}E_i$$

$$\frac{dE_i}{dt} = ke_i$$

Show that these equations use the method of steepest descent.
HINT: Take

$$d = \sum_{i=1}^{n} e_i^2$$

as the distance function.

8.16. Consider a simplified version of the intercept problem shown in Fig. P 8.16. At time $t = t_0$ a particle starts at rest, $r = 0$, $s = 0$, from point $x = 0$, $y = 0$. At time

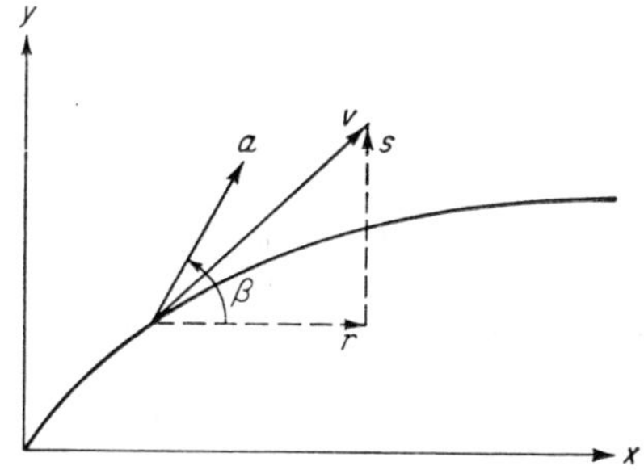

FIG. P 8.16

$t = T$ we want the particle to come as close as possible to the point $x = h$, $y = k$. We therefore choose as our distance function to be minimized

$$d = \phi[x(T), T] = \tfrac{1}{2}[(x - h)^2 + (y - k)^2] \qquad t = T$$

Assuming that the particle is a mass point, with drag and gravitational effects ignored, our equations of motion become

$$\frac{dr}{dt} = a \cos \beta$$

$$\frac{ds}{dt} = a \sin \beta$$

$$\frac{dx}{dt} = r$$

$$\frac{dy}{dt} = s$$

Take the magnitude of acceleration a as a constant and its angle β as a control variable. Set up the differential equations to determine the time-varying value of β by the method outlined in Sec. 8.9. Indicate values of all initial conditions.

Chapter 10

10.1. Using Fig. A8.1, draw a curve of d versus n for a confidence-interval size of ± 0.01.

10.2. Using the general technique developed in Sec. 10.2.3, find curves such as those of Figs. A8.1 and A8.2 for the case

$$p(x;\, \theta) = \tfrac{1}{2} e^{-(x-u)^2/2}$$

Let

$$\hat{\theta} = \bar{x} = \frac{\sum_{i=1}^{n} x_i}{n}$$

Then

$$h(\hat{\theta},\, \theta_0) = \sqrt{\frac{n}{2\pi}}\, e^{-(n/2)(\bar{x}-n)^2}$$

10.3. Choose a sample of 10 from a table of random numbers having a population of $N(0, 1)$. Calculate s and R_{10} and compare them as estimates of $\sigma = 1$.

10.4. Find the two numbers $r_{0.05/2}$ and $r_{(1-0.05)/2}$ in Eq. (10.6). Repeat for Eq. (10.9).

10.5. Suppose we assign a + sign every time the result of an analog-computer run yields a "positive voltage" and a − sign every time it yields a "negative voltage." Suppose positive and negative voltage are equally likely. If a given sequence of run yields a sequence of signs

$$+ \quad + \quad + \quad + \quad - \quad + \quad + \quad + \quad - \quad + \quad + \quad + \quad + \quad + \quad -$$

do we suspect for $\alpha = 0.05$ that the computer is malfunctioning?

10.6. Using the "runs test" in Prob. 10.5, do we suspect that the computer is malfunctioning?

10.7. Discuss the meaning of the term *design of experiments.*

10.8. Give examples where the techniques of Sec. 10.5 might be applicable other than in the area of optimum tolerance assignment.

Chapter 11

11.1. The white-noise output of a gaussian noise generator $z(t)$ is passed through first-order filters as shown in Fig. P 11.1.

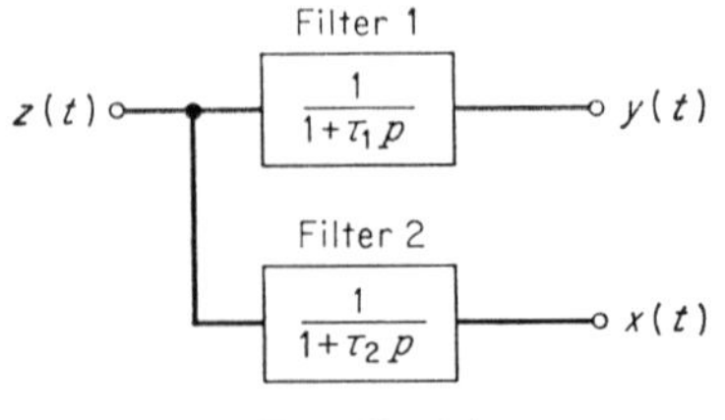

FIG. P 11.1

(*a*) Suppose we want two random initial conditions which have a correlation coefficient $\rho = 0.3$. Indicate the time constant of the filter if $x(t)$ and $y(t)$ are sampled at the same time. What would the time constant be if x is sampled 0.5 sec before y? Find the time constant if x is sampled 0.5 sec after y.

(*b*) Using the filter obtained in part *a*, what transformation should be made on $x(t)$ and $y(t)$ to yield two independent random samples at a given time?

(*c*) Suppose the time between samples is 10 sec. Let us designate the two variables which give the random samples in part *b* as $z_1(t)$ and $z_2(t)$. If t_1 is the time for one sample and $t_2 = t_1 + 10$, what is the correlation coefficient between $z_1(t_1)$ and $z_2(t_2)$?

(*d*) Suppose we want two independent noise sources to be used as problem forcing functions. Can we transform $x(t)$ and $y(t)$ to accomplish this? If so, indicate the transformation. If not, explain your answer.

HINT: For white noise with spectral density $S_z(f) = 1$,

$$E[x(t+\alpha)y(t)] = \int_{-\infty}^{t} h_1(t,\tau)h_2(t+\alpha,\tau)\,d\tau$$

where $h_1(t,\tau)$ is the weighting function for filter 1 and $h_2(t+\tau)$ is the weighting function for filter 2.

Thus

$$\rho(t) = \frac{E[x(t+\alpha)y(t)]}{\sigma_x\sigma_y}$$

where σ_x and σ_y are the standard deviations of x and y.

11.2. (*a*) x and y are stationary noise sources in Fig. P 11.2 which are uncorrelated and have the following spectral densities:

$$S_x(f) = \frac{1}{1+\tau_1^2w^2} \qquad \text{for } x$$

$$S_y(f) = \frac{1}{1+\tau_2^2w^2} \qquad \text{for } y$$

Find the spectral density of the voltage across C_1.

(*b*) x and y are stationary noise sources having the spectral densities indicated in part *a*. However, their correlation function is

$$R_{xy}(\tau) = e^{-\beta|\tau|}$$

Find the spectral density of the voltage across C_1.

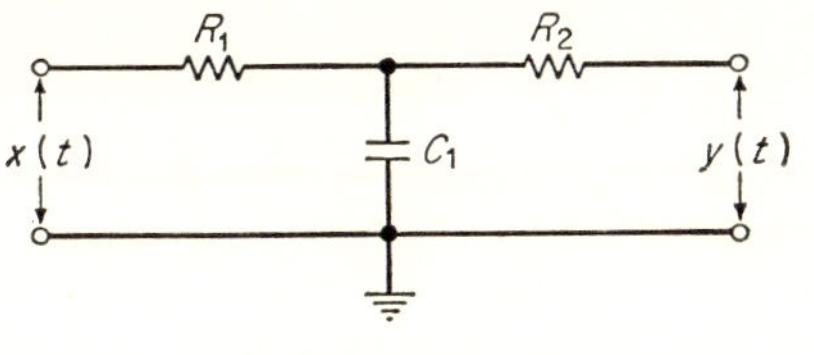

FIG. P 11.2

11.3. Suppose we want to sample three quantities a, b, and c with the following probabilities:

$$p(a) = \tfrac{1}{2} \qquad p(b) = \tfrac{1}{3} \qquad p(c) = \tfrac{1}{6}$$

We have a gaussian noise source with density

$$p(x) = \frac{1}{15\sqrt{2\pi}}\,e^{-x^2/30}$$

Set up a relay circuit and other elements for proper sampling.

11.4. Assume that k_1 and k_2 are set in Fig. P 11.4 so that, for $x = A \sin wt$,

$$E(w_1) = E(w_2) = A$$

Now let $x = f_2(t)$ be inserted, where x is a stationary noise source with probability density

$$p(x) = \frac{1}{\sqrt{2\pi}\,\sigma} e^{-x^2/2\sigma^2}$$

What will $E(w_1)$ and $E(w_2)$ read?

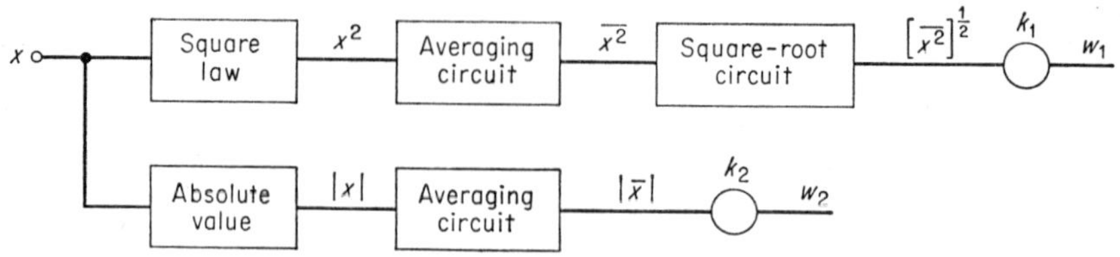

FIG. P 11.4

11.5. Suppose that in Eq. (11.92) the probability distribution is measured in the following two different ways:

(*a*) Each probability $\hat{p}_1$, $\hat{p}_2$, $\hat{p}_3$, and $\hat{p}_4$ is obtained in a separate set of computer runs and is statistically independent. Each $\hat{p}_i$ is the measured estimate of p_i.

(*b*) $\hat{p}$ is measured by counting the number of times the solution falls in the rectangle whose vertices are (x_1, y_1), (x_2, y_2), (x_3, y_3), and (x_4, y_4).

What techniques would you use to evaluate the confidence intervals for (*a*) and (*b*)? Assume that $\hat{p}_1$, $\hat{p}_2$, $\hat{p}_3$, and $\hat{p}_4$ are approximately normal in distribution.

11.6. A particular precision machine tool has a nut-and-bolt assembly which is the critical unit. If the nut and bolt have diameters that differ by 0.3 in. or more, the machine will chatter and perform defective work. Suppose that in the stock bin there are bolts with diameters 1.1, 1.2, 1.3, 1.3, 1.35, 1.35, 1.4, 1.4, 1.4, 1.4 and nuts with diameters 1.45, 1.45, 1.45, 1.5, 1.5, 1.5, 1.55, 1.55, 1.6, 1.6. If five machines are to be built with nuts and bolts chosen from stock with equal likelihood, show how the computer can determine the probability of having a defective machine. Why does a single sample require five computer runs? In the five runs that make up one sample are the solutions independent? Explain.

11.7. Assume that we have two noise sources with uniform distribution

$$p(x) = \tfrac{1}{2} \qquad -1 \le x \le 2$$

How could we use these noise sources and appropriate equipment so that the expected value of some random variable is a product of two variables?

11.8. We have a pinball machine which operates on the following principle. There are three receptacles A, B, C in a line, each of which contains three adjacent apertures. Thus we begin the game with nine open apertures. A ball is hurled, and has an equal likelihood of going through any aperture into a receptacle. After the ball passes through the aperture, the aperture closes. The next ball then has an equal likelihood of passing through the remaining eight apertures. This process is repeated for a total of five balls.

Assume that we have an original population of ten balls, of which five are white and five are black. Each successive ball is chosen at random, i.e., with equal likelihood, and is *not* replaced.

If only gaussian noise sources are available, program the computer to find the following probabilities:

(*a*) Three balls of one color in one of the receptacles.

(*b*) No balls in receptacle C.

For any number of samples n, how would you determine the confidence intervals for a given significance level?

11.9. Assume that we are measuring the standard deviation σ and have two repetitive computers with the following characteristics:

Computer A
1. 10 solutions per second
2. 5 per cent error in each measurement

Computer B
1. 1 solution per second
2. 1 per cent error in each measurement

Assume that all errors are of the same sign.

If we run both computers the same length of time, how much time is required so that the maximum difference between errorless estimates and those with error is the same for both computers?

11.10. In the text we show that, if the probability of occurrence of one event E_1 is p_1 and that of a second event E_2 is p_2, then the ordered pairs E_1E_2 and E_2E_1 have the same probability. Describe a computer mechanization which would record these ordered pairs and give two equally probable quantities a and b.

11.11. Assume that we have a system weighting function $h(t, \tau)$ and input white noise with gaussian distribution and spectral density N_0. Show how we can use the adjoint technique to determine the third and fourth moment.

11.12. In Fig. P 11.12a the switch S_1 is open for $t < 0$. At $t \geq 0$ the switch begins opening and closing in a periodic manner as shown in Fig. P 11.12b. The switch remains closed during the shaded portion.

(a) If $x(t)$ is a random noise source and $\bar{x}(t) = 10$ volts, set up the computer to determine $E[y(t)]$ at $t = 0.5$, 1.5, 2.25 sec.

(b) If $x(t)$ is a noise source with a spectral density $S_x(f) = 1/(1 + \tau_1{}^2\omega^2)$, determine $E[y^2(t)]$ at $t = 0.75$, 1.6, and 2.6 sec.

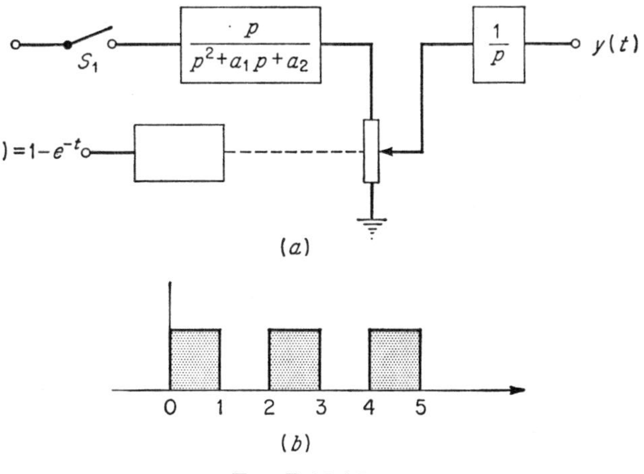

FIG. P 11.12

11.13. Given a forcing function with a correlation function

$$R(\tau) = A_1e^{\alpha_1|\tau|} + A_2e^{-\alpha_2|\tau|}$$

What transfer function is needed to make the input an equivalent white-noise source?

11.14. Assume that our gaussian noise generator has a mean of 10 volts and a variance of 4 volts. Our relay is fixed to switch at 3 volts. We want to program the relay so that $P_r(x + y \geq 3) = 0.67$, $P_r(x + y < 3) = 0.33$. Given the circuits in Fig. P 11.14*a* and *b*, what must be the values of k_1 and k_2 to satisfy these conditions? NOTE: k_1 and k_2 can be any finite positive or negative value.

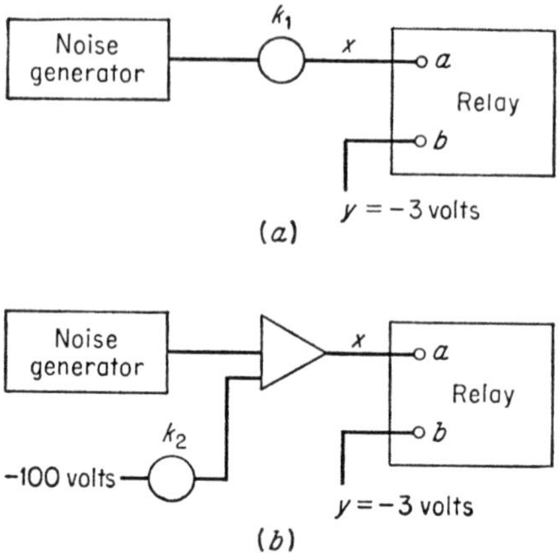

FIG. P 11.14

References

1. Polya, B.: "Mathematics and Plausible Reasoning," vol. I, "Induction and Analogy in Mathematics," Princeton University Press, Princeton, N.J., 1954.
2. Karplus, W. J., and W. W. Soroka: "Analog Methods," 2d ed., pp. 4–6, McGraw-Hill Book Company, Inc., New York, 1959.
3. Karplus, W. J.: "Analog Simulation," McGraw-Hill Book Company, Inc., New York, 1958.
4. Ehricke, Krafft A.: "Space Flight," vol. 1, "Environment and Celestial Mechanics," D. Van Nostrand Company, Inc., Princeton, N.J., 1960.
5. Scott, N. R.: "Analog and Digital Computer Technology," McGraw-Hill Book Company, Inc., New York, 1960.
6. Singer, C.: "A Short History of Scientific Ideas to 1900," pp. 236–242, Oxford University Press, Fair Lawn, N.J., 1959.
7. Sluchin, W.: "Minds and Machines," Penguin Books, Inc., Baltimore, 1960.
8. "Encyclopaedia Britannica," vol. 15, pp. 69–73, Encyclopaedia Britannica, Inc., Chicago, 1958.
9. "Encyclopaedia Britannica," vol. 4, pp. 548–554, Encyclopaedia Britannica, Inc., Chicago, 1958.
10. Price, D. J. de Solla: An Ancient Greek Computer, *Sci. Am.*, vol. 200, no. 6, pp. 60–66, June, 1959.
11. Kennedy, E. S.: Al-Kushi's Plate of Conjunctions, *Isis*, vol. 38, pp. 56–59, November, 1947.
12. Kennedy, E. S.: A Fifteenth Century Planetary Computer, I. Motion of the Sun and Moon in Longitude, *Isis*, vol. 41, pp. 180–183, April, 1950.
13. Kennedy, E. S.: A Fifteenth Century Planetary Computer, II. Longitudes, Distances, and Equations of the Planets, *Isis*, vol. 43, pp. 42–50, April, 1952.
14. Thomson, J.: On an Integrating Machine Having a New Kinetic Principle, *Proc. Roy. Soc.*, vol. 24, pp. 262–265, Feb. 3, 1876.
15. Thomson, W. (Lord Kelvin): On an Instrument for Calculating $\int \phi(x)\psi(x)\,dx$, The Integral of the Product of Two Given Functions, *Proc. Roy. Soc.*, vol. 24, pp. 266–269, Feb. 3, 1876.
16. Thomson, W. (Lord Kelvin): Mechanical Integration of the Linear Differential Equation of the Second Order with Variable Coefficients, *Proc. Roy. Soc.*, vol. 24, pp. 269–271, Feb. 3, 1876.
17. Thomson, W. (Lord Kelvin): Mechanical Integration of the General Linear Differential Equation of Any Order with Variable Coefficients, *Proc. Roy. Soc.*, vol. 24, pp. 271–275, Feb. 3, 1876.
18. Eterman, I. I.: "Analogue Computers," translated from the Russian by G. Segal, Pergamon Press, New York, 1960.
19. Bush, V.: Description of the First Differential Analyzer, *J. Franklin Inst.*, vol. 212, pp. 447–448, 1931.
20. Aiken, H. H., and G. M. Hopper: The Automatic Sequence Controlled Calculator, *Elec. Eng.*, vol. 65, pp. 384–391, 449–454, 522–528, 1946.
21. von Neumann, J.: "The Computer and the Brain," Yale University Press, New Haven, Conn., 1958.

22. Paynter, A. M. (ed.): "A Palimpsest on the Electronic Analog Art," George Philbrick Associates, Boston, 1955.
23. Hildebrand, F. B.: "Advanced Calculus for Engineers," Prentice-Hall, Inc., Englewood Cliffs, N.J., 1948.
24. Ince, E. L.: "Ordinary Differential Equations," Dover Publications, Inc., New York, 1956.
25. Ford, Lester R.: "Differential Equations," 2d ed., McGraw-Hill Book Company, Inc., New York, 1955.
26. Greenspan, Donald: "Theory and Solution of Ordinary Differential Equations," The Macmillan Company, New York, 1960.
27. Goursat, Edward: "A Course in Mathematical Analysis," vol. II, part II, "Differential Equations," translated by E. R. Hedrick and Otto Dunkel, Dover Publications, Inc., New York, 1959.
28. Struik, Dirk J.: "A Concise History of Mathematics," Dover Publications, Inc., New York, 1948.
29. Bell, Eric Temple: "The Development of Mathematics," 2d ed., pp. 400–409, McGraw-Hill Book Company, Inc., New York, 1945.
30. Sokolnikoff, I. S., and R. M. Redheffer: "Mathematics of Physics and Modern Engineering," pp. 95–99, McGraw-Hill Book Company, Inc., New York, 1958.
31. Gardner, M. F., and J. L. Barnes: "Transients in Linear Systems," John Wiley & Sons, Inc., New York, 1942.
32. Sneddon, Ian N.: "Fourier Transforms," McGraw-Hill Book Company, Inc., New York, 1951.
33. Ragazzini, John R., and Gene F. Franklin: "Sampled-data Control Systems," McGraw-Hill Book Company, Inc., New York, 1958.
34. Jury, Eliahu I.: "Sampled-data Control Systems," John Wiley & Sons, Inc., New York, 1958.
35. Truxal, J. G.: "Automatic Feedback Control System Synthesis," McGraw-Hill Book Company, Inc., New York, 1955.
36. Kamke, E.: "Differentialgleichungen: Lösungsmethoden und Lösungen," 3d ed., Chelsea Publishing Company, New York, 1944.
37. Stoker, J. J.: "Nonlinear Vibrations," Interscience Publishers, Inc., New York, 1950.
38. Cunningham, W. J.: "Introduction to Nonlinear Analysis," McGraw-Hill Book Company, Inc., New York, 1958.
39. Graham, D., and D. McRuer: "Analysis of Nonlinear Control Systems," John Wiley & Sons, Inc., New York, 1961.
40. Richards, Paul I.: "A Manual of Mathematical Physics," Pergamon Press, New York, 1959.
41. von Kármán, Theodore, and Maurice A. Biot: "Mathematical Methods in Engineering," 1st ed., McGraw-Hill Book Company, Inc., New York, 1940.
42. Andronow, A. A., and C. E. Chaikin: "Theory of Oscillations," Princeton University Press, Princeton, N.J., 1949.
43. Jackson, A. S.: "Analog Computation," McGraw-Hill Book Company, Inc., New York, 1960.
44. Rogers, A. E., and T. W. Connolly: "Analog Computation in Engineering Design," McGraw-Hill Book Company, Inc., New York, 1960.
45. Friedman, Bernard: "Principles and Techniques of Applied Mathematics," John Wiley & Sons, Inc., New York, 1956.
46. Fifer, Stanley: "Analogue Computation," 4 vols., McGraw-Hill Book Company, Inc., New York, 1961.
47. Laning, J. H., Jr., and R. H. Battin: An Application of Analog Computers to Problems of Statistical Analysis, pp. 79–88 in "Project Cyclone, Symposium II," sponsored by Reeves Instrument Corporation, New York, Apr. 28–May 2, 1952.

48. Laning, J. H., Jr., and R. H. Battin: "Random Processes in Automatic Control," McGraw-Hill Book Company, Inc., New York, 1956.
49. Grabbe, E. M., S. Ramo, and D. E. Wooldridge: "Handbook of Automation, Computation, and Control," vol. I, "Control Fundamentals," John Wiley & Sons, Inc., New York, 1958.
50. Churchill, R. V.: "Modern Operational Mathematics in Engineering," McGraw-Hill Book Company, Inc., New York, 1944.
51. Cooke, R. G.: "Linear Operators," St Martin's Press, Inc., New York, 1953.
52. Gille, J. C., M. J. Pélegrin, and P. Decaulne: "Feedback Control Systems," McGraw-Hill Book Company, Inc., New York, 1959.
53. Smith, O. J. M.: "Feedback Control Systems," McGraw-Hill Book Company, Inc., New York, 1958.
54. Konigsberg, R. L.: in L. Marton (ed.), "Advances in Electronics and Electronic Physics," vol. II, pp. 225–285, Academic Press Inc., New York, 1959.
55. Leitmann, G. (ed.): "Optimization Techniques," Academic Press Inc., New York, 1962.
56. Howe, R. M., and E. G. Gilbert: Trigonometric Resolution in Analog Computers by Means of Multiplier Elements, *IRE Trans. Electron. Computers*, vol. 6, pp. 86–92, June, 1957.
57. Lanczos, C.: "Applied Analysis," Prentice-Hall, Inc., Englewood Cliffs, N.J., 1956.
58. Korn, G. A., and T. M. Korn: "Mathematical Handbook for Scientists and Engineers," McGraw-Hill Book Company, Inc., New York, 1961.
59. Morrill, C. D., and R. V. Baum: Diode Limiters Simulate Mechanical Phenomena, *Electronics*, vol. 25, pp. 122–126, November, 1952.
60. Bennett, R. R.: The Generation of Straight Line Transfer Relationships, pp. 145–148 in "Project Cyclone, Symposium I on REAC Techniques," Reeves Instrument Corporation, under sponsorship of U.S. Navy Special Devices Center, New York, Mar. 15–16, 1951.
61. Burt, E. G. C., and O. H. Lange: Function Generators Based on Linear Interpolation with Applications to Analog Computing, *Proc. IEE*, Monograph 137 M, June, 1955.
62. Williams, R. W., and H. Marchant: Resistive Potentiometers as Function Generators, *Elec. Eng.*, vol. 30, no. 367, pp. 585–597, October, 1958.
63. Shen, D. W. C.: Approximating Nonlinear Functions by Shunt Loading Tapped Potentiometers in Analog Computing Machines, *Elec. Eng.*, vol. 29, no. 355, pp. 434–439, September, 1957.
64. Alt, F. L., and M. Rubinoff (eds.): "Advances in Computers," vol. 3, pp. 156–189, Academic Press Inc., New York, 1962.
65. Diendonné, J.: "Foundations of Modern Analysis," pp. 26–74, Academic Press Inc., New York, 1960.
66. Chestnut, H., A. Dabul, and D. Leiby: Analog Computer Study of Sampled-data Systems, *Proc. Computers Control Systems Conf.*, pp. 71–76, Atlantic City, N.J., Oct. 16–18, 1957; published by American Institute of Electrical Engineers, New York, May, 1958.
67. Meissinger, H. F.: Representation of Functions of Several Variables on REAC Equipment, pp. 125–130 in "Project Cyclone, Symposium I on REAC Techniques," Reeves Instrument Corporation, under sponsorship of U.S. Navy Special Devices Center, New York, Mar. 15–16, 1951.
68. Meissinger, H. F.: An Electronic Circuit for the Generation of Functions of Several Variables, *IRE Conv. Record*, pt. IV, 1955.
69. Levitt, J. R.: An Application of Numerical Mathematical Analysis to Analog Computation, *Proc. 2d Intern. Analog Computation Meetings*, pp. 397–399, Presses Académiques Européennes, Brussels, 1959.

70. Tuttle, D. F.: "Network Synthesis," vol. 1, John Wiley & Sons, Inc., New York, 1958.
71. Morrill, C. D.: A Sub Audio Time Delay Circuit, *IRE Trans. Electron. Computers*, vol. EC-3, no. 2, June, 1954.
72. Taylor, A. E.: "Advanced Calculus," Ginn and Company, Boston, 1955.
73. Courant, R.: "Differential and Integral Calculus," 2 vols., translated from German by J. McShane, Nordemann Publishing Company, New York, 1938.
74. Goldstein, H.: "Classical Mechanics," Addison-Wesley Publishing Company, Inc., Reading, Mass., 1959.
75. Greenwood, I. A., Jr., J. V. Holdam, Jr., and D. MacRae, Jr.: "Electronic Instruments," MIT Radiation Laboratory Series, vol. 21, McGraw-Hill Book Company, Inc., New York, 1948.
76. Mengel, A. S., and W. S. Melahn: Rand REAC Manual RM-525, RAND Corporation, Santa Monica, Calif., Dec. 1, 1950.
77. Synge, J. L., and B. A. Griffith: "Principles of Mechanics," 2d ed., McGraw-Hill Book Company, Inc., New York, 1949.
78. Mayer, A.: Rotations and Their Algebra, *SIAM Rev.*, vol. 2, no. 2, April, 1960.
79. Frazer, R. A., W. S. Duncan, and A. R. Collar: "Elementary Matrices and Some Applications to Dynamics and Differential Equations," The Macmillan Company, New York, 1946.
80. Guillemin, E. A.: "The Mathematics of Circuit Analysis," John Wiley & Sons, Inc., New York, 1949.
81. Pipes, L. A.: "Applied Mathematics for Engineers and Physicists," McGraw-Hill Book Company, Inc., New York, 1946.
82. Fitsner, L. N.: A Precision Multiplier, *Automation Remote Control* (English translation), January, 1959, pp. 56–62.
83. Erdélyi, A.: "Asymptotic Expansions," Dover Publications, Inc., New York, 1956.
84. Bush, L., and P. Orlando: A Perturbation Technique for Analog Computers, *IRE Trans. Electron. Computers*, vol. EC-8, no. 2, pp. 218–221, June, 1959.
85. Locke, A. S.: "Guidance," D. Van Nostrand Company, Inc., Princeton, N.J., 1955.
86. Meissinger, H. F.: The Use of Parameter Influence Coefficients in Computer Analysis of Dynamic Systems, *Proc. Western Joint Computer Conf.*, May 3–5, 1960.
87. Bellman, R. I.: "Adaptive Control Processes, A Guided Tour," Princeton University Press, Princeton, N.J., 1961.
88. Hancock, H.: "Theory of Maxima and Minima," Dover Publications, Inc., New York, 1960.
89. Brown, R. R.: A Generalized Computer Procedure for the Design of Optimum Systems, *Trans. AIEE, Commun. Electron.*, July, 1959, pp. 285–293.
90. Hestenes, M. R.: Iterative Computational Methods, *Commun. Pure Appl. Math.*, vol. 8, pp. 85–96, 1955.
91. Booth, A. D.: An Application of the Method of Steepest Descents to the Solution of Systems of Nonlinear Simultaneous Equations, *Quart. J. Mech. Appl. Math.*, vol. II, pt. 4, 1949.
92. Davies, O. L. (ed.): "Design and Analysis of Industrial Experiments," chap. XI, Determination of Optimum Conditions, pp. 475–579, Hafner Publishing Company, Inc., New York, 1956.
93. Courant, R.: Variational Methods for the Solution of Problems of Equilibrium and Vibrations, *Bull. Am. Math. Soc.*, vol. 49, pp. 1–23, January, 1943.
94. Box, G. E. P., and K. B. Wilson: On the Experimental Attainment of Optimum Conditions, *J. Roy. Statist. Soc.*, ser. B, vol. 13, no. 1, pp. 1–43, 1951.
95. Beckenbach, E. F. (ed.): "Modern Mathematics for the Engineer," McGraw-Hill Book Company, Inc., New York, 1956.

96. Levine, L., and H. F. Meissinger: An Automatic Analog Computer Method for Solving Polynomials and Finding Root Loci, *IRE Conv. Record*, pt. IV, pp. 164–172, 1957.
97. Ward, J. A.: The Downhill Method of Solving a Polynomial Equation, paper presented at 11th Annual Meeting of the ACM, Los Angeles, Calif., Aug. 27–29, 1956.
98. Rubin, A., R. R. Favreau, J. P. Landauer, and I. Hay: Plotting Results of a Quadratic Regression Analysis, *Analog Computer Appl. Bull.* 10, Electronic Associates, Inc., Princeton, N.J.
99. Neustadt, L. W.: Applications of Linear and Nonlinear Programming Techniques, presented at 3d International Conference on Analog Computations, Opatiza, Yugoslavia, Sept. 4–8, 1961.
100. Pyne, I. B.: Linear Programming on an Electronic Analog Computer, *Trans. AIEE, Commun. Electron.*, 1956, pp. 139–143.
101. Weinstock, R.: "Calculus of Variations," McGraw-Hill Book Company, Inc., New York, 1952.
102. Graves, R. L., and P. Wolfe (eds.): "Recent Advances in Mathematical Programming," McGraw-Hill Book Company, Inc., New York, 1963.
103. Feldbaum, A. A.: Automatic Optimalizer, *Automation Remote Control* (English translation), August, 1958, pp. 718–728.
104. Stakhovskii, R. I.: Twin-channel Automatic Optimalizer, *Automation Remote Control* (English translation), August, 1958, pp. 729–740.
105. Allen, D. N. de G.: "Relaxation Methods," McGraw-Hill Book Company, Inc., New York, 1954.
106. Bubb, F. W.: A Circuit for Generating Polynomials and Finding Their Zeros, *Proc. IRE*, vol. 39, pp. 1556–1561, 1951.
107. Brunner, W.: An Iteration Procedure for Parametric Model Building and Boundary Value Problems, *Proc. Western Joint Computer Conf.*, pp. 519–533, Los Angeles, Calif., May 9–11, 1961.
108. Moser, J. H., C. O. Reed, and H. L. Sellars: A Nonlinear Programming Technique for Analog Computation, presented at the Eastern Simulation Council, December, 1960 (available from Shell Oil Co., Houston Research Lab., Deer Park, Tex.).
109. Munson, J. K., and A. I. Rubin: Optimization by Random Search on the Analog Computer, National Simulation Conference, Dallas, Tex., Jan. 19–21, 1956.
110. Sarhan, A. E., and B. G. Greenberg: "Contributions to Order Statistics," John Wiley & Sons, Inc., New York, 1962.
111. deBruijn, N. G.: "Asymptotic Methods in Analysis," Interscience Publishers, Inc., New York, 1958.
112. Keifer, J., and J. Wolfowitz: Stochastic Estimation of the Maximum of a Regression Function, *Ann. Math. Statist.*, vol. 23, pp. 452–466, 1952.
113. Grabbe, E. M., S. Ramo, and D. E. Wooldridge (eds.): "Handbook of Automation, Computation, and Control," vol. 2, "Computers and Data Processing" (chap. 26, The Use of Noise and Statistical Techniques in Analog Computers, by H. Low), John Wiley & Sons, Inc., New York, 1957.
114. Middleton, D.: "An Introduction to Statistical Communication Theory," McGraw-Hill Book Company, Inc., New York, 1960.
115. Hald, A.: "Statistical Theory with Engineering Applications," John Wiley & Sons, Inc., New York, 1952.
116. Davenport, W. B., Jr., and W. L. Root: "Introduction to Random Signals and Noise," McGraw-Hill Book Company, Inc., New York, 1958.
117. Blackman, R. S., and J. W. Tukey: "The Measurement of Power Spectra," Dover Publications, Inc., New York, 1958.
118. Bendat, J.: "Principles and Applications of Random Noise Theory," John Wiley & Sons, Inc., New York, 1958.

119. Chang, S. S. L.: "Synthesis of Optimum Control Systems," McGraw-Hill Book Company, Inc., New York, 1961.
120. Wadsworth, G. P., and J. G. Bryan: "Introduction to Probability and Random Variables," McGraw-Hill Book Company, Inc., New York, 1960.
121. Vander Velde, W. E.: Make Statistical Studies on Analog Computers, *Control Eng.*, June, 1960, pp. 127–130.
122. Acton, F. S.: "Analysis of Straight-line Data," John Wiley & Sons, Inc., New York, 1960.
123. Korn, Granino: The Impact of Hybrid Analog-Digital Techniques on the Analog Computer Art, *Proc. IRE*, May, 1962, pp. 1077–1086.
124. Cramer, H.: "Mathematical Methods of Statistics," Princeton University Press, Princeton, N.J., 1957.
125. Hoel, P. G.: "Introduction to Mathematical Statistics," John Wiley & Sons, Inc., New York, 1954.
126. Williams, J. D.: "The Compleat Strategyst," McGraw-Hill Book Company, Inc., New York, 1954.
127. Siegel, S.: "Nonparametric Statistics," McGraw-Hill Book Company, Inc., New York, 1956.
128. Evans, D. H.: Optimum Tolerance Assignment to Yield Minimum Manufacturing Cost, *Bell System Tech. J.*, vol. 37, pp. 461–468, March, 1958.
129. Pike, E. W., and T. R. Silverberg: Assigning Tolerances for Maximum Economy, *Machine Design*, vol. 25, p. 139, September, 1953.
130. Davis, R. C.: A Statistical Method for Analyzing the Effects of Missile Guidance Systems Tolerance on Hit Probability, *Proc. 3d Exploratory Conf. Missile Model Design for Reliability Predictions*, pp. 85–96, White Sands Missile Range, April, 1959.
131. Cochran, W. G., and G. M. Cox: "Experimental Design," John Wiley & Sons, Inc., New York, 1950.
132. Hunter, J. S.: Designing and Interpreting Tests, *Control Eng.*, September, 1959, pp. 137–141.
133. Bekey, G. A.: Generalized Integration on the Analog Computer, *IRE Trans. Electron. Computers*, June, 1959, pp. 210–217.
134. Huskey, H. D., and G. A. Korn (eds.): "Computer Handbook," McGraw-Hill Book Company, Inc., New York, 1962.
135. Bose, A. G.: A Theory of Nonlinear Systems, *MIT Res. Lab. Electron. Tech. Rept.* 309, May 15, 1956.
136. Dixon, W. J., and F. J. Massey, Jr.: "Introduction to Statistical Analysis," 2d ed., McGraw-Hill Book Company, Inc., New York, 1957.
137. Neustadt, L. W.: A Method of Computing Eigenvectors and Eigenvalues on an Analog Computer, *Math. Tables Aids Comput.*, vol. 13, no. 67, pp. 194–201, 1959.
138. Schwartz, L.: "Théorie des distributions," 2 vols., Hermann & Cie, Paris, 1950.
139. Zadeh, L. A.: The Determination of the Impulsive Response of Variable Networks, *J. Appl. Phys.*, vol. 21, pp. 642–645, July, 1950.
140. Aseltine, J. A.: "Transform Method in Linear System Analysis," McGraw-Hill Book Company, Inc., New York, 1958.
141. Solodov, A. V.: On the Transformation of Initial Conditions on the Input of a Linear Time-varying System into an Equivalent Input Signal, *Automation Remote Control* (English translation from the Russian), vol. 19, no. 7, pp. 654–660, 1958.
142. Batkov, A. M.: On the Problem of Synthesis of Linear Dynamic Systems with Variable Parameters, *Automation Remote Control* (English translation from the Russian), vol. 19, no. 1, pp. 49–54, 1958.
143. Borskii, V.: On the Properties of Impulsive Response of Varying-parameter Systems, *Automation Remote Control* (English translation from the Russian), vol. 20, no. 7, pp. 848–855, 1959.

144. Miller, K. S.: Properties of Impulsive Responses and Green's Functions, *IRE Trans. Circuit Theory*, March, 1955, pp. 26–31.
145. Zadeh, L. A.: Time Varying Networks, I, *Proc. IRE*, October, 1961, pp. 1488–1503.
146. Bryson, A. E., W. F. Denham, F. J. Carroll, and K. Mikami: Determination of Lift Drag Programs to Minimize Re-entry Heating, *J. Aerospace Sci.*, vol. 29, no. 4, April, 1962.
147. Bryson, A. E., and W. F. Denham: A Steepest-ascent Method for Solving Optimum Programming Problems, *J. Appl. Mech.* (*Trans. ASME, ser. E*), vol. 29, no. 2, pp. 247–257, June, 1962.
148. Kelley, H. J., R. E. Kopp, and H. Moyer: Successive Approximation Techniques for Trajectory Optimization, *Proc. Inst. Aeron. Sci. Symp. Vehicle Systems Optimization*, Garden City, N.Y., Nov. 29–31, 1961.
149. McGhee, R. B., and A. Levine: Determination of Optimum Production Tolerances by Analog Simulation, *Proc. Eastern Joint Computer Conf.*, pp. 249–254, Boston, Mass., Dec. 1–3, 1959.

Index